du 23rd

||||||||||||||||||||||||||||||
D0849285

# Paul F. Aubin's
# Mastering Revit®
# Architecture 2009

# Paul F. Aubin's
# Mastering Revit®
# Architecture 2009

**PAUL F. AUBIN**

Autodesk

DELMAR
CENGAGE Learning

Australia • Brazil • Japan • Korea • Mexico • Singapore • Spain • United Kingdom • United States

**DELMAR**
CENGAGE Learning™

**Paul F. Aubin's Mastering Revit Architecture 2009**
Paul F. Aubin

Vice President, Career and Professional Editorial: David Garza

Director of Learning Solutions: Sandy Clark

Acquisitions Editor: John Fedor

Managing Editor: Larry Main

Senior Product Manager: John Fisher

Senior Editorial Assistant: Dawn Daugherty

Vice President, Career and Professional Marketing: Jennifer McAvey

Marketing Director: Deborah S. Yarnell

Marketing Manager: Jimmy Stephens

Marketing Specialist: Mark Pierro

Production Director: Wendy Troeger

Production Manager: Stacy Masucci

Content Project Manager: Angela Iula

Art Director: David Arsenault

Technology Project Manager: Christopher Catalina

Production Technology Analyst: Thomas Stover

For product information and technology assistance, contact us at
**Professional & Career Group Customer Support, 1-800-658-7450**

For permission to use material from this text or product, submit all requests online at **cengage.com/permissions.**
Further permission questions can be emailed to
**permissionrequest@cengage.com.**

Library of Congress Control Number: 2008926547

ISBN-13: 978-1-4354-0263-4

ISBN-10: 1-4354-0263-4

**Delmar**
5 Maxwell Drive
Clifton Park, NY 12065-2919
USA

Cengage Learning is a leading provider of customized learning solutions with office locations around the globe, including Singapore, the United Kingdom, Australia, Mexico, Brazil, and Japan. Locate your local office at:
**international.cengage.com/region**

Cengage Learning products are represented in Canada by Nelson Education, Ltd.

For your lifelong learning solutions, visit **delmar.cengage.com**

Visit our corporate website at **cengage.com.**

**NOTICE TO THE READER**

Publisher does not warrant or guarantee any of the products described herein or perform any independent analysis in connection with any of the product information contained herein. Publisher does not assume, and expressly disclaims, any obligation to obtain and include information other than that provided to it by the manufacturer. The reader is expressly warned to consider and adopt all safety precautions that might be indicated by the activities herein and to avoid all potential hazards. By following the instructions contain herein, the reader willingly assumes all risks in connection with such instructions. The publisher makes no representation or warranties of any kind, including but not limited to, the warranties of fitness for particular purpose or merchantability, nor are any such representations implied with respect to the material set forth herein, and the publisher takes no responsibility with respect to such material. The publisher shall not be liable for any special, consequential, or exemplary damages resulting, in whole or part, from the readers' use of, or reliance upon, this material.

Printed in Canada

1 2 3 4 5 XXX 10 09 08

# CONTENTS

## CHAPTER 5  COLUMN GRIDS AND STRUCTURAL LAYOUT ...... 246

## CHAPTER 13  CEILING PLANS AND INTERIOR ELEVATIONS .......................................... 731

# INTRODUCTION

## WELCOME

Within the pages of this book you will find a comprehensive introduction to the methods, philosophy, and procedures of the Revit Architecture software. Revit is an advanced and powerful architectural design and documentation software package. By following the detailed tutorials contained in this book, you will become immersed in its workings and functionality.

## WHO SHOULD READ THIS BOOK?

The primary audience for this book is users new to Revit Architecture. However, it is also appropriate for existing Revit users who wish to expand their knowledge. You need not be an experienced computer operator to use this book. Only basic knowledge of the Windows operating system and basic use of a mouse and keyboard are assumed. No prior computer aided design software knowledge is required. If part of your job requires that you design buildings and produce architectural construction documentation or design drawings, facilities layouts, or interior design studies and documentation, then this book is intended for you. Architects, interior designers, design build professionals, facilities planners, and building industry CAD professionals will benefit from the information contained within.

## FEATURES IN THIS EDITION

*Mastering Revit Architecture* is a concise manual focused squarely on the rationale and practicality of the Revit process. The book emphasizes the process of creating projects in Revit rather than a series of independent commands and tools. The goal of each lesson is to help readers complete building design projects successfully. Tools are introduced together in a focused process with a strong emphasis on "why" as well as "how." The text and exercises seek to give the reader a clear sense of the value of the tools, and a clear indication of each tool's potential. Mastering Revit is a resource designed to shorten your learning curve, raise your comfort level, and, most importantly, give you real-life, tested, and practical advice on the usage of the software to create architecture.

## WHAT YOU WILL FIND INSIDE

Section I of this book is focused on the underlying theory and user interface of Revit Architecture. The section is intended to get you acquainted with the software and put you in the proper mindset. Section II relies heavily on tutorial-based exercises to present the process of creating a building model in

Revit, relying on the software's powerful Building Information Modeling functionality. Two projects are developed concurrently throughout the tutorial section: one residential and one commercial. Detailed explanations are included throughout the tutorials to clearly identify why each step is employed. Annotation and other features specific to construction documentation are covered in Section III. Section IV contains appendices.

## WHAT YOU WON'T FIND INSIDE

This book is not a command reference. This book approaches the subject of learning Revit by both exposing conceptual aspects of the software and extensive tutorial coverage. No attempt is made to give a comprehensive explanation of every command or every method available to execute commands. Instead, explanations cover broad topics of how to perform various tasks in Revit, with specific examples coming from architectural practice. References are made within the text wherever appropriate to the extensive on-line help and reference materials available on the Web. The focus of this book is the design development and construction documentation phases of architectural design. Conceptual design tools are not extensively covered in this edition.

## STYLE CONVENTIONS

Style Conventions used in this text are as follows:

| Text | Revit Architecture |
| --- | --- |
| Step-by-step tutorials | 1. Perform these steps. |
| Menu picks | **View > Zoom > Zoom In Region** |
| On screen input | For the length type **10'-0"** **[3000]**. |
| Design Bar tools | On the Design Bar, click the **Wall** tool |
| Type names | From the Type Selector, choose the **Basic: Generic 12"** Wall Type |
| File and Directory names | *C:\MasterRevitBuilding\Chapter01\Sample File.rvt* |

## UNITS

This book references both imperial and metric units. Symbol names, scales, references, and measurements are given first in imperial units, and are then followed by the metric equivalent in square brackets [ ]. For example, when there are two versions of the same file, they will appear like this within the text:

*Curtain Wall Dbl Glass.rfa* [*M_Curtain Wall Dbl Glass.rfa*].

When the scale varies, a note like this will appear: **1/8"=1'-0"** **[1:100]**.

If a measurement must be input, the values will appear like this: **10'-0"** **[3000]**. Please note that in many cases, the closest logical corresponding metric value has been chosen, rather than a "direct" mathematical translation. For instance, 10'-0" in imperial drawings translates to 3048 millimeters; however, a value of 3000 will be used in most cases as a more logical value.

**Note:** Every attempt has been made to make these decisions in an informed manner. However, it is hoped that readers in countries where metric units are the standard will forgive the American author for any poor choices or translations made in this regard.

All project files are included in both imperial and metric units on the included CD ROM. See the "Files Included on the CD ROM" topic below for information on how to install the dataset in your preferred choice of units.

## HOW TO USE THIS BOOK

The order of chapters has been carefully thought out with the intention of following a logical flow and architectural process. If you are relatively new to Revit, it is recommended that you complete the entire book in order. However, if there are certain chapters that do not pertain to the type of work performed by you or your firm, feel free to skip those topics. However, bear in mind that not every procedure will be repeated in every chapter. For the best experience, it is recommended that you read the entire book, cover to cover. Most importantly, even after you have completed your initial pass of the tutorials in this book, keep *Mastering Revit Architecture* handy, as it will remain a valuable resource in the weeks and months to come.

## FILES INCLUDED ON THE CD ROM

Files used in the tutorials throughout this book, in various stages of completion, are located on the included CD ROM. Therefore, you will be able to load the file for a given chapter and begin working immediately. When you install the files from the CD, the files for all chapters will be installed automatically. The files will install into a folder on your C drive named *MRAC* by default. If you wish, you can choose a different root location such as your "My Documents" folder. Whatever location you choose, a folder named *MRAC* will be created. Inside this folder will be a folder for each chapter. Please note that in some cases a particular chapter or subfolder will not have any Revit files. This is usually indicated by a text file (TXT) within this folder. For example, the *Chapter01\Complete* folder contains no Revit files and instead contains a text document named: *There is no Complete version of Chapter 1.txt*. This text file simply explains that this folder was left empty intentionally.

## INSTALLING CD FILES

Locate the Mastering Revit CD ROM in the back cover of your book. Read the license agreement before breaking the seal to the CD. To install the dataset files, do the following:

1. Place the CD in your CD drive.

   An installer window should appear on screen after a moment or two.

2. To install the dataset files in imperial units, click the Imperial Dataset button. To install the dataset files in metric units, click the Metric Dataset button.

3. In the "WinZip Self-Extractor" dialog, click the Browse button to locate your desired installation folder (such as *My Documents*).

If you do not wish to change the location for the files, you can simply accept the default C Drive location. To do so, do not click Browse and continue to the next step.

4. Click the Unzip button to commence installation.

5. When all files are extracted, a dialog will appear. Click OK and then click Close to finish.

If you do not intend to perform the tutorials in certain chapters, it is OK to delete the files for those chapters. Simply delete the entire folder for the chapter(s) that you wish to skip. If you wish to install both the imperial and metric datasets, return to the installer and repeat the steps above for the other units. Installation requires approximately 275 MB of disk space per unit type (550 MB if you install both). If you install both datasets, some files will be the same. Click OK if WinZip asks to overwrite any files.

## KEEP YOUR SOFTWARE CURRENT

It is important to keep your software current. Be sure to check online at **www.autodesk. com** on a regular basis for the latest updates and service packs to the Revit Architecture software. Having the latest version installed will ensure that you benefit from the latest features and enhancements. If you are on the Autodesk Subscription program, you will be entitled to new releases as they become available. Visit the Autodesk Web site or talk to your local reseller for more information.

## WE WANT TO HEAR FROM YOU

We welcome your comments and suggestions regarding *Mastering Revit Architecture*. Please forward your comments and questions to:

The CADD Team
Cengage Learning
Executive Woods
5 Maxwell Drive
Clifton Park, NY 12065-8007
Web site: www.autodeskpress.com

## ABOUT THE AUTHOR

Paul F. Aubin is the author of several books on Revit Architecture and AutoCAD Architecture including *Mastering Revit Architecture*, *Mastering AutoCAD Architecture*, and *Autodesk Architectural Desktop: An Advanced Implementation Guide*. Paul has a background in the architectural profession spanning over nineteen years. These experiences include architectural design and production, CAD management, mentoring, and training. Paul is an independent consultant offering training and implementation services to architectural firms using Revit Architecture and AutoCAD Architecture. He is the moderator for *Cadalyst* magazine's online CAD questions forum and has spoken at Autodesk University (Autodesk's annual convention for users) for many years. The combination of his experiences in architectural practice, as a CAD manager and an instructor, give his writing and his classroom instruction a fresh and credible focus. Paul is an associate member of the AIA and is based in Chicago.

## DEDICATION

This book is dedicated to my mom. I know you can do it because you taught me how.

## ACKNOWLEDGEMENTS

The author would like to thank several people for their assistance and support throughout the writing of this book.

A very special thank you to David Mills and Christie Landry. It has been a long journey, but we have finally arrived at the destination. I couldn't have done it without your help.

Thanks to John Fisher, and all of the Delmar/Cengage team. It continues to be a pleasure to work with so dedicated a group of professionals.

Technical Contributors to portions of this text include Velina Mirincheva, Robert Guarcello Mencarini, Architect, AIA and Mark Schmieding. A special thanks to each of you. Additional contributions and quotations have been noted within the text. Thank you to those contributors as well.

A special acknowledgement is due the following instructors who reviewed the chapters in detail:

Matt Dillon–DC CADD Company

Mel Persin, Coordinator–Chicago Autodesk Revit Users Group

Stephen K. Stafford II–Stafford Consulting Services (Thanks for the Workset/Library analogy)

For taking the time to discuss this project personally and offer suggestions and feedback, thanks to Jeff Millett, AIA—Vice President and Director of Information Technology, Eddie Barnett—LEED, Interior Designer, Sarah Vekasy—LEED, Architect, Marc Gabriel—LEED, Architect, John Jackson—LEED, Architect and Marwan Bakri, Stubbins Associates, Boston, MA and also to Mark Dietrick—CIO and Senior Associate and Michael DeOrsey—Gradaute Architect of Burt Hill Kosar Rittlemann Associates, Boston, MA.

There are far too many folks at Autodesk to mention. Thanks to all of them, but in particular, Lillian Smith, David Mills, Kelcy Lemon, Jason Winstanley, Tatjana Dzambazova, David Conant, Matthew Jezyk, Steve Crotty, Erik Egbertson, Greg Demchak, Chico Membreno, Trey Klein, Michael Juros, Brian Fitzpatrick and all of the folks at Autodesk Tech Support.

I am ever grateful for blessings I have received in my many friends and family. Finally, I am most grateful for the constant love and support of my wife, Martha and our three wonderful children.

# SECTION I

# Introduction and Methodology

This section introduces the methodology of Revit Architecture. The concept of "Building Information Modeling" (BIM) is introduced and defined as are many other important topics and concepts. Within this section you will gain valuable experience using Revit Architecture and exploring its interface and overall conceptual underpinnings.

Section I is organized as follows:

**Quick Start**   General Overview
**Chapter I**    Conceptual Underpinnings of Revit Architecture
**Chapter 2**    Revit Architecture User Interface

# Quick Start

## General Revit Architecture Overview

### INTRODUCTION

This Quick Start provides a simple tutorial designed to give you a quick tour of some of the most common elements and features of Revit Architecture. You should be able to complete the entire exercise in one to two hours. At the completion of this tutorial, you will have experienced a first-hand look at what Revit Architecture has to offer.

### OBJECTIVES

In this chapter you will:

- Experience an overview of the software.
- Create your first Revit Architecture model.
- Receive a first-hand glimpse at many Revit Architecture tools and methods.
- Gain some basic experience with the Revit interface.

### CREATE A SMALL BUILDING

Let's get started using Revit Architecture right away. For the next several minutes, we will take a "whirlwind" tour of the Revit Architecture tool set. All of the tools covered in the following steps use simple, and often default, settings. The chapters that follow cover each of these tools and settings in detail. This book was authored using Microsoft Windows Vista Business, but the exercises and tutorials should perform equally well in Microsoft Windows XP Professional. Please refer to the Preface for complete details on prerequisites and assumptions.

### INSTALL THE CD FILES AND OPEN A PROJECT

The lessons that follow require the dataset included on the Mastering Revit Architecture CD ROM. If you have already installed all of the files from the CD, simply skip down to step 3 below to open the project. If you need to install the CD files, start at step 1.

1. Install the dataset files located on the Mastering Revit Architecture CD ROM.

Refer to "Files Included on the CD ROM" in the Preface for instructions on installing the dataset files included on the CD.

> 2. Launch Revit Architecture from the icon on your desktop or from the **Autodesk** group in **All Programs** on the Windows Start menu.

> **Tip:** In Windows Vista, you can click the Start button, and then begin typing **Revit** in the "Start Search" field.

> • If the New Features Workshop dialog appears, choose "Maybe later" and then click OK.

Revit Architecture presents a start page automatically upon launch.

> 3. Beneath the "Projects" heading, click the Browse icon (or, on the Standard toolbar, click the Open icon).

> **Tip:** The keyboard shortcut for Open is CTRL + O. **Open** is also located on the File menu.

> • In the "Open" dialog box, click the *My Documents* icon on the left side.

> • Double-click on the *MRAC* folder, and then the *Quick Start* folder.

If you installed the dataset files to a different location than the one listed here, use the "Look in" drop down list to browse to that location instead.

> 4. Double-click *Pavilion.rvt.*

You can also select it and then click the Open button (see Figure Q.1).

**Figure Q.1** *Open the Pavilion project to get started*

 **Note:** In this Quick Start tutorial, only one project has been provided and it uses Imperial units. The remainder of the book provides a Metric dataset as well.

The project will open in Revit Architecture with the last opened view visible on screen. In this case that is the *Level 1* floor plan view. This project has been started already and contains a Property Line element (dashed square) in the middle of the screen. There is also a Toposurface terrain model element in this file that represents the site for the building. Let's start by displaying this so we know where to place the Walls of our building.

## BEGIN A NEW MODEL

To get started, we need to begin with the basics: Walls, Doors, and Windows. These elements are the basic building blocks of any architectural model. Adding these elements in Revit Architecture is simple and straight forward.

## CREATE AN UNDERLAY

On the left side of the screen running vertically is a panel named: Project Browser. In it are listed several representations of our project including drawings, schedules, and sheets. Four floor plan views are provided here: *Level 1*, *Level 2*, *Roof*, and *Site*. The *Level 1* first floor plan view is bold indicating that it is the currently active view and open on screen. The screen will look something like Figure 2.1 in Chapter 2.

5. On the Project Browser, double-click to open the *Site* plan view.

Notice that the Site Plan includes contours and a shape in the middle of the plan representing the building footprint and its entrance patio.

- On the Project Browser, double-click *Level 1* to return to the first floor plan view.

We can display any one of the other levels (such as the *Site*) as an underlay to this view to help us coordinate elements at different levels.

6. On the Project Browser, right-click on the *Level 1* Floor Plan view and choose **Properties** (see Figure Q.2).

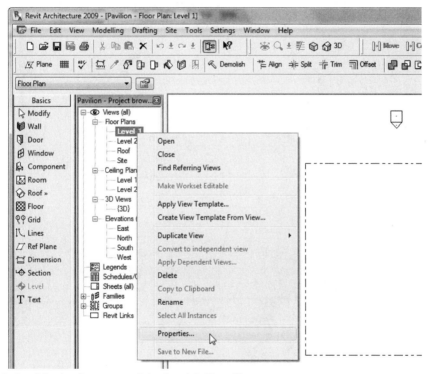

**Figure Q.2**   *Edit the Properties of the Level 1 Floor Plan view*

7. In the "Element Properties" dialog, beneath the "Graphics" grouping (near the bottom) click the word None next to "Underlay."

- Open the pop-up menu that appears and then choose **Site** (see Figure Q.3).

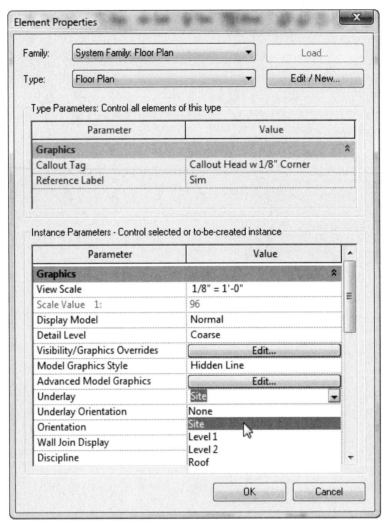

**Figure Q.3** *Assign the Site plan view as an underlay to the Level 1 plan*

- Click OK to dismiss the dialog and see the results.

Notice that only the patio and building footprint outline appeared. This is because they are the only parts of the Site that intersect the current level height (more on this later). Notice that they appear in 50% halftone gray as well. This reinforces visually that this is simply an underlay.

## CREATE WALLS

We begin our building model with some simple Walls.

Locate the Design Bar on the left side of the screen. The "Basics" tab should currently be active.

8. On the Design Bar, click the Basics tab and then click the **Wall** tool.

Several options will appear across the bar at the top of the screen just beneath the toolbar icons. This is called the Options Bar.

9. From the drop-down list on the left (known as the Type Selector) choose **Basic Wall : Generic - 8"**.

- From the "Loc Line" list, choose **Finish Face:Exterior**.

- On the right side of the Options Bar, click the rectangle icon (see Figure Q.4).

**Figure Q.4** *Pick the Wall tool and set it to draw Basic 8" Walls in a rectangular shape*

10. With the mouse pointer (now shaped like a small pencil) click the lower right corner of the gray shape on screen (see the left side of Figure Q.5).

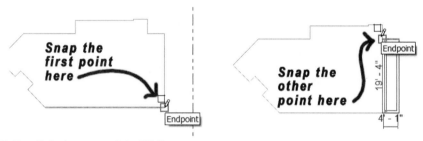

**Figure Q.5** *Click the start of the Walls*

- Pick the Endpoint of the short horizontal edge indicated on the right side of Figure Q.5.

You will now have four Walls on screen. However, the "room" they define is very narrow. We can easily adjust this. Before we can manipulate the Walls however, we must cancel the current Wall creation command.

11. On the Design Bar, click the **Modify** tool or press the ESC key twice.

**Note:** Either method can be used anytime in Revit Architecture to cancel the current command and return to the **Modify** (selection pointer) tool. In Revit Architecture there is always one active tool. The default tool is the "Modify" tool, which is really just the standard mouse pointer.

12. Click on the vertical Wall on the left to select it.

In Revit Architecture, objects turn red on screen when they are selected.

13. Click the mouse directly on the blue text of the dimension that appears between the selected Wall and the other vertical one (see Figure Q.6).

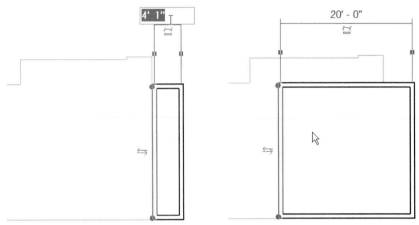

**Figure Q.6** *Click the text of a temporary dimension to edit it*

14. In the text field that appears, type **20** and then press ENTER.

Notice that the Wall moved to the new location as indicated by the value we input and that the two horizontal Walls stretched with it to remain attached. Please note that when you edit this way, the selected Wall moves. The dimensions that we used for this edit are referred to as "temporary dimensions."

15. On the Design Bar, click the **Wall** tool again.

- From the Type Selector choose **Basic Wall : Generic - 5"**.

- From the "Loc Line" list, choose **Wall Centerline**.

- Draw a vertical wall from top to bottom of the room, dividing it approximately into thirds (see Figure Q.7).

 **Note:** The exact dimensions are unimportant at this point, we will move the Wall next.

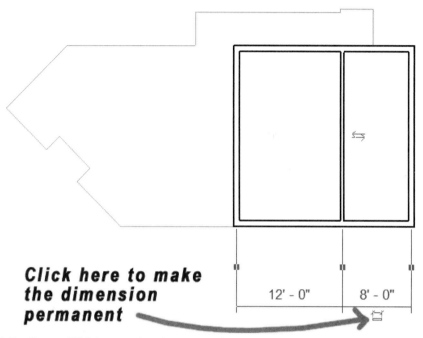

**Click here to make
the dimension
permanent**

**Figure Q.7**   *Draw a Wall in a random location in the space*

- Beneath the temporary dimension, click the small icon (indicated in Figure Q.7) to make the dimension permanent.

- On the Design Bar, click the **Modify** tool or press the ESC key twice.

Now that the dimension is permanent, notice that it remains on screen when the Wall is no longer selected.

16. Click to select the dimension.

- Click the small "EQ" (Equidistant) icon beneath the dimension (see Figure Q.8).

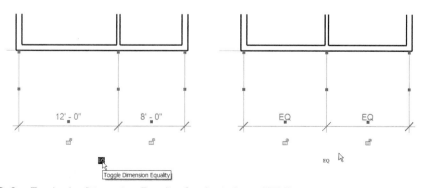

**Figure Q.8**   *Toggle the Dimension Equality for the indicated Walls*

**Note:** This icon is part of the Revit Architecture constraint system. The constraint system is used to "lock-in" design intent. This notion is an integral part of the underlying concepts inherent to Revit Architecture.

17. From the File menu, choose **Save**.

It is important to remember to save every so often to preserve your work. Revit Architecture is configured by default to remind you to save at regular intervals. You can edit the interval, but if the message asking you to save appears, you should always perform the save.

 **Note:** You can find more information and tutorials on working with Walls in Chapter 3.

## INSERT DOORS AND WINDOWS

Next we'll add some openings in our Walls.

18. On the Design Bar, click the Basics tab and then click the **Door** tool.

- Accept all of the defaults on the Options Bar.

- Move the pointer near the top horizontal Wall to begin placing the Door.

Move the mouse around without clicking it yet. Notice how the Door follows the cursor and also stays attached to the Wall as it does. This is because elements such as Doors are "hosted" elements. This Door is "hosted" by the wall. In other words, it is not possible to place a Door on its own (freestanding) without a Wall as a host. Also notice that moving the mouse from one side of the Wall to the other will flip the Door in or out relative to the Wall. The gray underlay we added previously indicates a patio shape to the left of the plan and wrapping around the top. Our first door will be out to that passageway along the top.

19. Position the mouse on the top Wall near the right side of the passageway so it swings out and then click (see Figure Q.9).

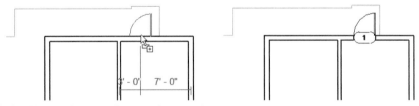

**Figure Q.9** *Place a Door to the outside near the passageway*

20. Place another near the top of the interior vertical Wall.

Notice that Door tags have automatically appeared and the numbers have filled in sequentially. We will learn more about tags later.

Sometimes after placing a Door, it is not positioned or oriented correctly. Just like the Walls above, we can select a Door, and then edit its temporary dimensions to move it to the desired location. There are also small flip control icons on the Door to control its orientation.

21. Click the Flip control to change one of the Door's orientation (see Figure Q.10). Repeat if desired on the other Door.

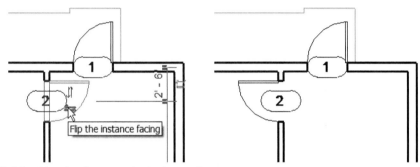

**Figure Q.10**   *Use the flip controls to change Door orientation*

Adding Windows works the same way as adding Doors.

22. On the Design Bar, click the Basics tab and then click the **Window** tool.

   • Accept all of the defaults on the Options Bar.

   • Move the pointer near the top horizontal Wall and move up then down.

Again notice how this controls the placement orientation of the Window. As with the Door, you can always flip it later if you make an error.

23. Place Windows in the two horizontal Walls only (see Figure Q.11).

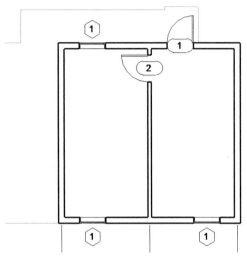

**Figure Q.11**   *Place Windows*

Let's place one more Door in the Wall at the left. For this door we will use a different type and to do so, we will load it from an external library.

24. On the Design Bar, click the Basics tab and then click the **Door** tool.

- On the Options Bar, click the Load button.

- Navigate to the folder where you installed the Mastering Revit Architecture CD ROM files.

- From there, double-click the *Library* folder, then the *Imperial Library* [*Metric Library*] sub-folder and then finally the *Doors* sub-folder.

- Select *Double-Glass 2.rfa* and then click Open.

This action loads the Door component into the currently active project making it available to place in the model. (This Door Family file is a copy of one provided with the out-of-the-box content in the Imperial install of the product).

- Place the Door in the center of the left Wall swinging out. Use the temporary dimensions to assist you in placement (see Figure Q.12).

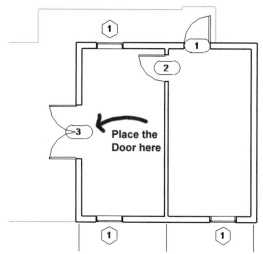

**Figure Q.12**  *Place a double entry Door*

- On the Design Bar, click the **Modify** tool or press the ESC key twice.

25. Save the project.

> **Note:** You can find more information and tutorials on working with Doors and Windows in Chapter 3.

## WORKING IN OTHER VIEWS

We can work in many types of views in Revit Architecture; not just floor plans. Our project includes elevation views and ceiling plan views already. We can also add section views and 3D views. Many other view types are also available and discussed in future chapters. One thing to keep in mind is that all of these views representing the project are contained in one Revit Architecture (RVT) project file.

### VIEW THE MODEL IN 3D

Opening a three-dimensional (3D) view will reveal that our Wall height needs adjustment.

1. On the Project Browser, double-click to open the {3D} 3D view.

This is the default three-dimensional view in Revit Architecture. You can modify it as you like or create others from it. We will use this one for our tutorial, but make some simple adjustments to its vantage point. Notice that the {3D} view is an isometric view of our building model. We can see the Walls, Doors, and Windows we added from a bird's eye vantage point. We also see the Toposurface terrain model that was included in this project. You can change the vantage point of a 3D view interactively on screen.

2. On the View toolbar, click the Steering Wheels icon. (You can also press F8).

A "steering wheel" control will appear on screen and follow the position of your cursor. Several types of 3D navigation are possible.

- Position the steering wheel over the middle of the model.

- Move your mouse to the left side of the steering wheel over the Orbit option.

- Click and drag the mouse in the view window to spin the model around interactively.

Drag side-to-side to move around the building.

Drag up or down to change height of the vantage point.

- Spin the model around so that the front double Door is visible (see Figure Q.13).

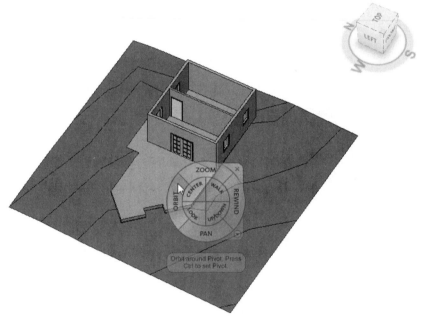

**Figure Q.13** *Dynamically Modify the {3D} view window*

Several other options are possible. For now just try Orbit, Zoom, and Pan. To use each option, click and drag. The position of the cursor becomes the center of the movement. Another option is the view cube in the upper right corner of the screen. You can click several hot spots on the cube to orient the model to that position.

**Tip:** You can also spin the model by holding down the SHIFT key and dragging with the middle wheel button on your mouse.

- On the Design Bar, click the **Modify** tool or press the ESC key.

## CREATE A SECTION VIEW

Let's create a section view to help us understand the relationships built into our Revit Architecture model very clearly as we make some simple edits.

3. On the Project Browser, double-click to open the *Level 1* floor plan view.

4. On the Design Bar, click the Basics tab and then click the **Section** tool.

- Click to the left of the double Door.

Move the pointer through the model to the right keeping the section line horizontal.

- Click outside the model to the right to complete the section line (see Figure Q.14).

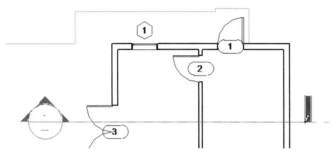

**Figure Q.14** *Cut a section through the model*

A section line, with a section head and tail will appear. Three green dashed lines with drag handles will also appear. The Section Head will currently be red indicating that it is selected.

5. Click next to the section line in the white area of the view background being careful not to click on any geometry.

**Tip:** This is a quick way to deselect the selected element(s). You can also just press the ESC key.

Notice that the Section Head now turns blue. This indicates that it has a linked view associated with it and acts as a "hyperlink" to the associated section view. In the Project Browser, notice that there is now a *Sections* category included in the list. Revit Architecture automatically creates such nodes in the Project Browser as needed.

- Double-click the blue Section Head to open the associated view.

You should now see the *Section 1* view on screen.

- On the Project Browser, double-click to open the *West* Elevation view.

We now have four views open. If you click on the Window menu, you will see all open views listed near the bottom of the menu. You can choose them off this list to bring them to the front of the pile, or you can simply double-click the view name in Project Browser again to display them. You can also tile them all on screen at once. Let's do that now.

6. From the Window menu, choose **Tile** (see Figure Q.15).

If the Recent Files window tiles with the others, minimize it (or close it) and then Tile the windows again.

- On the keyboard, type the letters ZA.

This is the keyboard shortcut for zoom all views to fit.

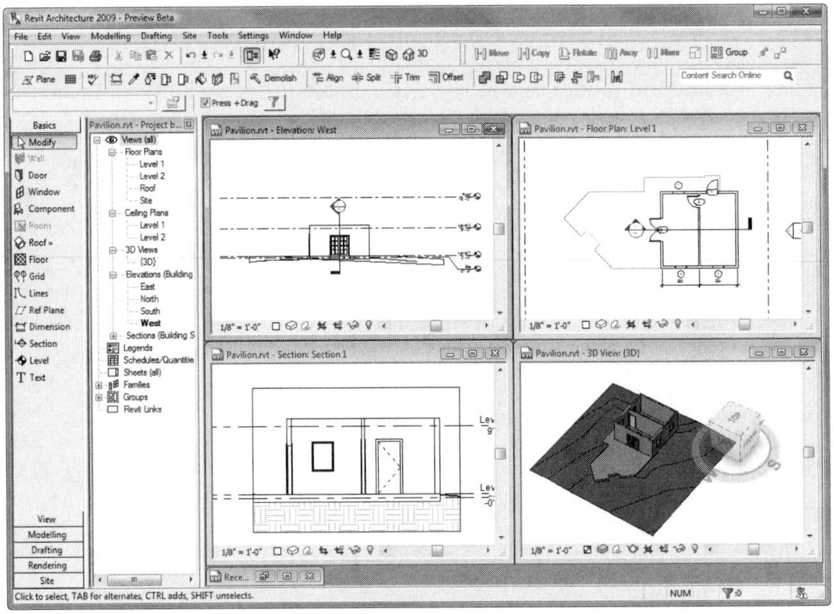

**Figure Q.15**  *Tile the views on screen to view them all at once*

**Note:** You can find more information and tutorials on working with views in Chapter 4.

## EDIT IN ANY VIEW

When you wish to edit a Revit Architecture model, you may perform the edits in *any* view. Changes will automatically be applied to *all* views. This is the power of Revit Architecture! You are describing a single virtual building model or *Building Information Model* (BIM) (see Chapter 1 for complete details). You can "view" it in an unlimited number of ways. Regardless of where you make the edit—plan, section, elevation, 3D, or even schedules, all views are completely coordinated. These graphical and tabular views are like reports of the data contained within the Building Information Model.

### EDITING LEVELS

With our current screen configuration, take a look at the elevation and section views in particular. This project has been set up to have two stories plus a roof. Currently our Walls only go up one story and in fact do not even coincide with the second floor level at all. Let's fix both problems.

1. Click in the floor plan view, *Level 1* to make it active.

2. Place your mouse pointer (the **Modify** tool) over one of the exterior Walls.

Notice the way that it highlights under the cursor. (If you move the mouse away without clicking, the Wall will no longer highlight.) This is called "pre-highlight" and is a useful aid to proper selection.

- Pre-highlight one exterior Wall—do not click yet.

- Press the TAB key.

Notice how all of the exterior Walls now pre-highlight.

- Click the left mouse button to select the pre-highlighted elements.

Notice how all four exterior Walls are now shaded red in all open views. This is called a chain selection.

3. On the Options Bar, click the Properties icon (see Figure Q.16).

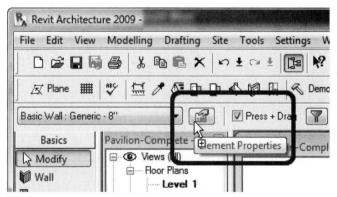

**Figure Q.16** *Click the Properties icon to access the "Element Properties" dialog*

- In the "Element Properties" dialog, from the "Top Constraint" list, choose **Up to level: Roof** and then click OK.

Notice how the Walls project up to the Roof Level line in all views. The Walls are now set relative to this Level in the project. If we were to change the height of the Roof Level, the Walls would also adjust accordingly. Let's try that now.

 **Note:** The section view will likely not show the top of the Walls as it is currently cropped to the first floor. You can adjust this with the double arrow drag control at the top of the Crop Boundary. Click the rectangular box surrounding the section. Click the control at the top edge of the box and drag it upward. The Walls should now show.

4. Click anywhere in the *West* elevation view.

Zoom In Region to get a better look if you need to (right-click to access Zoom In Region).

- Click to select the second floor Level line.

Notice the temporary dimensions that appear. Like the Walls and other elements drawn so far, we can edit the blue dimension value to move the Level lines to a new location. We will move both Level 2 and the Roof level; starting with Level 2.

- Click the blue text of the temporary dimension between Level 1 and Level 2, type **10** and press ENTER (see Figure Q.17).

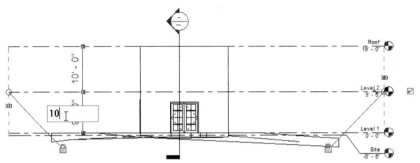

**Figure Q.17** *Move the Level line with the temporary dimension*

Notice that this moves Level 2 relative to Level 1, but that the distance between Level 2 and the Roof has changed as well.

- Repeat for the Roof—select the Roof Level line.

- Click the blue text of the temporary dimension between Level 2 and Roof, type **10** and press ENTER.

 **Note:** Be sure to use the temporary dimension between Level 2 and Roof. If you edit the height of the Roof level directly (the height shown on the level head symbol), you will need to type **20** instead.

Notice that not only does the Roof Level line move, but since we constrained the Walls to the Roof Level, the top edge of the Walls adjusts as well!

 **Note:** You can find more information and tutorials on working with Levels in Chapter 4.

## MODIFY A WINDOW

Let's edit a Window next. Again, we can edit in whatever view is convenient with confidence that the edit will appear in all appropriate views automatically.

5. Spin the 3D model view to show the north Wall.

You can hold down the SHIFT key and drag with the middle wheel button on your mouse, or click the small corner area across the top surface on the view cube.

6. Select the Window on the north Wall. (You can select it in any view—try plan).

- Click on the titlebar of the plan view, then the section view.

Not only is the Window highlighted red (indicating that it is selected) in all views, but in each of these orthographic views where it is visible, the temporary dimensions appear when the view is made active.

- Edit the temporary dimension values to move the Window.

Notice how the Window moves instantly in *all* views. When using Revit, you will never have to worry about chasing down a change in several different drawings to be certain that it has been coordinated everywhere. This will boost productivity and help reduce the number of costly change orders.

Move additional Windows in the same way if you wish.

At this point you may wish to line up the Window on the North Wall across the plan with the one on the South. You can do this and have Revit Architecture maintain the relationship with the Align tool.

7. Working in the plan view, click the Align tool on the Tools toolbar.

You first indicate the point of reference. We'll use the Window we just moved.

- Click near the center of the Window you just moved to set the point of alignment.

- Click near the center of the opposite Window to align it to the reference point (see Figure Q.18).

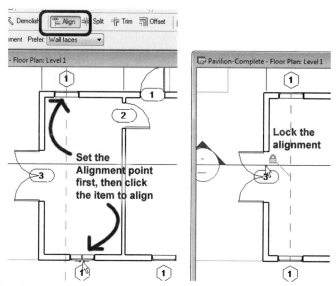

**Figure Q.18**   *Use the Align Tool to align Windows to one another*

- Click the lock icon to constrain the alignment of the two Windows together.

- On the Design Bar, click the **Modify** tool or press the ESC key twice.

If you now move either Window, they will move together. Try it!

## ADD OPENINGS ON THE SECOND FLOOR

Now that our Walls span the height of both floors, we should add some fenestration on the second floor.

8. On the Project Browser, double-click to open the *Level 2* floor plan view.

- Following the above procedures, add Windows and a Door to *Level 2* as shown in Figure Q.18.

**Tip:** You can start with the three Windows and Single-Flush exterior Door already on the *Level 1* plan. Copy these to the clipboard using CTRL + C or the command on the Edit menu. Open the *Level 2* plan and then choose the **Paste Aligned > Current View** command from the Edit menu. Add the remaining Windows using the Wall tool as outlined above.

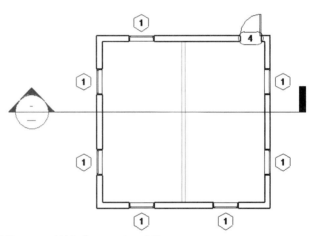

**Figure Q.19**  *Add Doors and Windows to Level 2*

Fine-tune the placement of any of these openings and feel free to repeat the alignment procedure on any of the pairs of Windows as well.

## ROUND OUT THE PROJECT

Our project needs more than just Walls, Doors, and Windows. Let's enclose it with a Floor and Roof and look at how to extract data from our model with Schedules.

### ADD A FLOOR

The second floor of our building will have an interior balcony on the right overlooking the space to the left.

1. On the Design Bar, click the Basics tab and then click the **Floor** tool.

When you click the Floor tool, the floor plan will turn gray. The Design Bar will also change placing the drawing editor into "Sketch mode." Sketch mode is a special two-dimensional drawing mode used when the element that you are creating has a shape that Revit Architecture cannot easily "guess." In this case, it would not be possible for Revit Architecture to assume the size and shape of the Floor that we want, so instead, we will sketch it. This is easy to do, given that we already have several Walls and can use them for reference.

On the Design Bar, the "Pick Walls" mode will already be enabled (selected).

- Click one of the horizontal Walls, then the other.
- Click the vertical exterior Wall on the right.

Notice that with each Wall you click a magenta sketch line will appear on the Wall. Also notice that the sketch will appear on either the inside face or the outside face depending on which side of the Wall the Pick Walls tool was on when you clicked the Wall. The Floor will only cover the right half of the plan, so for the last Wall, we will use the vertical one in the center rather than the exterior one on the left.

If the sketch lines appear on the out side face of a Wall use the Modify tool and select the sketch line. Then click the double arrow flip control to reposition the sketch line to the other face. Repeat as required to locate all three sketch lines on the inside face as indicated in Figure Q.20.

- Click on the vertical Wall in the center (see Figure Q.20).

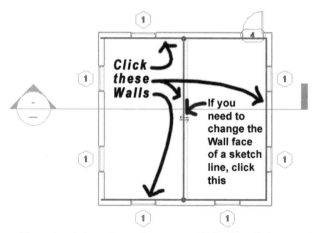

**Figure Q.20**  *Create Floor sketch lines from the existing Walls (Sketch lines in the figure enhanced for clarity)*

2. On the Tools toolbar click the Trim/Extend tool.

- Click the vertical sketch line in the center of the plan, and then click the right side of the horizontal line at the top.

**Note:** When using the Trim/Extend tool you always select the portion of the lines that you want to keep.

- Repeat by clicking the vertical again, then the right side of the horizontal one on the bottom (see Figure Q.21).

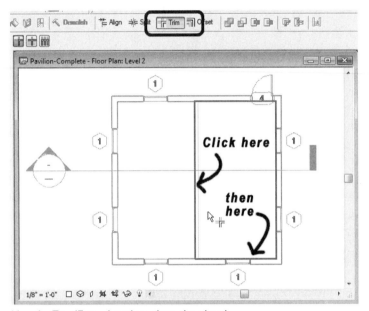

**Figure Q.21** *Use the Trim/Extend tool to close the sketch*

- On the Design Bar, click the Finish Sketch button.

- In the dialog that appears, click Yes.

3. On the Project Browser, double-click to open the {3D} view.

 **Note:** If you are still working with four tiled view windows, simply click the titlebar of the {3D} view to make it active. Double click the view's titlebar to maximize the view if desired.

- Spin the model around and angle down to see the new Floor.

You can also study it in the section view (see Figure Q.22).

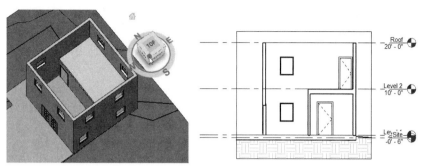

**Figure Q.22** *Study the new Floor in the {3D} and Section 1 views*

 **Note:** You can find more information and tutorials on working with Floors in Chapter 5.

## ADD A ROOF

We can sketch a roof in much the way as we sketched the Floor. There are a few Roof techniques available. For this exercise, we will make a Roof by Extrusion.

4. On the Project Browser, double-click to open the *West* elevation view.

 **Note:** If you are still working with four tiled view windows, simply click the titlebar of the *West* elevation view to make it active.

 **Note:** In the default Revit project template, North is toward the top of the screen in plan view. Therefore, the top elevation mark symbol is the North elevation, the bottom one is South, the right one is East, and the left one is West.

5. On the Design Bar, click the **Roof** tool. From the flyout that appears, choose **Roof by Extrusion**.

Revit Architecture will allow us to sketch a simple 2D shape that will become the shape of the Roof's section. The Roof will extrude this shape along a length we designate. To do this, Revit needs us to establish the plane in which we wish to work.

- In the "Work Plane" dialog, accept the defaults (Pick a plane) and click OK.

If you pause your mouse for a moment, a tooltip will appear that reads: "Pick a vertical plane." If you don't see a tooltip, the same message will appear in the Status Bar at the bottom of your screen.

- Click the Wall facing us (see Figure Q.23).

 **Tip:** To click the face of the wall, or any plane in Revit, you need to position the Pick Plane tool over the edge of the element. The perimeter of the plane will pre-highlight indicating which plane will be selected when you click.

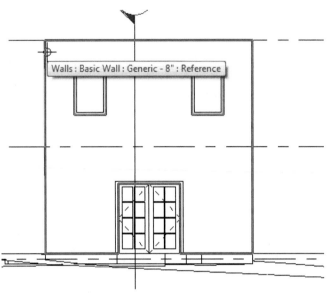

**Figure Q.23** *Click the Wall to set a Reference Plane*

- In the "Roof Reference Level and Offset" dialog that appears, verify that the Level is set to Roof with a 0 offset (this is the default) and then click OK.

We are placed into sketch mode as with the Floor element above. We will now create a curved shape for the Roof.

On the Design Bar, the **Lines** tool should be active.

- On the Options Bar, click the "Arc passing through three points" icon.
- Click the first point of the arc at the top left corner of the Wall.
- Set the next point about 6° below the right top corner (see Figure Q.24).

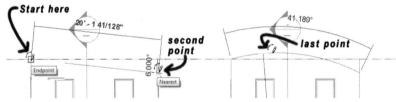

**Figure Q.24** *Create a three-point arc shape for the Roof Extrusion*

- Set the final point approximately where indicated in the figure. (It does not need to be precise).
- On the Design Bar, click the Finish Sketch button.
- On the Project Browser, double-click to open the {3D} view.

**Note:** If you are still working with four tiled view windows, simply click the titlebar of the {3D} view to make it active.

The Roof automatically spanned over the entire building model. However, its eaves are flush with the Walls and the Walls pass through the Roof. Let's add an overhang to the Roof and then we will fix the Walls.

6. Using the **Modify** tool select the Roof in the {3D} view.

Notice the small arrow handles pointing away from the Roof on two ends of the extrusion.

Click and drag each of these handles slightly away from the Walls to create an overhang (see Figure Q.25).

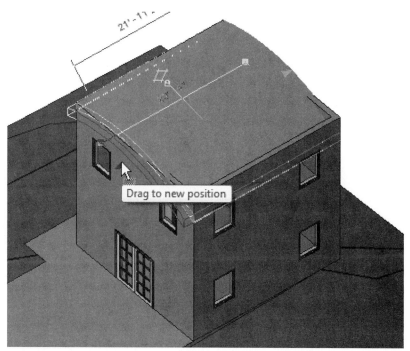

**Figure Q.25**   *Stretch Roof Extrusion to make overhangs*

The previous edit was simple because we were actually changing the overall length of the extrusion. To add overhangs in the other direction, we can edit the sketch again.

• With the Roof still selected, on the Options Bar, click the Edit button.

The sketch line will reappear and the Roof will temporarily disappear. You can switch back to the elevation view or edit the sketch directly in 3D. Remember, you can edit in any view and the change will occur in all views.

• Drag the ends of the line away from the Walls slightly as shown in Figure Q.26.

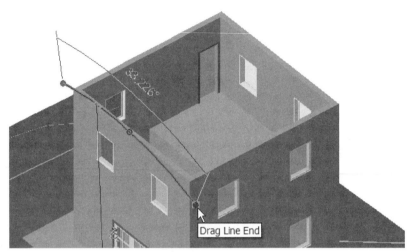

**Figure Q.26** *Edit the sketch line to re-shape the Roof to include overhangs*

- On the Design Bar, click the Finish Sketch button.

7. Use the **Modify** tool and the TAB select method above to chain-select all the exterior Walls.

- On the Options Bar (running horizontally across the top of the view window), click the Attach button.

- Click the Roof (see Figure Q.27).

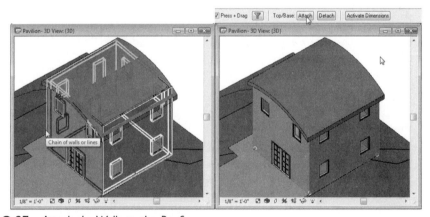

**Figure Q.27** *Attach the Walls to the Roof*

**Note:** If you edit the shape of the Roof, the Walls will remain attached. Try it out if you like. Select the Roof. On the Options bar click the Edit button and change the shape of the arc. Finish the sketch to see the change to the roof shape and how the Walls remain attached. Undo any change before continuing.

- Save the model.

 **Note:** You can find more information and tutorials on working with Roofs in Chapter 7.

## ADD A STAIR

We have no way to reach our second floor balcony. Let's add a Stair.

8. On the Project Browser, double-click to open the *Level 1* floor plan view.

 **Note:** Whether the view is open already or not, double-clicking its name on the Project Browser will make the view active.

9. On the Design Bar, click the Modeling tab and then click the ***Stairs*** tool.

The Stair will go to the north (top) side of the plan on the right of the building. There is a little bump out on the existing patio for this purpose.

- Click near the middle of the patio bump out and drag up (see Figure Q.28).

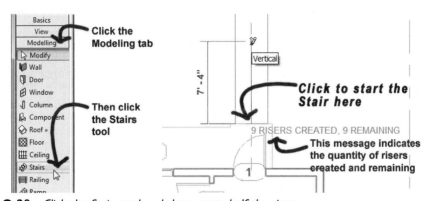

**Figure Q.28** *Click the Stairs tool and then create half the risers*

A small label will appear on screen indicating how many risers have been created and how many remain.

- Drag straight up until the gray label reads "9 Risers created, 9 remaining" and then click to create the first 9 risers.

- Click a point next to the first run of stairs at the location indicated in Figure Q.29.

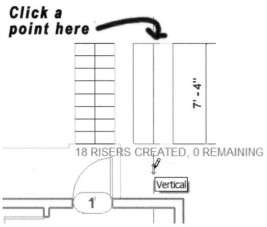

**Click a point here**

7' - 4"

18 RISERS CREATED, 0 REMAINING

Vertical

1'

**Figure Q.29**  *Create the remaining Risers*

- Drag straight down until the message indicates that zero risers remain and then click to create the remaining risers.

This will give us the basic Stair but it will not "hook up" with the second floor. We need to extend the top riser to make it a landing at the top.

10. Select the riser line at the bottom right (the last riser of the Stair) with the **Modify** tool.

- Drag it down until it snaps to the building.

- On the Tools toolbar, click the Split tool.

- Click on each of the green lines to split them where indicated in Figure Q.30.

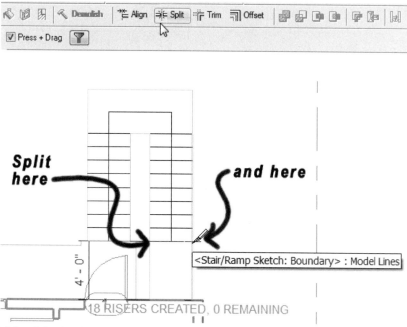

**Figure Q.30** *Split the Stringer Lines to make a landing*

The green lines represent the stringers of the Stair. It is necessary to split them (break them into two segments) so that the top part in this case can slope with the Stair and the bottom part can be flat and follow the landing. We also need to change the slope of these lines so they represent a landing and not additional stringers.

- Click the **Modify** tool.

- Select one of the stringer lines (the ones we just split) and on the Options Bar; choose **Flat** from the Slope list.

- Repeat for the other side (see Figure Q.31).

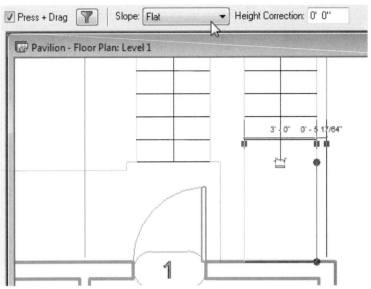

**Figure Q.31** *Set the slope of the sketch lines to Flat*

- On the Design Bar, click the Finish Sketch button.

When you are finished, repeat the process for the two Railings. Select one of the Railings and click the Edit button on the Options Bar. You do not need to split the sketch line, but you will need to set the slope to Flat. Finish the sketch and then do the other Railing.

11. On the Project Browser, double-click to open the {3D} view.

- Spin the model around to see the Stair (see Figure Q.32).

- Move the Door on the second floor if necessary.

**Figure Q.32** *Study the results in the {3D} view*

 **Note:** You can find more information and tutorials on working with Stairs and Railings in Chapter 6.

This completes the basic geometry of the model. We could add many more embellishments like railings on the patio and second floor balcony, adjustments to the windows and roof eaves, and the addition of materials to name a few. For now we'll finish this quick start tutorial by creating some construction documentation items like a Door and Window Schedule and some sheets for printing.

12. Save the project.

## CREATE A SCHEDULE

We can create automated schedules of anything in Revit Architecture. All we need to do is generate a Schedule view which, while not graphical like the plan, section and elevation views, are just like the other views in Revit Architecture. Plans, sections, elevations, and 3D views, etc. are graphical views. Schedules are tabular views. You can view information related to the model and even edit it directly from a schedule view.

13. On the Design Bar, click the view tab and then click the ***Schedules/Quantities*** tool.

The "New Schedule" dialog will appear.

- From the "Category" list, choose Doors and then click OK.

The "Schedule Properties" dialog will appear.

- In the "Schedule Properties" dialog, on the "Fields" tab, click Mark in the "Available Fields" list and then click the Add → button.

 **Note:** "Mark" represents the instance number of every Door element.

- Repeat for the following Fields: Level, Width, Height, Frame Type, Frame Material, Family and Type, and Comments (see Figure Q.33).

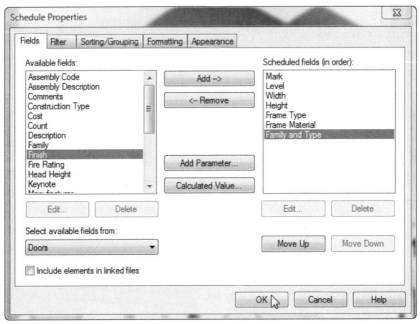

**Figure Q.33**  *Add Fields to the Door Schedule*

- Click OK to create the Schedule.

A Schedule view will appear on screen. The Schedule view appears much like a spreadsheet. Let's look at the Schedule tiled next to one of the floor plans. However, before we tile, let's close some of the other views.

If you still have the windows tiled on screen, maximize the current window. (use the small icon in the upper right corner of the window, or simply double-click the titlebar).

14. From the Window menu, choose **Close Hidden Windows**.

- On the Project Browser, double-click to open the *Level I* floor plan view.

- From the Window menu, choose **Tile**.

You should now have just the *Level 1* floor plan view and the *Door Schedule* view open on screen side by side.

15. In the Door Schedule view, click on Door number 3.

The Door number will highlight in the Schedule and the Door itself will highlight in the plan (see Figure Q.34).

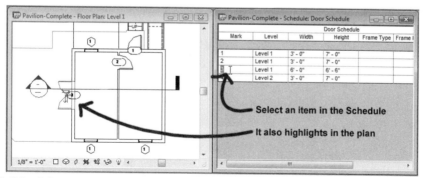

**Figure Q.34** *Selected elements highlight in graphical views and Schedules*

- Highlight the value in the Mark field, type **5** and then press ENTER.

Notice that the value changes in the Door Tag on the floor plan as well.

- For the same Door, click in the Family and Type column and choose **Double-Glass 2: 68" × 80"** from the pop-up list.

Notice that the size of the Door changes in both the Schedule and the floor plan. This is considered the correct way to change the size of a Door since most Doors use "Type-based" parameters to control their width. To illustrate this point, try an experiment on the other Doors in this project. Click in the Width column for Door 1. Note that it will again highlight in the plan as well. In the Width field, change the value to **2**. A message will appear stating: This change will be applied to all elements of type Single-Flush: 36" × 84". This means that *all* Doors of this Type will be affected when you click OK. Go ahead and click OK to see this. Notice that all single Doors in this project (including the one on the second floor) are now 2'-0". This is likely not the desired result, particularly since the Family and Type value still indicates that the name of this Door is Single-Flush: 36" × 84", which is likely to foster confusion. This topic will be discussed in more detail in later chapters. For now, it suffices to say that if you wish to change the width or height of the Door, that you should choose a different Door Family and/or Type instead rather than simply edit the dimensions directly in the schedule (unless you really want to edit all Doors at once). Experiment with other changes if you wish. Undo the change to the single Doors before continuing.

16. Create another Schedule view (Design Bar, click the **Schedule/Quantities** tool).

- In the "New Schedule" dialog, from the "Category" list, choose Windows and then click OK.

- In the "Schedule Properties" dialog, on the "Fields" tab, add the Mark, Level, Width, Height, Count and Comments fields.

- Click the Sorting/Grouping tab.

- For Sort by, choose Level and then place a checkmark in the "Footer" checkbox.

- At the bottom of the dialog, place a checkbox in the "Grand totals" checkbox and clear the "Itemize every instance" checkbox.

- Click OK to create the Schedule (see Figure Q.35).

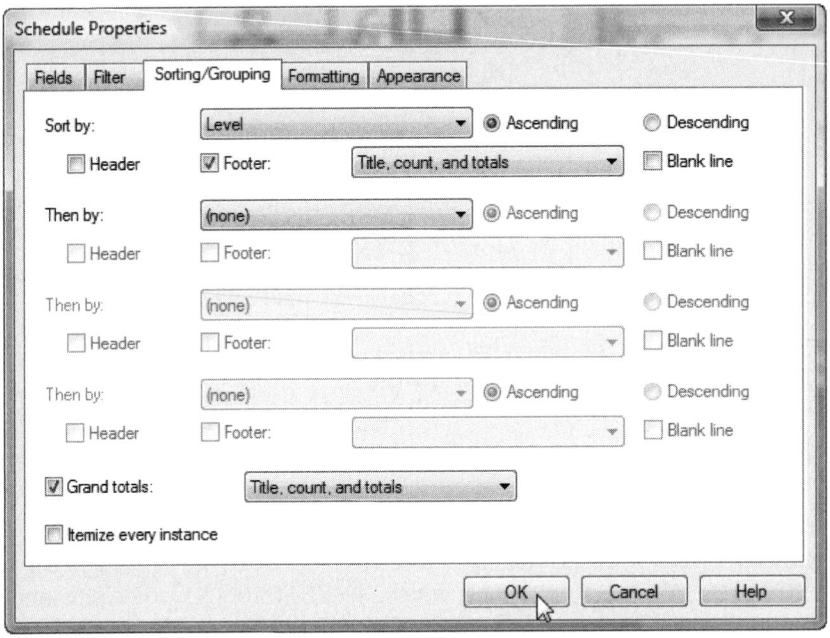

**Figure Q.35** *Configure the Window Schedule to sort and total the Windows in the project by Level*

17. In the *Level 1* floor plan view, add Windows to the right Wall.

Notice that the new Windows appear immediately in the Schedule.

18. Save the model.

**Note:** You can find more information and tutorials on working with Schedules in Chapter 11.

## PREPARING OUTPUT

At some point in every architectural project, you will need to output your designs and produce some form of deliverable. Currently the most common format for this is a collection of printed drawings. In Revit Architecture you use special "sheet" views for this purpose. These views emulate the final paper output and allow us to compose the completed sheets formatted the way we wish, complete with titleblocks.

## ADD A SHEET

While it is possible to print the views we already have, we will get more polished results by creating one or more sheet views. Sheet views are basically pieces of paper upon which we drag and drop the various views for printing. Both graphical (drawings) and tabular (schedule) views can be combined in sheet views.

1. On the Design Bar, click the view tab and then click the **Sheet** tool.

- In the "Select a Titleblock" dialog click OK.

- Double click the titlebar of the new window that appears to maximize it.

- From the view menu, choose **Zoom > Zoom To Fit**.

A blank sheet view with a Titleblock appears and is ready to receive views. There are two ways to do this. Right-click the sheet view name on the Project Browser and choose **Add View** to receive a dialog listing all available views, or simply drag and drop them from Project Browser. We'll use drag and drop here.

2. From the Project Browser, drag the *Level 1* floor plan view and drop it on the sheet.

An outline of the view will appear attached to the cursor. This is called the viewport boundary and it automatically sizes itself to the extents of the dragged view's contents. You can use this to place it on the sheet where desired.

- Position the view in the upper left corner of the sheet and then click to place it.

- Repeat the drag and drop process for each of the remaining floor plans.

It will be a close fit, but the first and second roof plans should fit across the top of the sheet.

- Drag each of the Schedule views to the sheet as well (see Figure Q.36).

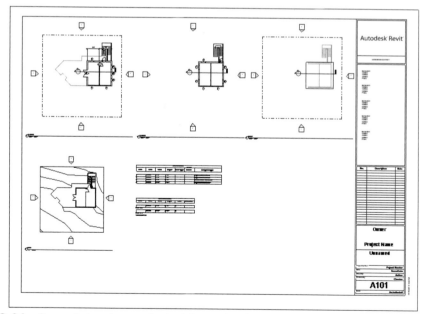

**Figure Q.36** *Drag all the plans and schedules to the sheet view and position them*

If you need to re-position a view after you drag it, you can click on it directly on the sheet with the **Modify** tool, and then drag it again to move it or use the arrow keys on your keyboard to nudge it slightly. When you drag the Site plan in, notice that it is a little smaller than the others. Each view has its own scale setting that is used when it is added to a sheet.

3.  Repeat the entire process to create another sheet and add the elevations and section views.

Suppose that you wish to change the scale of a view after it is added. For instance after adding all of the elevations and the section view to this new sheet, you may wish to enlarge the Section view.

4.  Select the Section view on the sheet, right-click and choose **Activate View**.

This makes the view editable as if you had opened it from the Project Browser and edited the original section view.

5.  Right-click in the Section view and choose **View Properties**.

    -   In the "Element Properties" dialog, beneath the "graphics" grouping, change the "View Scale" to **1/4"=1'-0"** and then click OK.

    -   Right click in the view again and choose **Deactivate View**.

    -   Reposition the view as required.

Notice that the change in scale only affected the graphics of the view. The graphics of the model elements such as walls, doors, and windows enlarged but the text and annotations remained the same size.

6.  Perform any other edits and explorations you wish.

7.  Save and close the project.

## GOING FURTHER

Feel free to print your two sheets out to your printer or plotter. You can also generate a "digital" plot using the **Publish DWF > 2D DWF** command on the File menu. This will create a compact potable file that can be opened and redlined on any computer using the freely available Design Review software from Autodesk. You can further edit and refine this model if you wish to add additional elements and annotations. Everything will remain coordinated and the sheets will automatically receive all the updates. Congratulations! You have completed your first Revit Architecture project! Your journey into Revit Architecture and Building Information Modeling awaits you in the coming chapters.

## SUMMARY

- Getting started with Revit Architecture is easy—click a tool on the Design Bar and place the item in the view window.

- Walls, Doors, and other elements interact with each other as you place them in the model.

- Relationships and constraints are maintained automatically as you work.

- Build or edit your model from any view and changes are fully coordinated in all views.

- Views include graphical representations like plans and sections and non-graphical tabular representations like Schedules.

- Drag views to sheets for printing.

# CHAPTER 1

# Conceptual Underpinnings of Revit Architecture

## INTRODUCTION

Revit Architecture is an object-based software package for architects and building designers. It facilitates the creation of a "Building Information Model" (BIM) in which plans, sections, elevations, 3D models, quantities and other data are fully coordinated and can be readily manipulated and accessed. From the BIM database, one can perform design tasks, query quantities and takeoffs and generate drawing sheets for construction documentation needs. The advantages of this approach are many. From a production point of view, it means less time drafting and coordinating building data, because all drawings and reports come from the same source model. If this model changes, all "Views", whether floor plans, sections, elevations, or schedules reflect the change immediately. To achieve this level of functionality, it is important to understand a bit about what it means to create and work within a Building Information Model. That is the primary goal of this chapter.

## OBJECTIVES

In this chapter, we will explore the meanings of parametric design, Building Information Modeling and take a high level look at the Revit Architecture software package. Working in a provided dataset, you will learn how to view a single model in many different ways that serve a variety of architectural drawing and documentation needs. Topics in this chapter include:

- Building Information Modeling
- The fully coordinated nature of a Revit Architecture Project File
- The basics of Revit Architecture Elements
- An introduction to Families and Types
- Core concepts within the context of a Project

## BUILDING INFORMATION MODELING

*In the Autodesk Revit Curriculum & Student Workbook, author Simon Greenwold concisely and elegantly presents the concept of Building Information Modeling. A portion of that text is reproduced here to help explain BIM and how it compares to more traditional computer aided design (or drafting) (CAD) technology. The following topic is excerpted from the above mentioned publication which is © 2005 Autodesk, Inc. It is used here with permission.*

Building Information Modeling [BIM] is a process that fundamentally changes the role of computation [and delineation] in architectural design. It means that rather than using a computer to help produce a series of drawings and schedules that together describe a building; you use the computer to produce a single, unified representation of the building so complete that it can generate all necessary documentation. The primitives from which you compose these models are not the same ones used in CAD (points, lines, curves). Instead you model with building components such as walls, doors, windows, ceilings, and roofs. The software you use to do this recognizes the form and behavior of these components, so it can ease much of the tedium of their manipulation. Walls, for instance, join and miter automatically, connecting structure layers to structure layers, and finish layers to finish layers.

Many of the advantages are obvious—for instance, changes made in elevation propagate automatically to every plan, section, callout, and rendering of the project. Other advantages are subtler and take some investigation to discover. The manipulation of parametric relationships to model coarsely and then refine is a technique that has more than one career's worth of depth to plumb.

BIM design marks a fundamental advance in computer-aided design. As the tools improve, ideas spread, and [practitioners] become versed in the principles, it is inevitable that just as traditional CAD has secured a deserved place in every office, so will BIM design.

### CAD Versus Building Information Modeling

**Modeling Is Not CAD**—BIM is entirely unlike the CAD tools that emerged over the last 50 years and are still in wide use in today's architectural profession. BIM methodologies, motivations, and principles represent a shift away from the kind of assisted drafting systems that CAD offers.

To arrive at a working definition of Building Information Modeling first requires an examination of the basic principles and assumptions behind this type of tool.

**Why Draw Anything Twice?**—You draw things multiple times for a variety of reasons. In the process of design refinement, you may want to use an old design as a template for a new one. You are always required to draw things multiple times to see them in different representations. Drawing a door in plan does not automatically place a door in your section. So the traditional CAD program requires that you draw the same door several times.

**Why Not Draw Anything Twice?**—There is more to the idea of not drawing anything than just saving time in your initial work of design representation. Suppose you have drawn a door in plan and have added that same door in two sections and one elevation. Now should you decide to move that door, you suddenly need to find every other representation of that door and change its location too. In a complicated set of drawings, the likelihood that you can find all the instances of that door the first time is slim unless you are using a good reference tracking system.

**Reference and Representation**—But doesn't reference tracking sound like something a computer ought to be good at? In fact, that's one of about three fundamental things a computer does at all. And it is exceedingly good at it. The basic principle of BIM design is that you are designing not by describing the forms of objects in a *specific* representation, but by placing a *reference* to an object into a holistic model. When you place a door into your plan, it automatically appears in any section or elevation or other plan in which it ought to be visible. When you move it, all these views update because there is only *one* instance of that door in the model. But there may be *many* representations of it.

**You Are Not Making Drawings**—That means that as you create a model in modeling software, you are not making a drawing. You are asked to specify only as much geometry as is necessary to locate and describe the building components you place into the model. For instance, once a wall exists, to place a door into it, you need to specify only the type of door it is (which automatically determines all its internal geometry) and how far along the wall it is to be placed. That same information places a door in as many drawings as there are. No further specification work is required.

That means that the act of placing a door into a model is not at all like drawing a door in plan or elevation, or even modeling it in 3D. You make no solids, draw no lines. You simply choose the type of door from a list and select the location in a wall. You don't draw it. A drawing is an artifact that can be automatically generated from the superior description you are making.

**Even More Than a 3D Model**—You are making a model—a *full* description of a building. This should *not* be confused with making a full three-dimensional (3D) model of a building. A 3D model is just another representation of a building model with the same incompleteness as a plan or section. A full 3D model can be cut to reveal the basic outlines for sections and plans, but there are drawing conventions in these representations that cannot be captured this way. How will a door swing be encoded into a 3D model? For a system to intelligently place a door swing into a plan but not into a 3D model, you need a high-level description of the building model separate from a 3D description of its form. This is the model in BIM design.

**Encoding of Design Intent**—This model encodes more than form; it encodes high-level design intent. A staircase is modeled not as a rising series of 3D solids, but as a staircase. That way if a level changes height, the stair automatically adjusts to the new criterion.

**Specification of Relationships and Behavior**—When a design changes, BIM software attempts to maintain design intent. The model implicitly encodes the behavior necessary to keep all relationships relative as the design evolves. Therefore the modeler is required to specify enough information that the system can apply the best changes to maintain design intent. When you move an object, it is placed at a location relative to specific data (often a floor level). When this [datum] moves, the object moves with it. This kind of relativity information is not necessary to add to CAD models, which are brittle to change.

**Objects and Parameters**—You may be troubled by the idea that the only doors you are allowed to place into a wall are the ones that appear on a predefined list. Doesn't this limit the range of possible doors? To allow variability in objects, they are created with a set of parameters that can take on arbitrary values. If you want to create a door that is nine feet high, it is only necessary to modify the height parameter of an existing door. Every object has parameters—doors, windows, walls, ceilings, roofs, floors, even drawings themselves. Some have fixed values, and some are modifiable. In advanced modeling you will also learn how to create custom object types with parameters of your choosing.

### How Do BIM Tools Differ from CAD Tools?

Clearly, because modeling is different from CAD, you are obliged to learn and use different tools.

**Modeling tools don't offer such low-level geometry options**—As a general rule modeling deals with higher-level operations than CAD does. You are placing and modifying entire objects rather than drawing and modifying sets of lines and points. Occasionally you must do this in BIM, but not frequently. Consequently, the geometry is generated from the model and is therefore not open to direct manipulation.

**Modeling tools are frustrating to people who really need CAD tools**—For users who are not skilled modelers, modeling can feel like a loss of control. This is much the same argument stick-shift car drivers make about control and feel for the road. But automatic transmission lets you eat a sandwich and drive, so the choice is yours. There are also ways to layer on low-level geometric control as a post-modeling operation, so you can regain control without destroying all the benefits of a full building model.

**Modeling entails a great deal of domain-specific knowledge**—Many of the operations in the creation of a Building Information Model have semantic content that comes directly from the architectural domain. The list of default door types is taken from a survey of the field. Whereas CAD gets its power from being entirely syntactical and agnostic to design intent, BIM design is the opposite. When you place a component in a model, you must tell the model what it is, *not* what it looks like.

**Or else requires you to build it in yourself**—Adding custom features and components to a BIM design is possible but requires more effort to specify than it does in CAD. Not only must geometry be specified, but also the meanings and relationships inherent in the geometry.

### Is Modeling Always Better Than CAD

As in anything, there are trade-offs.

**A model requires much more information**—A model comprises a great deal more information than CAD drafting. This information must come from somewhere. As you work in a modeling tool, it makes a huge number of simplifying assumptions to add all the necessary model information. For instance, as you lay down walls, they are all initially the same height. You can change these default values before object creation or later, but there are a near infinitude of parameters and possible values, so the program makes a great many assumptions as you work. The same thing happens whenever you read a sketch, in fact. That sketch does not contain enough information to fully determine a building. The viewer fills in the rest according to tacit assumptions.

**Flouting of convention makes for tough modeling**—This method works well when the building being modeled accords reasonably well with the assumptions the modeler is making. For instance, if the modeler makes an assumption that walls do not cant in or out but instead go straight up and down, that means that vertical angle does not need to be specified at the time of modeling. But if the designer wants tilted walls, it's going to require more work—potentially more work than it would to create these forms in a CAD program. It is even possible that the internal model that the software maintains does not have the flexibility to represent what the designer has in mind. Tilted walls may not even be representable in this piece of software. Then a workaround is required that is a compromise at best. Therefore unique designs are difficult to model.

**BIM Manager Note:** The author of this passage used the example of tilted walls simply to make the accompanying point. Tilted Walls are possible in Revit Architecture, but admittedly with a little more effort than common vertical ones.

**Whereas CAD doesn't care**—In CAD, geometry is geometry. CAD doesn't care what is or isn't a wall. You are still bound by the geometric limitations of the software (some CAD software supports nonuniform rational b-splines [NURBS] curves and surfaces, and others do not, for instance), but for the most part there is always a way to construct arbitrary forms if you want.

**Modeling can help project coordination**—Having a single unified description of a building can help coordinate a project. Because drawings *cannot* ever get out of sync, there is no need for concern that updates have not reached a certain party.

**A single model could drive the whole building lifecycle**—Increasingly, there is interest in the architectural community for designing the entire lifecycle of a building, which lasts much longer than the design and construction phases. A full building description is invaluable for such design. Energy use can be calculated from the building model, or security practices can be prototyped, for instance. Building systems can be made aware of their context in the whole structure.

**BIM potentially expands the role of the designer**—Clearly this has implications for the role of architects. They may become the designers of more than the building form, but also specifiers of use patterns and building services.

**Modeling may not save time while it's being learned**—It is likely that while designers are learning to model rather than to draft, the technique will not save time in an office. That is to be expected. The same is true of CAD. Switching offices from hand drafting to CAD occurred only as students became trained in CAD and did not therefore have to learn the techniques on the job. The same is likely to be true of BIM design. But students are beginning to learn it, so it is bound to enter offices soon.

**Potential hazards exist in BIM that do not exist in CAD**—Because design *intelligence* is embedded into a model, it is equally possible to embed design *stupidity*.

**BIM** *Manager Note:* For example, if the computer operator inputs a floor to floor height of only 4 feet [1200 millimeters], BIM software will not automatically correct or even flag this as an error. The knowledge of the architect is still the driving force behind the design intent that creates the model. Building Information Modeling software by itself does not replace the knowledge and experience of an Architect.

**Improperly structured models that look fine can be unusable**—It is possible to make a model that looks fine but is created in such a way that it is essentially unusable. For instance, it may be possible to create something that looks like a window out of a collection of extremely tiny walls. But then the program's rules for the behavior of windows would be wasted. Further, its rules for the behavior of walls would cause it to do the wrong things [with the "windows" modeled this way] when the design changed.

**What Is an Engineering Technique Doing in Architecture?** —BIM design comes from engineering techniques that have been refined for many years. Many forces are acting together to bring engineering methodologies like BIM into architecture. First, the computing power and the basic ability to use computers have become commonly available. Second, efficiencies of time and money are increasingly part of an architect's concern. BIM offers a possible edge in efficiency of design and construction.

## DEFINING BIM

This quoted passage does a wonderful job at outlining the high-level concepts involved in BIM and comparing and contrasting its tenets and techniques to traditional CAD and drafting methods. Using the points raised in this article, let's try to synthesize them into a definition of Building Information Modeling.

Unfortunately, Building Information Model(ing) is perhaps one of the most misunderstood terms in the architectural industry today. "BIM" is often assumed to be synonymous with simply generating a three-dimensional (3D) model of a building—whether that model has any useful non-graphical information or not, and regardless of the level to which the 3D model is detailed. Part of the problem is that Building Information Modeling *is* an evolving concept; one that will continue to change as the capabilities of technology and our own ability to manipulate technology improve. These issues make it difficult to formulate a simple definition for BIM. However, as the popularity of BIM is growing, reaching a consensus on its meaning and intent is increasingly important.

Summarizing all of the points made so far, the most important issue is that emphasis belongs on the "I—Information" in BIM. That information can be either graphical or non-graphical; either contained directly in the building model or accessible from the building model through linked data that is stored elsewhere. As has already been mentioned in the quoted passage above, when you exercise BIM, you are making a model which is a *full* description of a building—*not* just a 3D model. A data model is just as valid a model as a geometric model. (This is not a new concept in Architecture. Consider the existing requirement of both a set of drawings *and* a written specification to complete a construction documents package). Despite the importance of these distinctions, when we think of a Building Information Model, a three-dimensional geometric model of the building is often what comes to mind. So the first step to fully understanding BIM is to realize that "BIM" and "3D Model" is *not* the same thing.

In simple terms, a Building Information Model is a complete representation or depiction of a building that aids in its design, construction and potentially ongoing management. Such representation will often employ any combination of 3D graphics, 2D abstractions and/or non-graphical data as required for conveying full intent. The coordination and delivery of information and intent is the most important goal in BIM.

*Some of the content of the previous topic is paraphrased from Matt Dillon's Web Log (Blog). You can find the complete article at the following URL: http://modocrmadt.blogspot.com/2005/01/ bim-what-is-it-why-do-i-care-and-how.html Portions used here were used with permission.*

## REVIT ARCHITECTURE KEY CONCEPTS

So now that you've got a good idea of the BIM concept you may be wondering how it specifically relates to Revit Architecture. Even more importantly, you may also be wondering how BIM will improve the way you work. Throughout the course of this book, and even more as you begin working with Revit Architecture on your own projects, you will gain comfort and familiarity with key Revit Architecture concepts. In this topic, we will identify and describe some of the most important Revit Architecture concepts, including Revit Architecture Elements, Families & Types, and Editing Modes.

### ONE PROJECT FILE—EVERYTHING RELATES

In a Revit Architecture project (which is often contained within a single computer file), you will notice that no matter where in the project you work, no matter what kind of view you are working in, all changes occur immediately and all elements retain their relation-ships with each other. This complete *bi-directional* coordination is perhaps the most significant benefit to using Revit Architecture. You can make a change in any view (plan, section or schedule) with complete confidence that the change is instantly reflected throughout the entire Project File in all other views.

 **Note:** Changes to "Detail" views occur only in the edited view. Detailing is discussed in Chapter 10.

The most talked about examples of the bi-directional coordination usually have to do with the "physical" aspects of your building project. If you move a door in a plan view, for example, the same door will also move wherever it appears in an interior elevation or perspective view.

Another example is in the "informational" aspects of that door. If you go into a schedule view where that door is listed and change it from wood door to a glass door, not only will the calculations like quantities or costs for that door change in the schedule, but also the change will be reflected in all graphical views—for example, in shaded views, the door will now appear transparent. Likewise, if this data is linked to cost estimation or green building calculations, the change to a glass door type will have other important impacts as well. This example gives a good example of how the "I" in BIM often deserves more emphasis than the "M."

Another important aspect of Revit Architecture is the Project Browser. Every view, Family and Group is clearly listed and organized in the Project Browser. The advantage of this is that all pieces of the project are always accessible and neatly organized automatically. It is not possible for a team member to accidentally save project data in the wrong folder or location.

## REVIT ARCHITECTURE ELEMENTS

In Revit Architecture you create a representation, or model of your project, using three types of elements. An element in Revit Architecture is simply a discrete piece of data like an object or drawing sheet. The three basic types of elements are Model Elements (Walls, Doors, Roofs), view Elements (Plans, Sections, Schedules), and Annotation Elements (Tags, Text, Dimensions). Your graphical model primarily contains Model Elements. View Elements enable you to display, study, and edit the model in depictions that represent traditional architectural drawing types. All views automatically appear in logical categories in the Project Browser tree. To prepare views in Revit Architecture to appear on and print from sheets, Annotation Elements such as text, dimensions, and tags are used to notate and clarify the information shown in the various views. Specific architectural scale, level of detail, and other display characteristics are also the province of views. Views can be placed on sheets and plotted to produce presentation drawings or drawing sets.

An illustration of the major element types with examples and their relationship to Revit and each other appears in Figure 1.1.

# REVIT ELEMENTS

There are five types of elements. Each represents something fundamental to your project.

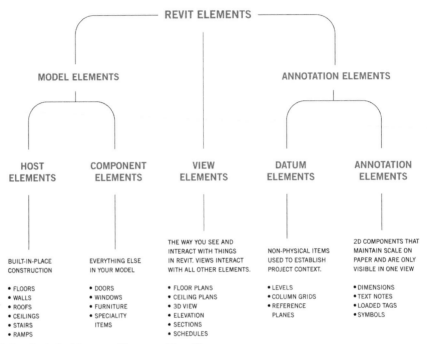

**Figure 1.1**    *Revit Architecture Elements Flow Chart*
*Image courtesy of Autodesk*

## Model Elements

You can think of Model Elements as the items you use to describe the physical aspects of your building project. These elements in Revit Architecture can be separated into two sub-categories (as shown in Figure 1.1)—Host Elements and Component Elements. Host Elements represent items that are constructed in place in the actual building like Walls and Roofs. Component Elements are placed or installed in the building like Doors and Furniture. The figure shows further examples of each of these element types. For the purpose of our discussion here, let's consider the most common element of each category: a Wall (Host Element) and a Door (Component Element).

**Table 1.1**   *Comparing Host and Component Elements*

| Host Elements | Component Elements |
|---|---|
| *Here are the similarities:* | |
| They are both considered Families. | |
| Wall – Basic | Door – Single-Flush |
| They can both have multiple Types. | |
| Generic – 4" Brick | 30" x 84" |
| Exterior – Brick on CMU | 36" x 80" |
| *Here are the differences:* | |
| Their Basic Role in Projects | |
| Define Spaces and Enclosures | Modify/Detail Spaces and Enclosures |
| Their Relationship in Hosting | |
| Can *Host* Components (A Wall can *Host* a Door) | Can be *Hosted* (A Door can be *Hosted* by a Wall) |
| They are saved in different places | |
| Saved in Project Files (*transferred* to new projects) | Saved in Family Files (*loaded* into new projects) |

The primary concept to understand about Model Elements is that they *are* the elements that you use to create a physical model that depicts your building, and these same elements are the ones that appear within *any* of the views of your project.

When adding a wall to your project, you start by deciding the type of Wall you are creating and in what location the Wall should appear. Determining how the lines and patterns that represent that wall will look on a particular drawing is handled automatically by the software. In this way, you are actually constructing a model of the required Walls rather than drafting a specific representation of them as you might in manual or CAD drafting.

 **Note:** The specific graphical settings can be modified manually if required in the Object Styles dialog box.

To get a complete listing of all of the Model Elements available in Revit Architecture, and how they will appear graphically in your project, open the Object Styles dialog box (Settings menu), and study the list on the "Model Objects" tab (see Figure 1.2).

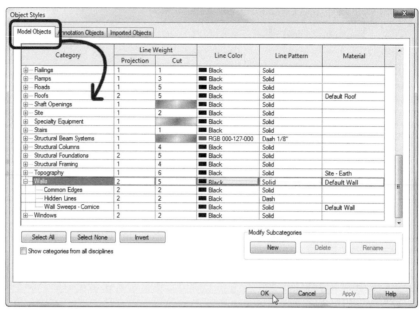

**Figure 1.2** *A list of Model Elements appears in the Object Styles dialog*

## View Elements

As you develop your project in Revit Architecture, you will work with and create several views. Every Revit Architecture project begins with at least some views already in the Project Browser. The specific views that are available are a function of the template project from which the project is created. (A project template provides the framework and settings for a new project. More information on project templates is available in Chapter 4). To open and work in a view, you simply double-click its name in the Project Browser. Views are available for every architectural drawing type traditionally included in Architectural documentation sets. Like the drawings they represent, views allow you to interface with your model and edit its contents and composition within a particular context like plan, elevation, or section. Unlike traditional drawings, (as mentioned above) an edit in one view is instantly reflected in all appropriate views throughout the project.

**Multiple Views-One Model**—Imagine two friends living on opposite sides of the same street. Let's assume that the street runs north-south and that one friend lives on the west side of the street while the other lives on the east side. If both friends were looking out their window at the same time, as a car was passing by on the street below traveling from the south to the north, which way would the friends say that the car was driving relative to their respective vantage point? The friend in the house on the west side would describe the car as traveling from his right to his left, while the friend on the east side would say the car traveled from her left to her right. Which friend was correct? What if both friends snapped a photo at the same time? The two photos would show a different "view" of the same car and its travel pattern. Upon comparing photos with one another, would the two friends describe the respective scenes as two different cars? Or would they rather describe them as two different aspects of the same car?

This hypothetical scenario illustrates how project views work in Revit Architecture. When working in Revit, we frequently switch from view to view to edit and create elements; and although the specific graphics displayed on screen may vary (like showing the driver or passenger side of the car in the scenario above), they convey aspects of the *same* model. Therefore, the specific view in which you make an edit is irrelevant. A change to the model occurs in only one place—the model. You can study the change from any number of vantage points as represented in various views.

To see a complete list of view types in Revit Architecture, you can look at the **View > New** Menu (see Figure 1.3).

 **Note:** Revit Architecture includes both graphical views like plans, sections, and elevations and tabular views like Schedules and Material Takeoffs. Both types provide the means to study and manipulate models. Examples occur throughout this book.

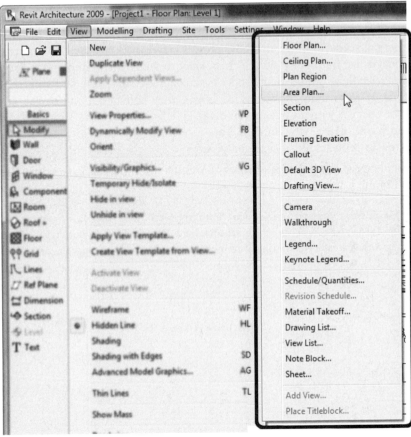

**Figure 1.3**  *Available view types are seen on the View > New menu*

## Annotation Elements

If you return to Figure 1.1, you will see that the overall Annotation Elements category of Revit Elements contains two sub-categories like the Model Elements category does. The first sub-category is Datum Elements and the other is simply named Annotation Elements again. While first category of datum elements does provide graphic annotations for views used as drawings, their behavior in terms of the project as a whole is very different from the latter category of annotation elements. This is because changes to datum elements affect the project as a whole, and will potentially cause changes to both model and view elements.

On the other hand, annotation elements *only* affect the views in which they are used. This makes their behavior unique in relation to the other types of Revit Architecture elements. Annotation elements include all of the text and other descriptive architectural symbology that is required on drawings to explain and clarify the intent of the graphics, such as tags, dimensions, and view-specific detail components.

Annotation elements are the closest thing to what you might consider "drafting" in Revit Architecture. In fact, most of the annotation elements tools are located on the Drafting tab of the Design Bar. This is because they rarely effect changes in other views, or to the project model as a whole. In some cases, annotation elements can be used to manipulate the objects from within the view where they are placed. Dimensions are one such example. Even elements in the so-called annotation category can have an impact on the model as a whole. We will see examples of dimensions used to manipulate model geometry in the lessons that follow.

Annotation elements include objects such as Dimensions, Text, Tags, and more. To get a complete listing of all of the annotation elements available in Revit Architecture, and how they will appear graphically in your project, open the Object Styles dialog box (Settings menu), and study the list on the "Annotation Objects" tab (see Figure 1.4).

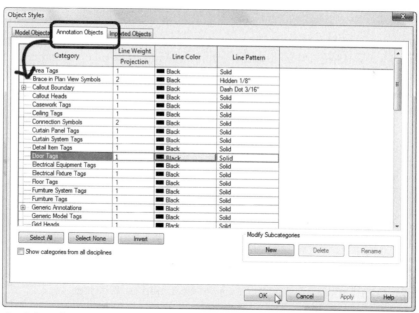

**Figure 1.4**  *A list of annotation elements can be seen in the Object Styles dialog*

### Components, Model Lines and Detail Lines

Most elements in Revit Architecture are purpose-built elements that have obvious functions based upon their namesakes like Wall, Door, Room Tag and Section. The function of each of these elements is easily inferred from their respective names. However, there are also more generic elements whose function is not as specific that serve important functions in Revit Architecture as well. These elements include items like components, text and lines. A Component is a model element that does not have a pre-defined function like a Door or a Window. Components are employed to create Furniture, Fixtures and other items that are placed in models. Components are based on predefined templates so that while there is no specific "Furniture" element per se, there is a Furniture category and Family Template. (The term Family will be discussed in more detail below).

Text and Lines on the other hand can actually be either model or annotation elements. The same distinction (discussed above) between model and annotation applies to these elements as well—Model Text and Model Lines are actually part of the model and used to represent real items in the model. For instance, if you want to create signage you create it with Model Text and Model Lines might be employed to create inlaid patterns on Walls or Floors that you don't find necessary to model three-dimensionally.

While the creation process and graphical appearance on screen of Model and Detail lines may appear very similar and therefore make it difficult to distinguish them from one another, they are in fact very different. For example, if you draw a floor pattern using Model Lines; it *will* appear in all appropriate views like the floor plan view and a 3D view of the space. Model Lines added to the surface of a Wall would appear in both elevation and section views as if you actually painted these lines on the surfaces of the model.

On the other hand, a Detail Line, like other annotation, will appear in *only* the view in which it is added. Adding a Detail Line is like drawing on a piece of paper covering the view of your project, and will not be added *physically* to your building model. It is treated like other annotation as a simple embellishment to that particular view only. The most common use of such embellishment would be on enlarged details created from the model. Rather than meticulously model components that would only be practical to show in large scale drawings of a design, Detail Lines can be employed to represent those elements that would otherwise take too much time and effort to model throughout and would also add unnecessary overhead to the model without a commensurate amount of benefit. Such items might include building paper in a wall or roof section, nails, screws or other fasteners, reinforcing, or even moldings and trim in some cases. It would certainly be possible to model any of these elements, but in many cases, the additional overhead and effort required to model them would not be justified. Understanding what *not* to model is perhaps even more important than knowing what or how to model. This is a very important issue and understanding it is critical to using Revit Architecture in the most efficient and practical manor. More information on this concept can be found in Chapter 10.

The Revit Architecture interface is covered in detail in the next chapter. However a point of clarification is useful here to understand when you get Model Lines and when you get Detail Lines in the interface. The Lines tool from the Basics and Modeling tabs of the Design Bar gives you Model Lines. To create a Detail Line, you must click on the Drafting tab of the Design Bar and choose the Detail Lines tool. In general, a tool labled simply "Lines" will be Model Lines, while Detail Lines will be labeled accordingly.

## FAMILIES & TYPES

One of the most common terms in Revit Architecture is the term "Family." A Family in Revit Architecture is an object designated for a particular purpose that has a specific collection of parameters and behaviors. Within the limits established by the Family and its parameters, an endless number of "Types" can be spawned. Where a Family establishes a set of available variable parameters, a Type is a specific version of the Family with actual values for each parameter. The term "Family" was selected to characterize objects in Revit Architecture that have an inherited-property relationship. The main idea comes from the notion of a Parent-Child relationship—which is a central idea in object-based computer programming.

In order to illustrate how this concept works in Revit Architecture, we will look at some common categories of Families and describe how they are created and used in a project. You will notice these follow along similar lines as the Elements hierarchy covered above, with the addition of an additional Element class called "System Families."

## Model Element Families

In Revit Architecture, everything in the software belongs to a Family. Each of the element types shown in Figure 1.1 above has a corresponding Family classification. Recall that the model element branch includes both Host and Component elements. This means that we have both Host Families and Component Families in Revit Architecture. The most common example of a Host Family is the Wall Family. The organization of Revit elements follows a hierarchical progression from global to specific parameters and can be referred to as a "Family Tree." Here is an example of a Wall Family Tree:

**Table 1.2** *Family Tree Examples*

| Family Tree Hierarchy | Sample Wall Family Tree |
|---|---|
| ElementElement Category | Walls |
| Family/System Family | Basic Wall |
| Type | Exterior – Brick on CMU |

You can also see the hierarchy interactively in a Revit project. To do so, expand the Families branch of the project Browser. Here is the Wall Family Tree illustrated in a Revit Architecture project (see Figure 1.5):

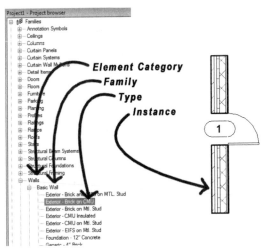

**Figure 1.5** *A typical Wall Family Tree viewed in the Project Browser*

Recalling our discussion of element categories above, a Wall Family is a host Family. Host elements represent "built in place" construction. (This includes items that are assembled from raw materials on the job site). Host Families are hard-coded into the software and cannot be created or edited by the user. Users can however add, edit, or delete Family Types associated with host Families. If the Wall Type that you wish to use is not present in the current project, you must either duplicate and modify a similar Type that is resident in the file or transfer one from another project. Wall Types and other host Element Types cannot be saved into their own individual files. Component Families include all of the component elements and as defined above. They represent items that are purchased and installed in a project (not assembled in place). Perhaps the most common component element Family is a Door Family. Here is an example of a Door Family tree:

**Table 1.3** *Family Tree Examples*

| Category | Item |
|---|---|
| Element Category | Doors |
| Family/System Family | Single-Flush |
| Type | 36" x 84" |
| Instance | Specific Instance of a Door selectable in the project |

Here is this Door Family tree illustrated in a Revit Architecture project (see Figure 1.6):

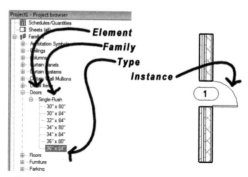

**Figure 1.6** *A typical Door Family tree viewed in the Project Browser*

Unlike host Families, component Families do not have to be resident in the project file before you can use them. You can load component Families from external files. Doing so will typically load all Types associated with that Family into the current project (unless the Family file has a "Type Catalog" associated with it like Structural shapes—see the online Help for more information). You can also create new Types within the project just as you can with host Families. Baring these subtle differences, host and component Families are very similar to one another. Families will be covered in more detail in the coming chapters.

### Annotation Families

Families are not limited to model elements. Every element in Revit Architecture belongs to a Family. Annotation Families are similar to model Families in that they are also objects that have a specific collection of parameters and behaviors. However, in this case, those parameters and behaviors are specific to annotation rather than the model. For example, there are Families for Dimensions, Text, Tags, and Datums. You can see most Annotation families loaded in your project by expanding the Annotation Symbols branch in the Project Browser (see Figure 1.7).

**Figure 1.7** *Annotation Element Families appear beneath Annotation Symbols on the Project Browser*

Many Annotation Families serve special purposes in a project. Some are simple symbols that you can add to any view. A Title block for instance is a special kind of Annotation Family that is used when you create a sheet. It can include text fields that automatically report project data and can also contain company logos and other graphics. Some Annotation Families, like Level Heads, Section Heads, and Elevation Heads are typically associated directly to some other view in the project and provide a means to navigate from one view to another. For example, add a Section line in a plan to indicate where the section is cut. Double-click the Section head to open the associated section view. Tags usually contain a symbol created from simple geometry and a label that reports a specific field or fields from an associated object. This might be the door number, room number, or area of a space, for example.

### System Families

Families even help to expand options about how to work with the software. One example of this is Browser Organization. You'll notice if you select the views branch of the Project Browser, the Type Selector field activates with the current Browser Organization Family Type. The default type is Browser:Views:all (see Figure 1.8).

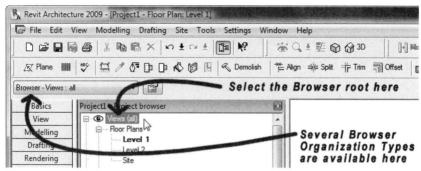

**Figure 1.8**    *Even the organization of the Project Browser is governed by a Family*

Like all Families, the Browser Family includes other Types. If you open the drop-down list (on the Type Selector) you will see the other Types in this Family. Many of the Types available in the list are quite useful depending on the kind of project on which you are working and/or the phase or composition of the design team (see Figure 1.9).

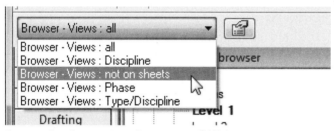

**Figure 1.9**    *The Browser Family contains other very useful Types*

If the Type of Browser Organization that you wish to use is not included in the Type Selector list, you can choose **Browser Organization** from the Settings menu to create new Types to help manage and display the views created in your Revit Architecture project (see Figure 1.10).

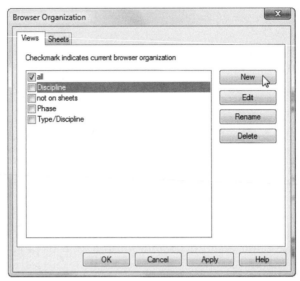

**Figure 1.10**  *The Browser Organization command allows you to modify or create Browser Organization Types*

Changing the Browser Organization can help you find a specific view you need particularly on large projects with many views. For example, if you are working in a project that will have many phases (existing, new, etc.) the Phase Type can be helpful. Discipline organization will sort the views by various disciplines like Architectural, Structural, or Mechanical. Not on sheets will hide all the views that have already been added to a sheet in the project, thus showing only those views not currently on a sheet. This is a handy way to be sure you don't forget to include an important view on a sheet in your document set. Should you wish to organize the browser in some other manner than those already included, you can use the New button in the Browser Organization dialog to create and customize your own.

## EXPLORE AN EXISTING PROJECT

We have covered a lot of important concepts in this chapter so far. To help reinforce and solidify the concepts presented let's open and explore a dataset provided with the files installed from the Mastering Revit Architecture CD ROM.

### INSTALL THE CD FILES AND OPEN A PROJECT

The lessons that follow require the dataset included on the Mastering Revit Architecture CD ROM. If you have already installed all of the files from the CD, simply skip down to step 3 below to open the project. If you need to install the CD files, start at step 1.

1. If you have not already done so, install the dataset files located on the Mastering Revit Architecture CD ROM.

Refer to "Files Included on the CD ROM" in the Preface for instructions on installing the dataset files included on the CD.

2. Launch Revit Architecture from the icon on your desktop or from the **Autodesk** group in **All Programs** on the Windows Start menu.

**Tip:** In Windows Vista, you can click the Start button, and then begin typing **Revit** in the "Start Search" field.

3. If the New Features Workshop dialog appears, choose "Maybe later" and then click OK.

4. On the Standard toolbar, click the Open icon.

**Tip:** The keyboard shortcut for Open is CTRL + O. **Open** is also located on the File menu.

- In the "Open" dialog box, click the *My Documents* icon on the left side.
- Double-click on the *MRAC* folder, and then the *Chapter01* folder.

If you installed the dataset files to a different location than the one listed here, use the "Look in" drop down list to browse to that location instead.

5. Double-click *MRAC Chapter01.rvt* to open the project.

You can also select it and then click the Open button.

**Note:** For this brief tutorial, only an Imperial units dataset has been provided. For the remainder of the book, Metric datasets are provided as well.

The project will open in Revit Architecture with the last opened view visible on screen. *The dataset for this chapter provided courtesy of Mark Schmieding.*

## GETTING ACQUAINTED WITH THE PROJECT

For this tutorial, we will explore a series of sheet views included in the project. A sheet view is a special kind of view that emulates a sheet of paper from which drawing sets can be printed to output devices. sheet views typically include a title block which includes project and drawing information.

Revit Architecture remembers the last view that was open when the project was saved. In this case, it is a three-dimensional aerial view of the entire project. This is a small one-floor project for a youth center. It includes offices, exam and counseling rooms, a multipurpose room, and media rooms. Let's take a closer look (see Figure 1.11).

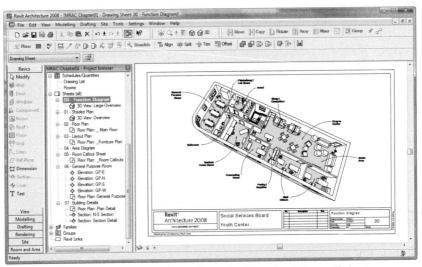

**Figure 1.11** *The Youth Center dataset shown from the "Function Diagram" sheet*

You can use the wheel on your mouse to zoom in and out in any view. You can hold the wheel in and drag to pan the screen. You can also use the scroll bars at the right and bottom for this purpose. If you do not have a wheel mouse, you should consider purchasing one. However, with or without a wheel mouse, you can use the commands on the **View > Zoom** menu to get a better look at the model in any view. Most of these commands will be available in all views, like Zoom To Fit (which fits the screen to the extent of the model) and Zoom In Region (which allows you to drag a rectangular region on screen to zoom). We also have the handy **Zoom > Sheet Size** available. When you are in a sheet, this will zoom the screen to the scale of the sheet and give you a good preview of how the sheet will look when printed. Each of the zoom commands has a command shortcut that you can execute via the keyboard. These shortcuts are two characters and you simply type both characters in succession to execute the appropriate command. For example, to issue Zoom to Fit, you can use the View menu, right-click in the drawing area, or simply type ZF. Commands with shortcuts have the shortcut listed next to the command on the right side of the menu.

6. From the View menu, choose **Zoom > Zoom In Region**.

You can also type ZR to issue this command.

• Drag a rectangular region around the upper left corner of the drawing.

• Hold in the wheel on the mouse and drag around to pan the model (see Figure 1.12).

If you prefer, or if you don't have a wheel, use the scroll bars instead.

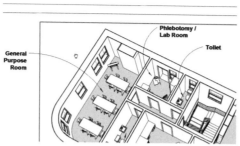

**Figure 1.12**  *Zoom the sheet to sheet Size and pan around to see it as it will print*

The image you see on screen is actually the view named *Large Overview* in the *3D Views* category of the project. It has been added to the current sheet and displays in a "Viewport."

7.  Zoom back out. The easiest way is to choose **Zoom To Fit** from the View menu. (Or type zF).

## UNDERSTANDING SCREEN TOOL TIPS

Let's return the screen to showing us the full image.

8.  Move your mouse pointer into the middle of the screen and pause it there—pause over the drawing, not a text note.

Do not click the mouse.

Notice how a rectangular border highlights around the image. As you pause the mouse, an onscreen tool tip should appear as well. In this case, this tip will read: Viewports : Viewport : No Title (see Figure 1.13).

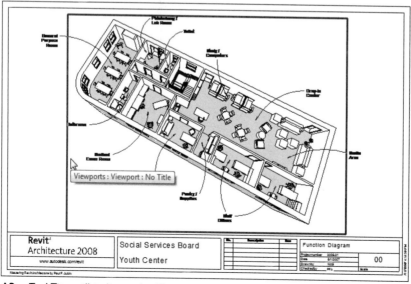

**Figure 1.13**  *Tool Tips will indicate the Element category, Family and Type*

The tool tip conveys three bits of information about the element highlighted—Element category : Family : Type. So in this case, the Element category is Viewports, the Family is Viewport and the Type is No Title. Now hover the Modify tool over a piece of text but do not click. This is called "Pre-highlighting." The tool tip for a piece of text will read—Text Notes : Text : 3D Notes. Here, Text Notes it the Element category, Text is the Family and 3D Notes is the Type.

Since sheets are primarily intended for printing, you do not initially see the elements within the model pre-highlighting. However, you can choose to "Activate" a Viewport that will give you access to the building model elements shown in the view. Editing them from a Viewport is no different than opening the view on Project Browser and editing them there; it is the same view either way. Let's take a look.

9. Pre-highlight the Viewport, and then click to select it this time.

- Right-click and choose **Activate View**.

Notice that the sheet title block and the text labels have grayed out. While they are still visible, this graying effect indicates that they are currently inactive.

- Move the mouse around the model.

Notice that the elements within the model now pre-highlight (see Figure 1.14).

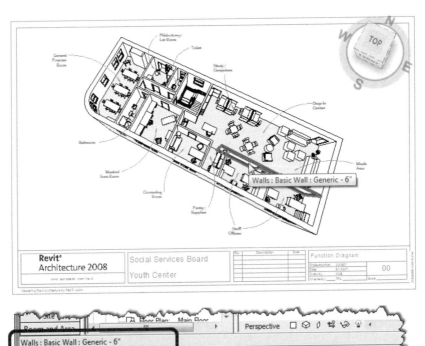

**Figure 1.14**  *Once the Viewport is activated, you can pre-highlight the elements in the model*

We will not actually edit any model objects in this view, but do take notice of the tool tips at this level. The interior partitions, for example, display Walls : Basic Wall : Generic – 6". The Element category is Walls, the Family is Basic Wall and the Type is Generic – 6".

In addition to the tooltips on screen, you will also notice the same information appears in the status bar at the lower left corner of the Revit Architecture screen. You can see an example of this at the bottom of Figure 1.14.

> 10. When you are done exploring in the model, right-click in the Viewport again and choose **Deactivate View**.

This returns you to the sheet and the elements in the view are no longer selectable.

## VIEWS AND DETAILING

We have discussed several distinctions between model and annotation in Revit Architecture. Using this dataset, let's explore these concepts a bit further.

> 11. On the Project Browser, beneath the *Views (all)* category, double-click to open the _ *Main Floor* plan view.

This is the basic floor plan view for this project.

> 12. On the Project Browser, double-click to open the *_Room Callouts* plan view.

This plan is very similar to the _ *Main Floor* view except that it also includes callouts around the General Purpose Room on the left and some elevation and section markers. A sheet has been provided showing each of these views.

> 13. On the Project Browser, double-click to open the *05 – Room Callout Sheet* sheet view.

Notice how the only visual difference here is that the plan appears on a title block sheet in this view.

> 14. On the Project Browser, double-click to open the *02 – Floor Plan* sheet view.

This is the sheet presentation of the _ *Main Floor* plan view. In other words, this sheet composes the _ *Main Floor* plan view on a title block for printing. You can easily see which views appear on a sheet in the Project Browser.

> 15. On the Project Browser, beneath the *Sheets* node, expand the tree (click the small plus (+) sign) beneath the *01 – Shaded Plan* sheet (see Figure 1.15).

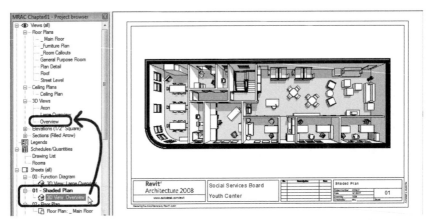

**Figure 1.15** *Expand the sheet entries in the Project Browser to see the Views they contain*

This provides an easy way to see which views inserted on particular sheets. Another useful tool noted above gives us a way to see which views have not yet been placed on sheets.

16. On the Project Browser, click on the top node labeled *Views (All)*.

- From the Type Selector, choose **Browser – Views : not on Sheets** (see Figure 1.16).

**Figure 1.16** *Change the Browser Organization to "not on Sheets"*

Notice that the list of views on the Project Browser filters to show only those views (in any category) that are not yet assigned to a sheet. In this particular project, there are only a couple views not yet on sheets, as you can see.

- From the Type Selector, return to **Browser – Views : all**.

17. On the Project Browser, double-click to return to the _ *Main Floor* plan view.

Suppose that we needed to create another floor plan that was similar to this one, but that was to convey a different type of information on the printed sheet or that we were planning to use simply as a convenient place in which to edit the model with no intention of using it to print. To achieve either goal, we simply duplicate an existing view.

- On the Project Browser, right-click the _ *Main Floor* plan view and choose **Duplicate View > Duplicate**.

PA new floor plan view named *Copy of _ Main Floor* will appear and become active. Notice that none of the room labels or dimensions were copied in this operation. This might be useful if you were creating a "working" view. A "working" view is intended as a view in which you manipulate the model only and do not plan to add to a sheet for printing. Bear in mind that nothing prevents the working view from being used on a sheet; rather it is simply not intended for that purpose by our project team. If we want to duplicate the view, including the tags and dimensions, we choose a different command.

- On the Project Browser, right-click the _ *Main Floor* plan view and choose **Duplicate View > Duplicate with Detailing**.

 **Note:** "Duplicate with Detailing" is short for "Duplicate with view specific detailing elements and annotation elements." Remember that the "Detailing" is being copied, while the model elements are simply being viewed.

Be sure to right-click on _ *Main Floor* and not *Copy of _ Main Floor* in this step. A new floor plan view named *Copy (2) of _ Main Floor* will appear and become active.

- Right-click *Copy (2) of _ Main Floor* and choose **Rename**.
- In the "rename View" dialog, type **Area Diagram** and then click OK.

18. With the CTRL key held down, select each of the Dimensions in the view.

- Press the DELETE key.

We do not need Dimensions for the new view we are creating. However, there is no way to duplicate only the room labels and not the dimensions, so simply deleting them achieves the result.

19. On the Design Bar, click the Drafting tab and then click the Color Scheme Legend tool.

If you do not see this tool, look for the More Tools > button. If your screen resolution is too low, some tools may be hidden and this will reveal them (see Figure 1.17).

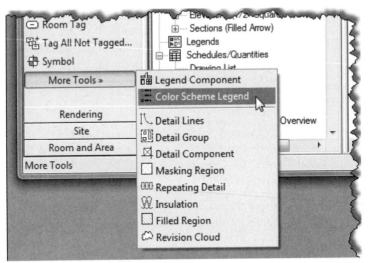

**Figure 1.17** *If you don't see all of the tools, look for them beneath the More Tools button*

A small square with a label will appear attached to the cursor.

- Click a point above the plan to place the Color Scheme Legend.

- In the dialog that appears, Scheme 1 will be selected. Accept this by clicking OK.

- Click on the Color Fill Legend and then drag the small round Control at the bottom to make the legend two columns (see Figure 1.18).

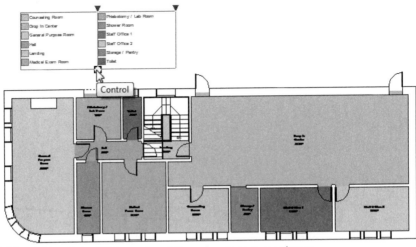

**Figure 1.18** *Add a Color Fill Legend and then drag it to two columns*

20. On the Project Browser, double-click to open the *04 – Area Diagram* sheet view.

A sheet appears on screen which does not yet have a drawing on it. Let's add our new shaded plan to this sheet.

- On the Project Browser, right-click the *04 – Area Diagram* sheet and choose **Add View**.

- From the "Views" dialog, choose *Floor Plan : Area Diagram* view and then click the Add View to Sheet button.

- Click to place the view on the sheet.

Notice that the view is a little too big for the sheet. We can adjust the scale of the view and it will update on the sheet.

21. On the Project Browser, reopen the *Area Diagram* view.

- At the bottom of the screen, choose **1/8"=1'-0"** from the scale pop-up menu (see Figure 1.19).

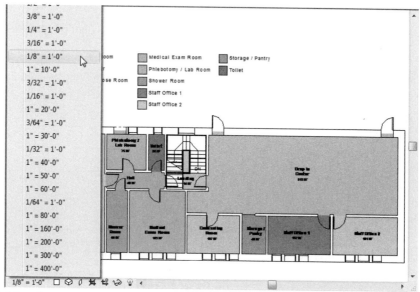

**Figure 1.19** *Change the scale of the view*

22. Return to the *04 – Area Diagram* sheet to see the change.

- To hide the titlebar under the viewport, select the viewport and then choose **Viewport : No Title** from the Type Selector.

You should also take a look at the *_Furniture Plan* floor plan view and the *03 – Layout Plan* sheet next. In this view and sheet, you will notice that the plan is displayed with furniture. Therefore creating plans with and without detailing (text and other annotation) is *not* the only way to vary the specifics of what we see. We can also control the visibility of each type of element in *any* Revit Architecture view. The visibility settings are a parameter of the view itself. This is how we can choose to display the furniture in the *_Furniture Plan* floor plan view and not display it in the _ *Main Floor* view. From the View menu, you can choose Visibility / Graphics to see a dialog listing all element categories and enabling you to turn on and off items per view. While we will discuss the specifics of this process in later chapters, the important point for this exercise is that this sort of control *is* possible and extremely useful. If you wish to explore the Visibility/Graphics Overrides dialog, please feel free to do so. Simply undo your changes before continuing with the lesson.

## EDIT IN ANY VIEW

Perhaps the most powerful feature of Revit Architecture is the ability to edit in any view and see the results instantly in all views.

23. On the Project Browser, double-click to open the *06 – General Purpose Room* sheet view.

This sheet shows the views that are associated to the callouts we saw on the *05 – Room Callout Sheet* sheet above.

- Select the plan view on the left, right-click and choose **Activate View**.

- On the Design Bar, click the Basics tab and then click the Window tool.

- On the Options Bar, clear the "Tag on Placement" checkbox.

- Click a point on the exterior Wall at the left to add a new Window (see Figure 1.20).

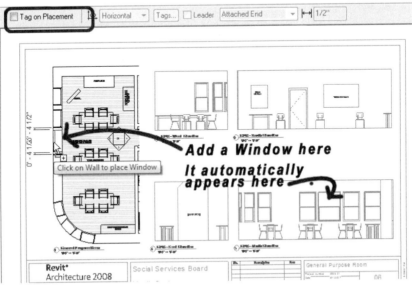

**Figure 1.20**  *Add a Window and it appears in all appropriate Views automatically*

- Right-click in the plan view again and choose **Deactivate View**.

## EXPLORE A DETAIL VIEW

As we have noted above, a Detail view is a little different than the other Views. Typically it will include a live view of the model—usually a callout of some part of a section or plan—and various types of annotation and other graphical embellishments on top. One such detail view has been included in this sample dataset.

24. On the Project Browser, expand (click the plus [+] sign) the *07 – Building Details* sheet view.

Beneath this sheet entry in the Project Browser will appear a listing of three views that are already placed on the sheet.

- Beneath the *07 – Building Details* sheet view entry, double-click to open the *Section : Section Detail* view.

- Pre-highlight some of the elements in this view (see Figure 1.21).

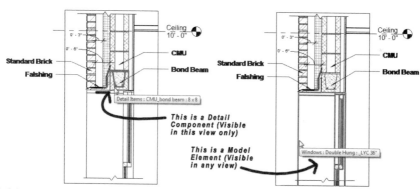

**Figure 1.21**  *Explore a Detail view—Note the combination of Detail and Model elements*

Notice that the Detail view contains both Model elements (which would appear in all Views) and Detail elements (which appear only in this view). Even though the Detail elements represent items like concrete blocks, brick, flashing and bond beams, the level of detail required in a construction detail is much higher than that required in nearly any other view. Therefore, these types of items are typically drawn as Detail elements on top of the Model view geometry in this way to keep overhead low and reduce the amount of time and effort required to build your overall model. Complete coverage of this technique can be found in Chapter 10.

Continue to explore in this dataset as much as you wish to get a better feel of how the various elements and Views in a Revit Architecture project interact. Close Revit Architecture when you are finished exploring. You do not need to save the file.

## SUMMARY

- Building Information Modeling is the process of creating an accurate representation of the building including its form, functions, and systems from which detailed and accurate information can be extracted.

- Using BIM successfully requires a firm understanding of the concepts and techniques enabling its use.

- In Revit Architecture, all views show the same source data stored in a single source model. If changes are made in one view, the change is immediately available in all views.

- Model elements are those items that represent real building components and are visible in all views.

- Annotation elements appear only in the view to which they are added and include notes, dimensions and detail embellishments.

- All items in Revit Architecture belong to Families.

- There are System Families, Model Families and Annotation Families.

- Model Families can be either Host or Component Families.

# CHAPTER 2

# Revit Architecture User Interface

## INTRODUCTION

This chapter is designed to acquaint you with the user interface and work environment of Revit Architecture. The Revit user interface is logically organized and easy to learn. In this chapter, we will explore its major features. While many features follow standard Microsoft Windows™ conventions, many aspects of the user Interface are unique to Revit Architecture. In this overview, our goal is to make you comfortable with all aspects of interacting with and receiving feedback from Revit Architecture. Many of the lessons that follow are descriptive in nature and some are tutorial-based. Feel free to follow along in Revit Architecture as you read the descriptions.

## OBJECTIVES

To get you quickly acquainted with the Revit Architecture user interface, topics we will explore include:

- An overview of the Revit Architecture user interface

- Interface terminology

- Working with the Design Bar and the Options Bar

- Moving around a Revit Architecture model

## UNIT CONVENTIONS

Throughout this book, Imperial units and files will be listed first, followed by metric in brackets, for example, **Imperial [Metric]**. See the Preface for complete details on style conventions used throughout this book.

Imperial dimensions throughout this text appear in the "Feet and Inch" format for clarity. However, when typing imperial values into Revit Architecture, neither the foot symbol (') (when typing whole feet) nor the hyphen separating the feet from inches (when typing both) is required. Therefore, to type values of whole feet, simply type the number. To type values of both feet and inches, type the number of feet with the foot symbol (') followed immediately by the number of inches; the inch symbol is not required. When typing only inches, the inch (") symbol *is* required unless you preface the value with a leading 0'. For example, 4'-0" can be typed in Revit Architecture as simply: **4** (or **48"**). To input four feet six inches, type: **4.5**, **4'6** or **54"**. You can also type **4 6** (that is **4** SPACE **6**). To type 10 inches, type: **10"** or **0'10** or **0 10** (that is **0** SPACE **10**). Hyphens are not required. When separating inches from fractions, use a SPACE. Consult Table 3.1 for additional examples.

**Table 2.1** *Acceptable Imperial Unit Input Formats*

| Value Required | Type This: | Or This |
|---|---|---|
| Four feet | **4** | **48"** |
| Six inches | **.5** or **0'6** or **0 6** | **6"** |
| Five feet six inches | **5'6** or **5.5** or **5 6** | **66"** |
| Four feet six and one half inches | **4'6 ½** or **4 6.5** or **4 6 1/2** | **54.5"** |

**Note:** Typing the foot (') mark is acceptable when typing whole feet as well; however, it is not required.

**Note:** Dimensions throughout this text are given in the Feet and Inch format for clarity. However, feel free to enter dimension values in whatever of the above acceptable formats you prefer. Eliminating the inch or foot marks where possible reduces keystrokes and is recommended despite their inclusion in this text.

If using metric units, all values in this text are in millimeters and can be typed in directly with no unit designation required. More information on style conventions used in this book can be found in the Preface.

## UNDERSTANDING THE USER INTERFACE

Revit Architecture offers a clean and streamlined work environment designed to put the tools and features that you need to use most often within easy reach. In addition to the many onscreen tools and controls, many of the most common tools also have keyboard shortcuts. The topic of shortcuts will be explored below. Figure 2.1 shows the Revit Architecture screen with each of the major interface elements labeled for your reference.

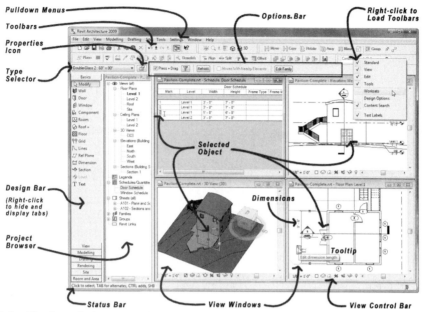

**Figure 2.1**   *The Revit Architecture User Interface*

Consistent with most Windows software, the Revit screen is framed with pull-down menus along the top edge, the Windows minimize, maximize, and close icons in the top right corner, a status bar along the bottom edge, and a variety of toolbars at the top edge of the screen. In addition to these Windows standards, the center of the Revit Architecture screen is divided into three major sections: the Design Bar, the Project Browser, and the Workspace. The Design Bar is organized into several tabs. Clicking on a tab reveals a collection of tools contained within. You can right-click the Design Bar to display hidden tabs. The Project Browser can be thought of as the "table of contents" for your Revit Architecture project. It reveals all of the various representations of your project data—referred to in Revit Architecture as "Views." Views in Revit Architecture can be graphical (drawings, sketches, diagrams) or non-graphical (schedules, legends, takeoffs) and offer the means to both query your project (output) and to manipulate and edit it (input). The Project Browser also lists the Families, Groups, and linked files that reside in the project. The Workspace is where view windows are displayed and manipulated. The Workspace can present one or more views of the project at the same time using the standard Windows minimize, maximize, and tile functionality on the Window menu. Finally, stretched across the top of the Design Bar, the Project Browser and the Workspace are the Type Selector menu, Properties icon, and the Options Bar. These controls are used to both create new objects and manipulate selected objects in the project (all of these items are labeled in Figure 2.1). Take some time to acquaint yourself with the various user interface elements. Be sure to right-click on each item to see additional context-sensitive menus.

## PULLDOWN MENUS

The most common way to issue commands in Revit Architecture is by choosing them from the pull-down menus across the top of the screen. This is consistent with the interface in most of today's software. In addition to the common "File," "Edit," "Window," and "Help" menus, Revit Architecture includes additional menus to categorize its collection of commands into logical menu structures (see Figure 2.2).

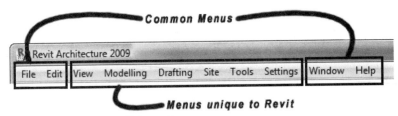

**Figure 2.2** *A look at the Revit pull-down Menus*

All Revit Architecture commands and functions are included in the menu structure. Most commands are accessible via other means as well, but the menus are a good place to look as you are learning Revit Architecture. In addition to the commands themselves, you can also use the pull-down menus as a way to discover if the particular command has a keyboard shortcut. Keyboard shortcuts are simple keystroke combinations that can be typed as an alternative way to issue a command (see Figure 2.3).

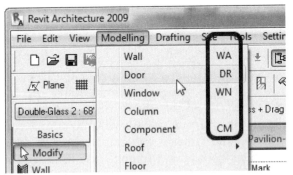

**Figure 2.3** *Keyboard shortcuts shown next to commands on the pull-down menus*

To use a keyboard shortcut, simply type the two letters on the keyboard in succession. There is no need to press ENTER following the keystrokes. You can find more details on available keyboard shortcuts in Appendix D. Next to some of the menu commands is a small arrow like the one next to the "Roof" command in Figure 2.3 and 2.4. This indicates that a cascading menu appears for this item. This is another standard Windows convention. Moving your cursor over this small arrow will cause the menu to "cascade" revealing a sub-menu (see Figure 2.4).

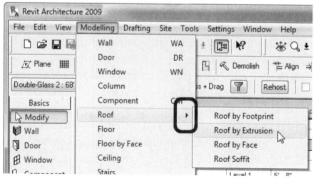

**Figure 2.4** *Example of a Cascading menu*

Another Windows convention supported by Revit Architecture is the ability to issue menu commands with the keyboard using the ALT key and a key letter from the desired command. To try this, press the ALT key. Doing so will place a small underline beneath one letter of every menu. Next, press the underlined key for the desired menu and then press the underlined key for the desired command. For example, to issue the Door command from the Modeling menu, you could press ALT + M THEN D (see Figure 2.5). If a cascading menu is involved, continue pressing the appropriate keystrokes to arrive at the desired command.

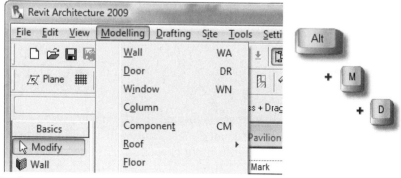

**Figure 2.5** *Press the ALT key to reveal menu shortcuts*

Menu items will appear grayed out if the particular command is not available in the current context. For instance, if you have a sheet view active on screen, most commands such as Wall, Door or Roof on the Modeling menu will not be available. Walls cannot be drawn in an elevation view and Levels cannot be drawn in a plan view. Similarly, items on the Edit menu will typically not be available unless an element in the model is selected (see below for more on selection).

## TOOLBARS

The toolbars in Revit Architecture are located just below the pull-down menus across the top of the screen. There are six toolbars available. You can choose which ones you wish to have displayed at any given time. Revit Architecture will remember your choices the next time you launch the program. Toolbars include commands that are typically used for editing elements in your models and modifying your work environment. To display or hide toolbars, right-click on any toolbar item for a context menu. You can also use the command on the Window menu as well. Displayed toolbars show in the context menu with a checkmark next to their name. By default the "Worksets" and "Design Options" toolbars are not displayed. If you have a large computer monitor, you can certainly choose to display these toolbars. However, you may wish to wait to enable them until you are ready to use their functions. Both of these topics are more advanced and covered in later chapters or the appendices. Figure 2.6 shows all of the Revit Architecture toolbars.

**Figure 2.6**  *Revit Architecture toolbars*

Toolbars can be "collapsed" if screen space is limited. To do this, click the small bar on the left side of the toolbar. In this state, a small double arrow will appear to indicate that some of the icons on the toolbar are hidden. Simply click this arrow to reveal a pop-up menu of the hidden commands (see Figure 2.7).

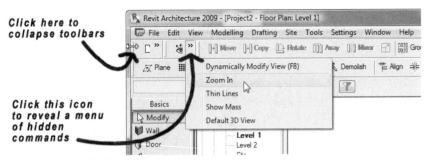

**Figure 2.7**  *The icons of collapsed toolbars can be accessed from a pop-up menu*

**Note:** Even though you can move and resize the toolbars and other interface elements, the physical arrangement of the screen reverts to the default setting every time you restart Revit Architecture.

Like menu items above, toolbar icons will appear grayed out if the particular command is not available in the current context. For instance, like the Edit menu, icons on the Edit toolbar will not be available unless an element in the model is selected (see below for more on selection).

## DESIGN BAR

The Design Bar is located vertically along the left side of the screen. Think of the Design Bar as an organized toolbox for the most common Revit Architecture functions, particularly those involving adding or creating elements and views in your Revit projects. Look to the toolbars for tools to edit *existing* elements in your project; look to the Design Bar for commands to create geometry and views. As mentioned above, all functions on the Design Bar can also be accessed through the pull-down menus, and often by keyboard shortcuts.

The Design Bar is organized into groups called "tabs." Each tab has a name that generally categorizes all commands contained on that tab. There are several tabs; some examples are shown in Figure 2.8.

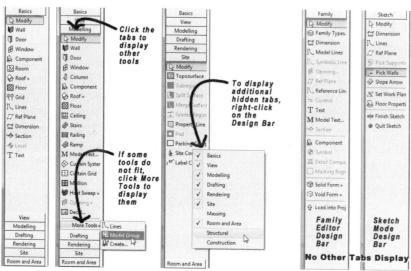

**Figure 2.8**    *The Design Bar*

As with toolbars, there is a fixed quantity of Design Bar tabs included with the software and you can choose to hide or display each tab. By default, most tabs are displayed. You can display the hidden ones with a right-click (see the middle of Figure 2.8). You can also use the command on the Window menu as well. To access a command from the Design

Bar, first click on the tab to reveal its tools (only one tab can be displayed at a time; the other tabs will collapse automatically). Next click on the tool you wish to execute.

At low screen resolution (e.g. 1024×768 or lower) some of the Design Bar tools may be cropped at the bottom of the screen. In such a case, any missing tools will be accessible via a "More Tools" button displayed at the bottom of the list. You can see an example of this in Figure 2.8.

If you have difficulty reading the full names of some of the tools, you can widen the Design Bar. To do this, place your cursor over the border line on the right edge of the Design Bar and drag it to the right. Please note that this width change does not carry over from session to session.

The Design Bar has some specialized working modes. Two such modes are the Family editor and Sketch mode. (These modes will be covered in detail in the appropriate chapters.) When you enter one of these modes, the Design Bar will change to show you the mode you are in as well as what functions are available to you in that mode. Also, all other Design Bar tabs will disappear while in the special mode (see the right side of Figure 2.8).

Like menu and toolbar items above, Design Bar tools will appear grayed out if the particular command is not available in the current context. For instance, if you have a Sheet open and active on screen, most commands such as Wall, Door or Roof on the Modeling Design Bar tab will not be available. Similarly, many of the items on the View tab will typically not be available unless a viewport is activated on the Sheet.

## PROJECT BROWSER

Menus, Toolbars, and the Design Bar are a fixed part of the Revit Architecture user interface. While we can manipulate which of these items displays and how, they remain consistent regardless of the project that we happen to be editing. The Project Browser on the other hand is very specific to the currently open project. When you open a Revit Architecture project (a file with an RVT extension) its contents will be displayed in the Project Browser. Think of the Project Browser as the table of contents for your project. It is the primary organizational tool for a Revit Architecture project. It is typically docked between the Design Bar and the view windows near the left side of the screen (see Figure 2.9).

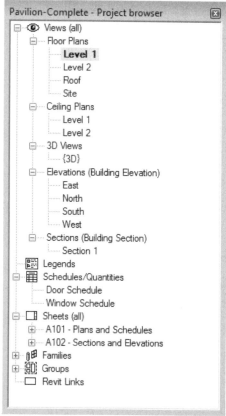

**Figure 2.9**　*The Project Browser*

You can re-size and move the Project Browser depending on your screen resolution and space needs. However, please be aware that when you launch a Revit session, the Project Browser will reset to the default location and width. While it is possible to close the Project Browser, it is *not* recommended. Since the Project Browser is the primary means of interacting with and navigating between your project's Views, closing it makes these functions difficult and inefficient. Should you inadvertently close the Project Browser you can restore it by clicking on the Project Browser icon on the toolbar (see Figure 2.10). You can restore it with the command on the Window menu as well.

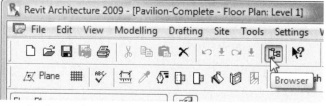

**Figure 2.10**　*Restore the Project Browser using the icon*

Like the Design Bar, longer names in the Project Browser can get truncated. It is therefore recommend that you widen the Project Browser window by stretching the right edge as much as your screen size will permit and to the extent of you own personal preferences. This will make it easier to read the full names of your Views. However, if you are not able to widen the Project Browser or you prefer not to, you can simply hover your mouse over a View to see a tooltip of its full name (see Figure 2.11).

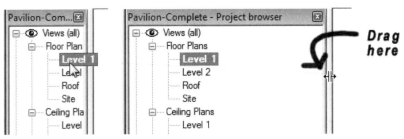

**Figure 2.11**  *Partially hidden names can be seen with tooltips*

The Project Browser enables you to view and work with two distinct parts of your project: views of your model and elements within your model. Views are listed at the top in the *Views, Legends, Schedule/Quantities,* and *Sheets* nodes. In the previous chapter, we explored this part of the Project Browser and its organization options. Please refer to that discussion for more details. To interact directly with elements within your model from the Project Browser, you will work within the *Families, Groups,* and the *Revit Links* nodes of the Project Browser tree. Every Family, Group, and Revit file link in a project is listed among these items. From these nodes, you can edit existing items, or even create new ones (see Figure 2.12).

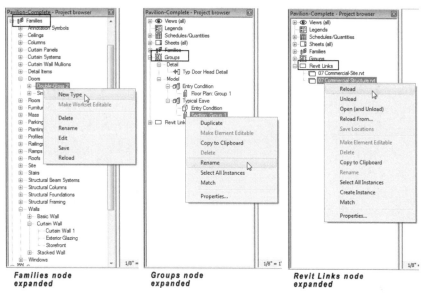

Families node
expanded

Groups node
expanded

Revit Links node
expanded

**Figure 2.12** *Understanding the various nodes of the Project Browser*

Right-click options are sometimes available on each node and sub-branch of the Project Browser. Please take a moment to right-click on each item and study the menus and commands that are available. If you right-click an item like the *Families* node, and no menu appears, this simply indicates that there are no commands associated with that node or branch of the Project Browser.

In the previous chapter, we explored the *Views* node of the Project Browser within the context of a provided dataset. Here we will reiterate certain behaviors that have more to do with understanding the user interface associated with Project Browser. Beneath the *Views* node of the tree, certain categories will appear automatically as various views are added to the project. For instance, a *Floor Plans* node is automatically created to house the various floor plan views of your project. Likewise, "*Elevation*" and "*Section*" nodes will appear as their view types are added. If you delete all views in a particular category, the category itself will also disappear. You cannot make your own categories. These are created and maintained automatically by the software.

The name of the currently active view will appear bold in the Project Browser. You can open any view by simply double-clicking on its name in the Project Browser. This is the most common function of the Project Browser. In addition you can right-click views in the Browser to access commands specific to that item (see below for more on right-clicking). It is also important to note that more than one item may be selected on the Project Browser at the same time. To select multiple views, select the first view, then hold down the CTRL key and select additional views. Once you have more than one View selected, you may right-click on any of the highlighted view names to access a menu that will apply to the entire selection of views. This might be useful if you wish to make a global change such as editing the scale of several views at once (see Figure 2.13). (Scale would be accessed from the **Properties** item on the right-click.)

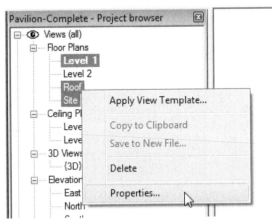

**Figure 2.13**  *Select multiple items in the Project Browser with the* CTRL *key*

Another important aspect of the Project Browser is its presentation of the project's sheets. Although sheets are technically classified as views in Revit Architecture they have some unique properties not shared by other views. They are also presented in their own node in the Project Browser. Expand the *Sheets* node to see the sheet views in your project. Since sheets can actually have other views placed on them, a plus (+) will appear next to the sheet name. Expanding this will reveal the names of the views that have been placed upon the sheet. A sheet appearing in the Project Browser without a plus sign indicates that the sheet does not yet contain references to any other views. Like other views, you can open a sheet view by double-clicking on its name in the Project Browser. In addition, you can open referenced views placed on sheets by double-clicking their names beneath the expanded sheet name (see Figure 2.14).

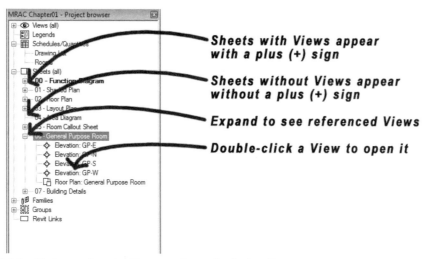

**Figure 2.14** *Understanding the Sheets node on the Project Browser*

When using these techniques to navigate your project, recall from the previous chapter our discussion of changing the way that the Project Browser is organized. In that exercise, we noted how we could organize our views by discipline, phase or even by those not yet placed on sheets. Look back to Figure 1.9 and the accompanying discussion for more details.

When you choose to organize the Browser using the "Not on Sheets" type, you do not loose anything. You simple change the way things are displayed. If you wish to work on a view that is already on a sheet, you first expand the sheet on which it is placed, and then double-click the view name beneath it. If the view is not yet on a sheet, it will be listed beneath the *Views* node at the top of the Browser instead. Once you add a view to a sheet, it will disappear from the *Views* node and appear instead beneath its host sheet. If you wish to experiment with this, try reopening the dataset from Chapter 1 and explore further.

In addition to all of the project views accessible from the Project Browser, many other resources such as Families, Groups, and Revit Links are also accessible from Project Browser. The *Families, Groups,* and *Revit Links* nodes of the Project Browser show all Families and Groups in a project file: both the ones already used in the project's geometry and those that have not been used yet. Families, Groups and Revit Links are defined and discussed in more detail in later chapters. This section will focus only on their placement and access via the Project Browser.

Navigating the *Families, Groups,* and *Revit Links* nodes works the same as navigating the *Views* node. If you expand an item, you will see Families, Groups, or Links within that category. The items shown are those that are already loaded and therefore part of the current project. You can further expand a Family itself to reveal the Types that it contains (see Figure 2.15).

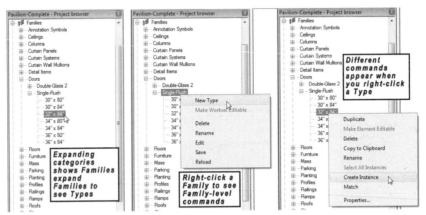

**Figure 2.15** *Expanding an item in the Families node of the Project Browser*

As you can see in the figure, you will get different options depending on whether you right-click on the Family itself, or its Types listed beneath it.

Groups and Revit Links behave similarly to Families in the Project Browser. Groups appear in two main categories: Detail and Model Groups. A Detail Group contains detail components that are not part of the model. They can be inserted into detail views. Model Groups contain model elements and can be inserted directly into the model. Detail Groups can also be associated with Model Groups and will appear nested beneath their parent Group in the Project Browser tree. To see if a Group has an associated Attached Detail Group expand the Group in the list. Linked files exhibit similar behavior. For example, if a linked file has linked files attached to it, they will appear nested beneath their parent file (see Figure 2.16).

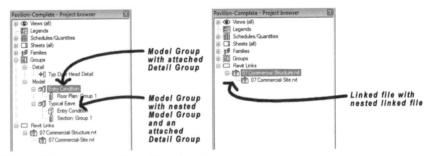

**Figure 2.16** *Right-click a Family or Type for options*

To place an instance of a Family, Group, or Revit Link in the model, you can right-click and choose **Create Instance** or you can simply drag it from the Project Browser to the View window. Be sure you perform this action on a Type, not the Family itself.

## TYPE SELECTOR / PROPERTIES ICON

Just above the Design Bar and beneath the toolbars on the left side of the Options Bar is the Type Selector. The Type Selector is a drop-down list of Types available for whatever item is currently selected in your project. If you have nothing selected the Type Selector will be grayed out and unavailable. As its name implies, the Type Selector is used to select a Type for an element in your project. Whenever you create elements, you will choose an appropriate Type from this list. You can also use the Type Selector to change the Type of elements already in the model. Simply select an element and then choose a different Type from the list (see Figure 2.17).

 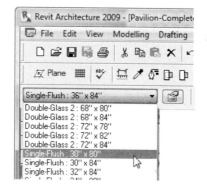

**Figure 2.17** *The Type Selector and Properties Icon*

Next to the Type Selector is the Properties Icon. You can use the Properties icon to call the "Element Properties" dialog. This will allow you to explore or edit the parameters of the selected element. Every Element in Revit Architecture, from Model Elements like Walls and Floors, to annotation Elements like text, tags, and Levels, and even Views themselves have instance and/or type properties. You will access these properties frequently from this dialog. From the "Element Properties" dialog, you can use the Edit/New button to access the "Type Properties" dialog to explore or edit the parameters of the element's Type (see Figure 2.18).

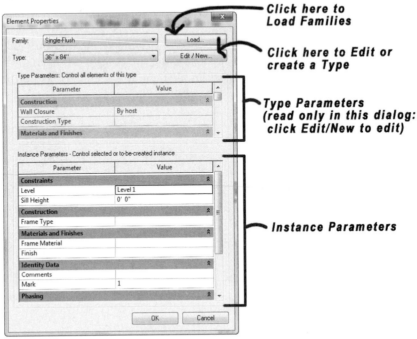

**Figure 2.18** *Element Properties dialog*

You will interact with both the Type Selector and the Properties icon when creating or modifying elements. When you are creating an element, you first choose the tool for the item you wish to create on the Design Bar. Next choose a Type from the Type Selector and if necessary click the Properties icon to view or edit additional parameters. If you are modifying an existing element, simply select the element (or elements) and then choose the desired Type from the Type Selector or click the Properties icon to edit parameters in the "Element Properties" dialog.

When you select multiple elements, the Type Selector displays the common Type of the group of items you have selected. If the selection of items is the same kind of element (all Doors for example) but are not currently the same Type, then the Type Selector will remain active, but will display a blank entry (all white) rather than show a Type. If the Type Selector box appears grayed out, this indicates that you have made a selection of dissimilar elements (like a Wall and a Door), and you cannot make changes to them using the Type Selector or Properties icon.

## OPTIONS BAR

The Options Bar is located directly to the right of the Properties Button, and runs horizontally across the screen. The Options Bar is constantly changing, depending on what element you have selected or what operation is being performed. In this way, you will only see those options relating to the current situation. This helps simplify the interface and keep tasks more focused (see Figure 2.19).

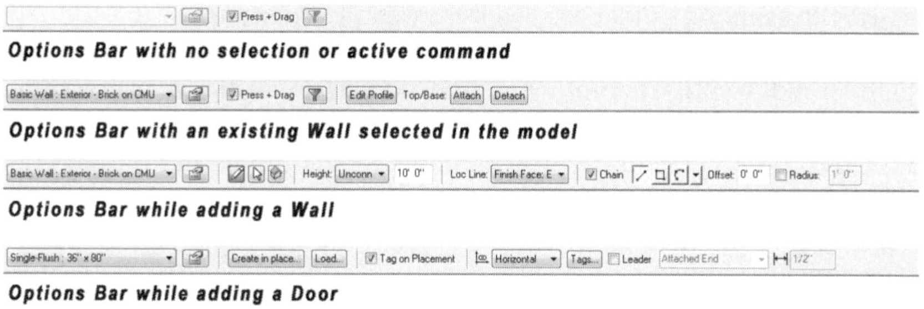

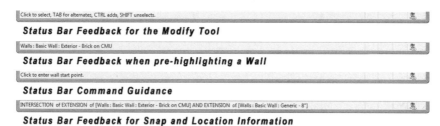

**Figure 2.19**    *The Options Bar constantly adjusts to show options appropriate to the current task*

You will interact with the Options Bar constantly as you work in Revit Architecture. Keep this in mind and always remember to look there for context-specific options relating to the task at hand.

## THE WORKSPACE (VIEW WINDOW DISPLAY AREA)

The most interactive portion of the interface is the workspace where you work with the views of your building model. When you open a view, it appears in a window in the workspace area. At least one view window must be open to work in a project. You can however open several views for a project at one time. If the view windows are maximized, you can see which are opened on the Window menu. You can also choose to tile the Windows so that they appear side by side in the Workspace. This command is also on the Window menu. If you have many views open at once, tiling can make the actual windows very small and hard to work with. To prevent this, you can cascade them or maximize them instead. Be sure to close windows when you no longer need them. An easy way to do this is to maximize the current view window and then choose **Close Hidden Windows** from the Window menu.

## STATUS BAR

The Status Bar is the gray bar along the bottom edge of the Revit Architecture screen. If you glance down at the Status Bar, you will notice a constant readout of feedback appears there. In some cases, the information provided prompts and clues as to what actions are required within a particular command, in other cases; the feedback may simply describe an action taking place or an element beneath the cursor (see Figure 2.20 for examples).

**Figure 2.20**    *The Status Bar provides ongoing feedback and guidance as you work*

In many cases, the same or similar feedback is available on screen in the form of tool tips. The extent to which tooltips appear is controlled by a setting that you can modify. You can opt for a high level of tooltip prompting, a moderate level or none at all. To edit the degree of Tooltip Assistance, choose **Options** from the Settings menu and edit the "Tooltip Assistance" item on the General tab. You cannot edit the Status Bar messages in any way.

## VIEW CONTROL BAR

Every graphical view window has a View Control Bar located at the bottom edge of the window. When windows are maximized, this will appear directly above the Status Bar adjacent to the horizontal scroll bar. If the view windows are tiled, it will appear in the lower left corner of the window (see Figure 2.21). The View Control bar serves two purposes—it displays at a glance the most common view settings of the window and provides a simple and convenient way to change them if required.

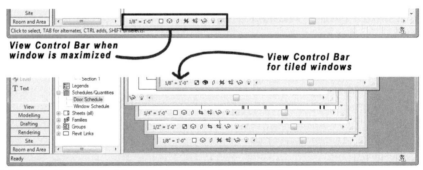

**Figure 2.21** *The View Control Bar maximized (top) and tiled (bottom)*

Like so many other aspects of Revit Architecture, the View Control Bar is context sensitive. Notice the window second from the top in Figure 2.21 has a different View Control Bar than the others (containing only the Hide/Isolate settings). This window is a sheet view and sheet views do not have as many settings available as other views. Click on any of the icons on the View Control Bar to access a pop-up menu of available choices. For example, to change the scale of a view, click on the scale item, or pick the Detail Level icon to change the level of detail in which the view is being displayed (see Figure 2.22).

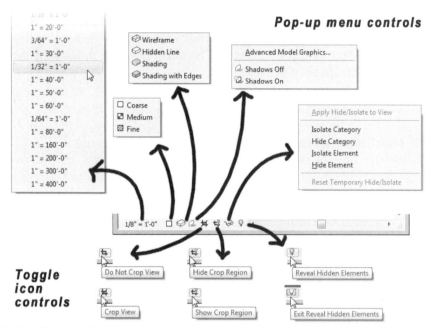

**Figure 2.22** *The View Control Bar provides quick access to the most common View settings*

All of the settings, except Hide/Isolate, on the View Control Bar are also accessible from the view's "Element Properties" dialog. To access the view's "Element Properties" dialog, right click in the view window with no objects selected and choose **View Properties**. You can also right-click on its name in the Project Browser and choose **Properties**.

The last two icons on the right are used to hide and show elements in your model to make it easier to see things as you work. In Revit Architecture, there are two ways to hide objects. You can use the icon on the Temporary Hide/Isolate View Control Bar (looks like sunglasses) to hide elements temporarily. You can also now hide elements permanently. To use the temporary Hide/Isolate tools, select an element or elements in the model and then click the Hide/Isolate icon and choose an option. The **Hide Element** option temporarily makes the selected element(s) invisible. The **Isolate Element** option leaves the selected elements visible and hides everything else. There are also options to hide or isolate the entire category of objects based upon the category of objects you have selected. For example, if you select a Door in the model and then choose **Hide Category**, all Doors in the model will temporarily hide. When you use the temporary Hide/Isolate command, a cyan border will appear around the current view window until the mode is disabled.

The important thing to know about using the Temporary Hide/Isolate function in a view, is that the changes it makes to the view window will not be saved outside the current work session. Furthermore, if you were to print or export the current view, Hide/Isolate settings will be ignored. If you want to hide an object permanently on screen or when printing, use the "Hide in View" command on the right-click menu instead. This command is new in 2008. To see hidden elements, both temporary and permanent, you click the Reveal

Hidden Elements toggle icon (light bulb on the far right). If there are temporarily hidden elements in the current view, this mode will reveal them in a cyan color. Permanently hidden elements will be revealed in red. A red border will appear around the current view window as long as this mode is active.

Remember also that while the View Control Bar shows the most common View Properties, right-clicking and choosing the **View Properties** command will give a dialog with a complete list of properties—many more than on the View Control Bar. If you wish to make the Hide/Isolate settings permanent for instance, you can edit the View Properties and then click the Edit button next to "Visibility." This will call the "View/Visibility Graphics" dialog where you can modify the display of certain Element Categories for that View permanently.

## RIGHT-CLICKING

Like most Windows software, in Revit Architecture, you can right-click on almost anything and receive a context-sensitive menu. In fact, we have already seen examples of this in the previous chapters and in this chapter. Right-click menus are loaded with functionality and will be used extensively.

 **Tip:** As a general rule of thumb, "When in doubt, right-click."

The next several figures highlight some of the more common right-click menus you will encounter in Revit Architecture. Take a moment to experiment with right-clicking in each section of the user interface. You will also discover the typical Windows right-click menus (for Cut, Copy, Paste and Select All) appear in all text fields and other similar contexts.

### RIGHT-CLICK ON TOOLBARS

As mentioned above, to load or hide toolbars, right-click in the toolbar area. You simply move your mouse over any visible toolbar icon, and then right-click. This will make a menu appear (see Figure 2.1 above). Toolbars that are *currently* displayed on screen will have a checkmark next to them. Toolbars that are *not* displayed will have no checkmark. Many toolbars also have text labels. These labels provide a text description of the tool name. There is an option to toggle these text labels on and off. Choose an item without a checkmark (such as Worksets or Design Options) to display it. The toolbar will appear on screen. Right-click any icon on the newly loaded toolbar and choose it again to hide it.

### RIGHT-CLICK ON THE DESIGN BAR

The Design Bar was covered in detail above. It is not necessary to display all Design Bar tabs at once. Right-click to decide which tabs you wish to see and those you wish to hide. Just like toolbars, tabs that are displayed will have a checkmark, and those that are hidden will not have a checkmark (see Figure 2.8 above). Right-click on the Design Bar again and choose a checked item to hide that tab. Whatever tabs you choose to display will remain displayed the next time you launch Revit Architecture.

## RIGHT-CLICK ITEMS IN PROJECT BROWSER

Items on Project Browser often have right-click options as well. Context menus are not available for every node of the tree; however, you can always right-click directly on a View to receive a context menu of options (examples can be seen in Figures 2.12, 13, and 15 above). A menu will appear with several options. Some of these items have been used already in the previous chapter and above.

## RIGHT-CLICK IN THE VIEW WINDOW

When you right-click in the workspace view window, you get a different menu depending on whether there are any objects selected or not. If there are no objects selected, the menu contains basic zooming and scrolling commands, the View Properties command, and the Find Referring Views command. The "Find Referring Views" command is particularly useful. When choosing this command, a dialog will appear like the one shown in Figure 2.23.

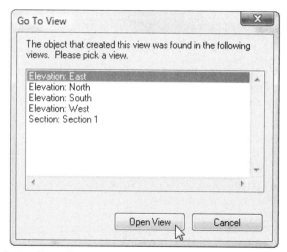

**Figure 2.23** *Find Referring Views locates all views that refer to the current view*

You can choose any view listed in the dialog and then click the Open View button. The figure illustrates an example of running the command from a plan view. If you run the command from a section or elevation view, a similar dialog will appear listing different views including plans and reflected ceiling plans.

When you right-click with one or more objects selected on screen, you get a different menu than if there is no selection active. However, some commands appear in either case. One such command is the **View Properties** command. This command calls the properties dialog for the view itself. So if you are looking at a floor plan view, you will get a dialog allowing you to edit the properties of the floor plan such as its scale and display overrides. If you have an element selected, and you wish to edit the properties of the selected element, you choose the **Element Properties** command on the right-click menu. This command will appear directly beneath the **View Properties** command (see Figure 2.24).

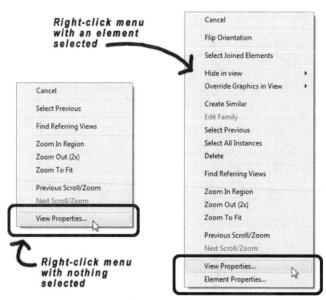

**Right-click menu
with an element
selected**

Cancel

Flip Orientation

Select Joined Elements

Hide in view ▸
Override Graphics in View ▸

Create Similar
Edit Family
Select Previous
Select All Instances
Delete

Find Referring Views

Zoom In Region
Zoom Out (2x)
Zoom To Fit

Previous Scroll/Zoom
Next Scroll/Zoom

View Properties...
Element Properties...

Cancel

Select Previous

Find Referring Views

Zoom In Region
Zoom Out (2x)
Zoom To Fit

Previous Scroll/Zoom
Next Scroll/Zoom

View Properties...

**Right-click menu
with nothing
selected**

**Figure 2.24** *The View Properties command appears with or without a selection. Element Properties appears only with a selection*

Notice that all of the commands from the no selection menu (on the left) also appear in the menu that appears with a selection. They are simply accompanied by several commands appropriate to the element you have selected. (The figure shows how the menu looks when a Wall is selected.) You can cancel an active command by choosing Cancel from the menu, or by simply pressing the ESC key.

## NAVIGATING IN VIEWS

When working with models on a computer screen, you need more than the standard Windows scroll bars to navigate your model. You will need to change the magnification of the model and frequently scroll other parts of the model into view on the limited screen space available. The act of changing the magnification of the screen is referred to as "zooming" and moving the image on screen within the borders of the view window to see parts off screen is referred to as "scrolling."

### USING A WHEEL MOUSE TO NAVIGATE

If you have a mouse with a middle wheel button, Revit Architecture provides instant zooming and scrolling using the wheel! If you don't have a wheel mouse, this might be a good time to get one. This modest investment in hardware will pay for itself in time saved and increased productivity by the end of the first day of usage. Using the wheel you have the following benefits:

- To **Zoom**—Roll the wheel. Roll up to zoom in (closer), roll down to zoom out (farther away). Move the pointer before you roll to control the center of the zooming.

- To **Scroll**—Push and hold the wheel down and then drag the mouse.
- To **Dynamic Zoom**—Hold down the CTRL key and then drag the wheel (this is similar to rolling the wheel, except the zoom is centered).
- To **Dynamic Spin**—Hold down the SHIFT key and then drag the wheel (3D Views only).

## ZOOM

In addition to the wheel mouse, Revit Architecture provides three basic zooming methods.

**Zoom In Region**—This command will enlarge an area of the model that you designate by dragging a box around the area on screen.

**Zoom Out (2x)**—This is also a zoom in (or magnification) command that simply doubles the size of the image on screen.

**Zoom To Fit**—Most often this is a zoom out (or reduction in magnification) command. The function of this command is to fit the entire extents of model and any annotation objects into the available view window space on screen. A similar command is available on the View menu: **Zoom All to Fit**. This command performs the "Zoom To Fit" command in all of the open view windows rather than just the active one.

You can find these commands in several locations. They are located on the **View > Zoom** menu, on an icon on the View toolbar (both shown in Figure 2.25), the right-click menu (shown above in Figure 2.24) and by using keyboard shortcuts (as seen next to the menu commands in Figure 2.25).

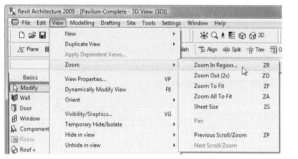

**Figure 2.25** *Zoom commands are available from various places in the interface*

## SCROLL

Scroll bars are located on the bottom and right sides of each View window. You can use these to scroll the window. You can also use the wheel mouse. If you have zoomed or scrolled the window, you can choose the previous and next scroll options from the same Zoom menus shown in Figure 2.25.

## SPIN (AND MORE)

If you are working in a three-dimensional View, you can also "spin" the model. This allows you to study the model from all directions and heights. To spin the model, you use the

"Dynamically Modify View" icon on the View toolbar (or the SHIFT + wheel shortcut mentioned above). This command is covered in detail in Chapter 3.

## SELECTION METHODS

Before elements can be modified in Revit they must be selected. There are various methods used to select elements, some of which are similar to other software. When modifying elements in a model, the basic workflow is as follows:

1. Select the Element(s) to be manipulated with the Modify tool.

2. Issue a command to perform on the selected element(s).

3. Indicate on screen where and how to perform the command.

To summarize this workflow, choose (select) *what* you want to modify and then indicate *how* you want to modify it—to this selection of elements I wish to perform this action.

### CREATING A SELECTION SET:

A selection set is one or more selected elements in Revit Architecture. As you move the Modify tool around a view window you will notice that any elements available for selection will temporarily highlight while under the cursor—this is know as "pre-highlighting." The purpose of pre-highlighting is to preview what will be selected if you click the mouse. This is particularly helpful when working in a complex model with many elements close together. Use the pre-highlighting (and often the TAB key—see below) as a tool to assist you in accurate selection.

The simplest way to select an element is to click on it. When you select an element it will display in red on screen—this indicates that the element(s) is selected. Once selected, an element will remain selected until you deselect it. You can deselect elements in three ways: selecting another element will automatically deselect the current selection set (unless you hold down the CTRL key) and become the new selection set. You can deselect all elements without creating a new selection set by clicking on a blank portion of the screen (where there are no elements) or by pressing the ESC key. (You can also right-click and choose **Cancel** as noted above.)

To create a selection set containing more than one element, use the following techniques:

- Hold the CTRL key down while clicking on another element. The new element will be added to the current selection set (and highlight red as well).

- Hold the SHIFT key down while clicking on a selected element (highlighted red) to remove the element from the current selection set (it will no longer be highlighted in red).

If you accidentally pick an object without holding down the CTRL key you will lose your selection set (which will be replaced with only the one element just picked). You can restore your previous selection set by right-clicking in the view window and then choosing **Select Previous**.

**Note:** You can also select previous by holding down the CTRL key and then pressing the LEFT ARROW key.

Either action will restore your previous selection without having to start all over again.

### SELECTION BOXES:

Even with the CTRL and SHIFT keys, making selections of multiple elements can be time consuming. The easiest way to create a large selection set is by using a selection box. To create a selection box with the Modify tool, click down with the left mouse button next to an element, hold the button down and drag a rectangular box around the elements you wish to select. Elements will pre-highlight as you make the selection box.

The direction in which you drag the selection box determines which specific elements are selected. If you create your selection box by dragging the Modify tool from left to right on the screen, the edge of the selection box will appear solid as you drag and only elements *completely* within the box will be selected. If you create your selection box by dragging from right to left on the screen, the edges of the selection box will appear dashed and any element completely or partially included in the box will be selected (see Figure 2.26).

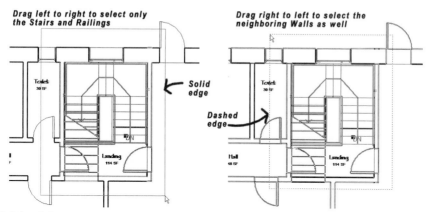

**Figure 2.26**   *Selection Windows vary with the direction you drag*

You can always combine methods—first create a selection box, and then use the CTRL and SHIFT keys to add and remove from the basic selection.

### FILTER SELECTION

Another approach to building a large selection set is to deliberately select too many elements and then use the Filter function to remove items of particular categories from the selection. The Filter Selection icon is located on the Options Bar. Remember, when using the Filter Selection tool, you always start with a selection that includes *more* than the items you want, and you then filter *out* the undesired elements by category (see Figure 2.27).

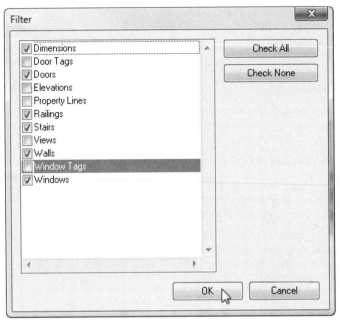

**Figure 2.27** *Use the Filter Selection dialog to remove categories of elements from the selection set*

Each of the selection methods will require a little practice to become second nature to you. Take the time to practice each one since selection is such an important and frequent part of nearly all tasks in Revit Architecture.

## THE ALMIGHTY TAB

You will find that the TAB key is probably the most used and most useful of all the keys on the keyboard when working in Revit. As mentioned in the previous topic, being able to quickly add or remove from a selection set can greatly increase your efficiency during a work session. In general, you can think of the TAB key as a "toggle switch," allowing you to cycle through potential selections. When you attempt to pre-highlight a particular element on screen, sometimes a neighboring element will pre-highlight instead. No matter how subtly you move your mouse in an attempt to capture the desired element, it often proves difficult or impossible to capture the right one. The TAB key provides the solution to this situation. Use it to cycle through adjacent or stacked items to highlight the particular element you desire.

 **Tip:** Like our right-click rule of thumb above, "When in doubt, TAB."

While there are dozens of examples of using the TAB key in a Revit Architecture session, a few examples are presented here that will give you an idea of where using the TAB key proves most handy.

## PRE-HIGHLIGHTING ELEMENTS FOR SELECTION

If the tool, whether it is the Modify, Dimension, Split, or another tool, is near two or more Elements in a view window, you can either move your mouse around the view to pre-highlight the different objects for selection, or use the TAB key to pre-highlight each element in succession. Each time you press TAB, a different element will pre-highlight until all items have been cycled through. Once you have tabbed through all elements, the cycling will repeat (see Figure 2.28).

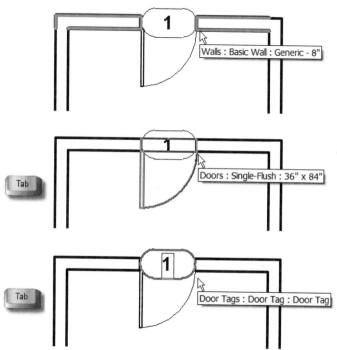

**Figure 2.28** *Use the tab to cycle through and pre-highlight nearby elements*

When the element that you wish to select is pre-highlighted, click the mouse as normal to select it.

## DURING DIMENSIONING

When you are using existing geometry for reference points such as when adding dimensions, using the align command or tracing a background, you can use the tab key to cycle through possible reference points (see Figure 2.29).

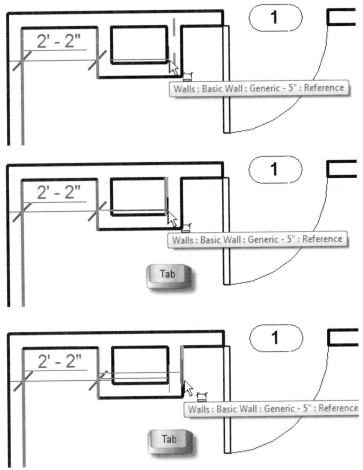

**Figure 2.29**  *Use the tab to cycle through dimension reference points*

## CHAIN SELECTION

Chain selection is a very powerful feature of Revit Architecture. Chain selection works with Walls or Lines. Like an actual chain, chain selection highlights all of the Walls or Lines that touch one another end to end. To make a chain selection, you first pre-highlight a single Wall or Line. Next you press the TAB key and in so doing, any walls or lines that form a chain (i.e. are located end-to-end) with that wall or line are pre-highlighted together. Finally, you click the mouse to make the selection with a single mouse-click (see Figure 2.30).

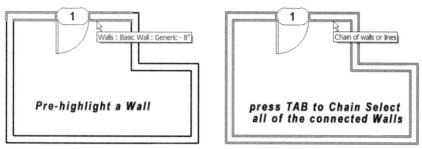

**Figure 2.30**   *Use the tab key to chain select a collection of Walls*

There are dozens of additional ways to use the TAB key in Revit Architecture. One of the easiest ways things to do is simply try tabbing in various situations while you work in the software. Other examples will be presented throughout this book in the tutorials that follow. For more discussion on the topic, consult the Revit Architecture online help.

## SETTINGS

While not strictly related to interface, this section covers some of the settings you can use to manipulate workspace preferences. In some cases, the settings apply to your Revit Architecture application in general (in other words, the setting will persist regardless of the project you open) and in other cases the setting is saved with the project and can vary from one project to the next.

Examples of global settings include the User name, the Tooltip Assistance or the Path settings. Any of these settings applies to your installation of Revit Architecture not to a particular project file. On the other hand, Units settings apply to only the current project file. So if you change the Project Units from Feet-Inches to Inches, this would apply only to the current project file.

While there are many Options and Project Settings available, we will only look at a few of the more common ones here.

### THE GENERAL TAB

From the Settings menu, choose the **Options** command to open the "Options" dialog and view or edit the settings therein. The dialog is divided into five tabs, each containing several related options (see Figure 2.31).

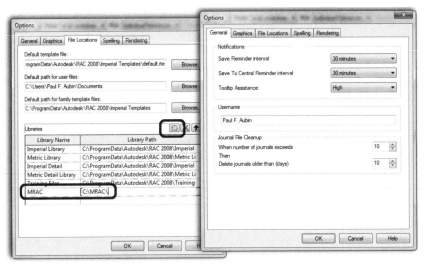

**Figure 2.31** *The Options dialog box*

**Save Reminder Interval**—The default setting is 30 minutes. This means that every 30 minutes during your work session, Revit will prompt you to save your file. You can change the increment to suit your preference.

 **Tip:** Regardless of your Save Reminder frequency, you should make it a habit to save your work often.

**Tooltip Assistance**—Use this option to control how much feedback and assistance is given via on screen tooltips.

**Username**—By default, this is set to your Windows login name. The only time this is important is if you are working with Worksets. Worksets enable you to work with teams on the same project. Refer to Appendix A for more information on using Worksets. The important thing to remember about Username is that Revit will only allow you to open and work in a Workset-enabled local file if your Username matches the Username that was active when that file was saved.

## FILE LOCATIONS

The File Locations tab of the "Options" dialog is where you configure the hard drive and/or network locations of the template and library files used by Revit Architecture.

**Default template file**—Project Template files are discussed in detail in Chapter 4. Use the setting here to point to your firm's preferred project template file.

**Default path for user files**—If you have a specific location on your computer that you typically keep your project files, you can browse to that location here. The folder written here will be used by Revit when the Open dialog is used.

**Libraries**—This is a very handy feature. Each entry in this list will show up as an icon in the Open and Save dialog boxes. In this way, you can jump from one library location to another with a single click.

 **Tip:** Add a Library pointer for the folder in which you installed your Mastering Revit Architecture CD ROM files. In this way, you can navigate to this location quickly at the start of each tutorial.

## TEMPORARY DIMENSIONS

Although you can toggle the reference point used by temporary dimensions using the TAB key, it is often more efficient to change the default behavior instead. Defaults for Temporary Dimensions can be configured for Walls, Doors, and Windows. Depending on your preferences, you can have the Temporary Dimensions default to either the centers (default) or edges of objects. From the Settings menu, choose **Temporary Dimensions** (see Figure 2.32).

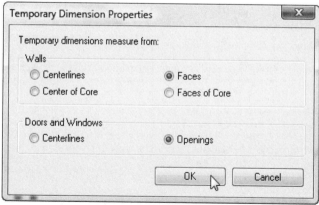

**Figure 2.32**   *Configure temporary dimension behavior*

Choose your preference for both the "Walls" and "Doors and Windows" settings. Revit Architecture defaults to centerlines for both. For Walls, you can dimension faces, centerlines and/or Wall Cores. For Doors and Windows choose between centerlines and openings. The author recommends Faces for Walls and Openings for Doors and Windows.

## SUMMARY

Now that you have completed the Quick Start Tutorial, read the high level overview of Building Information Modeling in Chapter 1 and have been guided through an overview of the user interface in this chapter, you are now ready to begin creating your first Revit projects. In the chapters that follow, we will build two projects from scratch: one residential and one commercial. In this chapter you have learned:

- The User Interface of Revit Architecture uses many common Windows conventions and many unique ones as well.

- The Design Bar along the left side of the screen contains several common Tools organized into Tabs.

- The Options Bar changes to reflect the current tool or command.

- One or more views of the project can be opened at one time in the Workspace area.

- The Project Browser, along the left side, contains all of the views of the project as well as the Families, Groups, and Linked files.

- The View Control Bar can be used to manipulate the display settings of the active View.

- The Status Bar along the bottom of the screen shows prompts for the current command.

- Many commands are available from the context-sensitive right-click menus.

- The easiest way to navigate a view is using the wheel of your mouse and the appropriate modifier keys.

- Roll the wheel to zoom, drag the wheel to scroll, hold the shift key and drag to spin the 3D model.

- Most editing commands require element selection.

- Drag left to right to select all elements within the selection box, right to left to select all those in contact with the box.

- The tab key is used to cycle through potential selections when more than one is possible.

- Use the tab key to "Chain" select Walls or Lines.

- Configure settings in the Options and Temporary Dimensions dialogs to suit your personal preferences.

## SECTION II

# Create the Building Model

The first step toward reaping the benefits of Building Information Modeling is to construct a virtual building model in the software. The tools provided for this purpose in Revit Architecture are many and varied. In this section we will explore the many building modeling tools available such as Walls, Doors, Windows, Columns, Beams, Stairs, Railings, Roofs, Floors, and Curtain Walls. In Chapter 10, we will take a detailed look at Families and the Family Editor.

Section II is organized as follows:

# CHAPTER 3

# Creating a Building Layout

## INTRODUCTION

Revit Architecture can be used successfully in all types of architectural projects and within all project phases. Most of our attention in this book is given to the design development and construction documentation phases. The tutorial exercises in this book will explore two building types concurrently, starting at different points in the project cycle. This will give a sense of the multiple ways you can approach the design process while keeping some variety in the lessons. Don't become limited to the techniques covered here. The aim of the tutorials is to get you thinking in the right direction. Exploration and playtime are highly encouraged.

Architectural projects start in many different ways. You may be commissioned to design a new building from its early conception all the way through construction and beyond, or you may be asked to join a project already in progress as a member of a larger team. Likewise, design data comes in many forms. If the project is new construction, you may have only the site plan and some other contextual data. If the project is a renovation, or if you join a project already in progress, you may use or be given existing design data in formats that vary from traditional hand-drawn sketches to digitally created images and models saved in all manor of software formats.

The first two chapters were intended to get you comfortable with the theoretical underpinnings of Revit Architecture. Now that you have the correct mind-set and a level of comfort with the user interface, get ready to roll up your sleeves—it is time to begin creating our first model.

## OBJECTIVES

Throughout the course of the following hands-on tutorials, we will lay out the existing conditions for our residential project. In this chapter, we will explore the various techniques for adding and modifying Walls, Doors, and Windows. In addition, we will add plumbing fixtures and other elements to make the layout more complete. After completing this chapter you will know how to:

- Add and modify Walls
- Explore Wall properties
- Add and modify Doors and Windows

- Assign Phase parameters to model components
- Add plumbing fixtures
- Add an In-Place Family

## WORKING WITH WALLS

Basic object creation in Revit Architecture involves the simple process of choosing a tool from the Design Bar and then designating the desired settings on the Options Bar. We will begin by working with Walls. You can add Walls point by point or enable the "Chain" option to create them in series (each segment beginning where the previous one ended). Walls will "join" automatically with intersecting Walls at corners. Walls have many parameters such as length, height, and type; can have custom shapes and profiles; and can receive (i.e., host) Doors and Windows by automatically creating openings for them.

### BASIC WALL OPTIONS

Since Walls are the basic building blocks of any building, we will start with them. We will create a new temporary project and sketch some Walls to get comfortable with the various options. Later we will create one of the actual projects that will be used throughout the rest of the book.

### CREATE A NEW PROJECT

We will begin our work in a new project created from the Revit Architecture default template file.

1. Launch Revit Architecture from the icon on your desktop or from the *Autodesk >
Revit Architecture* group in *All Programs* on the Windows Start menu.

 **Tip:** In Windows Vista, you can click the Start button, and then begin typing **Revit** in the "Start Search" field. After a couple letters, Revit Architecture should appear near the top of the list. Click it to launch to program.

- If the New Features Workshop dialog appears, choose "Maybe later" and then click OK.

Revit Architecture launches to an empty project based on your default template. Let's be sure that we are starting from the out-of-the-box default template. First we will close the empty project that opened automatically and then create a new one based on the default template file.

2. From the File menu, choose **New > Project**.

- In the New Project dialog, in the "Template File" area, be sure that the lower radio button is selected (the one that lists a template file).

- Verify that the *default.rte* [*DefaultMetric.rte*] template file is listed in the text field.

The default template file name and location for the Imperial template file is:

*C:\ProgramData\Autodesk\RAC 2009\Imperial Templates\default.rte*

The default template file name and location for the Metric template file is:

[*C:\ProgramData\Autodesk\RAC 2009\Metric Templates\DefaultMetric.rte*]

**Note:** In Windows XP, the folder is: *C:\Documents and Settings\All Users\Application Data\Autodesk\RAC 2009\Imperial Templates\*. [*C:\Documents and Settings\All Users\ Application Data\Autodesk\RAC 2009\Metric Templates\*].

If the template shown on your screen does not match the location and name listed above, click the Browse button to locate it before continuing.

**Note:** If you are in a country for which your version of Revit Architecture does not include these template files, they have both been provided with the files from the Mastering Revit Architecture CD ROM. Please browse to the *\MRAC\Templates* folder to locate them.

- In the "Create New" area, verify that the Project radio button is selected and then click OK (see Figure 3.1).

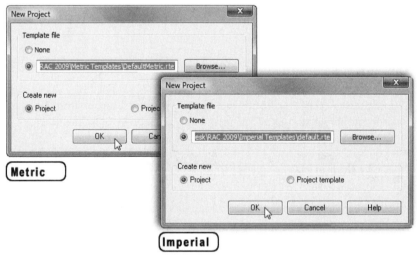

**Figure 3.1** *Create a new project based upon the default template file*

Depending on your system's settings, the same result can normally be achieved by simply clicking the NEW icon on the Standard toolbar. The NEW icon automatically creates a new project using the default template file. To configure the default template file, choose **Options** from the Settings menu. Click the File Locations tab and then edit the location in the Default template file field.

## GETTING STARTED WITH WALLS

Nearly every building component in Revit Architecture is created following the same basic procedure. Locate the tool for the object you wish to create on the Design Bar (on the left side of the screen, see Chapter 2 for more information). Use the Options bar to configure the settings of the object as you create it. Use your mouse to create the object in the current view.

3. On the Design Bar, click the Basics tab and then click the **Wall** tool (see Figure 3.2).

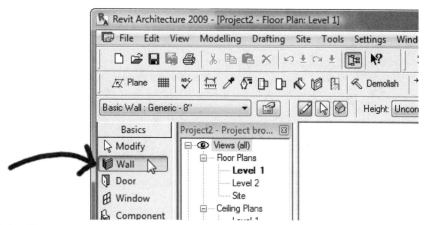

**Figure 3.2** *The Wall tool on the Basics tab of the Design Bar*

 **Note:** If the Basics Design Bar is not showing on your screen, choose Design Bars from the Window menu to retrieve it or right-click on the Design Bar to display it. See Chapter 2 for more information.

The pointer will change to a pencil cursor and the Status Bar prompt will read: "Click to enter wall start point."

- From the Type Selector, choose **Basic Wall: Generic - 8"** [**Basic Wall: Generic – 200mm**].

The following list explains the major fields and controls shown on the Options Bar when you are adding a Wall (see Figure 3.3).

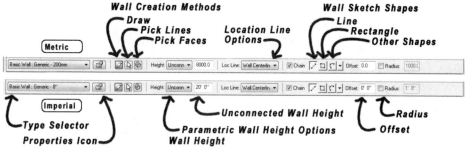

**Figure 3.3** *The Options Bar for adding Walls*

- **Type Selector**—Use this drop-down to choose from a list of Wall types in the current file. The specific list is populated by the template we used to create the project. Always choose a type from this list before configuring other options.

**BIM** *Manager Note:* Walls are a System Family. System Families are the elements in Revit Architecture with the most built-in parameters. (They typically represent building components that are constructed on-site like floors or roofs as opposed to those created in a factory or shop and delivered to the site and installed like doors and light fixtures.) System Families cannot be edited outside of the Project Editor. Furthermore, the System Family itself cannot be renamed, deleted, or edited. System Families have Types like component Families and user edits are limited to the System Family's Types. In order to automatically have customized System Family Types available to new projects, you must add and edit them in your project template files. You can create your own project templates by choosing the "Project Template" radio button when creating a new project. You can also transfer Wall Types, other System Family Types and general system settings from existing projects to your current project by opening both projects and then choosing the **Transfer Project Standards** from the File menu. For more information, search for "Transfer Project Standards" in the online help.

- **Properties icon**—Click this icon to open the Element Properties dialog (see Figure 3.14 below for an example). The Element Properties dialog gives you access to both "Type" and "Instance" parameters. Type parameters apply to all Walls modeled in the project of the same type, Instance parameters apply only to the selected Wall(s).

- **Wall Creation Methods**—Walls can be drawn on-screen or created from existing model components. There are three creation methods, shown by their icons: Draw, Pick Lines, and Pick Faces. When you choose Draw, you can create Walls on-screen in a variety of shapes. When you choose the Pick Lines option, the Wall Shapes icons disappear (since the shape of the Wall will be determined by the lines you pick on screen). Likewise, even fewer options will be available when you choose the Pick Faces option. Here Walls are created directly from the faces of 3D geometry.

- **Height**—There are two options for Height; a drop down list of parametric height options and a text field (available when "Unconnected" height is chosen). When you choose "Unconnected" you are able to simply type in a fixed height for the Wall. The heights of Walls can also be connected to the project's levels instead.

- **Loc Line**—Short for "Location Line." The Location Line is the point within the width of the Wall that is used as a reference. Choices include: Wall Centerline, Interior and Exterior Finish Faces and several "Core" options. The Core of the Wall will be discussed in detail in later chapters.

- **Chain**—This option creates Walls in a sequence automatically joined end to end. If you deselect the Chain option, you will need to indicate the start and end point of each Wall segment you draw. With Chain enabled, the end of your first Wall will automatically become the start point of the next Wall and so on. This option is selected by default.

- **Wall Sketch Shapes**—These options are only available when you choose the "Draw" icon from the creation methods (see above). There are three basic sketch shape icons in this section: Line, Rectangle, and Arc. Several additional options are accessible from the drop down menu icon next to the arc icon. This reveals a collection of other options, such as Polygon, Circle, and several Arc types.

- **Offset**—Use this field to input a dimension. The Walls you create will be placed parallel (offset) to the points you click at a distance equal to this value.

- **Radius**—This option can be used to create rounded corner joins as walls are created.

Now that we have an overview of the available options, let's see some of them in practice.

- Verify that both the "Draw" and the "Line" icons are enabled and then click anywhere on screen to place the first point of the Wall.

  Move your mouse to the right, keep it horizontal, but don't click yet (see the top of Figure 3.4).

Notice the dimension on screen as you move the mouse. As you move your pointer, this dimension automatically snaps to whole unit increments. Depending on the size and resolution of your screen, the exact increment may vary. For example, if you are using Imperial units, the increment is likely 1'-0" and the increment for Metric is likely 100mm.

4. Roll the wheel of your mouse up a few clicks to zoom in a bit and continue to move the pointer left or right (see the bottom of Figure 3.4).

 **Note:** If you do not have a wheel mouse, you might want to consider purchasing one. You can use the Zoom commands on the View menu instead of the wheel, but the wheel makes quick zooming much easier. Refer to Chapter 2 for more information.

As you zoom in, the increment of the dimension will reduce. Depending on the unit type

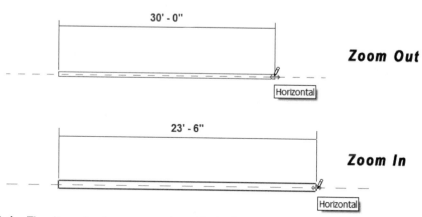

**Figure 3.4** *The dimension increment varies with the level of zoom, (The exact values on your screen may not match the figure)*

you are using, you may need to zoom in or out to see this. This is easy to do with the wheel. Simply roll the wheel up or down to zoom in or out. If you zoom off screen, keep the command active, press and hold the wheel button in and then drag. This will scroll the screen. You can also access zoom commands on the View menu, Tool bar, right-click menu, and by pressing the F8 key to use the Dynamic View dialog.

- Click to set the other point of the Wall.

  A single Wall segment will be created.

Often you will want to create more than one Wall segment, each beginning where the previous one ended. You can achieve this with the "Chain" option on the Options Bar.

5. On the Options Bar, verify that there is a checkmark in the "Chain" checkbox to enable this option.

• Click any point on screen to begin placing the next Wall.

• Click two more points at any locations on screen (see Figure 3.5).

  Notice that the corner where the two Walls meet has formed a clean intersection (called a "Wall Join").

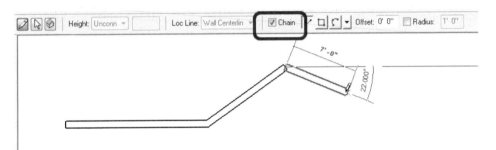

**Figure 3.5** *Use the Chain option to create continuous Wall segments*

6. Click the Arc icon.

The default *Arc* tool first places each of the endpoints and then adds the intermediate point last to define the degree of curvature. Other Arc options are also available if you click the small arrow icon (next to the Arc icon) to reveal the drop down list of Wall shapes available (see the "Wall Shapes" bullet point above).

• Move the pointer in any direction and click to place the end point of the Arc.

• Following the cue at the Status Bar, click to place the intermediate point of the Arc.

  Continue to experiment by adding additional Wall segments as desired.

7. To finish adding Walls, click the **Modify** tool on the Basics Design Bar (or press the ESC key twice).

Think of these few Walls as a simple "warm up" exercise. Drawing them helped us explore some of the basic Wall options. However, these Walls have been placed a bit too randomly to be useful. Let's delete them now and create some new ones.

## USING A CROSSING SELECTION BOX

Before we can delete the existing Walls, we need to select them. There are many ways to select objects in Revit Architecture. Rather than learn them all at once, let's take them one at a time as our current need dictates. A convenient way to select multiple objects at one time is with the Window and Crossing selection methods. In either technique, you create a box by selecting two opposite corners with your mouse. To create a Window selection, click and drag from left to right. To make a Crossing selection, click and drag the opposite direction—from right to left. A Window selection selects only those items completely surrounded by the box, while a Crossing selects anything touched by or within the box. This was discussed in brief in Chapter 2. Let's review it now.

8.  Click a point below and to the right of the Walls you have on screen.

- Hold down the mouse button and drag up and to the left far enough to touch all objects with the dashed (Crossing) selection box (see Figure 3.6).

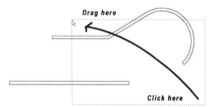

**Figure 3.6**   *Make a Crossing selection by dragging from right to left*

- When all of the Walls highlight, release the mouse button.

All of the Walls will turn red to indicate that they are now selected. Once you have a selection of objects, you can manipulate their properties, move, rotate, or mirror them or you can delete them.

9.  Press the DELETE key on your keyboard to delete the Walls.

## ADDING WALLS WITH THE RECTANGLE OPTION

Several of the Wall shape icons that are available create Walls that form closed geometric shapes like rectangles, circles, and polygons. Often using these is the quickest way to layout these more regular shapes. Let's try the rectangle option now.

10.  On the Basics Design Bar, click the **Wall** tool.

- On the Options Bar, verify that **Basic Wall: Generic - 8"** [*Basic Wall: Generic – 200mm*] is chosen from the Type Selector.

- Verify also that the Draw icon is active. If it is not, click it now.

- Accept the defaults for Height and Location Line and then click the Rectangle shape icon (see Figure 3.7).

**Figure 3.7**   *Using the rectangle draw option for Walls*

Our project contains four elevation markers. Let's zoom the screen to fit these to the screen so they are visible.

11. On the keyboard, type **ZF**.

 **Note:** If you prefer, you can also choose **Zoom > Zoom To Fit** from the View menu. However, where available, the keyboard shortcuts like the one suggested in the previous step are usually quicker once you learn them. Menu commands that have keyboard shortcuts list them to the right of the menu pick. Several examples of this are seen on the **View > Zoom** menu.

The mouse pointer will show a small rectangle next to it indicating that we are in rectangle drawing mode.

12. Click a point within the upper left region of the space surrounded by the elevation markers.

• Move the mouse down and to the right and watch the values of the dimensions.

13. Click a point in the lower right region of the space surrounded by the elevation markers.

Notice that the temporary dimensions continue to display on the Walls just drawn. Furthermore, these dimensions appear in blue. The blue color indicates that their values may be edited dynamically; or put another way, blue means "interactive."

14. Click on the blue value of the horizontal dimension.

The value will become an editable text field.

• Type a new value into this field and then press ENTER. (The exact value is unimportant—see Figure 3.8).

 **Note:** The value on your screen will vary from that shown in the figure.

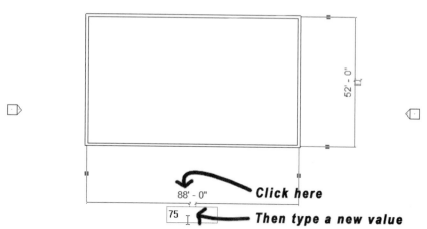

**Figure 3.8**  *Use the temporary dimensions to edit the locations of the Walls*

15. The Wall command is still active. Using the same technique, draw another rectangle overlapping the first.

Notice that the temporary dimensions now reference points on the first rectangle to points on the new one. You can edit these values in the same way that we edited the ones above. You can also move the witness lines of the dimensions to gain more control over their exact locations. We will explore this below. If you wish, try some of the other sketch shapes like circle or polygon. In the case of the polygon, additional controls will appear on the Options Bar.

16. When you are finished exploring Walls, choose **Close** from the File menu to close the current project. When prompted to save, choose No.

## CREATE AN EXISTING CONDITIONS LAYOUT

Now that we have practiced adding a few Walls and seen some of the options available while doing so, let's begin creating an actual model. In this book we will follow two projects from the early schematic phase through to the construction document phase. We will start with the first floor existing conditions for our residential project. The residential project is an 800 ft² [75 m²] residential addition. This project will require a little bit of demolition and new construction and will require plans, sections, elevations, details, and schedules. In this project we will explore the Phasing tools in Revit—Demolition, Existing, and New Construction will be articulated later in the tutorial. This will give us the required separation between construction phases of the project. The completed files for the residential project are available in the *Chapter03\Complete* folder with the files installed from the CD. You can open the completed version at any time to compare it to your progress.

## CREATE A NEW RESIDENTIAL PROJECT

We will use the same template file that we used above to begin our residential model.

1. Create a new project file using the *default.rte* [*DefaultMetric.rte*] template file as we did in the "Create a New Project" heading above.

Be sure that the *Level 1* floor plan view is open on screen. You can see this indicated in bold (under *Views (all) > Floor Plans*) on the Project Browser and in the title bar of the Revit Architecture window.

2. On the Design Bar, click the Basics tab and then click the **Wall** tool (shown in Figure 3.2 above).

• From the Type Selector, choose **Basic Wall: Generic - 12"** [**Basic Wall: Generic – 300mm**].

• Verify also that the Draw icon is active. If it is not, click it now.

• For Height choose **Unconnected** and set the value to **18'-0"** [**5500**].

**Tip:** If you are using Imperial units, simply type "18" and then press ENTER. No unit symbol or zero inches is necessary. Also note that to enter a value like "18'-2 1/2" you can type it in several ways. For example you can type 18' 2 1/2" or 18' 2.5" or 18 2 1/2 or 18 2.5 or 18.20833. Refer to the "Unit Conventions" topic in Chapter 2 for more information.

• For the Location Line, choose **Finish Face: Exterior** and then click the Rectangle sketch icon (see Figure 3.9).

**Figure 3.9** *Set the Options for the exterior Walls of the residential project*

3. Click two opposite corners on screen within the space bounded by the elevation markers (the exact size is not important for initial placement).

Notice that even though we have chosen a Location Line of Finish Face: Exterior, that the temporary dimensions still have witness lines at the centerlines of the Walls. This is simply a default behavior independent of the Wall's individual location line setting. If you want to input a value for the dimension based upon the face of the Walls, you can simply move the witness lines.

4. Click the small blue square handle on one of the horizontal dimension's witness lines. Zoom into the small blue square to see this better.

Notice that the witness line moves to one of the Wall faces. If you click it again, it will move again, this time to the opposite face. One more click returns it to the centerline (see Figure 3.10).

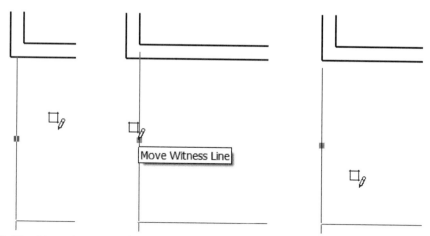

**Figure 3.10** *Move the witness lines of the temporary dimension*

- Click the witness line handle (of the horizontal dimension) until both sides reference the outside edges.

- Click the blue value of the horizontal dimension and then type: **33'-0"** [**10000**] and then press ENTER.

5. Repeat this process on the vertical dimension making the outside face to face dimension equal to: **24'-0"** [**73000**].

**Tip:** If your witness lines are in the wrong locations when you edit the temporary dimension, simply repeat the process to move the witness lines and then repeat the dimension edit process.

**Caution:** Be careful not to click the small "permanent" dimension icon when editing the dimension values. Clicking this icon will make the temporary dimension a "permanent dimension" (it will remain in the current view even after the associated Wall is deselected). If this happens, simply select the dimension and delete it.

6. On the Design Bar, click the Modify tool, or press ESC twice to complete the operation.

You should now have four walls in a rectangular configuration measuring 33'-0" × 24'-0" [10000 × 73000] outside dimensions.

7. From the File menu, choose **Save**.

- In the Save As dialog, navigate to a location on your local system where you wish to save the project—your *Documents* folder for example. (*My Documents* in Windows XP.)

- For the File name, type: **03 Residential** and then click the Save button.

## OFFSET WALLS

Let's begin adding the interior partitions. There are several techniques that we could employ to do this. In this sequence, we will use the *Offset* tool. The *Offset* tool moves or copies the selected objects parallel to themselves by an amount that you input. This tool is located on the Tools toolbar or the Tools menu. The Tools toolbar is displayed by default and therefore should already be displayed on your screen. If you have closed this toolbar, please re-display it now. Refer to Chapter 2 for more information on loading toolbars.

8. On the Tools toolbar, click the Offset tool.

 **Tip:** The shortcut for offset is **OF**. **Offset** is also located on the Tools menu.

The Options Bar for Offset has two modes: Graphical and Numerical. When you choose Graphical, the numeric input field is disabled and you use a temporary dimension on screen to indicate the distance of the offset. When you choose the Numerical option, you input the offset distance first, and then use the pointer on screen to indicate the side of the offset. In both cases, you can enable the "Copy" checkbox, which will create a copy of the object as it offsets. If you disable this option, the object you offset will be moved parallel to itself by the amount you indicate.

- On the Options Bar, verify that Numerical is chosen (if it is not, choose it now).

- Zoom in closer if necessary.

- In the Offset field on the Options Bar, type **12'-5 1/2" [3794]** and verify that "Copy" is selected.

  The mouse pointer will change to an Offset cursor and the Status Bar will prompt for a selection.

- In the workspace, move the Offset cursor over the top edge of the bottom horizontal Wall (see Figure 3.11).

  A dashed line will appear indicating the location of the Offset. If you move your mouse up and down this line will shift up and down as well.

As in most places in Revit Architecture, if you press the TAB key, the pre-highlighted selection will cycle to the next available option. In this case, all four Walls will pre-highlight and the offset result would be a concentric ring inside or outside the building. Go ahead and try it if you like. Press the TAB key once, and then move the pointer to indicate outside or inside offset and then click. Be sure to undo (Edit menu) after this experimentation.

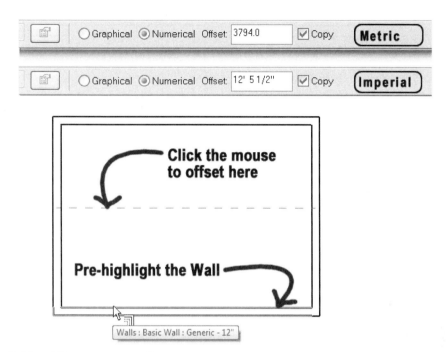

**Figure 3.11**   *Offsetting a new wall*

- When the dashed line appears above the Wall (inside the house) click the mouse.

  A new Wall will appear inside the house.

- On the Design Bar, click the Modify tool, or press the ESC key twice to complete the operation.

## COPYING WALLS

We can also copy Walls. You select the Wall to copy first and then click the Copy tool.

9. Select the vertical Wall on the left.

- On the Edit toolbar, click the Copy tool.

> **Tip:** The keyboard shortcut for Copy is **CO**. **Copy** is also located on the Edit menu.

With the Copy tool you input the distance of the offset on the screen. At the Status Bar, a prompt reads "Click to enter move start point."

- Click a point anywhere on screen. (The exact start point is not important, but for convenience, you can click directly on the Wall). Begin moving the Wall to the right, type **12'-8 1/2" [3818]** and then press ENTER for the Offset distance this time (see Figure 3.12).

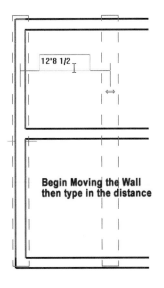 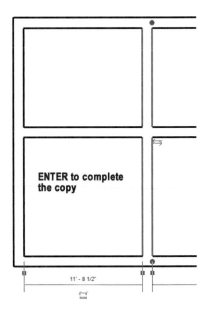

12'8 1/2

Begin Moving the Wall
then type in the distance

ENTER to complete
the copy

11' - 8 1/2"

**Figure 3.12**  *Offset a vertical Wall inside*

- Save your project.
10. On the Design Bar, click the **Modify** tool or press the ESC key twice.

## CHANGING A WALL'S PARAMETERS

One of the features that makes Revit Architecture such a powerful tool is the ability to change an object's parameters at any time, as design needs change. Let's take a look at modifying some of the walls as we continue with the layout of the first floor existing conditions for the residential project.

11. Select the two internal Walls created in the previous steps.

To select the Walls, you can either click just inside the house near the lower right corner and then drag up and to the left until both Walls highlight or you can hold down the CTRL key and click each of the Walls one at a time (see Figure 3.13).

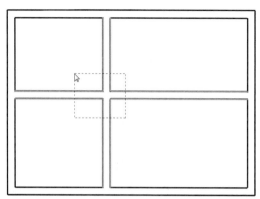

**Figure 3.13** *Select the interior Walls*

12. On the Options Bar, click the Properties icon (or right-click, and choose **Element Properties**).

    The Element Properties dialog will appear showing both the Type and Instance parameters for the selected Walls.

    • In the Element Properties dialog, at the top, choose **Generic – 5" [Interior - 135mm Partition (2-hr)]** from the Type list.

    • In the Instance Parameters area, within the Constraints grouping, choose **Wall Centerline** for the Location Line (see Figure 3.14).

**Tip:** To make this change, click in the Location Line field, and then click the small down arrow that appears to choose the Wall Centerline option from the popup menu.

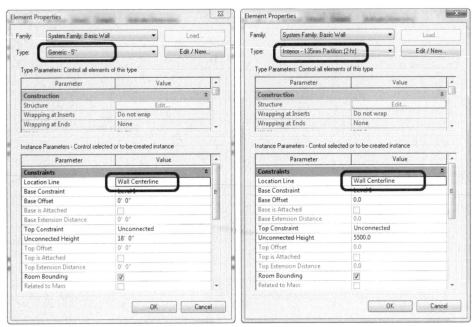

**Figure 3.14** *Change the parameters of the interior Walls*

13. Click OK to dismiss the Element Properties dialog.

    Notice the change to the Walls.

Let's verify that everything is in the correct place. As noted above, we are laying out the existing conditions of our residential project. Therefore, the locations of the Walls that we just placed are based on field dimensions. The room in the lower left corner of the plan is supposed to be 11'-6" [3500] × 11'-9" [3658]. We can verify these dimensions and easily make adjustments as necessary.

## ADJUST WALL LOCATIONS

In the "Create a New Residential Project" heading above, we learned to use the temporary dimensions and the handles on the witness lines to edit the location of Walls. We can use the same technique to verify our dimensions now.

14. Click on the interior horizontal Wall.

The temporary dimensions will appear indicating its current location relative to the other Walls. However, depending on the settings of your system, the dimensions may be from the centerlines. The desired size of the room that we are verifying are to the inside faces of the Walls (as you might expect from field dimensions). Therefore, we could use the witness line technique covered above to adjust the dimension points, but this is not persistent and the next time we would need to move the witness lines again. It will be better to change the default behavior of the dimensions from now on. We do this with the **Temporary Dimensions** command from the Settings menu.

15. From the Settings menu, choose **Temporary Dimensions**.

- In the Walls area, click Faces and then click OK (see Figure 3.15).

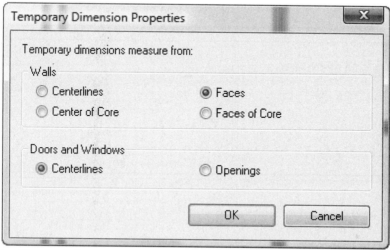

**Figure 3.15**    *Change the Temporary Dimensions to measure from Wall faces*

16. Click the horizontal interior Wall again.

The dimension from the inside face of the bottom exterior Wall to the inside face of the selected Wall should be 11'-5 1/2" [3658].

17. Click on the vertical interior Wall.

This time, the dimension from the inside face of the left exterior Wall to the inside face of the selected Wall should read 11'-8 1/2" [3650]. However, this value is not correct; the distance should be 11'-6" [3500]. Despite best efforts to transfer field measurements to our building model, an error crept into our calculations. Fortunately, errors like this are very easy to correct in Revit Architecture.

- Click the blue text of the left-hand dimension and type the correct value **11'-6"** [3500] (see Figure 3.16).

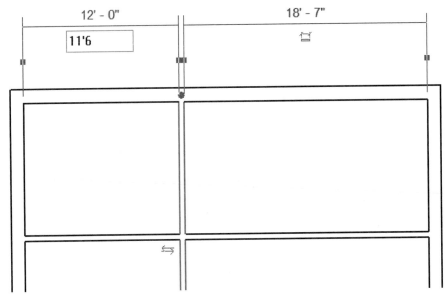

**Figure 3.16** *Edit the Temporary Dimension to move the Wall to the correct location*

18. Save the project.

## SKETCHING WALLS

The offset and copy techniques covered above are effective ways to add new Walls based upon Walls already existing in the model. In many cases however, it is easier to sketch the new Walls and then use the temporary dimensions to position them correctly. Let's try that technique next.

19. On the Design Bar, click the **Wall** tool.

- From the Type Selector, choose **Generic – 5" [Interior - 135mm Partition (2-hr)]**.

- Verify that the Draw icon is selected.

- For Location Line, choose Wall Centerline.

- For the Sketch Shape, choose Line and make sure that Chain in *not* selected.

20. Move the mouse over the exterior vertical Wall on the left.

Notice the temporary dimension that appears.

- When the value of the dimension above the interior horizontal Wall is about **4'-0"** **[1200]** click to set the first point.

- Move the pointer horizontally to the right and click to set the second point just past the middle of the plan (see Figure 3.17).

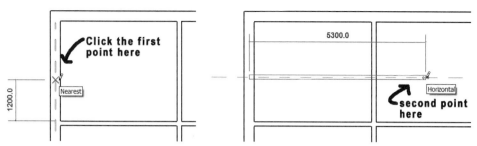

**Figure 3.17** *Sketch a Wall segment above the horizontal interior Wall*

 **Note:** The figure shows Metric units, however, the exact locations of both clicks is not critical since we will use temporary dimensions to edit them next. Use the figure to achieve approximate placement.

Several temporary dimensions will remain on screen after the placement of the Wall.

21. Click the blue text of the vertical temporary dimension between the new Wall that you just created and the other horizontal interior Wall.

   - Type **3'-10" [1170]** and then press ENTER.

 **Note:** This is a face to face dimension. Once you edited the Temporary Dimension setting above, it should remain persistent.

22. Using the same process, create a vertical Wall approximately **3'-0" [900]** to the left of the existing interior vertical Wall.

   - Start at the lower exterior Wall and draw up until it intersects with the upper horizontal interior Wall (the one drawn in the last step).

   - Using the temporary dimensions, edit the face to face distance between the two Walls to be **2'-6" [750]** (see Figure 3.18).

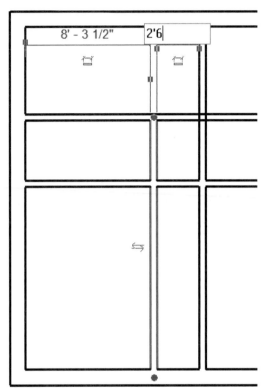

**Figure 3.18**  *Sketch Walls and use temporary dimensions to fine-tune their placement*

**Note:** On the Design Bar, click the **Modify** tool or press the ESC key twice. Manipulate the Walls with Control Handles. The two Walls that we just added frame out a small closet in the middle of the plan. We can use a few techniques to clean up the plan a bit and remove unnecessary Wall segments. We will look at some control handle editing techniques next.

23. Select the vertical interior Wall that you just completed (the one on the left).

At each end of the Wall is a small round blue handle.

- Click and drag the handle at the bottom upward.

- Using the temporary guidelines that appear, drag up and snap to the horizontal Wall (see Figure 3.19).

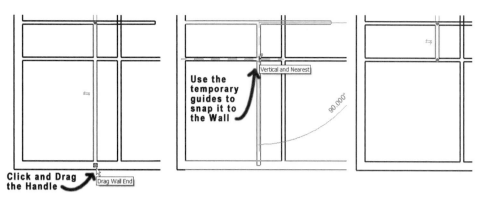

**Click and Drag the Handle**

**Figure 3.19**   *Drag the Wall End Handle up to the intersection with the horizontal Wall*

## TRIM WALLS

While the control handles are quick and easy, another common way to edit a plan layout is using the Trim/Extend tool.

24. On the Tools toolbar, click the ***Trim/Extend*** tool.

**Tip:** The keyboard shortcut for Trim/Extend is **TR**. **Trim/Extend** is also located on the Tools menu.

The ***Trim/Extend*** tool has three modes on the Options Bar. The "Trim/Extend to Corner" mode works on two Walls or sketch lines. It will create a clean corner from the two segments by either lengthening or shortening the selected segments. The second mode is "Trim/Extend Single Element." This mode will lengthen or shorten the selected segment to a designated boundary edge. The third option, "Trim/Extend Multiple Elements" works the same way except that multiple Walls or Lines can be lengthened or shortened to the same boundary.

25. Make sure that the first mode is selected (Trim/Extend to Corner) and then click the vertical Wall that was just edited with the handle.

**Note:** Pay attention to the prompt at the Status Bar, in this case instructing you to select the first Wall that you want to Trim or Extend. Please note that you are further prompted to select the portion that you wish to keep.

26. Next click the horizontal Wall to the right of the intersection (see Figure 3.20).

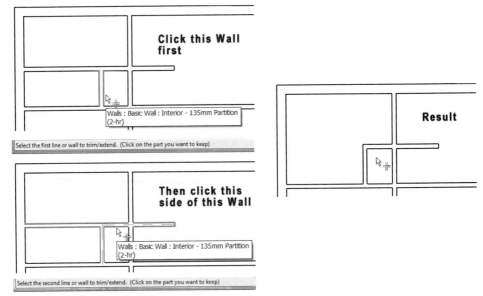

**Figure 3.20** *Using Trim/Extend to create the closet corner*

27. On the Design Bar, click the **Modify** tool or press the ESC key twice.

## ADDING WALLS USING THE PICK LINES MODE

Let's continue adding to our floor plan layout by using a technique that combines features of both the previous techniques. We will add Walls using the **Wall** tool and its "Pick Lines" mode. We will use this in conjunction with the Offset option on the Options Bar to place the Wall close to where we need it.

28. On the Design Bar, click the **Wall** tool.

   - From the Type Selector, choose **Generic – 5" [Interior - 135mm Partition (2-hr)]**.
   - Click the Pick Lines icon to activate this mode.
   - For Location Line, choose Wall Centerline.
   - In the Offset field, type **6'-0" [1800]**.

29. Move the mouse next to the right side of the interior vertical Wall in the middle of the plan.

   The right face of the Wall will pre-highlight.

30. Be sure that the temporary guideline indicates a new Wall to the right and then click.

31. On the Design Bar, click the **Modify** tool or press the ESC key twice.

32. Click on the new Wall, and edit the dimension between this Wall and the one it was offset from to **6'-6" [1983]** (see Figure 3.21).

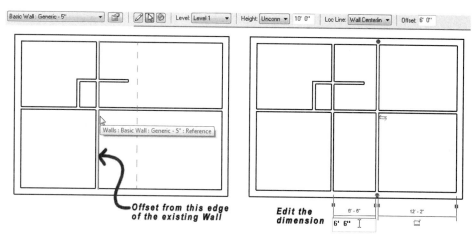

**Offset from this edge
of the existing Wall**

**Edit the
dimension**

**Figure 3.21**    *Adding a Wall using the Pick Lines mode*

If you want to calculate the exact offset to use instead of editing the temporary dimension, you can do so. The Offset value will be measured from the line you pick to the Location Line of the new Wall—in this case Wall Centerline. Therefore, half of the Wall's width would have to be added to the dimension. Despite the relative ease of this calculation, it is typically easier to place the Wall close and then edit the dimension to the exact value as we have done here.

## TRIM/EXTEND SINGLE ELEMENT MODE

Let's Trim a few more Wall segments.

     33. On the Tools toolbar, click the ***Trim/Extend*** tool (or type **TR**).

       • On the Options Bar, click the "Trim/Extend Single Element" mode icon.

In this mode you first select the element to use as a boundary and then the click the Wall or line you wish to extend or trim.

      **Tip:** Remember to read the prompts on the Status Bar.

       • Click the vertical interior Wall on the right as the boundary.

       • Click the short horizontal Wall at the top of the closet to extend (see Figure 3.22).

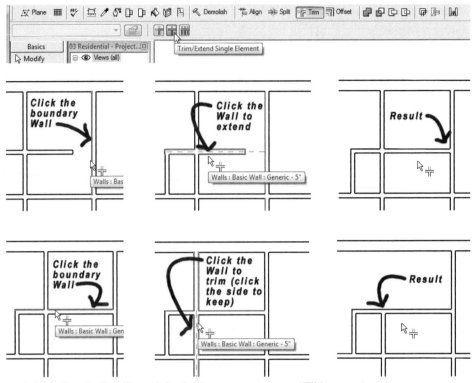

**Figure 3.22** *Use the Trim/Extend Single Element to create a "T" Intersection*

34. Use the same technique to trim away the vertical Wall at the top of the closet (Remember to click the piece you want to keep—read the Status Bar—see the lower portion of Figure 3.22).

35. On the Design Bar, click the **Modify** tool or press the ESC key twice.

## USING THE SPLIT AND MOVE TOOLS

The layout is coming along. Next let's focus on the stair hallway.

36. On the Tools toolbar, click the **Split** tool.

**Tip:** The keyboard shortcut for Split is **SL**. **Split Walls and Lines** is also located on the Tools menu.

• Place the cursor (now shaped like a knife) over the long horizontal Wall in the middle of the plan where it intersects the left vertical Wall in the middle.

The vertical Wall should highlight.

• Press the ESC key a few times until the horizontal Wall highlights.

- When the horizontal Wall is highlighted and small vertical line appears at the cursor, click the mouse (see Figure 3.23).

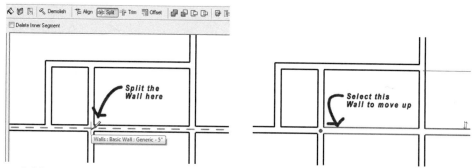

**Figure 3.23** *Move the Wall up and then edit the dimension to the correct value*

- On the Design Bar, click the **Modify** tool or press the ESC key twice.

37. Click the right side of the Wall we just split.

Notice how the Wall is now two segments. Use Split anytime you need to break a single Wall into two parts.

38. With the Wall still selected, click the **Move** tool on the Edit toolbar.

 **Tip:** The keyboard shortcut for Move is **MV**. **Move** is also located on the Edit menu.

- Pick a start point on screen, such as the Endpoint at the top of the Wall.
- Begin moving the mouse up (do not click) and then type **10" [250]**.

If you are working in Imperial units, be sure to type the inch symbol ("). If you prefer, you can also zoom in and snap to this value as you move the mouse. In this case, simply click the mouse when the right value appears on screen. Or, you can also move randomly at first, and then use the temporary dimension to edit it to the correct value. The exact technique you use is up to you, as long as the Wall moves up 10" [250]. Select the long vertical Wall on the left and then click the Copy tool on the Edit toolbar.

- Pick a start point on the selected Wall and then move to the right.
- Move the mouse to the right **3'-6" [1067]** and click to place the copy.

 **Tip:** You can begin moving the mouse to the right, type the desired value, and then press ENTER.

- Verify that the face to face dimension between the new Wall and the original is **3'-1" [942]**, if it is not, edit the temporary dimensions (see the left panel of Figure 3.24).

- Using the middle panel of Figure 3.24 as a guide, Trim the two new Walls to form a corner.

- Using the right panel of Figure 3.24 as a guide, drag the handle up and edit the temporary dimension as shown: **8'-3"** [**2500**].

39. Save your Project.

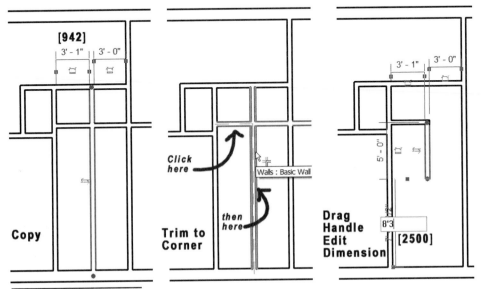

**Figure 3.24**   *Copy, Trim, and Edit Walls for the hallway stairs*

## SKETCH THE REMAINING WALLS

At the top middle of the plan is a small half-bath. Let's sketch these Walls in to complete the Wall layout of our first floor existing conditions.

40. On the Design Bar, click the **Wall** tool.

- From the Type Selector, choose **Generic – 5"** [**Interior - 135mm Partition (2-hr)**].

- Click the Draw icon to enable this mode.

- For Location Line, choose Wall Centerline.

- For the Sketch Shape, choose Line.

- Place a checkmark in the Chain checkbox.

- Verify that Offset is set to zero.

41. Working at the top middle of the plan, sketch two Wall segments to create a small half-bath.

- Edit the temporary dimensions to make the inside clear dimensions of the half-bath **4'-4"** [**1300**] × **3'-6"** [**1050**] (see Figure 3.25).

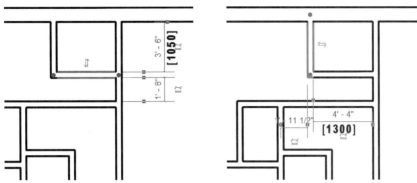

**Figure 3.25**   *Sketch the small half-bath*

42. Extend the vertical Wall in the middle of the stair hallway up to the half-bath.

- Use the **Split** tool with the "Delete Inner Segment" option on the Options Bar to remove the small piece of Wall shown in the middle panel of Figure 3.26.

- Use the **Trim/Extend** tool to complete the layout as shown in the right panel of Figure 3.26.

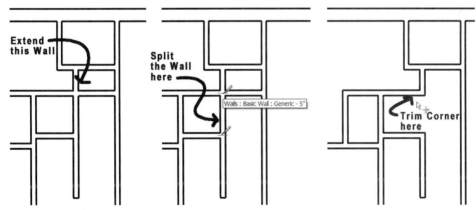

**Figure 3.26**   *Complete the Wall Layout using Trim/Extend and Split*

43. Save the Project.

## WORKING WITH PHASING

We have completed the layout of existing walls on the first floor of the house. Assigning the Walls to a particular construction phase will help us distinguish them as existing construction as the project progresses. Revit Architecture includes robust Phasing tools allowing us to quickly and easily designate these Walls as existing construction. Every element in the model has two phasing parameters: "Phase Created" and "Phase Demolished." In this way you can track the life span of any element in the model. In the case of the Walls that we have drawn here, all of them are "existing construction" and none will be demolished. We will be demolishing some of the Windows and Doors later on in the chapter.

## ASSIGN WALLS TO A CONSTRUCTION PHASE

1. Using a window selection (click and drag to surround all Walls), select all of the Walls in the model.

2. On the Options Bar, click the Properties icon (or right-click and choose **Element Properties**).

   The Element Properties dialog will appear.

   • Beneath the Phasing grouping, for "Phase Created," choose **Existing**.

   • For the "Phase Demolished," verify that **None** is chosen (see Figure 3.27).

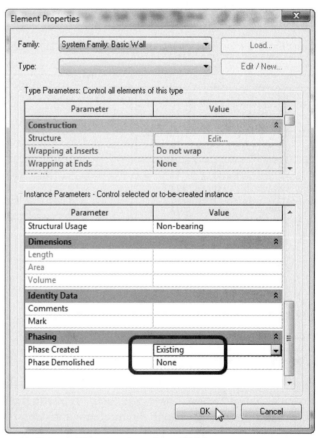

**Figure 3.27** *Assign the Existing Phase to the selected Walls*

   • Click OK to close the Element Properties dialog.

3. Deselect all of the Walls to see the change.

> You can deselect the Walls by simply clicking in the white space next to the Walls. You can also click the **Modify** tool on the Design Bar or press the ESC key twice.

The Walls will display in a lighter lineweight, and will also be colored gray. This indicates that they are existing construction. Assigning a phase to an element automatically assigns a graphic override to it. You can control the settings of these overrides as well as edit and add phases in the Phasing dialog. To open this dialog, choose **Phases** from the Settings menu. Feel free to explore this dialog now if you wish. A good example of a typical task that you could perform in this dialog would be to set up phases of construction in a large project. For example, you could create a "Foundations and Caissons" phase, "Phase 1" and "Phase 2 New Construction" phases. For our residential project, the out-of-the-box phases of "Existing" and "New Construction" are sufficient. Therefore, if you make any changes in the Phases dialog, please do not apply them.

## EXPLORING THE VIEW PROPERTIES

At this time, it is appropriate to explore Phasing a bit deeper to get a full understanding of the tools available. Many of the changes we are about to make are not required by our residential project, therefore we will undo the next few steps after we finish exploring.

1. On the Design Bar, click the **Wall** tool.

- Accept all the default settings and draw a single Wall segment anywhere in the model, but touching one of the existing Walls.

- On the Design Bar, click the **Modify** tool or press the ESC key twice.

Notice that the new Wall came in as New Construction. Notice also that it did join with the existing Wall. Elements you create automatically use the current phase assigned to the view. In this case, while we changed the phase of the Walls, the *Level 1* floor plan view is still assigned to New Construction.

2. Select any Existing (phase) Wall and then click the Properties icon (on the Options Bar).

- From the Phase Demolished list, choose **New Construction** and then click OK.

Notice that the Wall is now displayed as dashed.

3. Select a different Existing (phase) Wall and then click the Properties icon.

- From the Phase Demolished list, choose **Existing** and then click OK.

Notice that this time, the Wall has disappeared. This is because each view has its own display parameters with "Phase" and "Phase Filter" being two of them. In this case, we have the *Level 1* view open on screen. You can verify this on the title bar at the top of the Revit Architecture window and by noting that *Level 1* is bold in the Project Browser. Let's change the current phase of this view to Existing to see how the display changes.

4. Make sure that there are no elements selected. Right-click in the Workspace and then choose **View Properties**.

**Tip:** The keyboard shortcut for View Properties is **VP**. **View Properties** is also located on the View menu.

The Element Properties dialog will appear.

Notice that the same "Element Properties" dialog appears. However, this time the "element" in question is the view named: "Level 1" rather than a selection of Wall elements. Regardless, to Revit Architecture both items are "Elements" that have properties we can manipulate.

- Beneath the Phasing grouping, for "Phase," choose **Existing**.

- Click OK to see the change.

This change makes the Existing construction phase the current phase for the *Level 1* view. Therefore, the view will now show only items that existed in that phase. This means that no elements assigned to the New Construction phase will show, and the Wall that we specified as demolished in New Construction will now appear solid again. Furthermore, the Wall that we set to demolish in the Existing Phase now appears again, but it is dashed and cross hatched. This style indicates "temporary" construction, which occurs when an element is both created and demolished in the same phase.

5. Draw another Wall.

- Select the new Wall and edit its Properties.

Notice that the newly drawn Wall was automatically assigned to the Existing phase. Any elements that we draw will automatically be added to the current phase of the view. We will work with this technique below when we add Doors and Windows.

In addition to the current Phase setting, we can also assign a "Phase Filter" in the View Properties dialog. A Phase Filter does not change the active phase, rather it changes which elements display and how they appear in the current view based on their phase designations. By default, all views are set to "Show All." Several Phase Filters are included in the default templates from which our project was created. For example, if you choose None, all elements will display (from all phases) and they will display the same regardless of phase setting. There are several possibilities available; some are more useful when there are more phases than just Existing and New. Explore them now to get a sense of how they behave. Please note that the Phase Filter operates on the view's currently active phase. Therefore, choosing "Show Previous Phase," for example, will make everything disappear since there is currently no phase before "Existing" in the current project. Like Phases, Phase Filters can be edited in the Phasing dialog. To open this dialog, choose **Phases** from the Settings menu.

6. Return to the View Properties dialog for the Level 1 floor plan view. Make the **Existing** Phase active.

- Return the Phase Filter to **Show All**.

- Select any demolished Walls, edit their **Element Properties**, and change the Phase Demolished to **None**.

- Delete any extraneous Walls that you drew.

7. Return to the View Properties dialog for the Level 1 floor plan view. Make the **New Construction** Phase active.

- Select any demolished Walls, edit their **Element Properties**, and return the Phase Demolished to **None**.

- Delete any extraneous Walls that you drew.

It may be possible to click the undo icon several times to restore the drawing to the state before we began demolishing and adding extraneous Walls. When you have finished exploring the various view properties and phasing options, and restored your model, it should look like Figure 3.28.

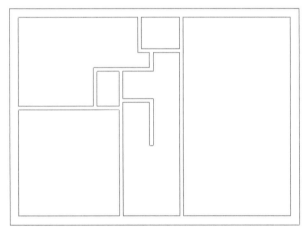

**Figure 3.28**   *The Residential Level 1 Wall layout*

8. Be sure that New Construction is the view's active Phase and then save the project.

## WORKING WITH DOORS AND WINDOWS

Continuing with our layout of the residential existing conditions, let's add some Doors and Windows. Doors and Windows automatically interact with Walls when inserted to create the opening and attach themselves to the Wall in a parametric way.

### ADD A DOOR

Doors are "Wall-hosted" elements. This means that Doors must be placed within Walls. Like Walls, Doors have both Type and Instance parameters. Several Door Types have been included in the out-of-the-box template file from which our residential project was based.

1. Continuing in the *Level 1* view of the Residential project, zoom out to see the entire plan layout.

You can do this with the mouse wheel or the commands on the **View > Zoom** menu.

2. On the Design Bar, click the Door tool (see Figure 3.29).

**Figure 3.29** *The Door tool on the Basics tab of the Design Bar*

Much like the *Wall* tool, when you click the *Door* tool, several options appear on the Options Bar. The pointer for Doors will have a small (+) sign next to the cursor and the Status Bar prompt will read: "Click on a Wall to place Door."

- From the Type Selector, choose Single Flush: **36" × 80"** [*M_Single-Flush : 0915 × 2032mm*].

The following list explains the major fields and controls shown on the Options Bar when you are adding a Door (see Figure 3.30).

**Figure 3.30** *The Options Bar for adding Doors*

- **Type Selector**—Use this drop-down to choose from a list of Door types in the current file. The specific list is populated by the door Families that are already loaded into the template we used to create the project. Always choose a type from this list before configuring other options. You can use the Load button (to the right of the Type Selector) to load additional Families and Types from an external library.

- **Properties icon**—Click this icon to open the Element Properties dialog. The Element Properties dialog gives you access to both "Type" and "Instance" parameters. Just like Walls, Type parameters apply to all Doors of the same type, Instance parameters apply only the selected Door(s).

- **Create in place**—Click to create a new in-place Door Family. In Place Families are used to create non-typical "one off" Families. In most cases, it would be rare to use In Place Families for Doors.

**Note:** In-place Families are not designed to be moved, copied, rotated, etc. They are meant to be used only once. If you need to use it more than once within this project or in a different project a regular door family should be created in the Family Editor, saved to a library and then loaded into your project as needed. The Family Editor will be explored in Chapter 9.

- **Load**—Noted in the Type Selector above, use this button to access the external libraries and load Door Families and Types not already available in the current project.

- **Tag on Placement**—Check this box to add a Door Tag as Doors are added to the project. Even if you deselect this option, you can still add Tags to Doors later.

- **Tag Orientation**—When "Tag on Placement" is selected, this option sets the orientation of the Tag as either horizontal or vertical.

- **Tags**—Opens the Tag dialog where you can see which Tags are currently loaded in your project and also choose to load additional Tags from a library file.

- **Leader**—Check this to add a Leader to your Tags as they are placed.

- **Leader Offset**—When you enable the Leader option, this field controls the offset of the Leader from the Tag.

As you move your mouse pointer around on screen, a Door will only appear when you move the pointer over a Wall. If you are unhappy with the direction of the door swing, press the SPACE BAR to flip it before you click to place the Door.

3. On the Options Bar, remove the checkmark from the "Tag on Placement" checkbox.

4. Move the mouse to the horizontal exterior Wall at the top right side of the plan.

- Position the Door roughly in the center of the top exterior Wall of the room on the right and then click the mouse (see Figure 3.31).

  As with Walls, temporary dimensions will guide your placement.

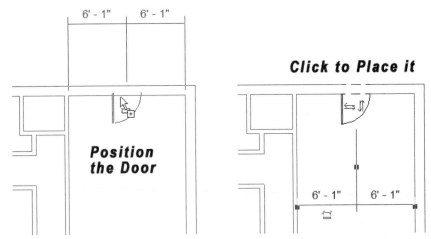

**Figure 3.31** *Click to place the first Door*

Notice that the Door appears in the drawing and cuts a hole in the Wall. However, notice that the hole in the Wall is filled with dashed lines. This is because we set the Phase of the view back to "New Construction" above. Therefore, Revit Architecture is showing this Door as being a new Door placed into an existing Wall. This requires the opening for the Door to be shown as demolition while the Door appears as new construction.

    5. On the Design Bar, click the Modify tool or press the ESC key twice.

### CHANGE THE PHASE FILTER

Let's try repeating the Phase Filter exercise above to see the different ways that this condition will display in each phase.

    6. Make sure that there are no elements selected, right-click in the Workspace and then choose **View Properties**.

       The Element Properties dialog will appear.

      • Beneath the Phasing grouping, for "Phase Filter," choose **Show Previous + Demo**.

      • Click OK to see the change (see the top of Figure 3.32).

With this phase filter active, you only see the previous phase (Existing in this case since it is the phase that occurs before new construction) and any demolition. In this case, the only demolition is the opening for the Door.

    7. Edit the View Properties again.

      • Beneath the Phasing grouping, for "Phase Filter," choose **Show Complete**.

      • Click OK to see the change (see the bottom of Figure 3.32).

This filter shows what the view would look like when the final phase is complete. So in this case, all the Walls appear bold and there is no demolition shown.

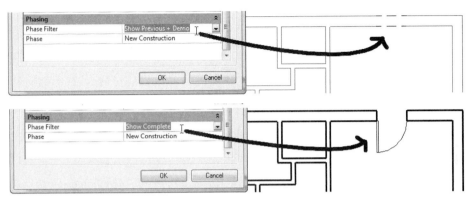

**Figure 3.32**   *Change the Phase Filter to view the model at different points in time*

8. Return to View Properties dialog (type **VP**) and choose **Show All** for the Phase Filter.

The power and potential of the Phasing parameters was seen when we explored these options with just Walls. Now that we have added a Door, we can truly see the full potential of these tools. If this Door truly were a new Door being added to these existing Walls, all of these graphical behaviors would be managed for us automatically by Revit Architecture simply by assigning the Door to the New Construction Phase parameter as we have done here. It turns out that this Door is actually an existing Door. Therefore, we need to change its Phase parameter to make it display properly.

9. Select the Door in the model and then on the Options Bar, click the Properties icon (you can also right-click and choose **Element Properties**).

• Beneath the Phasing grouping, for "Phase Created," choose **Existing**.

• Click OK to dismiss the Element Properties dialog.

The Door now displays the same as the Wall in which it is inserted and the dashed demolition lines no longer display. This is because the Door and Wall now belong to the same Phase, therefore there is no demolition required. Since we are going to add several more existing Doors, let's change the view's active Phase to Existing to save us the trouble of having to edit phase property of the Doors (and Windows below) later.

10. Return to View Properties dialog (type **VP**) and choose **Existing** for the Phase.

The Walls will turn bolder to reflect this change.

 **Note:** A typical set of construction documents requires existing conditions, demolition, and new construction drawings. In Revit Architecture this is easily achieved by duplicating the views (plans, sections, and or elevations, even schedules) and editing the views' Phase and Phase Filter parameters to display the correct data. Refer to the "Create an Existing Conditions View" heading in Chapter 6 for an example of this.

## PLACE A DOOR WITH TEMPORARY DIMENSIONS

Let's add several more existing Doors to our model. For the next several Doors, it will be easier to place them if the temporary dimensions are set to the openings rather than the centers. (This is similar to the change we made for Walls above).

11. On the Settings menu, choose **Temporary Dimensions**.

- In the Temporary Dimension Properties dialog, in the Doors and Windows area, choose Openings and then click OK.

12. On the Design Bar, click the Door tool.

- From the Type Selector, choose **Single Flush: 30" × 80"** [**M_Single-Flush : 0762 × 2032mm**].

- Verify that "Tag on Placement" is *not* selected.

- Move the cursor to the upper left corner of the plan and position it so that the Door is being added to the topmost horizontal exterior Wall.

- When the temporary dimension reads 2'-0" [600] from the upper left corner, click the mouse.

  As before, the temporary dimensions will remain on screen until you cancel the command or place another Door.

13. Click the blue value of the temporary dimension on the left and type **2'-4"** [**762**] and then press ENTER (see Figure 3.33).

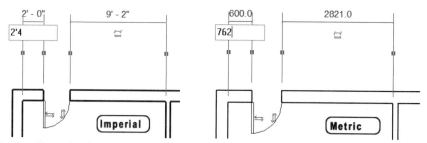

**Figure 3.33** *Place the Door in the approximate location, then use the temporary dimensions to fine-tune placement*

The Door will shift the indicated amount.

## LOAD A DOOR FAMILY

The next Door that we are going to add is a bi-fold door for the small closet in the middle of the plan. There is no bi-fold type available in the current project. Therefore, we will use the Load function to access the Revit Architecture library and load some bi-fold Door Types.

> You should still be in the Door command. If you have canceled it, click the **Door** tool on the Design Bar.

As noted, if you open the Type Selector, there are only Single Flush types loaded.

14. On the Options Bar, click the Load button.

- In the Open dialog, click the *Imperial Library* [*Metric Library*] icon on the left side.

- On the right side, double-click the *Doors* folder, choose *Bifold-2 Panel.rfa* [*M_Bifold-2 Panel.rfa*] and then click Open.

    There will be a pause while Revit loads the Family and its Door Types. If during this process a Save Reminder appears, click Dismiss for now.

- From the Type Selector, choose Bifold-2 Panel : **30" × 80"** [**M_Bifold-2 Panel : 0762 × 2032mm**].

 **Note:** When you open the Type Selector, you will notice several Bifold Types are now loaded into the project.

- Verify that "Tag on Placement" is *not* selected.

15. Move the pointer over the left vertical Wall of the closet in the middle of the plan.

- Using the temporary dimensions, get the Door centered on the vertical Wall.

    Do NOT click the mouse to place the door yet.

- Slowly move the mouse left to the right.

Notice that you can control whether the Door swings into or out of the closet with the mouse, but not which side of the opening (up or down in this case) that it swings. Take note of the Status Bar and in some cases a tip that appears on screen. There it notes that you can use the SPACE BAR to flip the swing.

- Press the SPACE BAR to flip the swing.

- Press it again to flip back. (see Figure 3.34).

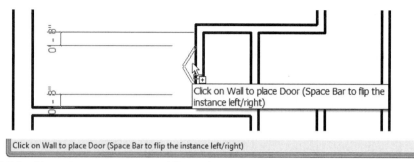

**Figure 3.34** *Move the mouse to control Door placement, press the SPACE BAR to flip the Door*

- When the Door opens to the lower portion of the Wall, click the mouse

Don't worry if you added the Door with the wrong swing. We can easily edit this after the Door is placed.

16. On the Design Bar, click the Modify tool or press the ESC key twice.

17. Save your project.

## FLIP A DOOR WITH FLIP ARROWS

To flip a Door after it is placed, use its handles.

18. Click on the hinged Door at the top right, (the first Door placed in the previous steps above).

Notice there are two sets of small arrow handles, one group horizontal and the other vertical. Use these handles to flip the Door swing.

19. Click either one of the Flip handles.

It will pre-highlight and a tip will appear to indicate its function (see Figure 3.35).

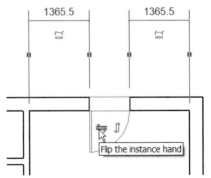

**Figure 3.35** *Flip the instance hand of a Door*

Repeat this process on other Doors if you wish.

## CHANGE THE DOOR SIZE

Door size is governed by a Door's Type. If you wish to change the size of Door in your model, simply choose a different Type. If you wish to use a size that is not on the list, you must edit the Door Family and create a new Type with a different size. This process will be covered below.

20. Select the Door in the top right of the plan.

- On the Options Bar, from the Type Selector, choose a new size, such as **Single-Flush : 34" x 80" [M_Single-Flush : 0864 × 2032mm]**.

21. Click next to the drawing to deselect the Door (or press ESC).

You can edit other Instance or Type parameters for the selected element (Door in this case) by clicking the Properties icon on the Options Bar. For example, we used this technique above to set the Phase Created parameter of our first Door.

## EDIT DOOR PLACEMENT WITH DIMENSIONS

We have seen several examples so far of the use of temporary dimensions to control the placement of elements in the model. We can also create permanent dimensions which give us even greater flexibility and control over element placement.

**Note:** The following examples would be more appropriate in a New Construction model since it is unlikely that we would need to maintain constraints in an existing layout. These tools are covered here merely for the educational value, not as a recommendation of their usage for existing conditions plans.

22. Select the Door in the top left side of the plan.

    Two temporary dimensions will appear; one on either side—take notice of the small blue icon that looks like a dimension itself (see Figure 3.36).

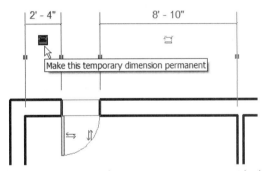

**Figure 3.36**   *You can convert a temporary dimension to permanent with the handle*

23. On the left side dimension, click this icon to make the dimension permanent.

When you do this, a permanent dimension will appear. It will be colored black. There are four basic colors used in Revit Architecture to indicate various states of selection and/or if the element is editable. Here is a summary of those colors and what they signify:

- **Black**—The element is not selected

- **50% screened**—The element is pre-highlighted (caused by the mouse passing over it, or the TAB key being used to cycle to the element).

- **Red**—The element is selected.

- **Blue**—The element is interactive (for example, temporary dimensions and handles).

**Note:** Selection color can be customized. On the Settings menu, choose **Options**. On the Graphics tab of the Options dialog, change the colors to suit your preference. Element colors can be customized in the Object Styles dialog from the Settings menu.

When you make the temporary dimension permanent, the new dimension appears in black. The new permanent dimension is a new element in the current view (*Level 1*).

Notice that the Door remains selected, and that the temporary dimension also remains but is directly underneath the new permanent dimension. To access the options of the new permanent dimension, we must select it instead of the Door.

 **Note:** Dimensions are annotation elements. Therefore, they are view-specific—meaning that they appear only in the view in which they are added. Walls and Doors on the other hand are model elements and appear in all views in which they would normally be seen. However, although the display of dimensions are view specific their constraints are not. If the dimension is locked the constraint is enforced throughout the model.

24. Click on the new permanent dimension.

    The Door will deselect (turning black) and the permanent dimension will select (turning red) instead.

    • Click the small padlock icon (see the left side of Figure 3.37).

    The padlock icon will "close" to indicate that the dimension is now constrained.

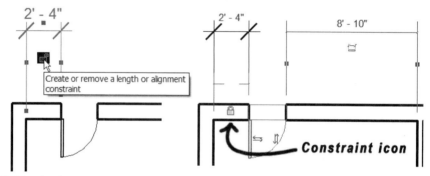

**Figure 3.37** *Apply a constraint to the Door position using the padlock on the permanent dimension*

25. Click on the Door to select it.

    The permanent dimension will deselect (turning black) and the Door will select (turning red) instead.

Notice the padlock icon that displays beneath the dimension (see the right side of Figure 3.37). If you click the blue dimension value and attempt to edit it, an error message will display. The only way that you could edit the value would be to first remove the constraint. You can do this by clicking the closed padlock icon. Let's test our new constraint with a quick experiment. We will undo it when we are finished.

26. Click on the left vertical exterior Wall.

    • Move the pointer over the selected Wall (the pointer changes to a small move cursor), click and drag the Wall to the left.

    The exact amount is not important.

Notice that the Wall moved, and the Door moved and maintained its dimension as well.

- Undo the change. (Click the Undo icon or press CTRL + Z).

27. Try the same experiment with the right vertical exterior Wall. Move it to the right.

Notice that the Door on that side did not move. It has no constraints applied to it.

- Undo the change.

## APPLYING AN EQUALITY CONSTRAINT

Let's apply another kind of constraint to a different Door. The Door that we have on the right should be centered in the room. We can add a permanent dimension with an equality constraint to maintain this relationship automatically for us.

28. On the Design Bar, click the Dimension tool.

We are going to place the dimension ourselves this time because we want more control over the specific dimension points. When you place a dimension, pre-highlight the elements that you wish to dimension. If the wrong element pre-highlights, use the TAB key to cycle to the one you want.

29. Move the mouse over the right exterior vertical Wall.

    The Wall centerline will pre-highlight.

- Press TAB and repeat until the inner edge of the Wall pre-highlights and then click.

- Click the center of the Door next.

    The center of the Door should pre-highlight, when it does, click the mouse. Otherwise, use the TAB to cycle to the center first.

- Set the final point at the inside face of the vertical Wall on the other side of the room—remember the TAB (see Figure 3.38).

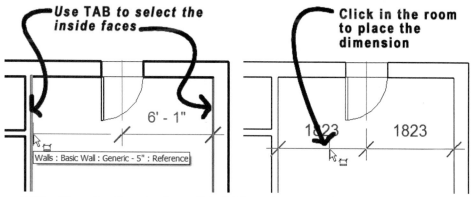

**Figure 3.38** *Dimension from the faces of the Walls to the center of the Door using* TAB *to cycle to the correct points*

30. To place the dimension, click anywhere in the blank white space of the room.

31. Press ESC twice to exit the dimension command.

32. Click on the new dimension to select it.

Notice the small "EQ" icon with a line though it (see the left side of Figure 3.39). This indicates that the permanent dimension in *not* set to maintain an equal distance. If you click this icon, you enable the equality constraint and force the elements attached to the dimension to remain equally spaced.

- Click the EQ icon to enable the equal constraint.

Notice that the Dimension Equality toggle icon no longer has the slash through it (see the right side of Figure 3.39).

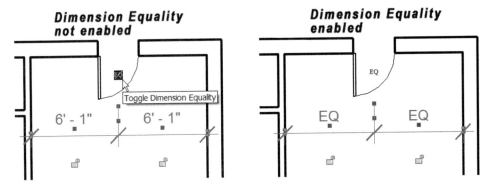

**Figure 3.39** *Set the dimension to an equal constraint*

33. Repeat the experiment above, and move the exterior Wall.

Notice that the Door remains constrained to the center of the room.

- Undo the Wall move.

34. Save the Project.

 **Tip:** Use CTRL + S to save quickly.

## ADD A NEW DOOR SIZE

The Door to the small half-bath at the top of the plan is 2'-4" [710]. This size is not available in our current list of Door Types. We could try loading the required size from an external library as we did above, but in this case it is just as easy to create a new size. To do this, we need to choose a type that is close to the one that we want, then duplicate, and edit it.

35. On the Design Bar, click the Door tool.

- From the Type Selector, choose *Single Flush: 30" × 80"* [*M_Single-Flush : 0762 × 2032mm*].

Don't place one yet.

• Click the Properties icon.

At the top of the Element Properties dialog, you can see that the Door Family is **Single Flush** [**M_Single-Flush**] and the Type is **30″ × 80″** [**0762 × 2032mm**]. The Family list includes all of the Families loaded in the current project. This currently includes only the single flush and bi-fold Families. Next to the Family list, is the Load button. You can use this button to load additional Families from external libraries as we did above. The Type list includes all of the Types available for the selected Family. If you scan that list, you will not find the size we need which is 28″ × 80″ [0710 × 2032mm]. Next to the Type list is the Edit/New button. You use this button to edit the currently selected Type or to create a new Type by duplicating the current one. Let's duplicate the current Type and modify the size.

36. Next to the Type list, click the Edit/New button.

**Tip:** A shortcut to this is to press ALT + E.

The Type Properties dialog will appear.

At this point, we could edit the Type parameters of the selected Door Type. The change would affect all Doors of this Type in the entire project. It is pretty important to understand this distinction between "Type" and "Instance" parameters. Rather than edit the existing Type, which would give us the size we need but would also reduce the size of the Doors that he have already inserted, let's copy this Type and edit the new copy.

• Next to the Type list, click the Duplicate button.

**Tip:** A shortcut to this is to press ALT + D.

A new Name dialog will appear suggesting the name: **Single Flush: 30″ × 80″ (2)** [**M_Single-Flush : 0762 × 2032mm (2)**].

It is a good idea to change this name. Let's follow the convention established with the out-of-the-box Families and name this Type after its size.

• Change the name to Single Flush: **28″ × 80″** [**M_Single-Flush : 0710 × 2032mm**] and then click OK (see Figure 3.40).

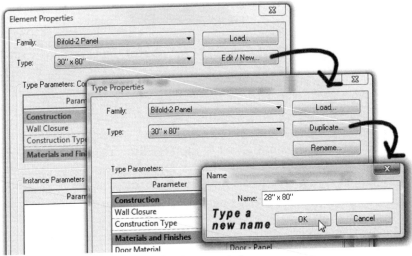

**Figure 3.40** *Create a new Type by Duplicating an existing one*

Now that we have created and named a new Type, let's edit it to the size we need.

37. Beneath the Dimensions grouping, for "Width," type **2'-4"** **[762]** and then click OK two times.

We are now ready to add our new Door Type.

38. On the Options Bar, verify that "Tag on Placement" is not selected.

- Add a Door to the small half-bath at the top of the plan.

- Use the SPACE BAR to swing the door in to the right (see Figure 3.41).

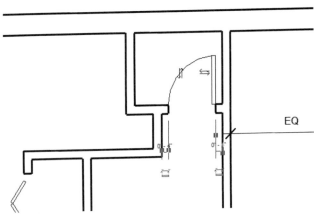

**Figure 3.41** *Add a Door in the new Type to the half-bath*

## ADD THE REMAINING DOORS

39. Using Figure 3.42 as a guide and the techniques covered above, place the remaining doors in the plan.

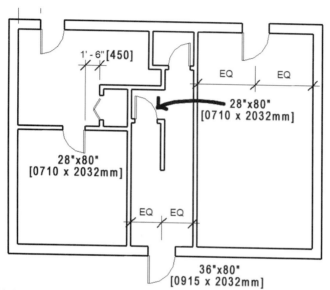

**Figure 3.42** *Add the remaining Doors as shown*

 **Hint:** You can select a Door, right-click, and choose **Create Similar** as a way of quickly adding Doors with similar parameters.

40. Save the project.

## ADD WINDOWS

Working with Windows is nearly identical to working with Doors. Many of the parameters are the same and placement and manipulation of Windows works the same as with Doors.

41. On the Design Bar, click the Window tool.

All of the options on the Options Bar for Windows match those of Doors. For detailed descriptions of each of these options, see Figure 3.30 above and the accompanying descriptions.

The only Window Family loaded in the current project is the *Fixed* Family. This house has double hung windows. Let's see what we have in the library.

42. On the Options Bar, click the Load button.

- In the Open dialog, click the *Imperial Library* [*Metric Library*] icon on the left side.

- On the right side, double-click the *Windows* folder, choose *Double Hung.rfa* [*M_Double Hung.rfa*] and then click Open.

The size that we need is not included on the list. So as we did above for the Door, we will create a new type.

- From the Type Selector, choose Double Hung : **36" × 48"** [**M_Double Hung : 0915 × 1220mm**].

43. Click the Properties icon.

- Next to the Type list, click the Edit/New button (or press ALT + E).
- Next to the Type list, click the Duplicate button (or press ALT + D).
- Change the name to **Double Hung : 36" × 56"** [**M_Double Hung : 0915 × 1422mm**] and then click OK.
- Beneath the Dimensions grouping, for "Height," type **4'-8"** [**1422**] and then click OK two times.

44. Using Figure 3.43 as a guide and the techniques covered above, place the Windows indicated.

- On the Options Bar, remove the checkmark from the "Tag on Placement" checkbox.
- Use the new Type that you just created for all of these Windows.

Note that the Window in the lower left corner is placed in at the midpoint of the Wall. The temporary dimensions should automatically snap to a centered configuration as you move the mouse nearby. This can also be done easily by using clicking the control on the temporary dimension to make them permanent dimensions and then clicking the EQ constraint.

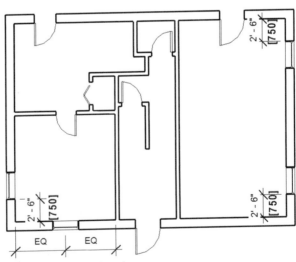

**Figure 3.43** *Add several Windows to the plan*

45. Using the process covered here, create two new Window Types:

- Double Hung : **36" × 32"** [**M_Double Hung : 0915 × 0800mm**]
- Double Hung : **24" × 32"** [**M_Double Hung : 0600 × 0800mm**]

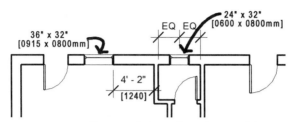

**Figure 3.44**   *Create two new custom Types and then place the remaining two Windows*

46. Place the remaining two windows as shown in Figure 3.44.

47. Save the project.

## ADD CASED OPENINGS

There are several passages between the hallway and the neighboring rooms that are simply cased openings. You place these the same way as Doors—in fact they are made from Door Families, but they are categorized as Components so that they will not appear in Door Schedules. The element includes the casing and a hole in the Wall, but no Door panel.

48. On the Design Bar, click the Basics tab and then click the Component tool.

Components are used to represent many things including furniture, fixtures, parking spaces, and nearly anything else you can imagine. In this case, the Cased Opening components that we need are not currently loaded in the model.

- On the Design Bar, click the Load button.
- In the Open dialog, click the *Imperial Library* [*Metric Library*] icon on the left side.
- On the right side, double-click the *Doors* folder, choose *Opening-Cased.rfa* [*M_Opening-Cased.rfa*] and then click Open.

 **Note:** Remember, the Cased Opening Families are stored in the Doors folder, and are essentially Door elements, but they have been categorized as Components (Generic Models) instead because typically you would not want cased openings to appear on Door Schedules.

- From the Type Selector, choose **Cased Opening : 30" × 80"** [**M_Opening-Cased : 0762 × 2032mm**].
- Click on right-most interior vertical Wall (see Figure 3.45)

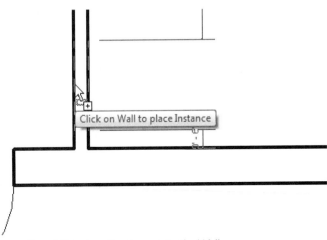

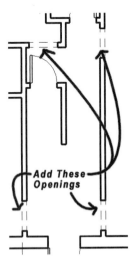

**Figure 3.45** *Place a Cased Opening Component in the Wall*

Watch the Status Bar or the tip that appears on screen. You are prompted to select a Wall.

- Using the temporary dimension place it **6" [150]** from the intersection with the bottom Wall.

 **Tip:** You can use temporary dimensions to assist you after placement if you have trouble getting it in the correct spot.

49. Using Figure 3.46 as a guide, place three more Openings of the same size.

**Figure 3.46** *Add Openings to the hallway*

After placement, you can click on a Cased Opening and move it or edit it just like the Doors and Windows.

## LOAD A CUSTOM FAMILY

There is one additional Window in the existing house that we need to add. In the front of the house in the living room (right side at the bottom) is a picture window comprised of a fixed Window flanked by two double-hung units. We could add this as three Windows simply enough. To save time, this configuration of three Windows has been pre-assembled and saved as a Family file and included with the Chapter 3 files from the Mastering Revit Architecture CD ROM. In this sequence, we will load this Family into our current project and add it to the front of the existing house.

50. On the File menu, choose **Load from Library > Load Family**.

   - In the "Open" dialog, browse to the location where you installed the CD ROM files and open the *MRAC\Chapter03* folder. Choose *Existing Living Room Front Window.rfa* [*Existing Living Room Front Window-Metric.rfa*] and then click Open.

The Family will load into the current project. This step simply loaded the Family into the current project. Now that it is loaded, we can use the ***Window*** tool to add this Window to the model. In previous examples, we used the "Load" button on the Options Bar as we were adding the Window to load the Window and place it in a single operation. The same process could be used here. The approach shown is merely an alternative.

51. On the Design Bar, click the Window tool.

The Type Selector should default to the newly loaded Family.

   - Verify that the Type Selector, reads: **Existing Living Room Front Window** [***Existing Living Room Front Window-Metric***].

   - Add the Window to the bottom horizontal Wall in the room at the right (see Figure 3.47).

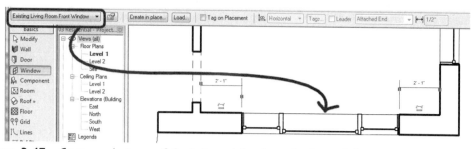

**Figure 3.47**  *Create an Instance of the imported Family at the front of the house*

This Family is a very simple one that contains three Window elements. This can be thought of as a "nested" Family because the Family file contains reference to other Families. We will explore Families in more detail in Chapter 9.

52. Save the project.

## MODIFYING DOOR SIZE AND TYPE

Modifying Doors and Windows is easy. We have already covered most of the basic techniques above. It turns out that the Door in the top right of the plan (the first one we added) is supposed to be a double door. Furthermore, it is actually a French door leading out to an existing patio. Let's make this edit to our model.

53. Select the Single Hinged door in the top right corner of the plan (the first one we added).

- On the Options Bar, click the Properties icon.

- Next to the Family list, click the Load button.

- In the Open dialog, click the *Imperial Library* [*Metric Library*] icon on the left side.

- On the right side, double-click the *Doors* folder, choose *Double-Glass 2.rfa* [*M_Double-Glass 2.rfa*] and then click Open.

- From the Type list, choose **68" x 80"** [**1730 × 2032mm**].

- Click OK.

The new Door Type has replaced the previous one and remains nicely centered since we added that constraint to this Door earlier. In plan view, it is not obvious what affect choosing a glass Door Type had. To see the glass, we will have to view the model in 3D.

 **Note:** You can use this technique for Windows and other elements as well.

## VIEWING THE MODEL IN 3D

We have spent the entire chapter working in a single plan view. However, as we saw in previous chapters, you can add or edit elements in any view you wish and then view them or edit them in elevation, section, and/or three-dimensional views. Revit Architecture maintains all of the parameters of our building model including those required to view it in 3D, even if we don't explicitly ask it to!

### OPEN THE DEFAULT THREE-DIMENSIONAL VIEW

Let's have a look at how the model is shaping up in the third dimension.

1. On the View toolbar, click the Default 3D View icon.

 **Tip:** If you prefer, you can choose **New > Default 3D View** from the View menu instead.

A new view will appear on screen named: *3D Views : {3D}*. This name is assigned automatically to the default 3D view in Revit. The default 3D view is an axonometric of the complete building model looking from the south-east. This view appears in the Project Browser beneath the *3D Views* category. From the default view, we can see the ganged Window that we added above in the front of the house. Even though the graphics in the

Floor Plan view indicated that the middle Window was different than the two flanking Windows, from the 3D vantage point it is now very clear that the middle Window is fixed while the two flanking ones are double-hung.

You can dynamically scroll and/or zoom any view as we have already seen. In a 3D view, you can also "spin" the view. When you do this, you can change the angle and height from which we are viewing the model and virtually spin it in all three dimensions.

2. On the View toolbar, click the Steering Wheels icon (see Figure 3.48).

 **Tip:** The keyboard shortcut for Steering Wheels is the F8 key.

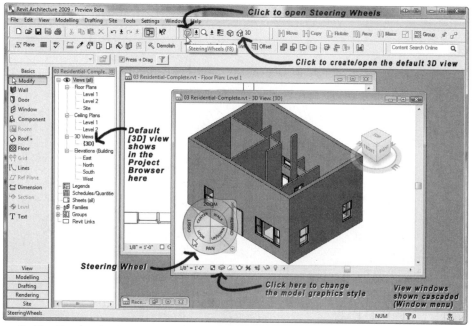

**Figure 3.48** *Display the Default 3D view and manipulate it dynamically*

Note the Steering Wheel appears on screen and follows the movement of the cursor. The Steering Wheel has several controls. Among them are: Scroll, Zoom, and Orbit. By now you should already be comfortable with scrolling and zooming. The corresponding tools on the Steering Wheel behave no differently than they did elsewhere. Orbit is only available in a 3D view. In other words, you can open the Steering Wheel from any view such as the Floor Plan view *Level 1* that we have worked in throughout the chapter. When you do, only Zoom and Pan will appear. Let's give Orbit a try in our 3D view.

• Move your mouse pointer over the Orbit portion of the Steering Wheel; click and hold down the mouse button and then drag to orbit.

The pointer will change to an orbit icon to indicate that this mode is active. A Pivot point icon will also appear on screen.

- In the middle of the Steering Wheel, move your mouse pointer over the Center portion; click and drag to place a new pivot point.

In the Orbit mode, drag side to side to rotate the model and drag up or down to tilt the vantage point up or down. The Look mode functions similarly. Give it a try.

3. Spin the model around so that you can see the back of the house.

- Use the Scroll and Zoom controls to fine-tune the view to your liking.

4. On the View Control Bar (bottom left corner of each view window) click the Model Graphic Style icon and choose **Shading w/Edges** from the pop-up menu that appears (see Figure 3.49).

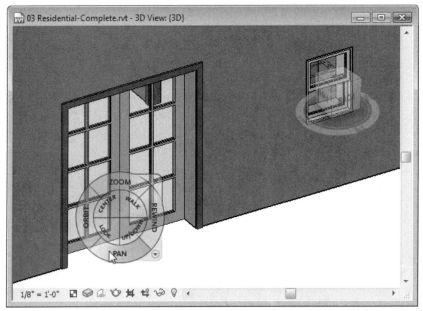

**Figure 3.49**  *Use the Steering Wheels to scroll, zoom, or orbit the model*

 **Tip:** The dynamic viewing functions can also be performed with a wheel mouse and the CTRL and SHIFT keys. Drag with the wheel pushed in to scroll, or simply roll the wheel to zoom. Hold down the SHIFT key and drag with the wheel to orbit. Your specific mouse and mouse driver will determine the exact behavior. Mouse drivers by many manufacturers allow customization of specific button functions. Sorry, 3D Connexion devices are not yet supported.

**Tip:** The keyboard shortcut for Shading with Edges is **SD**. The keyboard shortcut for Hidden Line is **HL**. Both commands are also located on the View menu.

The default three-dimensional view: *{3D}* is always available. You could certainly use this view exclusively and simply orbit, scroll, and zoom the view as needed each time you displayed it. However, once you get a 3D view displaying just the way you like, to convey a piece of information about the project in a certain way, you are encouraged to save the view. The new view will be added to the Project Browser beneath the 3D views category along with any other 3D views your project has. To do this, right-click on *{3D}* beneath 3D views in the Project Browser and choose **Rename**. A new view will appear in the Project Browser and the name on the title bar will change to reflect this. If you then click the Default 3D view icon on the view toolbar again, a new default *{3D}* view will be created. If you click the 3D view icon and the default *{3D}* view already exists it will open this existing view instead of creating a new one.

5. Save your modified 3D view.

   • Right-click the name {3D} beneath *3D Views* on the Project Browser and choose **Rename**.

   • Give the new view a unique name such as "**Rear Existing French Door**".

## EDIT IN ANY VIEW

From our explorations in 3D, you may have noticed that many of the interior Walls are the wrong height. We can correct this easily.

6. Close your "Rear Existing French Door" view and then click the Default 3D view icon to create a new {3D} view.

   • From the Window menu, choose Tile

**Tip:** The keyboard shortcut for Tile is **WT**.

   • Spin the view down slightly so that you can clearly see the heights of the interior partitions.

   • From the View menu, choose **Zoom > Zoom All To Fit**.

**Tip:** The keyboard shortcut for Zoom All To Fit is **ZA**.

7. Activate the View: *Floor Plan:Level 1* by clicking any where in the view or the view's title bar and then use the Modify tool to create a selection window, click within the exterior walls and drag from right to left to select all interior Walls.

Be sure to select only the interior Walls and not any of the exterior ones. Do not worry about selecting Doors and Openings. We will remove them from the selection next.

All of the selected elements will highlight in red in both views (see Figure 3.50).

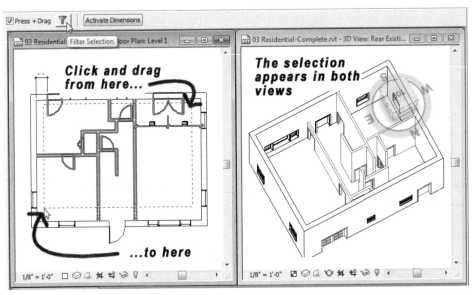

**Figure 3.50** *Select all interior Walls*

8. On the Options Bar, click the Filter Selection icon.

- In the Filter Selection dialog, remove the checkmarks from all boxes *except* "Walls" and then click OK (shown in the top of Figure 3.50).

  Now only the Walls will be highlighted.

9. On the Options Bar, click the Properties icon.

  The Element Properties dialog will appear.

- Beneath the Constraints grouping, for "Top Constraint," choose **Up to level: Level 2** and then click OK (see Figure 3.51).

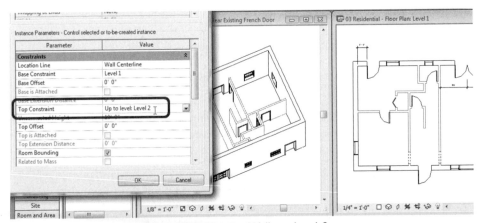

**Figure 3.51** *Set the Top Constraint of the interior Walls to Level 2*

Note that even though we made the change from a 2D floor plan view, the edit appeared in the 3D view immediately. In Revit Architecture, every view is *always* up to date. There is no need to coordinate or refresh anything. Regardless of the view you are working in, all edits apply directly to the model. You simply choose the most convenient and logical view in which to work and Revit takes care of the rest.

In this exercise, we have constrained the top of the interior Walls to the height of the second floor. It turns out that the default template from which we started our project included two levels. However the height of those levels does not match the existing conditions of our residential project. We will adjust the height of the levels and in doing so, all of these interior partitions will automatically adjust with the level height. This is the primary reason to apply such a constraint.

10. On the Project Browser, double-click the *East* elevation view to open it.

    Your windows should still be tiled on screen.

    • Position the view so that you can clearly see Level heads on the right side and the 3D view in the background.

11. Click on Level 2 to select it.

    • Click again directly on the blue dimension text beneath the Level 2 label.

    The text will highlight—ready to receive input.

    Change the value to **9'-0" [1800]** and then press ENTER (see Figure 3.52).

    Notice the immediate change to all of the interior Walls in the 3D view.

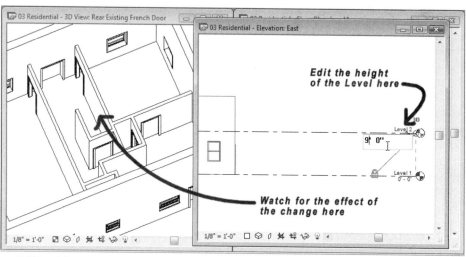

**Figure 3.52**  *Adjust the height of Level 2 and see the immediate effect on the interior Walls*

12. Maximize the *Floor Plan:Level 1* view

13. From the Window menu, choose **Close Hidden Windows**.

14. Save your project.

## ADDING PLUMBING FIXTURES

The small half-bath in the top of our plan could use some fixtures. The library of components provided with Revit includes a variety of items such as furniture, toilets, trees, parking lot layouts, equipment, electrical fixture symbols, targets, tags, and much more. All of these items are Families and many have special behaviors and parameters appropriate to the object that they represent. In this exercise, we will simply load the Families we need and insert them in the model much like we did for the ganged Window above. In later chapters we will learn how to manipulate and create our own Families.

### ADD COMPONENTS TO THE MODEL

You add Components like plumbing fixtures and furniture in nearly the same way as Walls, Doors, and Windows. Each Family may vary slightly depending on the parameters built into it. However, the basic process is the same: choose the *Component* tool from the Design Bar, choose an existing Type from the Type Selector, or click the Load button to access the library. Then load the Component in the model and make your placements. You can also use the **Load from Library** command on the File menu as an alternative as noted above.

1. On the Design Bar, click the Basics tab and then click the **Component** tool.

   If you open the Type Selector, you will note that no plumbing fixtures currently appear on the list.

As we did above for Doors and Windows, we will simply load the items we need from the library.

2. On the Options Bar, click the Load button.

   • In the Open dialog, click the *Imperial Library* [*Metric Library*] icon on the left side.

   • On the right side, double-click the *Plumbing Fixtures* folder.

You will note that two Family files exist for domestic toilets, one for 2D and the other for 3D. The reason for this is that in many cases, you will not need to see plumbing fixtures in 3D. By using a 2D symbol in those cases, you can reduce demand on computer resources in the project file. In reality, in a model the size of the one that we are building here, there would be no noticeable difference using either Family. However, let's begin developing good BIM habits right away. Since we have no need for a 3D toilet in this particular project, let's load the 2D Family for this exercise. Like all Families, you can swap out the 2D Component later with a 3D one should project needs change.

**Note:** The name of the 3D Family here is actually misleading. If you were to choose this Family, you would notice that a 2D symbol would appear in all floor plan views and the 3D would only appear in 3D views. The Family file is composed of both 2D and 3D representations of the toilet and Revit Architecture knows automatically when to display which one.

**CAD Manager Note:**    When you create a Building Information Model, it is important to realize that "Model" does not always mean "3D." Nor does "Model" imply that every facet and screw of an item should be painstakingly represented in graphical form. Always remember to strike a balance between what is conveyed graphically and what is conveyed by other means such as with attached data parameters—this is the "Information" part of BIM. In many cases, an "Information" model is much more practical and useful to a project team than a "3D" model. There are many types of models. Statisticians and economists refer to their spreadsheets as "models." Meteorologists refer to their predictions and weather simulations as "models." As Architects, we tend to only think "3D" when the word model is mentioned. Learning to embrace BIM means understanding that a model is not always 3D, and that the "I" is just as important as, and sometimes more important than the "M." A really good BIM will include both a graphical and data model tightly integrated with one another.

- Choose *Toilet-Domestic-2D.rfa* [*M_Toilet-Domestic-2D.rfa*] and then click Open.

3. Zoom in on the half-bath at the top of the plan.

- Move the pointer around the four Walls of the half-bath.

Notice how the Toilet automatically orients itself to the Walls (see Figure 3.53). This is because the 2D toilet family was made in a wall-hosted Family template file. The Doors and Windows we added already were also hosted Families, but they also cut holes in the Walls. Naturally the toilet does not need to cut the Wall, but requiring the attachment is a nice feature. It is also possible to create a toilet Family that is not hosted by a wall, if it is needed.

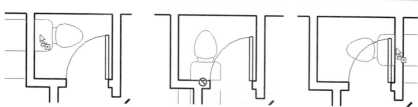

**Figure 3.53**    *The Toilet Family is wall-hosted, so it automatically attaches to the Walls*

- Click a point on the left vertical Wall to place the toilet in the model.

   This toilet room is clearly not up to code! Just place the toilet very close to the intersection with the exterior Wall. We still need room for a sink.

4. On the Options Bar, click the Load button again.

- Return to the same location and Open the *Sink-Single-2D.rfa* [*M_Sink-Single-2D.rfa*] Family this time.

Move the pointer around again and notice that this Component is also hosted. However, it will be too big for the space that we have available. Sometimes what we find in the field does not meet current building codes. The sink in this existing half-bath is *very* small.

5. On the Options Bar, click the Properties icon.

- Next to the Type list, click the Edit/New button (or press ALT + E).

- Next to the Type list, click the Duplicate button (or press ALT + D).

- Change the name to *18" × 15"* [*450 × 375mm*] and then click OK.

6. Beneath the Dimensions grouping, for "Depth," type **1'-3"** [**375**]

**Tip:** Remember, if you are using Imperial units, you can simply type **1 3** with a space between the numbers and Revit Architecture will interpret this as 1'-3".

- Beneath the Dimensions grouping, for "Width," type **1'-6"** [**450**] and then click OK two times.

- Place the sink next to the toilet. Use the temporary dimensions to fine-tune the placement of both the sink and the toilet.

It will be a tight fit. You will need to set the values of the temporary dimensions to about **1" [25]** between the Walls and the Components to get everything to fit (see Figure 3.54).

**Tip:** You can also use the Nudge tool to move the toilet and or sink. Select the element to nudge and then use the arrow keys to move it. The element will move approximately 2mm on screen. The more you are zoomed into the view the smaller the distance each press of the arrow key will move the object.

7. On the Design Bar, click the Modify tool or press the ESC key twice.

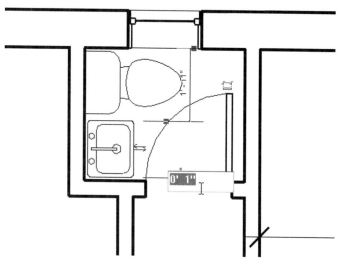

**Figure 3.54** *Use the Temporary Dimensions to fine-tune placement of the Components*

8. Save the project.

## CREATE AN IN-PLACE FAMILY

The first floor existing conditions plan is nearly finished. We still need to add the fireplace in the living room. While it would be possible to create a Fireplace Family in our library for use in any project, this would only make sense if we used the same fireplace design often. In this case, we will create the exact fireplace we need for this project directly in-place. This is called an "In-Place Family."

**Note:** In-Place Families are not designed to be moved, copied, rotated, etc. They are meant to be used only once. If you need to use it more than once within this project or in a different project a regular family should be created in the Family Editor, saved to a library and then loaded into your project as needed. The Family Editor will be explored in Chapter 9.

### CREATE THE FAMILY AND CHOOSE A CATEGORY

To get started, we will create a new In-Place Family and assign it to a pre-defined category.

1. Zoom in on the middle of the right vertical exterior Wall.

   This is where our fireplace will go.

2. On the Design Bar, click the Modeling tab and then click the Create tool.

**Tip:** Create is short for "Create In-Place Family." **Create** is also located on the Modeling menu.

- In the "Family Category and Parameters" dialog, choose Generic Models and then click OK (see Figure 3.55).

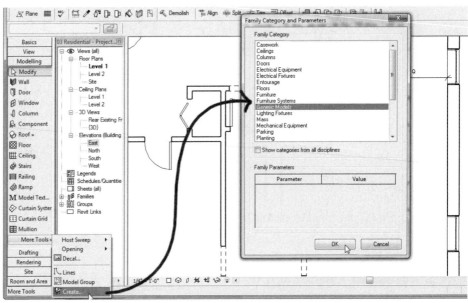

**Figure 3.55** *Create an In-Place Family and choose its category*

The Family Category list is a fixed list built into the software. When you create a Family, you must assign it one of these categories. The Family you create will gain the characteristics of the category to which assign it. We chose "Generic Models" here because our fireplace does not fit neatly into any of the other categories. This is sort of a "catch all" category. We will not benefit from many specialized parameters that might be available from others, but our existing fireplace has few specialized needs.

- In the Name dialog, type: **Existing Fireplace** and then click OK.

You are now in "In-Place Family Editing" mode. The Design Bar will change to a collection of In-Place Family Editing tools and all of the other tabs will disappear (see Figure 3.56).

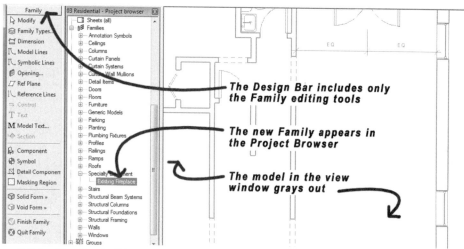

The Design Bar includes only the Family editing tools

The new Family appears in the Project Browser

The model in the view window grays out

**Figure 3.56** *The Family Editor mode is enabled when you create a new Family*

Many common tools can be found on the Design Bar while in Family editing mode. Two tools that we have not seen yet are **Solid Form** and **Void Form**. We use these tools to sculpt the form of our Family. You will also note that you can insert Components, Symbols, Dimensions, and Detail Components into Families. Some of these items have been explored above, others we will explore in the chapters that follow.

## ADDING REFERENCE PLANES

When you construct complex geometry, it is sometimes useful to have guidelines to assist in locating elements. Reference Planes are used for this purpose in Revit Architecture. You sketch a Reference Plane similar to the way you sketch Walls or lines. You can snap and constrain other elements to Reference Planes, making them useful tools for design layout. You can add Reference Planes in any orthographic view of the model. Reference Planes do not show in 3D. In this example, we will add them within our In-Place Family. When you add them in this way, the Reference Planes will become part of the In-Place Family and will be visible only when editing the In-Place Family.

3. On the Design Bar, click the Ref Plane tool.

- Click a point inside the large room on the right just above the Window.

- Move the pointer horizontally to the right past the exterior Wall and then click outside.

  The exact locations of either click are not critical.

A small Reference Plane (green dashed line with round blue handles at the ends) will appear.

- Edit the temporary dimension from the bottom horizontal Wall to **7'-11"** [**2400**] (see the left side of Figure 3.57).

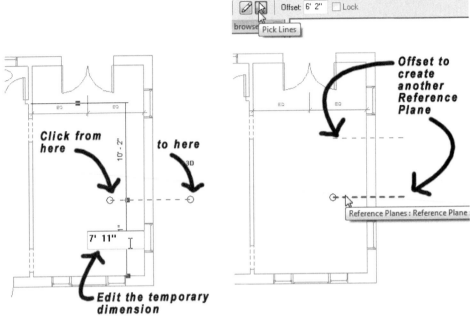

**Figure 3.57** *Create two horizontal Reference Planes*

- On the Options Bar, click the Pick Lines icon.

- Type **7'-2"** [**1900**] in the Offset field.

- Pre-highlight the first Reference Plane and move the mouse so that the offset line appears above.

- Click to create the new Reference Plane (see the right side of Figure 3.57).

Repeat the process to create two more vertical Reference Planes. These will frame out the rectangular footprint of the fireplace.

4. With the Reference Plane tool still active, on the Options Bar, click the Draw icon.

- Type **4"** [**100**] in the Offset field.

- Snap to the endpoint of the lower Window on the inside edge of the Wall.

- Snap to the endpoint of the upper Window on the inside edge of the Wall (see the left side of Figure 3.58).

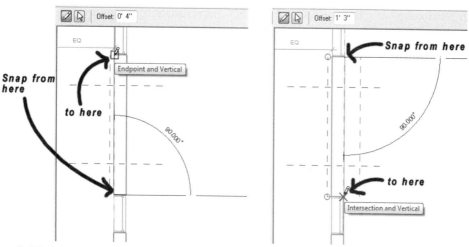

**Figure 3.58**   *Create vertical Reference Planes relative to the horizontal ones*

5. Change the Offset, to **1'-3"** [**380**].

- Snap to the endpoint of the upper Window on the outside edge of the Wall.

- Snap to the endpoint of the lower Window on the outside edge of the Wall (see the right side of Figure 3.58).

Figure 3.59 shows the completed Reference Plane layout. The dimensions in the figure are for your reference. They do not need to be added to the model.

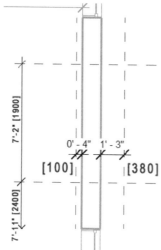

**Figure 3.59**   *The completed Reference Plane layout*

We now have four Reference Planes that we can use to guide the creation of our fireplace's form. It is not required that you use Reference Planes for this. However, they do make it easier particularly when you begin creating more advanced Families with constraints. Complete details will be discussed in Chapter 9.

## CREATE A SOLID FORM

Using our Reference Planes as a guide, let's create the overall mass of the fireplace.

6. On the Design Bar, click the Solid Form tool, and then from the flyout menu that appears, choose **Solid Extrusion** (see Figure 3.60).

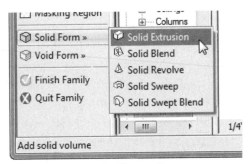

**Figure 3.60**  *Create a Solid Extrusion from the Design Bar*

This places the Design Bar in Sketch mode. The tools change accordingly.

7. On the Options Bar, in the "Depth" field, type **9'-0" [2750]**.

• Verify that the Draw icon is active. If it is not, push it in now.

• For the sketch shape, click the Rectangle icon.

8. Snap to the intersection of two of the Reference Planes and then snap to an opposite intersection to define the rectangular shape (see Figure 3.61).

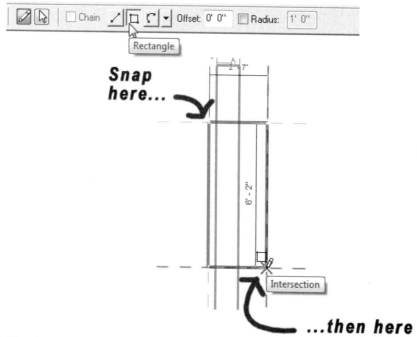

**Figure 3.61**   *Creating the shape of the fireplace*

An open padlock icon will appear on each side of the shape. If we wanted to constrain the sides to the Reference Planes, we could click these locks. There is no need for explicit (deliberate) constraints here so we will not do that for this exercise. Revit does interpret the location of this solid as your design intent and internally uses "implied constraints." You do not need to do anything to set implied constraints. It is simply part of the parametric nature of the software. If you add a(n) (explicit) constraint it will override and remove any implied constraints that may conflict.

> 9. On the Design Bar, click the Finish Sketch button.

This gives us our basic fireplace mass. We now need to carve out the firebox.

## CREATE A VOID FORM

Using the same basic process, we can create a Void form that will carve away from the solid form in our Family giving us the firebox opening.

> 10. On the Design Bar, click the Void Form tool, and then from the flyout menu that appears, choose **Void Extrusion**.
>
>    This places the Design Bar in Sketch mode again. The tools change accordingly.
>
> 11. On the Options Bar, in the "Depth" field, type **4'-0"** [**1200**].
>
> • Click the Pick Lines icon.
>
> • Click the left vertical edge of the Solid Extrusion.

A magenta sketch line will appear along this edge.

12. On the Options Bar, change the Offset value to **1'-1" [325]**

- Click the left edge of the Solid Extrusion to create a magenta sketch line within the fireplace structure.

- Change the Offset value to **1'-5" [430]** and click the top edge to create a sketch line below it.

- Click the bottom edge of the Solid Extrusion to create a sketch line above it (see Figure 3.62).

The side of the offset will pre-highlight before you click so that you can be sure to offset the sketch line to the correct side. Notice that the lines automatically trim when you offset more lines. If yours do not trim exactly the way they are illustrated in the figure, you can simply use the Trim/Extend tool to clean up the sketch.

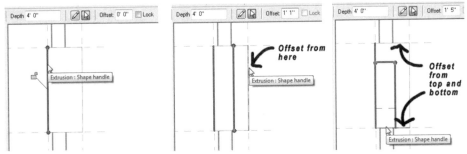

**Figure 3.62** *Offset sketch lines to form the firebox shape*

Now we will use the **_Trim/Extend_** tool (the same one we used for Walls at the start of the chapter) to cleanup the sketch.

13. On the Tools toolbar, click the Trim/Extend tool (or Type **TR**).

- On the Options Bar, click the first mode (Trim/Extend to Corner).

- Trim any segments as required to make a rectangular shape.

 **Note:** Select the portion of the sketch line that you wish to keep.

- On the Design Bar, click the **Modify** tool or press the ESC key to finish trimming (see the left side of Figure 3.63).

14. Click the lower horizontal sketch line and drag the handle on the right up till the temporary dimension reads 20° (see the middle of Figure 3.63).

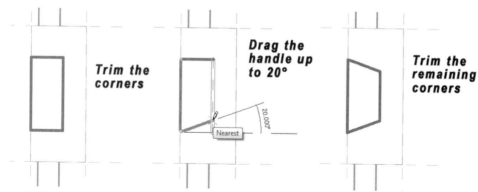

**Figure 3.63**   *Edit the offset sketch lines to finalize the shape*

- Repeat by stretching the top line down 20°

- Use Trim/Extend once more to cleanup the remaining corners (see the right side of Figure 3.63).

15. On the Design Bar, click the Finish Sketch button.

16. On the Design Bar, click the Finish Family button.

## JOIN THE FIREPLACE WITH THE WALL

The Fireplace Family is finished but it is modeled inside of the Wall making both the Wall and the Fireplace hard to read. Let's fix this.

17. On the Tools toolbar, click the Split tool (or type **SL**).

- On the Options Bar, place a checkmark in the "Delete Inner Segment" checkbox.

- Split the exterior vertical Wall on both sides of the fireplace (see Figure 3.64).

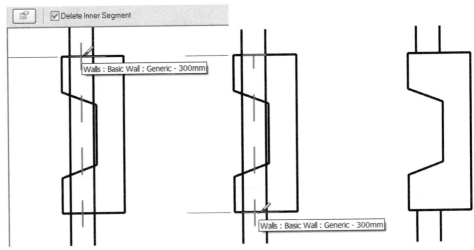

**Figure 3.64** *Split the exterior Wall*

- On the Design Bar, click the **Modify** tool or press the ESC key twice.

This is close to what we want but let's make one more edit.

18. On the Tools toolbar, click the Join Geometry icon (see the top of Figure 3.65).

- Click one of the exterior Walls.

- Then click the Fireplace to join them (see Figure 3.65).

 **Tip:** Remember to watch the Status Bar for detailed prompts.

- Repeat for the other Wall.

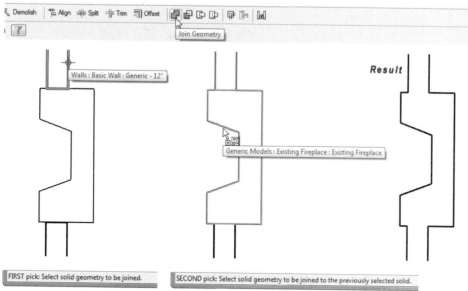

**Figure 3.65** *Use Join Geometry to join the Walls to the Fireplace*

19. On the Design Bar, click the Modify tool or press the ESC key twice to cancel the Join command.

20. On the View toolbar, click the Default 3D view icon.

- Use the techniques covered above and spin the model around so that you can see the Fireplace.

We modeled the fireplace a bit to short. However, for now we will leave this alone. In later chapters we will address the height of the fireplace as well as how it changes on the second floor. The fireplace could also use a mantel and a hearth. However, because there will be no new work done in the living room of this project and therefore no sections or elevations are needed of the fireplace, that extra level of detail is unnecessary for this tutorial. What we have created works well for floor plans. If you wish to try it anyway for the practice, feel free. Select the fireplace, and then on the Options Bar, click the Edit button. This will return you to the In-Place Family editor where you can add these accoutrements using additional solids.

## RESET THE CURRENT PHASE

Congratulations! Our work on residential project first floor existing conditions layout is complete for now (see Figure 3.66). We still need to add the Stairs to this model. However, Stairs will be covered in a dedicated chapter. Therefore, we will save our layout without the Stairs for now.

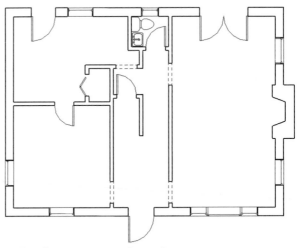

**Figure 3.66** *The final first floor existing conditions layout*

21. Make sure that there are no elements selected, right-click in the Workspace and then choose **View Properties** (or type **VP**).

- Beneath the Phasing grouping, for "Phase," choose **New Construction**.

- Verify that "Phase Filter" is set to **Show All**.

- Click OK to see the change.

 **Note:** Later in Chapter 6, we will actually duplicate this view and create a permanent Existing Conditions view. For now, we have simply returned the view to its Phase settings at the start of the chapter.

22. Save the project.

## SUMMARY

- The basic process for adding elements in Revit Architecture is to choose a tool on the Design Bar, choose a Type from the Type Selector, set additional options on the Options Bar and then click to add the item in the workspace.

- Walls can be added one segment at a time or chained to draw them end to end.

- Assign elements to construction Phases to show Existing, Demolition, and New Construction.

- Doors, Windows, and Openings automatically "cut" a hole in, and remain attached to, the receiving Wall.

- Add Walls, Doors, and Windows quickly, and modify their properties to add detail later.

- Use Trim/Extend, Split, Offset, Move, and Copy to quickly layout a series of Walls.

- Door, Window, and Component Types can be included with the project template or loaded as needed from library files.

- It is not always necessary to model all elements in 3D. 2D Components can save overhead and computer resources when 3D is not required.

- View the model interactively in 3D, plan, section, etc. to study and investigate possibilities in the design.

- Edit in any view and see the change immediately in all views.

- Use In-Place Families to model custom or project-specific components directly in the model where they are required.

# CHAPTER 4

# Setting up Project Levels and Views

## INTRODUCTION

One of the most important concepts to grasp early in your understanding of Revit Architecture and its implementation of the Building Information Modeling (BIM) concept is the concept of "Views." When you create a Building Information Model, you are creating a unified model that represents all aspects of a project using both graphics (the "M" in BIM) and associated data (the "I" in BIM). The way that you create, manipulate, and extract data from this model is with one or more views. A wide variety of view types is possible in Revit. Among these types are floor plan views, elevation views, 3D views, sheet views schedules, and legends. Each view presents different vantage point on the building model filtered to a particular need or perspective. For example, a floor plan view takes a slice of a building at a particular height and presents it graphically on screen using conventional two-dimensional architectural abstraction to present the data. On the other hand, a schedule view presents a collection of common data extracted from the building model presented in tabular format.

Views are not just for display of our model on screen; views are also used to create and edit data and graphics in the building model. When you create a new project, the program does not presume to know the type of building you wish to create or the kinds of views you intend to use to view and edit it. Some basic views are common to nearly all projects like floor plans, but mostly you must create the views you need in your projects. To save time and ensure consistency from one project to the next, Project Templates are available when creating new projects. A Project Template is simply a Revit project that has been pre-configured to contain the most common and useful views, Families and settings required to start a new project. Revit ships with several sample Project Templates and you can customize and save your own templates based upon your preferred standards.

## OBJECTIVES

In this chapter, we will explore Project Templates and Project views. First we will explore the many out-of-the-box Project Templates at your disposal. Then we will choose a Project Template from which to begin our second project—the commercial project. We'll add some basic geometry to the project for use as a backdrop to explore the many different types of views available in Revit Architecture. After completing this chapter you will know how to:

- Open and explore several Project Templates
- Create a new project from a template
- Set up preliminary views
- Understand switching between and working with views
- Work with sheet views
- Print a digital cartoon set

## UNDERSTANDING PROJECT TEMPLATES

Project template files provide a means to quickly apply project setup information, enforce company standards and project-specific settings, and save time. The concept is not unlike other popular Windows software packages such as Microsoft Word. Please note that template files apply only at the time of project creation. Projects do *not* remain linked to the template. However, you can transfer project standards from one project to another later, as required. A project template is simply a project file preconfigured for a particular type of building or task that has been saved in the template format. Template files have a RTE extension and are available when you choose the **New > Project** command from the File menu. In addition to the time saved when creating projects, templates help to ensure file consistency and office standards by giving all projects the same basic starting point. Several pre-made template files ship with the product and are ready to use. However, because office standards and project-specific needs vary, feel free to modify the default templates to better suit your needs. The exact composition of the template used to create a project is not as important as ensuring that templates are used consistently on all new projects. Table 4-1 lists some of the Revit Architecture project templates included with the software.

 **Note:** The table lists imperial and metric templates typically installed in the United States. When you install Revit, you are given the option to install several other templates and content files appropriate to other parts of the world. The exact items you have available may therefore vary from the table.

**Table 4.1** *Out-of-the-Box Project Templates*

| Template File Name | Drawing Units | Rounding | Description |
| --- | --- | --- | --- |
| **Imperial Templates** | | | |
| *default.rte* | Feet and fractional inches | To the nearest 1/32" | Default template for Imperial Units. |
| *Commercial-Default.rte* | Feet and fractional inches | To the nearest 1/32" | Intended for Commercial Projects |
| *Construction-Default.rte* | Feet and fractional inches | To the nearest 1/32" | Intended for Construction Projects |

**Table 4.1**   *Out-of-the-Box Project Templates (continued)*

| Template File Name | Drawing Units | Rounding | Description |
|---|---|---|---|
| *Residential-Default.rte* | Feet and fractional inches | To the nearest 1/32" | Intended for Residential Pr |
| **Metric Templates** | | | |
| *DefaultMetric.rte* | Millimeters | 0 decimal places | Default template for Metri Units. |
| *DefaultUS-Canada.rte* | Millimeters | 0 decimal places | Default template for Metri Units in the United States Canada. |
| *Construction-DefaultMetric.rte* | Millimeters | 0 decimal places | Intended for Construction Projects |
| *Construction-DefaultUS-Canada.rte* | Millimeters | 0 decimal places | Intended for Construction Projects in the United Stat Canada |

**Caution:** *Please avoid creating projects without a template, because doing so requires an enormous amount of user configuration before serious work can begin.*

**BIM** **Manager Note:** Revit Architecture saves virtually all data and configuration within the project file. This includes Families, settings and Levels and views. Therefore, template files provide an excellent tool for promoting and maintaining office standards. Several sample template files ship with the product to help you get started. These include project templates tailored to different types of projects such as construction or commercial in both Imperial and Metric units. Refer to the table above for more examples. You will likely find one of these templates suitable for your firm's needs directly or useable as a good starting point for customization. When you establish and configure office standards, it is highly recommended that you start with one of these default templates. Modify them to suit individual project or office-wide needs. Families, title blocks, and other resources can also be stored in separate library files stored on the network or hard drive. Externally saved resources can be accessed using the Load button on the Options Bar or within the Element Properties dialog or from the **Load From Library** command on the File menu. This approach can help keep the template file size smaller by including only those items needed in all (or most) projects, while still providing a central repository for additional office standard items. This method also provides additional ongoing flexibility because new items can be easily added to the office standard library. In addition, Revit Architecture provides the **Transfer Project Standards** command for copying System Family items such as Wall Families and other System Families from one project to another. To use this tool, first open the source project (the one with the standards that you want to copy), and then open the project to which you wish to transfer. Execute this command from the File menu, and check the boxes for the items you wish to transfer in the "Select Items To Copy" dialog. Some firms even do this using a simple copy and paste. In this case, the firm will create a library file containing instances of all the office standard elements. For example, there might be one of each Wall Type inserted into the library project. To access one of these Wall Types, you would open the library file, copy the instance of the Wall you want, and then simply paste it into your active project.

The best way to demonstrate the importance of using project templates is to create a project with some of them.

## CREATE A NEW PROJECT WITH THE DEFAULT TEMPLATE

Let's create a new project using the default template and explore some of its settings.

1. Launch Revit Architecture from the icon on your desktop or from the *Autodesk > Revit Architecture* group in *All Programs* on the Windows Start menu.

 **Tip:** In Windows Vista, you can click the Start button, and then begin typing **Revit** in the "Start Search field. After a couple letters, Revit Architecture should appear near the top of the list. Click it to launch to program.

- If the New Features Workshop dialog appears, choose "Maybe later" and then click OK.

Revit Architecture launches to the Recent Files home page.

2. On the home page, click the New link in the "Projects" column (or from the File menu, choose **New > Project**).

- In the New Project dialog, in the "Template File" area, be sure that the lower radio button is selected (the one that lists a template file).

- Click the Browse button.

The default template file name and location for the Imperial template file is:

*C:\ProgramData\Autodesk\RAC 2009\Imperial Templates\default.rte*

The default template file name and location for the Metric template file is:

*[C:\ProgramData\Autodesk\RAC 2009\Metric Templates\DefaultMetric.rte]*

 **Note:** In Windows XP, the folder is: *C:\Documents and Settings\All Users\Application Data\Autodesk\RAC 2009\Imperial Templates\. [C:\Documents and Settings\All Users\Application Data\Autodesk\RAC 2009\Metric Templates\].*

- Browse to this location and then select the *default.rte [DefaultMetric.rte]* template file and then click Open (see Figure 4.1).

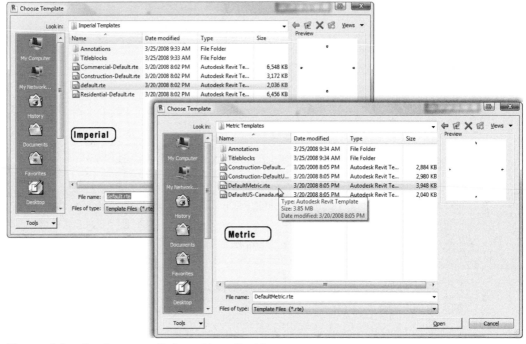

**Figure 4.1**  *Creating a new Project with no Template*

 **Note:** If your version of Revit Architecture does not include the template files cited here, both have been provided on the Mastering Revit Architecture CD ROM. Please browse to the *Templates* folder in the location where you installed the CD files to locate them.

- In the "Create New" area, verify that Project is selected and then click OK.

You may notice that if you are using Metric units that there is also a *DefaultUS-Canada.rte* template. If you wish, you can use this template instead of the default Metric template. The differences between the two are subtle. Both the *default.rte* Imperial and the *DefaultUS-Canada.rte* Metric templates use the same (square) Elevation tags. The *DefaultMetric.rte* template uses a round Elevation tag (see Figure 4.2).

**default.rte &**
**DefaultUS-Canada.rte**                    **DefaultMetric.rte**

**Figure 4.2**  *The shape of the Elevation tag varies in the North American and non-North American templates*

**BIM** *Manager Note:* You can change the shape of the elevation tags from square to circle if you wish. To do so, select one, edit its properties, then click the Edit/New button. In the Type Properties dialog, you can change the Elevation Tag setting. Click OK twice to update all tags. If you want to make the change permanent, re-save your default template file this way.

> 3. From the Settings menu, choose **Materials** (see Figure 4.3).

Some of the language used in the Materials and other settings varies between the North American and non-North American templates.

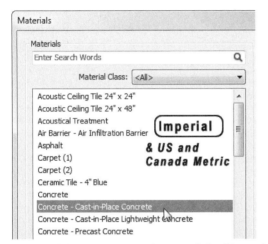 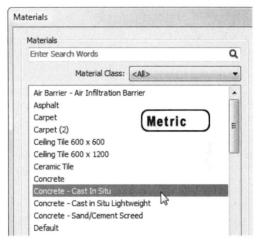

**Figure 4.3** *The names of some of the Materials vary*

- Close the Materials dialog when you are finished exploring.

Despite these minor differences, the three "Default" templates are virtually the same. They all have the same Levels and views pre-configured. They all have the same basic Families loaded and the other settings are similar but with unit-specific or regional vernacular differences (like "Cast-in-Place vs. "Cast In Situ").

> 4. On the Settings menu, choose Project Units.

**Tip:** The keyboard shortcut for Project Units is **UN**.

- Explore the various settings without making any permanent changes. Refer to the table above for the default settings of each Template.

- Click Cancel when finished.

> 5. On the Project Browser, under Views (all), expand Floor Plans, Ceiling Plans and Elevations (Building Elevation).

- There will be three Floor Plan views: *Level 1*, *Level 2* and *Site*

- There will be two Ceiling Plan views: *Level 1* and *Level 2*.

- And under Elevations, there are four Elevation views: *East*, *North*, *South* and *West* (see Figure 4.4).

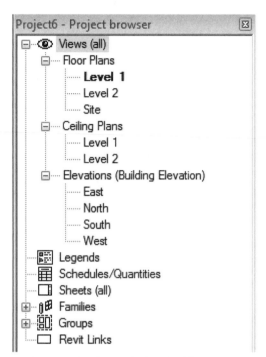

**Figure 4.4**  *There are several pre-made plan and elevation Views*

6. On the Design Bar, click the Basics tab and then click the **Wall** tool.

- On the Options Bar, open the Type Selector.

There will be several Wall Types already in the project (see Figure 4.5).

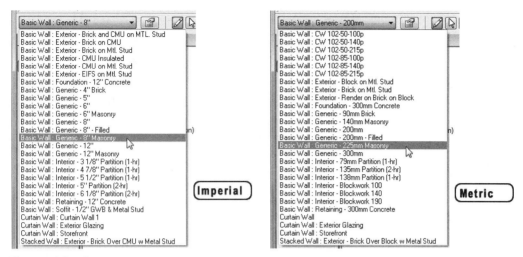

**Figure 4.5**  *Several Types are present in the project*

- Select any Wall Type and draw a short segment of Wall.

- On the Design Bar click the ***Modify*** tool or press ESC twice.

Section, Elevation and Level Heads are loaded in the project (however, as noted above, they may vary slightly in Imperial and Metric).

7. On the Design Bar, click the ***Section*** tool.

- Click a point anywhere on screen, and then drag to the right and click again (see Figure 4.6).

Notice the Section Head and the Section Tail that appear.

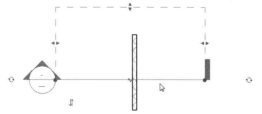

**Figure 4.6**  *Section Lines are pre-configured with a section head and tail*

8. On the Project Browser, expand *Elevations (Building Elevation)* and then double click *East*.

- Zoom in on the Level Heads to the right (see Figure 4.7).

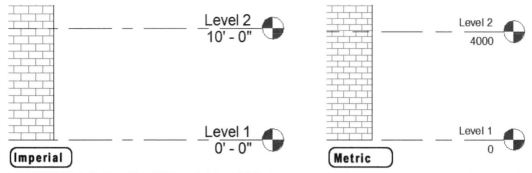

**Figure 4.7** *Levels have Level Heads*

You can see what other tags are loaded in the project by choosing one of the options from the **Settings** > **View Tags** menu.

## CREATE A PROJECT FROM THE CONSTRUCTION AND STRUCTURAL TEMPLATES

Let's repeat the exploration process in some of the other provided project templates.

    9. From the File menu, choose **Close**.

    • When prompted to save changes, click No.

    10. From the File menu, choose **New > Project**.

    • In the New Project dialog, in the "Template File" area, be sure that the lower radio button is selected (the one that lists a template file).

    • Click the Browse button.

As above, the dialog should open directly to your default template folder. If it does not, browse to the template location listed at the start of the tutorial above.

    • Select the *Construction-Default.rte* [*Construction-DefaultMetric.rte*] template file and then click Open.

    • In the "Create New" area, verify that Project is selected and then click OK.

The first thing you should notice about this template is that there are many more views in the Project Browser; particularly beneath the *Schedules/Quantities* node (see Figure 4.8).

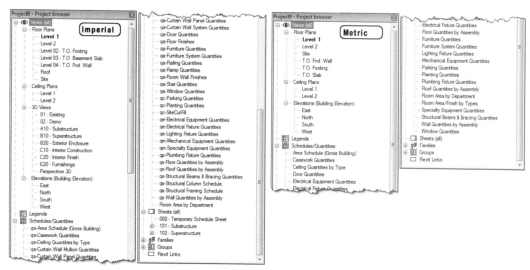

**Figure 4.8** *The Construction template is pre-loaded with dozens of schedules and views*

Spend some time in this template exploring the floor plan, ceiling plan and elevation views like we did above. The same four elevation views are provided here. There are a few additional floor plan views however, and if you open one of the elevation views, you will notice that there are Levels defined for the foundation and footing. The most significant addition to this template that was not included in the Default template is the inclusion of dozens of schedule views. As you can see from the names of these schedules, you can determine quantities for nearly every element in the project.

11. On the Project Browser, double-click the floor plan View: *Level 1* to make it active.

   • Create a Wall anywhere on the screen. Click **Modify** or press ESC twice when done.

12. On the Project Browser, double-click the *qs-Wall Quantities by Assembly* [*Wall Quantities by Assembly*] Schedule view to open it (see Figure 4.9).

**Imperial**

| | | | qs-Wall Quantities by Assembly | | |
| | | | Calculated To Butt-End Dimensions | | |
| Assembly Code | Assembly Description | Wall Assembly | Area | Volume | |
| B2010 | Exterior Walls | Generic - 8" | 480.00 | 320.00 | 4: |
| Grand total: 1 | | | 480.00 | 320.00 | |

**Metric**

| | | | Wall Quantities by Assembly | | | |
| | | | | | Calculated To Butt-End Dimensio | |
| Assembly Code | Assembly Description | Wall Assembly | Length - Center To Center | Width | Area | Volume |
| 6000 | | | | | | |
| | | Generic - 200mm | 6000 | 200 | 48 m² | 9.60 m³ |
| 6000: 1 | | | | | 48 m² | 9.60 m³ |

**Figure 4.9** *Schedules will populate automatically as elements are added to the model*

Take a little time exploring these schedules now if you like. Add a few more Walls, or some Doors and then open some of the schedules to see how they update. Schedules will be covered in detail in later chapters. The exact list and names of schedules in the Imperial and Metric templates vary. Therefore, despite your units preference, you may wish to open both templates and look at them.

**BIM** *Manager Note:* The Imperial template provides some sheets as well. In particular, there is a sheet named *000 – Temporary Schedule Sheet*. This sheet has each of the Schedules inserted onto it. There is a piece of text noting "Temporary schedule sheet to allow easy copy and pasting into other projects." This sheet remains part of this template, but is no longer required. In earlier versions of Revit, you could not easily copy and paste a schedule between projects; it could only be done if it was placed on a sheet. This limitation no longer exists. You can now simply select a schedule from the list in Project Browser and then choose copy and paste commands from the Edit menu. You can still copy and paste from a sheet, so the presence of the *000 – Temporary Schedule Sheet* can still be useful.

13. On the Project Browser, expand *Elevations (Building Elevation)* and then double click *East*.

   - Zoom in on the Level Heads to the right (see Figure 4.10).

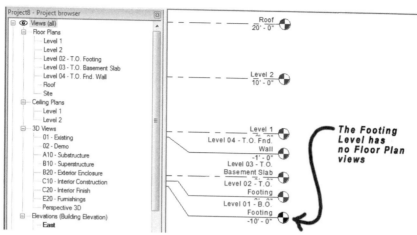

**Figure 4.10** *Blue level heads indicate an associated floor plan view while black level heads have no plan views*

If you compare the list of floor plan views in the Project Browser with the level heads in the elevation, you will notice that there is no "Footing" floor plan. Further, the Footing level head is colored black rather than blue like the others. The color blue on screen indicates interactivity. We have seen this in previous chapters with the temporary dimensions. Here, when you see a tag like the level heads colored blue, it is also interactive. Double-click on the level head to open the associated floor plan view. Think of it like a hyperlink in a web browser. The same behavior is true with elevation tags and section tags. If the symbol is black, it means that it is just annotation and there is no associated view.

## OTHER TEMPLATES

The imperial content contains a *Commercial-Default.rte* and a *Residential-Default.rte* template as well. You can create a project from each of these and explore them the same way. Even if you do not have imperial templates installed, you can find copies of these templates with the files installed from the CD.

## STARTING A PROJECT WITHOUT A TEMPLATE (NOT RECOMMENDED)

It is possible to begin a project without a template; doing so is not recommended, however. After exploring several of the provided templates in the exercise above, you should be getting a sense of the variety of settings and elements that are resident in a typical template. If you start a project without one, you are forced to either configure/create all of these items on your own manually as needed, or import them from other projects or libraries. While it is possible to do this, the amount of extra work that it adds to your project makes it an ill-advised approach to creating a new project. If you have any doubt as to the validity of these claims, try it out for yourself to see.

14. Create a New Project in Revit Architecture but in the New Project dialog, choose None for "Template file" and then click OK. When prompted for "Initial Units" make your choice of either Imperial or Metric.

15. Repeat the process that was followed above to explore the template.

- Check the Project Browser, check the Families and Types. Check the Materials.

- Add a Section or Elevation and notice that they do not even include Section or Elevation Tags.

16. Close the Project without saving when you are satisfied that you understand the differences between it and a Project created from a template.

## CREATE YOUR OWN TEMPLATE

Hopefully, you are beginning to see the benefits to starting new projects with a template. There is really no compelling reason to begin projects any other way. As you work through the exercises in the coming chapters, you will certainly discover areas where the default templates could be enhanced and improved. Make note of these observations as you go. When you are ready, try your hand at creating your own template file. The basic steps are simple:

- From the File menu, choose **New > Project**.

- Load the existing template that is most similar to the one you wish to create.

- Choose the "Project Template" option in the New Project dialog.

- Edit any settings as you see fit. (Change Settings, add or delete views, load or delete Families, etc.)

- Add, edit, or delete Materials

- Configure settings on the Settings menu such as units, fill patterns, line styles, objects styles, etc.

- Choose **File > Save As**.

- From the "Save as type" list, choose **Template Files (*.rte)**.

- Be sure to browse to your *Template* folder, type a name for your new template and then click the Save button (see Figure 4.11).

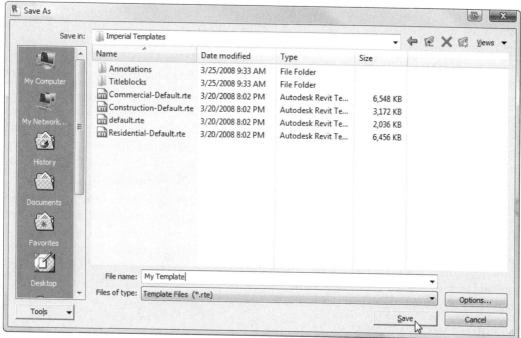

**Figure 4.11**    *Creating a new Template*

**BIM** *Manager Note:*    The templates shipping with Revit Architecture are ready to use straight out-of-the-box. They are excellent starting points for developing your own office-standard project template file(s). It is highly recommended that you become comfortable with the provided offerings and then if necessary, customize them to meet your firm's specific needs. Consider including those settings and views that people will use most frequently. Once you have a standard template in place, it is imperative that all users be required to use it. In some cases, a project will have special needs not addressed by this office standard template. In such cases, project-based derivatives can be created.

## SETTING UP A COMMERCIAL PROJECT

Now that we have seen some of the features and benefits of creating projects from templates, let's create an actual project that we will follow (as well as the one started in the last chapter) throughout the remainder of the book. This project is a 30,000 SF [2,800 SM] small commercial office building. The project is mostly core and shell with some build out occurring on one of the tenant floors. The tutorials that follow walk through the startup of the commercial project and the setup of several views including: plans, elevations, sections and schedules. At the end of the chapter, we will generate sheet views and print a cartoon set. The completed files for the project are available with the files you installed from the CD in the *Chapter04\Complete* folder on your hard drive.

Getting started with the commercial project gives us a nice practical exercise to further the goals of this chapter: namely the understanding of Project Templates and views. Even though we have explored many templates thus far, it might be difficult to decide exactly which one we should use to get started. In the remainder of this chapter, we will create a new project with the default template. We will adjust the project levels to suit the needs of our commercial building and then begin creating views that we will need. In some cases, we will create these items ourselves, and in others we will borrow them from additional templates as appropriate.

Programmatic and preliminary design can originate in a variety of forms, such as hand-drawn sketches, AutoCAD files, SketchUp models or other CAD files. Both electronic files and scans of paper sketches can be imported into Revit and used as the basis for preliminary design layout. Revit Architecture also contains its own collection of massing and preliminary design tools (often referred to as "Building Maker") so that you can also choose to do your preliminary explorations directly within Revit.

Regardless of the source of preliminary design data, take the opportunity early in the project cycle to develop a digital "cartoon set." Like a traditional cartoon set, a digital one is simply a collection of preliminary sheets that will allow you to assess the quantity and composition of each of the drawings and schedules required in the final document package. This will assist the team in allocating resources to the project. Remember, like everything else in Revit Architecture, the cartoon set will evolve and update live as the project progresses. The goal at this early stage is to assist with planning and gain a jump-start on production.

## WORKING WITH LEVELS

Before we start our commercial project, close all projects that you currently have open (**File > Close**). If it is easier, you can choose **File > Exit** to quit Revit Architecture and then launch a fresh session. If you do this, be sure to close the empty project that loads automatically upon launch. In either case, be sure that all projects are closed before proceeding. (You do not need to save anything from the previous exercises).

## ADDING AND MODIFYING LEVELS

1. From the File menu, choose **New > Project**.

- In the New Project dialog, in the "Template File" area, be sure that the lower radio button is selected (the one that lists a template file).

- Verify that the *default.rte* [*DefaultMetric.rte*] template file is listed in the text field (use the Browse button if necessary to choose it).

 **Note:** If you are in a country for which your version of Revit Architecture does not include these template files, they have both been provided with the files from the "Mastering Revit Architecture" CD ROM. Please browse to the \*Template* folder to locate them.

- In the "Create New" area, verify that Project is selected and then click OK.

By now we have started a few projects this way, and the familiar *Level 1*, *Level 2* and *Site* plan views will show on the Project Browser. Normally, the names of the Levels in the project will also be used as the names of the Floor Plan views. As our first task in our new commercial project, let's modify and add to the project Levels.

A Level is a horizontal datum element used to represent each of the actual floor levels or other significant vertical reference points in the building. Levels are used to organize projects vertically, in the Z axis. Typically you will add a Level for each story of the building ("First Floor," "Second Floor," "Roof" etc). In addition Levels can also be used to define other meaningful horizontal planes such as the "Top of Structure" or the "Bottom of Footing." We saw examples of this in several of the sample templates. For our commercial project, we have four stories, a roof and the grade level.

Take a look at the Basics tab of the Design Bar. Notice that the **Level** tool is grayed out. This is because the currently active view: *Level 1*, is a floor plan view. You cannot add or edit Levels in a floor plan view. To do so, you must switch to an elevation or section view.

> 2. On the Project Browser, expand *Elevations (Building Elevation)* and then double-click *South* (see Figure 4.12).

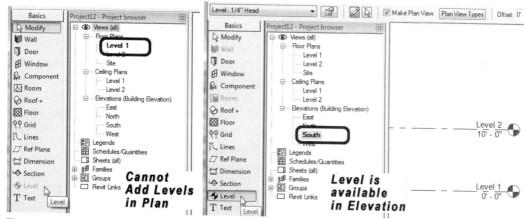

**Figure 4.12** *Levels cannot be added or edited in Plan views*

In this view, you can see two Levels: Level 1 and Level 2.

> **Note:** Notice that even though there is a "*Site*" Floor Plan view in the Project Browser, that there is *not* a "Site" Level. The *Site* view is actually associated to Level 1. In other words, there can be multiple plan views associated with the same Level. In addition, as we saw above, you can also have Levels that have no associated plan views. We will explore this further below.

> 3. Zoom in on the Level Heads at the right side of the screen so that you can see them clearly.

Most elements in Revit Architecture will show a Tool Tip when you pass the mouse over them. Please note that you can turn these tips off in the Options dialog from the Settings menu.

- Move your pointer over the Level line to see a Tool Tip.
- Move your pointer over the Level Head to see a different tip (see Figure 4.13).

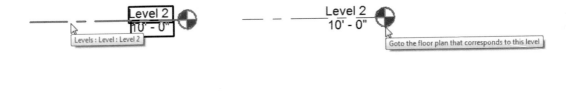

Figure 4.13  *Level Heads can be used to navigate to the associated Plan view*

 **Tip:** In general, the color blue in Revit Architecture indicates that the item is interactive. In the case of Level, Section, and Elevation tags, double-click to open the associated view. In the case of blue text, labels, or dimensions; click to edit the value.

To select the Level, click the dashed level line as shown on the left side of Figure 4.13. To open the associated floor plan view for the level, double-click the blue Level Head (or right-click and choose **Go to Floor Plan**) as shown on the right side. This is analogous to clicking on a link on a web page (except you must double-click in Revit) and it opens the associated view. If the Level Head is black, that means that there is no associated view. The Level by itself is a datum for documentation and modeling purposes only and should not be confused with a plan view, which is one of many ways to view and interact with your Revit model. Levels typically have associated floor plan views, but they are not required.

4. Select Level 2 (click the dashed line, not the Level Head symbol).

When you select the Level, it will highlight in red and several additional drag controls and handles appear. As you move your pointer (referred to as the *Modify* tool) over each one (don't click them yet, simply hover the pointer over each one), the handle will temporarily highlight and the tool tip will appear (see Figure 4.14).

Moving left to right in Figure 4.14, the following briefly describes each control.

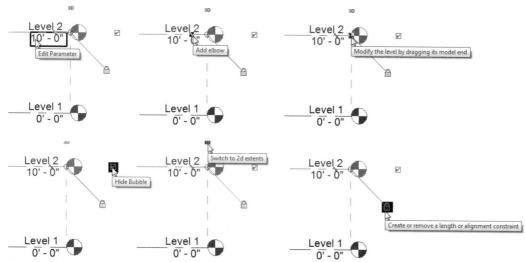

**Figure 4.14**  *Levels have many control handles and drag points*

- **Edit Parameter**—Click the blue text to rename the Level or edit its height. If you rename a Level, you will be prompted to also rename associated plan views. You can accept or reject this suggestion. This means that the Level and its associated views can have different names if you wish.

**BIM** *Manager Note:*   The height is measured from an "Elevation Base" which is a Type parameter of the Level System Family. The Elevation Base is the zero point from which the Level heights are measured. The Elevation Base can be either "Project" or "Shared." With Elevation Base set to Project the values will be relative to the origin of the project (this is at Level 1 in the default template). If you choose the "Shared" Elevation Base, values will report relative to a shared origin. This could be the height at Sea Level or some other appropriate datum height. You can find more information on this topic in the on-line help.

- **Add elbow**—This small "squiggle-shaped" handle creates an elbow in the Level line. This is useful when the annotation of two Level Heads overlap one another in a particular view. Click this handle to create the elbow, and then drag the resultant drag handles to your liking.

- **Modify the Level Drag Control**—This round handle is used to drag the extent of the Level. If the length or alignment constraint parameter is also active, dragging one Level will affect the extent of the other constrained Levels as well.

- **Hide/Show Bubble**—Use this control to hide or show the Level Head bubble at either end of the Level line. This control toggles from hide to show.

- **2D/3D Extents Control**—When 3D Extents are enabled, editing the extent of the Level line in one view affects all views in which the Level appears (and is also set to 3D Extents). If you toggle this to 2D Extents, dragging the extent of the Level line affects only the current view.

- **Length and Alignment Constraint**—This padlock icon is used to constrain the length and extents of one Level line to the others nearby. This is useful to keep all of your Level lines lined up with one another.

5. Click anywhere next to the Level (where there are no objects) to deselect it (or press the ESC key).

6. On the Design Bar, click the Basics tab and then click the **Level** tool.

- Zoom out so that you can see the entire length of the Levels.

- Move your mouse above the left ends of the existing levels.

A temporary dimension will appear with an alignment vector above the top Level.

- When the temporary dimension reads 12'-0" [3600], click the mouse to set the first point (see the left side of Figure 4.15).

**Line up the first point with the left side**

**Line up the other point with the right side**

**Figure 4.15** Add a Level aligned with the existing ones

**Tip:** Remember, you can zoom in to change the temporary dimension snap increment if necessary. Also, you can simply type the value that you wish while the temporary dimension is active and then press ENTER to apply it. Finally, like other tools that we have already seen in Revit Architecture, you can place the Level at any dimension and then edit the value of the temporary dimension later to the correct value.

- Move the pointer over to the right side; when an alignment vector appears above the existing Level Heads, click to set the other point (see the right side of Figure 4.15).

A new Level is added as well as two new plan views: a *Level 3* Floor Plan view and a *Level 3* Ceiling Plan view (see Figure 4.16). To control whether adding a new Level also creates new plan views, use the "Make Plan view" checkbox on the Options Bar. Furthermore, if you click the "Plan view Types" button, you can control which type of plan view(s) are created. By default, a floor plan and a ceiling plan view are created. Notice also that the "Length and Alignment" constraint is also automatically applied as you add Levels.

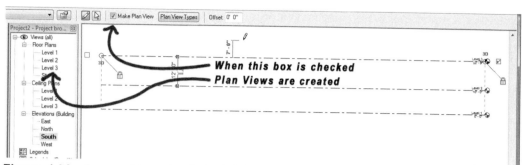

**Figure 4.16**  *As you add new Levels, by default new floor plan and ceiling views are created*

The Level command should still be active. If you canceled it, click the **Level** tool on the Design Bar to restart it.

    7. On the Options Bar, click the Pick Lines icon. In the Offset field, type: **12'-0" [3600]**.

      • Place your mouse over the Level 3 line, and when the green offset line appears above, click to create a new Level 4.

      • Repeat once more to create an additional Level above Level 4.

You should now have five Levels. Our project already has a *Site* plan view. If you edit its properties, you will see that the associated Level is actually Level 1 though. There is nothing wrong with that approach, but sometimes it is preferable to have dedicated Level for the site plan. To do this, we must first delete the existing site plan and recreate it with a new Level.

    8. On the Project Browser, right-click the *Site* floor plan and choose **Delete**.

    9. On the Design Bar, click the Basics tab and then click the **Level** tool.

      • Click the Plan View Types button.

      • Click on Ceiling Plan to deselect it (you only want Floor Plan highlighted).

      • On the Options Bar, click the Pick Lines icon. In the Offset field, type: **3'-0" [900]**.

      • Offset a Level beneath Level 1 (see Figure 4.17).

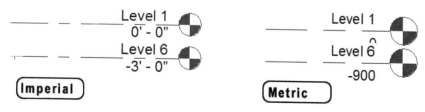

**Figure 4.17**  *Add the final Level below Level 1 without creating associated plan views*

      • On the Design Bar, click the **Modify** tool or press the ESC key twice.

Note that a new Level 6 has appeared in only the *Floor Plans* category.

## RENAMING LEVELS

We can accept the default level names as they appear, or we can change them to something more suitable for our specific project. The names of Levels can be anything that makes sense to the project team. Levels in the model will appear in their correct physical locations; however, in the Project Browser they will sort alphabetically. Therefore, on larger projects with many levels, it might make more sense to use a naming scheme with a numerical prefix such as 01 Level, 02 Level, etc. (In projects with more than 10 stories, you should consider placing a leading zero in the level names so that Level 10 does not inadvertently sort before Level 1). We'll just rename the top and bottom levels in this project to something a bit more descriptive.

10. Select Level 5 (the one at the top).

- Click on the blue text of the name to edit it. Type: **Roof** for the new name.

- A dialog will appear asking you if you wish to rename the corresponding plan views, click Yes (see Figure 4.18).

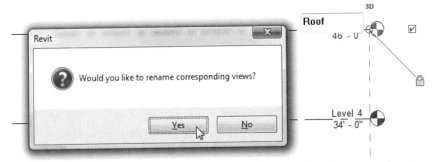

**Figure 4.18**   *Rename the top Level to Roof and accept the renaming of corresponding views*

When you choose "Yes" in this dialog, the two associated plan views: *Level 5* floor plan and *Level 5* ceiling plan, become "*Roof.*"

11. Repeat this process on Level 6 (the one at the bottom). Rename it to **Street Level**.

- In the dialog that appears, answer No this time.

  You do not need to name the Levels and the views the same. In this case, it will be more useful to call out the level on the elevations and sections as "Street Level" but the plan associated to this level will be our site plan. Therefore, in this case, it is better to name the Level and the plan view separately.

12. Right-click Level 6 beneath *Floor Plans* on the Project Browser and choose **Rename**.

- Name the view: **Site Plan** and then click OK.

- In the elevation view on screen, double-click the Street Level head.

Notice that this opens the *Site Plan* view. So while the names are no longer the same, they are still linked to one another. Look carefully at this view. When you create a floor plan view, certain defaults are automatically applied to it like display settings and scale for example. Typically, a site plan is drawn at a scale smaller than the default 1/8"=1'-0" [1:100] used here. While we could manually configure all of the various settings needed to make this view a more suitable site plan, we can do this more efficiently by applying a view template.

13. From the View menu, choose **Apply View Template**.

- In the "Select View Template" dialog, select Site Plan from the list and then click OK (see Figure 4.19).

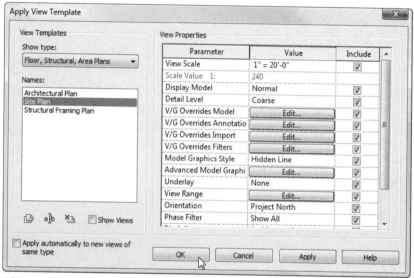

**Figure 4.19** *Apply a view template to the Site Plan view*

The most obvious change will be the increase in size of the elevation tags. This reflects that the scale of the view has changed. There are other settings that were changed as well. These will become more evident as the plan progresses.

14. From the File menu, choose **Save**.

- Browse to the *Chapter04* folder, give the project a name such as **MRAC-Commercial** and then click Save.

**BIM** *Manager Note:* You can create your Levels and plan views at the same time as we did here. You can also remove the option to create plan views completely when new Levels are added. You may choose to do this to create datum views to demarcate Top of Steel, Top of Plate, or Bottom of Footing, etc. New plan views can be created and associated to any and all existing Levels at any time. To add a view to an existing Level, choose the **View > New > Floor Plan** or **View > New > Ceiling Plan** option. Be sure to choose an appropriate scale before clicking OK. If the Level you select already has a view associated to it, clear the "Do not duplicate existing views" checkbox. Otherwise, a new view will not be created. You can also right-click and existing view at any time in the Project Browser and duplicate it directly.

## WORKING WITH SITE TOOLS

Now that we have created and named all of our Levels, we can move on to adding some basic building elements to our model. Having some simple geometry in the file will make planning views and setting up preliminary sheets easier. Since the site plan is already open, this is a good place to start.

### CREATE A TOPOSURFACE

The first thing we need in a site plan is our site! Revit Architecture has some simple site tools that enable us to build a topographical surface upon which our building model can sit. We can also add planting, parking, and other site accoutrements. You can build a Toposurface two ways: by manually placing points, or by importing site data from external files. Let's start by creating a simple surface.

1. On the Design Bar, click the Site tab.

   If you do not see this tab, right-click on the Design Bar to load it.

2. Click the *Toposurface* tool.

The Design Bar will change to sketch mode and only the Toposurface tools will be displayed. The view window will halftone the existing items in the view (such as the elevation tags).

Be sure the *Point* tool is selected.

- On the Options Bar, set the Elevation field to **0**.

- Click four or five points across the top of the screen slightly above the top elevation marker.

- Change the Elevation to **-1'-0"** [**-300**]. Click four or five more points running horizontally beneath the elevation marker.

- Change the Elevation to **-2'-0"** [**-600**]. Click the next set of points horizontally beneath the east and west elevation markers.

- Change the Elevation to **-3'-0"** [**-900**]. Click a final set of points horizontally beneath the south elevation marker (see Figure 4.20).

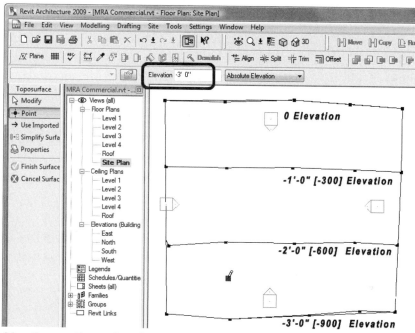

**Figure 4.20**   *Create a Toposurface from sketched points*

3. On the Design Bar, click the Finish Surface button.

- Open the *East* elevation view.

Zoom in as necessary and study the results. Notice that the surface appears with a gentle slope and an earth pattern applied where the grade is cut (see Figure 4.21).

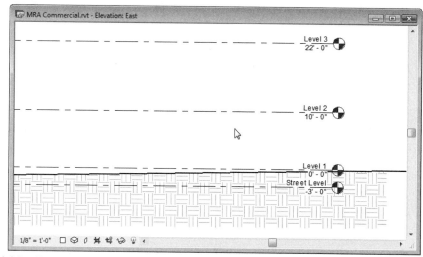

**Figure 4.21**   *Finish the sketch and view the Toposurface in elevation*

Like any sketch-based object in Revit, you can select the Toposurface at any time. On the Options Bar, click the Edit button. This will return you to sketch mode where you can select and modify the points from which the topo is derived. You can move the points, change their elevations, add new points, and delete existing points. If you choose to add a new point, you add it at an absolute elevation like the ones we added above, or when editing an existing surface, you can also choose to make them relative to the surface. You will find this on the Options Bar after you click the *Point* tool. Feel free to experiment with this Toposurface as much as you like before continuing. We will be deleting it in favor of a new one we will create from an imported file. So have some fun exploring the options and don't worry about "messing" this one up.

## IMPORT SITE DATA

Frequently you will receive site plan data from outside firms in AutoCAD DWG or Microstation DGN format. Revit Architecture readily imports files saved in either format and others as well. The linework in those files can be used to create your Toposurface. (In order for this to work correctly, the linework in the file has to be drawn at the correct z-height corresponding to the actual contour level you wish to create). Let's import some contour lines from a DWG file and generate a new Toposurface.

Make sure that the *Site Plan* view is open. If you switched to another view above, double-click *Site Plan* on Project Browser now.

> 4. Select the Toposurface created above and delete it.

> 5. On the File menu, choose **Import/Link > CAD Formats**.

In the Import/Link dialog, you should automatically be viewing the *Chapter04* folder, if not, you will need to browse there.

> • Select (do not double-click) the *Commercial-Site.dwg* [*Commercial-Site-Metric.dwg*] file (don't click Open yet).

Several options appear at the bottom of the dialog. Imported files can be linked or imported. In the "Import or Link" area, if you choose the "Link" option, a live link to the DWG file will be created. If the original file should be changed in its host environment, Revit Architecture will be able to re-import the changes. The "Current view only" option will import the file into the active Revit view only. (However, if you wish to use the imported file to generate a Toposurface, do not use this option as you will not be able to select the contours of the imported file while using this option).

Most CAD files use layers or levels to organize the data they contain. These layers/levels can be interpreted in the incoming file. If you wish to import only certain layers in the DWG file, you can choose an option next to the Layers list. Most DWG or DGN files are saved in multiple colors. The options under "Layer/Level Colors" allow you to control how this color data is handled on import. If the file you are importing has special scaling needs, you can choose options under Scaling. In most cases you will want to initially leave this set to "Auto-Detect."

For "Positioning" there are several options. "Center to Center" is the simplest option. It simply matches the geometric center of the imported file to the geometric center of your active Revit Architecture view. If the file is a one-time import and you are reasonably certain that you will not need to import additional files, this can be the most convenient option. As a matter of best practice, "Origin to origin" is often preferred. When you allow Revit Architecture to align the origin of the DWG or DGN file to the Revit Architecture model origin, you can later import additional DWG or DGN files based upon the same origin point and be certain that they will automatically align with the existing geometry properly. The only problem with the origin to origin option is that your Revit project may not be built to match the origin in the incoming file. In this case, you can use another option and then synchronize the coordinates between your project and the link with the shared coordinates feature. This is the approach we'll explore below.

If none of these methods meets your needs, "Manually place" is available allowing you to use the mouse pointer to place the imported file in any location that you like. This would not typically be the best option for site plan data, but may be fine for smaller components like furniture or building components designed outside of Revit Architecture.

- In the "Import or Link" area, place a checkmark in the "Link" box.

- In the "Layer/Level Colors" area, choose **Preserve**.

- In the "Positioning" area, choose **Center to Center** and then click Open (see Figure 4.22).

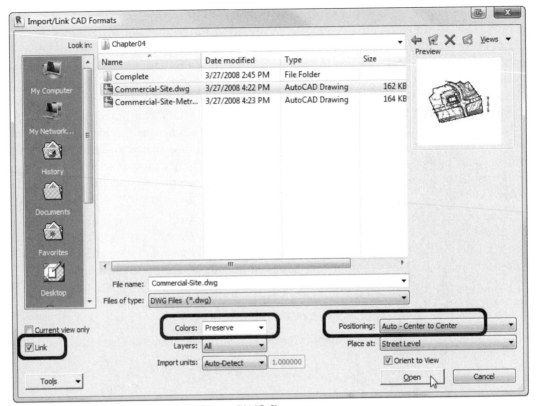

**Figure 4.22**  *Import the Site data from a DWG file*

Notice that the file contains several contour lines and a blue rectangle slightly off center and rotated relative to this view. While we could build our model to match the orientation of the imported file (True North), it will be more convenient to rotate and reposition the file to better suit our project template setup (Project North).

6. Select the linked drawing file. (You can click on it anywhere and the entire file will highlight).

- On the Edit toolbar, click the **Rotate** tool.

A "Center of Rotation" control will appear on screen. To change the center of the rotation, we drag this icon to the desired position. To do this, click and hold down the mouse on the icon, and then drag it to the new location and release. The motion is click, drag, release; not click, click.

- Drag the Center of Rotation control icon and snap it to the endpoint at the corner of the building footprint (see the left side of Figure 4.23).

If you dragged it correctly, moving your mouse should reveal a line spinning about the corner of the building footprint.

- Move the mouse to the opposite endpoint and then click (see the middle of Figure 4.23).

- Finally, move the mouse to the right. It will snap vertically. When it does, click to finish the rotation (see the right side of Figure 4.23).

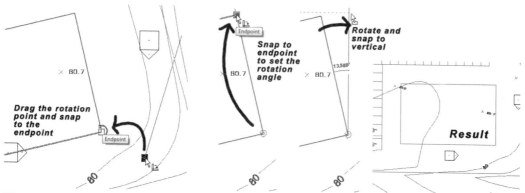

**Figure 4.23**   *Rotate the linked site file to make the building footprint horizontal*

Next we'll move the file both in plan and elevation to finish the positioning.

7. Select the linked file again.

- On the Edit toolbar, click the **Move** tool.

- Click anywhere for the start point and then click a new point such that the rectangular building footprint ends up centered between the four elevation tags.

The exact amount of the movement is not critical. Just get it close.

Finally, if we look at one of the elevation views, it will become clear that there is one more adjustment required. If you look at the labels on the contour lines, you will see that near the center of the building, the contours are at 81'-0". This means that in order for our building to sit on the site properly, we must move the linked file down.

8. In an elevation view, select the linked file again.

- On the Edit toolbar, click the **Move** tool.

- Click anywhere for the start point.

- Move the mouse straight down, type **81'-0"** [**24300**] and then press ENTER.

The contours should now be overlapping Level 1 and Street Level (see Figure 4.24).

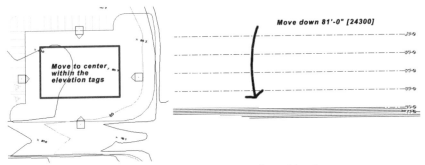

**Figure 4.24** *Move the file in plan and elevation to properly position it*

### SAVING SHARED COORDINATES

If this file were to later change and require reloading, we want to be sure that it reloads in the same relative location. To do this, we can save a shared location for the file.

9. Select the linked file on screen.

- On the Options Bar, click the Properties icon.

- Beneath the Other grouping, next to Shared Location, click the <Not Shared> button.

The "Share Coordinates" dialog lists two ways that the coordinate systems can be reconciled. The two methods are very similar and differ only in which file will be recorded as the "predominant" file. In each case, both the host file and the linked file must be saved with the shared coordinate information. Publishing makes the host file predominant while acquiring makes the linked file predominant. We will accept the default to of Publish.

- At the bottom of the dialog, click the Change button next to Default Location.

Revit projects can have one or more saved locations within them. This is useful when the same building model must be repeated on a site, such as a multi-building campus of condominium buildings. In this case, we have only one building and could simply accept the default location name. However, it is good practice to get in the habit of renaming the default location so that later you can use this to verify that you have in fact reconciled the coordinates.

- In the "Manage Place and Locations" dialog, click the Rename button and change the name to **Commercial Site** (see Figure 4.25).

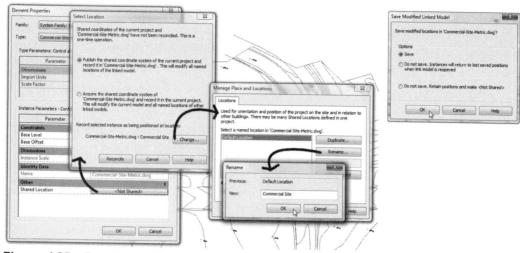

**Figure 4.25** *Establish the Shared Coordinates and then save the files*

- Click OK two times.
- In the "Select Location" dialog, click the Reconcile button.
- Finally in the "Element Properties" dialog, click OK.

10. Save the project.

- In the "Save Modified Linked Model" dialog, accept the Save option, and then click OK (see the right side of Figure 4.25).

We are saving both the host file and the linked file here.

To understand what we have accomplished, let's try an experiment.

11. Select and drag the linked file to move it.

A warning will appear on screen. You can ignore the warning, but reading the message reveals that if we reload the linked file without first saving the new location, the file will return to its original location.

Click OK to dismiss the warning dialog.

- From the File menu, choose **Manage Links**.
- On the CAD Formats tab, click the *Commercial-Site.dwg [Commercial-Site-Metric.dwg]* file and then click the Reload button.
- Click OK to return to the view window and see the results.

Notice that the file returns to its centered location. This method makes it simple for you to coordinate files with different coordinate systems and maintain the proper level of coordination between them. The shared coordinates ensure that files from each consultant and team member will remain properly aligned.

12. Save the project.

## BUILD A TOPOSURFACE FROM IMPORTED DATA

Now that we have imported the 3D contour line data from the DWG file, we can use it to create a Toposurface instead of the manual points we sketched above.

Make sure the *Site Plan* view is open.

13. On the Design Bar, click the Site tab and then click the **Toposurface** tool.

The Design Bar changes to show only the Toposurface tools again.

- On the Design Bar, click the → **Use Imported > Import Instance**.

- Click anywhere on the imported DWG to select it.

- In the "Add Points from Selected Layers" dialog, click the "Check None" button.

- Place a checkmark in only the "C-Site-Cntr" and the "C-Site-Cntr-Intm" checkboxes to select only those two layers (see Figure 4.26).

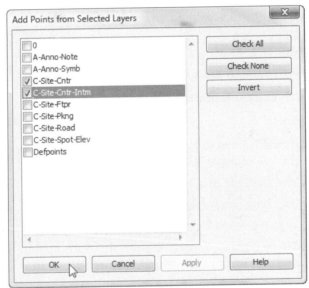

**Figure 4.26** *Choose a Layer from which to create the Toposurface*

- Click OK to create the Toposurface.

Several points will be extracted from the geometry on the selected layer and from those points a Toposurface will be created. The resulting surface is similar to the one we created above.

- On the Design Bar, click the **Finish Surface** button to exit sketch mode and complete the Toposurface.

14. On the Project Browser, double-click the *East* elevation view.

Notice that the profile of the terrain is again similar to the surface we created above. This Toposurface simply has more points and does a better job suggesting the roads that surround the building site. Either method showcased here is appropriate for creating Toposurfaces in your own projects. If you receive a site plan file, it is usually easier to use the linked file. Otherwise, you can quickly create a suitable site for your building model with the point sketching method as well.

## ADD A BUILDING PAD

Let's add a Building Pad. A Building Pad adds a simple level surface that cuts into the terrain model as appropriate to suggest the required excavation. It will be easier to do this without having the model shaded.

15. On the Project Browser, double-click to open the *Site Plan* view.

The site plan data imported from the DWG file includes a rectangle that approximates the rough footprint of the building. We can use this to assist us in sketching the Building Pad. However, at the moment the Toposurface is concealing the linked file.

- On the View Control Bar, click the Model Graphics pop-up and choose **Wireframe**.

16. On the Design Bar, click the *Pad* tool.

- On the Design Bar (now in Sketch mode) click the *Lines* tool.

- On the Options Bar, click the Pick Lines icon.

- In the view window, position the pointer over one edge of the rectangle in the middle of the site (it should pre-highlight) and then press the TAB key.

- When all four sides of the rectangle pre-highlight, click the mouse to create sketch lines (see Figure 4.27).

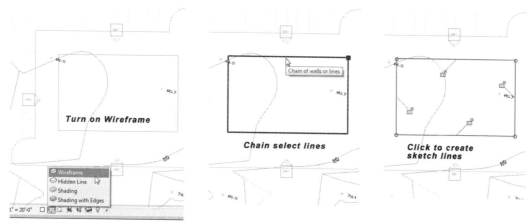

**Figure 4.27**   *Use the tab key to chain select and create Sketch Lines*

- On the Design Bar, click the *Pad Properties* button.

- In the Element Properties dialog, for the "Height Offset From Level" parameter, input **-4'-0" [-1200]** and then click OK.

- On the Design Bar, click the **Finish Sketch** button.

17. On the View toolbar, click the Default 3D View icon.

- Zoom as required to see the Pad and its relationship to the Toposurface.

18. Save the Project.

## ROUGH OUT THE BUILDING FORM

Since our goal in this chapter continues to be the overall setup of our commercial project, we will take this opportunity to add some simple geometry to the project to suggest building form. The purpose is not to arrive at complete design solution, but rather to provide enough of a building form to give the various views and sheets that we will establish below some context. The tasks that follow are therefore simple suggestions of project setup workflow. Feel free to vary the process as appropriate in your own projects.

### ADDING WALLS AND CONSTRAINING WALLS TO LEVELS

Now that we have our site contours, a Toposurface and a building Pad, we can add some geometry to rough out the overall form of our building. At this early stage of the project, some simple Generic Walls will be sufficient. As the design evolves the wall types can be changed and edited to suit our needs.

1. On the Project Browser, double-click the Level 1 floor plan view.

- Zoom so you can see the rectangle in the middle of the site plan.

2. On the Design Bar, click the Basics tab and then click the **Wall** tool.

- From the Type Selector, choose **Basic Wall : Generic - 12"** [**Basic Wall : Generic - 300mm**]

- On the Options Bar, click the Pick Lines icon.

- Change the Level to **Street Level**, and then from the Height list, choose **Roof**.

- For the Loc Line choose **Finish Face: Exterior** (see Figure 4.28).

**Figure 4.28**  *Set the options to create new Walls*

3. In the view window, position the pointer over one edge of the rectangle in the middle of the site to make it pre-highlight.

Look carefully at the way that the edge pre-highlights. A dashed line will appear parallel to the edge on one side or the other. If you move your pointer slightly, the dashed line will shift to the other side of the edge. This indicates the side on which the Wall will be created.

- When the dashed line is on the inside of the rectangle, press the TAB key.

  All four edges should pre-highlight with dashed green lines to the inside.

- When all four sides of the rectangle pre-highlight, (and the dashed lines are on the inside) click the mouse to create Walls.

4. On the Project Browser, double-click to open the *West* elevation view.

- Zoom in near the bottom of the building

The outside edge of the Walls should be aligned to the outside edge of the Pad. You can temporarily switch to wireframe to check this (see the View Control Bar topic in Chapter 2 for more information on switching display modes). If your Walls were created to the outside of the Pad and are not flush, you can either Undo, or return to the Level 1 plan view and Change the Wall orientations with the flip handle on each Wall.

5. On the Project Browser, beneath 3D Views, double-click on {*3D*} to return to the default 3D view.

  If you prefer, simply click the Default 3D View icon again.

- On the View Control Bar, click the Model Graphics Style icon and choose **Shading with Edges**.

We now have a big hollow box sitting on our terrain. Let's add a simple Roof.

## ADD A SIMPLE ROOF

6. On the Design Bar, click the *Roof* tool and then choose **Roof by Footprint**.

A message will appear requesting that we select the desired Level to build the Roof. This message appears because we are in a 3D view. Naturally we will want to create the Roof at the Roof Level.

- In the "Lowest Level Notice" dialog, choose **Roof** from the drop-down list and then click Yes (see Figure 4.29).

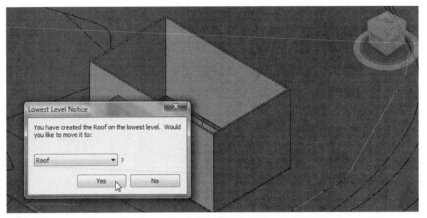

**Figure 4.29** *When creating a Roof in 3D, clarify the correct creation Level*

- On the Design Bar (now in Sketch mode) verify that **Pick Walls** is selected.
- On the Options Bar, clear the checkmark from the "Defines Slope" checkbox.
- Using the TAB key, chain select all four Walls and then click to create sketch lines.
- Zoom in as required and verify that the sketch lines are on the inside edge of the Walls. If they are not, click the small flip control to shift the sketch lines.
- On the Design Bar, choose Finish Roof.

## ADJUST WALL HEIGHT

Our goal at this stage is to create a very rough model that we can use to help setup and understand the views in our project. However, one quick edit to the Walls is immediately obvious. Let's adjust the Wall height to suggest a parapet at the Roof.

7. Chain select all four Walls and then click the Properties icon (on the Options Bar).

- For the Top Offset parameter, type **4'-0" [1200]** and then click OK.
- On the Design Bar, click the **Modify** tool or press the ESC key twice.

8. Save the Project.

## ADD A MASS

In the coming chapters, the design of both our projects will evolve. At this stage of our commercial project we have little information other than the number of levels and site data that we have already incorporated into the project and few notions of the building's overall form. It is desired that some sort of special treatment be given to the front façade. We don't have the specifics about this treatment yet but at this stage can make a few assumptions. To do this, we will sketch out a simple Massing element on the front façade of our building to suggest that some design element will later occur in this location. You can design the entire form of the building using Masses or just a portion of it as we are doing here.

9. On the Project Browser, double-click to open the *Level 1* floor plan view.

10. On the Design Bar, click the Massing tab and then click the **Create Mass** tool.

If you do not see the Massing tab, right-click on the Design Bar and choose **Massing** to load it.

A message will appear indicating that the "Show Mass" mode will be enabled. Massing tools are meant as design tools and as such are not visible by default. We can enable their visibility in any view that we wish. This message is simply a courtesy indicating that Massing display can be enabled temporarily for us. There is also an icon on the toolbar next to the Default 3D View icon that toggles massing display on and off.

- In the message dialog, click OK after you have read it.

- In the Name dialog, type: **Front Façade** for the name of the Mass that we are creating.

- On the Design Bar (now in Massing mode) click the **Solid Form** tool and then choose **Solid Extrusion**.

- On the Design Bar (now in Sketch mode) click the **Lines** tool.

- Sketch the shape indicated in Figure 4.30 using the dimensions indicated.

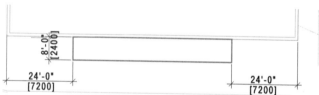

**Figure 4.30** *Sketch the footprint of the Mass*

- On the Design Bar, click the **Extrusion Properties** button.

- In the Element Properties dialog, in the Extrusion End field, type **40'-0" [12,000]** and then click OK.

- On the Design Bar, click the **Finish Sketch** button, and then click the **Finish Mass** button.

Masses are intended as a quick way to create and study building form. They are primarily a design tool but have some functions that allow them to easily transition into design development. In particular, looking at the Design Bar, you will notice tools to create Walls, Roofs, Floors and Curtain Systems on the faces of a Mass. We will try a few of these tools here. We will work with this Mass element more in Chapter 9.

11. On the Project Browser, beneath 3D Views, double-click on {3D} to return to the default 3D view (see Figure 4.31).

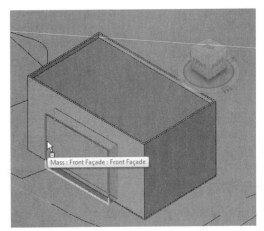 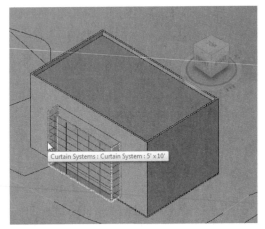

**Figure 4.31** *The completed Mass at the front of the building*

12. On the Design Bar, click the **Curtain System by Face** tool.

- At the Status Bar, you are prompted to select a face. Click each of the three sides of the Mass.

- With the three surfaces selected (highlighted red) click the Create System button on the Options Bar.

- On the Design Bar, click the **Roof by Face** tool.

- At the Status Bar, you are prompted to select a face. Click the top surface of the Mass.

- On the Options Bar, click the Create Roof button.

This quick example shows how you can use the Massing tools in a schematic design stage to suggest design elements in which you are considering. One you have completed working with the Mass, you can toggle their display off again. Since we have created other surfaces relative to our Mass, they will continue to display. If we later edit the Mass, a remake option appears on the Options Bar for the Walls, Curtain Systems, Floor, and Roofs that we create (refer to the "Editing Massing Elements" topic in Chapter 9 for an example). This makes it easy to keep your Mass model and building model coordinated. You can hide the underlying Mass using the icon on the toolbar next to the Default 3D View icon. This will turn Masses off in the project. Later, if you wish to edit the Mass, simply toggle them on again with this icon and make your edits.

## WORKING WITH ELEVATION VIEWS

The default templates that we used to create our project already included four building elevations. As our project progresses, we can work with these elevations, delete them, and/or add additional ones. Elevation views are vertical slices through the building model cut from a certain point and projected orthographically. Level heads will display automatically in elevation views and elevation view tags will appear by default in all plan views to indicate where the elevations are cut. The four default elevation views are sufficient to our needs at this time but we will make a series of adjustments to them to fine-tune them.

## ADJUSTING ELEVATION TAGS

We have added enough geometry to our model to begin studying the effectiveness of our elevations. As we study them, we can see that it might be beneficial to adjust a few things.

1. On the Project Browser, double-click each elevation view in succession.

Zoom and pan within each elevation to study them carefully. As you open each elevation, take note of the edges of the Toposurface. Most of the surface appears in section with a bold top edge and an earth fill pattern beneath. This result is due to the locations of the elevation tags in the plan views. Let's take a closer look at the elevation tags to get a better understanding of their behavior.

2. On the Project Browser, double-click to open the *Level 1* plan view.

- Zoom in on the bottom elevation tag.

- Move the mouse pointer over the tag.

Notice that it is actually made of two pieces. The elevation tag is the square [round] shape, and the arrow (triangle) portion indicates the viewing direction and extent of the associated elevation view (see Figure 4.32).

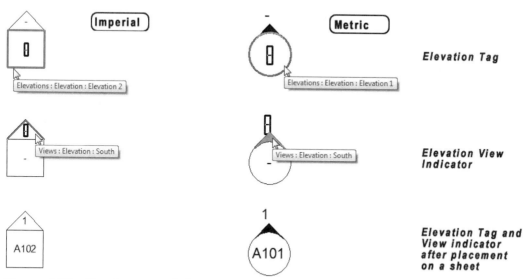

**Figure 4.32** *Elevation tags and elevation view indicators*

A small dash appears within each tag. These are text labels that fill in automatically when the elevation view is placed on a sheet. If you want to see this, you can add a sheet to the project. Choose **New > Sheet** from the View menu. Next, drag the *South* elevation view from Project Browser and drop it on the sheet. Return to the *Level 1* plan and note that the values in the elevation tag will update automatically. (An example is shown at the bottom of the figure). We will add sheets later in this chapter, so you do not have to create them now.

- Select the Elevation Tag (square in Imperial and round in Metric).

By default one elevation arrow is active per elevation tag. Each elevation tag can have up to four elevation arrows and corresponding elevation views. Three additional elevation view Indicator arrows will appear temporarily as long as the Tag is selected. Small checkboxes appear that you can use to add new elevation views. If you remove a checkmark, the associated elevation view will be deleted (see the left side of Figure 4.33). A prompt will appear to warn you of this. To test this out, check one of the empty boxes, then note the addition of a new elevation view on the Project Browser. Remove the checkmark from the same box and a warning will appear indicating that this new view will be deleted. There is also a rotate handle that appears when the tag is selected. This will rotate the entire elevation tag and all of its associated views. As the elevation tag is rotated it will snap to 90° increments and perpendicular to adjacent walls and other geometry. Try it out if you like, just be sure to undo before continuing.

- Deselect the Elevation Tag and then select the Elevation view Indicator arrow.

When this arrow is selected, it turns blue instead of the customary red. You can double-click the blue arrow to open the associated elevation view. First select the arrow, then double-click it if you want to open the view. Selecting the elevation arrow also reveals a line perpendicular to the direction of the arrow with a drag control at either end (see the right side of Figure 4.33).

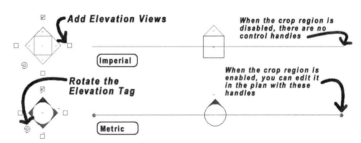

**Figure 4.33** *Handles and drag controls on the elevation tags and elevation view Indicators*

This line indicates the extent of the cut plane of the elevation view. An elevation view is very similar to a section cut at this line. Objects crossing this line will show as cut in the corresponding view. We have seen this above with the Toposurface. Objects behind this line will not be shown in the elevation view. Objects in front of the line appear in projection. The extents of the elevation line can actually crop the view. To do this, you must enable the crop view parameter in the associated view properties. Let's explore a few of these parameters.

3. Type **ZF** on the keyboard, or choose **Zoom > Zoom to Fit** from the View menu.

4. Select both parts of the south elevation Tag (the Tag and the arrow for the bottom-most one pointing up).

   • Drag it to up into the building just above the front façade (see the left side of Figure 4.34).

If you tile both the plan and elevation views side by side on the screen at once while performing these steps, you can see the elevation view change instantly as you move the tag around.

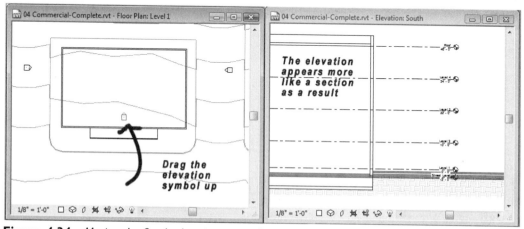

**Figure 4.34**   *Moving the South elevation tag and arrow inside the building results in a section*

   • Undo the change or simply move the symbol back to its original location.

   • Nudge each elevation view tag closer to the building on all four sides.

   • On the Design Bar, click the **Modify** tool or press the ESC key.

## ADJUSTING ELEVATION CROPPING

The elevation views by default do not use or show the crop regions. This means that the view continues to dynamically resize itself as the building geometry grows and shrinks. If you wish to limit the size of the elevation, you can display the crop region and then resize it to crop away the unwanted portion of the view. In this case, it is not really necessary to do so, but let's quickly look at the process in case you need to make such an adjustment in a future project.

5. Open the *South* elevation view.

   • On the View Control Bar, click the Crop view icon and the Show Crop Region icon.

   • Select the Crop Region.

   • Drag the handles to crop away the unwanted portion of the view.

   • Click the Hide Crop Region icon (see Figure 4.35).

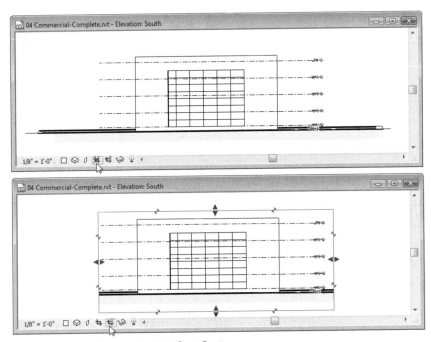

**Figure 4.35** *Showing and adjusting the Crop Region*

If you like the results, repeat the process on the North elevation. Otherwise, click the Do Not Crop View icon to turn it off and return to the full un-cropped view. Please note that the "Crop View" icon toggles to become the "Do Not Crop View" icon after you click it and likewise the "Show Crop Region" icon toggles to become the "Hide Crop Region" icon.

## ADJUSTING LEVEL HEADS

Now that we have moved and resized the elevations, all of the views will display correctly. However, the Level Heads still need adjustment.

6. On the Project Browser, double-click each elevation view in succession.

In some views the Levels are a little off-center or a bit too long. We can select and move them the like any other Revit element. However, if we use the drag technique, we can accidentally move them vertically as well which would affect our Level heights. To move them precisely left or right only, use the Move tool. You can also drag the length of the level lines to adjust them as well. We will use this approach here.

7. On the Project Browser, double-click to open the *West* elevation view.

- Click on one of the Level lines.

- Click and drag the small hollow circle handle at the end of the Level line and stretch the end closer to the elevation (see Figure 4.36).

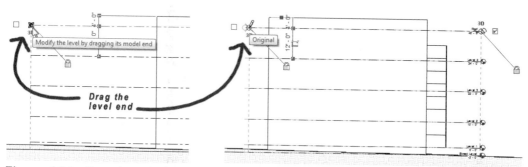

**Figure 4.36**   *Drag the end of the Level Heads to center them on the Elevation*

The precise amount of the stretch does not matter. You simply want the end result to have the Level heads comfortably centered on the elevation.

8. Repeat this process in the *North* elevation view if necessary.

Notice that you only need to move the Levels in two elevation views. Since elevations (like all Revit Architecture views) are live views of the model, changes you made in the *West* view can be seen automatically in the *East* view and likewise the changes in the *North* view appear automatically in the *South*. You will also notice that when you drag one Level line, all of them will adjust together. This is because they are all constrained to one another (see above).

9. On the Design Bar, click the **Modify** tool or press the ESC key.

10. Save the Project.

It may be necessary to adjust the extent of the Levels in the *North/South* elevations as well. This edit is likely necessary only if you are working in Metric units as the default Metric template from which we began the project has smaller Level extents than the Imperial template.

## EDIT LEVEL HEIGHTS

Earlier when we set up all of the Levels, we neglected to adjust the default height between Level 1 and Level 2. By default in the Imperial template it is 10'-0" and it is 4000mm in the Metric template. However, for this project, the heights of all four Levels should be the same at 12'-0" [3600]. The change was postponed till now so that we can get a sense of the power of some of the simple constraints that we have already built into our model.

11. On the Project Browser, double-click to open the West elevation view.

• Select Level 2.

Notice the temporary dimensions that appear. You can certainly edit those values to move the Levels to their correct height. However, the problem with that technique is that you must move the Levels one at a time. A better approach is to use the Move tool or use the "Activate Dimensions" Options Bar option for a multiple selection.

12. Select the Level 2, Level 3, Level 4 and the Roof Level.

**Tip:** Remember, you can hold down the CTRL key and click each one, or drag a window from right to left touching each Level you wish to select.

- On the Options Bar, click the "Activate Dimensions" button.
- In the temporary dimension that appears between Level I and Level 2, type **12'-0"** [**3600**] (see Figure 4.37).
- Watch the top edge of the Wall as you press ENTER.

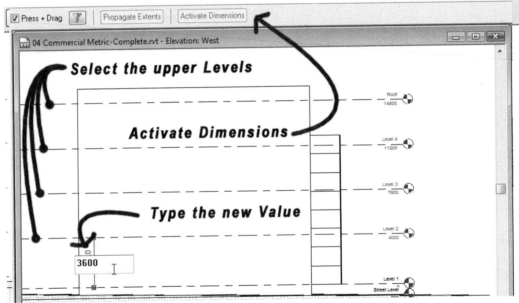

**Figure 4.37** *Activate Dimensions on a multiple selection to move them together*

You should notice that the top edge of the Wall moves with the Levels. This is because earlier, we constrained the top of the Wall to the Roof Level. This is another example of the power of the parametric relationships between elements in a Revit model.

## CREATING SECTION VIEWS

Unlike elevations, section views are not included in the default template. Section views are also vertical cuts through the building model. They typically show the building in a cutaway fashion in an orthographic view. Level Heads also appear in Sections automatically. Adding Sections to our model is a simple task.

### CREATE A SECTION LINE AND ASSOCIATED VIEW

Let's add two building sections, one longitudinal and one transverse.

1. On the Project Browser, double-click to open the *Level I* floor plan view.

2. On the Design Bar, click the Basics tab and then click the **Section** tool.

- Click the first point on the outside left of the building near the middle.

- Move the pointer horizontally to the right and then click the second point outside the building to the right (see Figure 4.38).

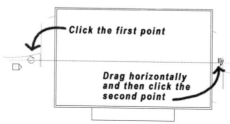

**Figure 4.38**  *Click two points to create a Section Line*

A section line will appear with several Drag controls and Handles on it. Like the other tags and symbols that we have seen so far, simply hover the Modify tool (mouse pointer) over each control; the handle will temporarily highlight and the tool tip will appear indicating the function (see Figure 4.39). Brief descriptions of each control follow.

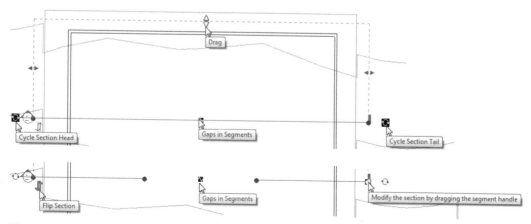

**Figure 4.39**  *Section Line Drag Controls and Handles*

- **Cycle Section Head/Tail**—These controls appear at each end; one for the head and another for the tail. Each time you click one of these handles, the Section head or tail will cycle to a different symbol.

- **Flip Section**—Click this control to flip the Section Line and symbols to look the opposite way.

- **Segment Drag Handle**—This handle (round dot) controls the length of the section line and moves the section heads with it. Dragging this handle does not change the size of the section crop box.

- **Drag**—Drag Handles appear on three sides of the Section Crop Box. Use them to define the precise extent of the associated Section view.

- **Gaps in Segments**—This control breaks the Section Line and removes the inner segment. You can then drag the exact length of each end segment.

Notice that when you add the Section marker, a new "Sections" branch appears in the Project Browser and that an associated Section view appears on this branch.

- Using the Drag Handles, make any desired adjustments to the extent of the section box or the position of the section heads.

- If you wish, click the "Gaps in Segments" handle to remove the middle section of the Section Line.

- On the Project Browser, expand *Sections (Building Section)*.

- Right click *Section I* and choose **Rename**.

- Type **Longitudinal** and then click OK.

- Double-click *Longitudinal* to open the view.

**Tip:** You can also double-click the blue Section Head in any plan view to jump to the associated Section view.

Notice the box that surrounds the section in the view. This is the Crop Region for this section and it matches the size of the section box that we created in plan. This is functionally identical to the crop region we manipulated for the elevations above. By default crop regions are turned on in section views.

3. Repeat the steps above to create a vertical section looking to the right.

- Rename the view to **Transverse**.

Make any additional adjustments that you wish. When you open other views (like plans and elevations), the section lines will appear automatically. You can make the same sort of adjustments to them in each view: add gaps to the line, drag the Section Head to a different location. These adjustments are view specific. The same Section marker in the other views can look different. The location of the Section marker line (perpendicular to its length) is in the exact same place. Or you can adjust the size of the Crop Region from any view. This is not view specific. If you adjust the Crop Region or the location of the Section marker in any view, it will move accordingly in all other views.

4. Save the project.

## SCHEDULE VIEWS

We have now created and worked with several types of views. While floor plans, ceiling plans, sections and elevations may be the most common views that our projects will contain, Revit Architecture includes over a dozen distinct view types. (You can see the complete list on the **View > New** menu). One of the most powerful view types at our disposal is the schedule view. A schedule view is not a drawing, but rather a non-graphical view of a collection of related data extracted from the model and presented in tabular format. You do not need to "draw" and manually compile your schedules when you work in Revit Architecture. Working them up in Excel is not required. You simply create a schedule view directly in the software by indicating which data fields you wish to include and how you wish it to be formatted and the data required to populate the schedule will be extracted from the model and tabulated into a list view.

## ADD A SCHEDULE VIEW

In this sequence, we will add a few simple schedule views to our project. We will not get into a detailed explanation of scheduling features at this time. Detailed information on schedules is found in Chapter 11.

 1. On the Design Bar, click the View tab and then click the **Schedule/Quantities** tool.

 • In the "New Schedule" dialog, choose Walls from the Category list and then click OK (see Figure 4.40).

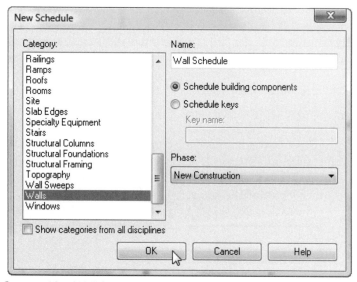

**Figure 4.40** *Create a New Wall Schedule*

 • In the "Available fields" column, select "Type Mark" and then click the Add → button.

 • Repeat this process for the "Length," "Width," "Family and Type" and "Comments" fields (see Figure 4.41).

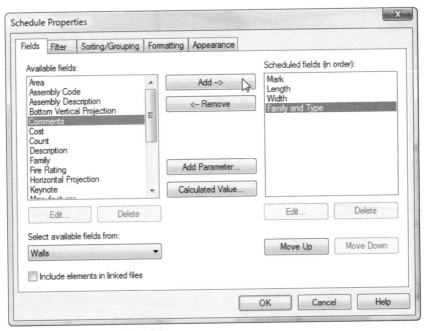

**Figure 4.41** *Add Fields to the Schedule*

- Click OK to complete field selection and open the Schedule view.

The *Wall Schedule* view will appear on screen with each of the four Walls that we currently have in our project listed. The Walls do not yet have "Type Marks" or "Comments" so these fields are empty. Later, as we edit the data parameters of these Walls, these fields will update to reflect the latest information. The Schedule view also appears in the "Schedules/Quantities" node of the Project Browser (see Figure 4.42). You can open it from there any time just like the other views of the project that we explored above.

| | | Wall Schedule | | |
|---|---|---|---|---|
| Mark | Length | Width | Family and Type | Comments |
| | 102' - 8" | 1' - 0" | Basic Wall: Generic - 12" | |
| | 62' - 10 1/16" | 1' - 0" | Basic Wall: Generic - 12" | |
| | 102' - 8" | 1' - 0" | Basic Wall: Generic - 12" | |
| | 62' - 10 1/16" | 1' - 0" | Basic Wall: Generic - 12" | |

Project Browser items:
04 Commercial-Comp...
- Views (all)
- Legends
- Schedules/Quantitie
  - **Wall Schedule**
- Sheets (all)
- Families
- Groups
- Revit Links

**Figure 4.42** *The Wall Schedule appears in the Project Browser and opens on screen*

## IMPORTING SCHEDULE VIEWS FROM OTHER PROJECTS

When we began this chapter, we opened several out-of-the-box Revit Architecture template projects. It was noted above that starting a project from a template was the preferred method of beginning a Revit Architecture project. In some cases, you will begin a project with one template and then later realize that a particular Family, Type or view required already exists in another template file or project. Rather than re-create the item, it is easier to borrow the element from the other project. In this topic, we will create a new project from the *Commercial-Default.rte* template file (installed with the out-of-the-box Imperial templates) and borrow some Schedule views from that project to use in our own commercial project.

 **Note:** The *Commercial-Default.rte* template file is installed with the out-of-the-box Imperial templates. If you did not install the Imperial templates, or your version of Revit Architecture does not give you access to this file, it has been provided for you in the *Templates* folder with the files installed from the Mastering Revit Architecture CDROM.

2. On the File menu, choose **New > Project**.

- In the New Project dialog, in the "Template File" area, be sure that the lower radio button is selected (the one that lists a template file) and then click the Browse button.

- Browse to the *Imperial Templates* folder.

- Select the *Commercial-Default.rte* template file and then click Open.

Take a look around this project. Many of the views that we have spent time configuring above have been included in this project template. In particular, you will note that the Level structure of this project includes Levels for such datums as bottom and top of footing. Levels such as this are not required, but can be helpful when creating and presenting your model. In addition to the Levels, this project has some useful Schedule views and a large collection of Sheets. We are going to borrow the Schedules to use in our project and then we'll have a look at some of the Sheets.

3. On the Project Browser, expand *Schedules/Quantities*.

- Double-click on *Door Schedule*.

Take note of the formatting of the schedule view. Several fields appear, and some columns contain headers. As noted above, we will learn how to build such a schedule in Chapter 11, for now, it will be useful to simply copy this view with all its formatting and headers for use in our project.

- Right-click on *Door Schedule* and choose **Copy to Clipboard**.

- From the Window menu, choose **MRAC-Commercial – Floor Plan:Level I**.

This will return you to your original project so that you can paste the Schedule.

4. From the Edit menu, choose **Paste from Clipboard** (or press CTRL + v).

On the Project Browser, beneath *Schedules/Quantities*, the pasted *Door Schedule* view will appear.

5. On the Project Browser, beneath Schedules/Quantities, double-click the *Door Schedule* view to open it.

Notice that this Schedule contains all of the pre-formatted columns, but lists no Doors. This is because our project does not yet contain any Doors.

## ADD A DOOR TO THE PROJECT

When you add elements to the project, they will automatically appear in *all* appropriate views—including the Schedule views. Let's add a Door to our project to see this.

6. On the Project Browser, double-click to open the *Level 1* floor plan view.

- Zoom in on the middle of the top horizontal Wall in the plan.

- On the Design Bar, click the Basics tab and then click the **Door** tool.

- On the Type Selector, choose **Single-Flush : 36" × 84" [M_Single-Flush : 0915 × 2134mm]** from the list.

Make sure Tag on Placement is selected.

- Add a Door to the middle of the Wall.

- On the Design Bar, click the **Modify** tool or press the ESC key twice.

7. On the Project Browser, double-click to open the *Door Schedule* view.

Notice that the Door also appears in the Schedule.

8. Click in the Door field of the Schedule.

- Change the door number to **101** and then press enter.

- Return to the *Level 1* plan view and confirm that the door tag reflects the new number (see Figure 4.43).

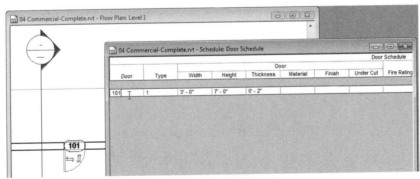

**Figure 4.43** *A change in any view is reflected automatically in all views*

In Revit Architecture, a change made in one view applies in all associated views. You are not changing a graphic representation in a drawing. You are actually editing the parameters of an element in your virtual building model. This change must therefore be reflected in all views that show that component of the model—regardless if they are graphical views or tabular views.

9. Return to the other project and repeat the process to copy and paste the other two schedule views (*Room Schedule* and *Window Schedule*) to your *MRAC Commercial* project.

Be sure to choose a floor plan view from the Window menu before pasting. The Paste command will not be available if you open a schedule view first.

**BIM** *Manager Note:*    If you use the same Schedule views (or any views) repeatedly from one project to the next, you should create your own project template that already includes the views you require.

10. Save the project.

## SHEET VIEWS AND THE CARTOON SET

A sheet is special type of view that is intended to compose and print your document set—they are printing views. When you create a sheet, you can add a title block to it, configure the print setup to match the paper size and other required printing parameters and add viewports containing the various views of your project. Since most architectural projects will require printed output (often at several points) in the life of the project, it is a good idea to set up a set of sheet views to create professional looking results.

Project submission time is usually a high-stress endeavor characterized by panic as project team members scramble to assemble the required sheets and print everything required for a particular submission. This frenzy often occurs at the last minute, which contributes to the stressful environment. One simple way to help alleviate this situation is to create the sheet views early in the project so that they are ready to print at any time. You can also consider adding a number of pre-configured sheet views to your office-standard project template. This will ensure consistency from one project to the next and simplify project setup.

### CREATE A FLOOR PLAN SHEET

1. On the Project Browser, right-click *Sheets* and choose **New Sheet**.

• In the "Select a Titleblock" dialog, accept the default and click OK (see Figure 4.44).

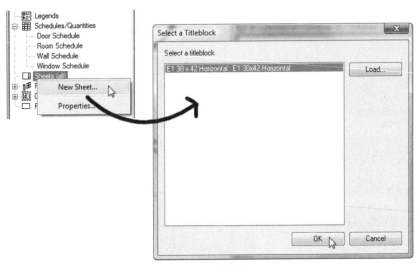

**Figure 4.44** *Create a new sheet and choose a titleblock*

A plus (+) sign will appear next to the *Sheets* node indicating that a new sheet has been added.

2. Expand the *Sheets* node.

The sheet will have a default name and number that we can change.

- Right-click on the sheet and choose **Rename**.

- Change the sheet Number to **A101**, the Name to **Floor Plans,** and then click OK (see Figure 4.45).

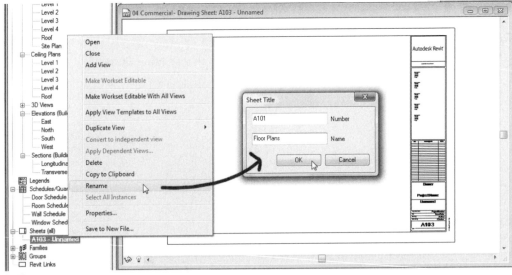

**Figure 4.45** *Rename the new sheet*

**Tip:** If you prefer, you can rename and re-number Sheets directly in the titleblock. Click on the titleblock to highlight it and then click the blue text to edit the value. The result will be the same.

Make sure that the *A101 – Floor Plans* sheet is open. It should have opened automatically when it was created.

- From the Project Browser, drag the *Level 1* floor plan view and drop it on the sheet.

- Zoom in on the view title beneath the floor plan on the sheet.

Notice that the name of the drawing is "Level 1." While this is logical, and in some cases desirable, chances are we would like the name of this drawing on the printed Sheet to be something else. This is easily changed.

3. On the Project Browser, right-click the *Level 1* view and choose **Properties**.

- In the Element Properties dialog, type **First Floor Plan** in the "Title on Sheet" parameter field (see Figure 4.46).

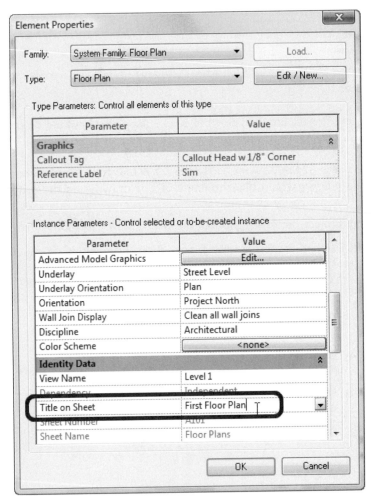

**Figure 4.46** *Edit the value of the Title on the Sheet*

- Click OK to complete the change (see Figure 4.47).

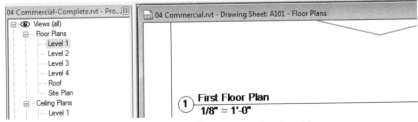

**Figure 4.47** *The title on the sheet can be different than the view Name*

Notice that the number in the view title tag has automatically filled in with the number 1. This is because this is the first (and currently only) view that we dragged to this sheet. Revit will automatically enumerate the annotation of all views as you drag drawings onto sheet views.

4. Save the project.

## EDIT THE CROP BOX

The *Level 1* plan (First Floor Plan) contains an underlay of the site plan (added above). While it might be desirable for some of the site to show on the First Floor Plan, we are not likely to want to show the entire site plan, but rather a small portion of just around the perimeter of the First Floor Plan. To do this, we need to enable the Crop Region of the *Level 1* view.

5. On the sheet, right-click on the viewport (showing the first floor plan) and choose **Activate View**.

This command allows us to edit the view directly in context with the sheet. This can be helpful when you want to see the borders of the sheet as you make changes to be certain that everything fits properly and as expected. When you deactivate the viewport, you can move it around to compose the layout of the Sheet.

On the View Control Bar, the crop icons we used above will become available.

• Click the Show Crop Region icon.

A rectangular Crop Region will now surround the plan. (You may need to zoom out a bit to see it).

• With the Modify tool (mouse pointer), click on the Crop Region boundary to select it.

Several drag handles will appear on each side. The triangular shaped ones at the midpoints can be used to edit the size of the Crop Region. We will drag these to make the Crop Region smaller for this view.

• Drag each of the drag controls in toward the plan until only a small portion of the site plan shows on all sides of the plan (see Figure 4.48).

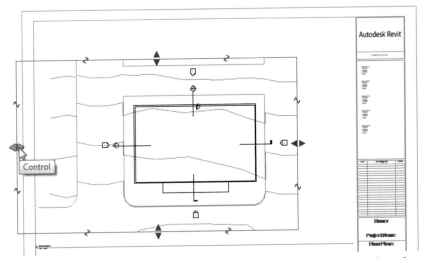

**Figure 4.48** *Drag the extents of the Crop Region to show only a small portion of the Site*

Make sure that you do not crop away the elevation tags; leave them visible.

- Right-click anywhere inside the viewport and choose **Deactivate View**.

The view title beneath the plan now needs adjustment. You can move it independently of the viewport by clicking on it with the Modify tool and dragging it to a new location. To move it with the viewport, select and move the viewport and the titlemark automatically moves with it. To resize the view title line, select the viewport first, then stretch the handles.

- Move and resize the view title as required.

- Move the viewport to the upper left corner of the Sheet (see Figure 4.49).

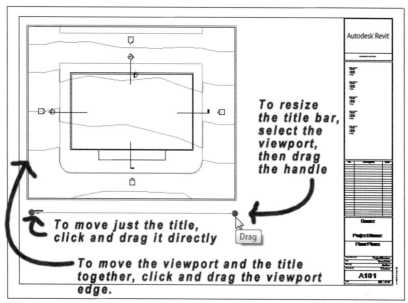

To resize
the title bar,
select the
viewport,
then drag
the handle

To move just the title,
click and drag it directly

To move the viewport and the title
together, click and drag the viewport
edge.

**Figure 4.49** *Finalize the size and position of the viewport and its view title*

Take note of the Sheet on the Project Browser. A small plus (+) sign will appear next to a sheet when there are views placed on it. You can click the plus sign to see each view listed beneath the sheet. You can double-click on views from here or the standard location in Project Browser to open them.

### CREATE THE REMAINING FLOOR PLAN SHEET FILES

Let's create the remaining floor plan sheets.

6. Repeating the process outlined above, create Sheet *A102 – Floor Plans*.

**Tip:** You can rename and renumber a sheet directly by selecting the titleblock and then clicking on the blue text labels.

- Create each of the additional Sheets listed in the "Drag to Sheet" column of Table 4.2.

7. On the Project Browser, right-click the *Level 2* floor plan and choose **Properties**.

- Change the "Title on Sheet" parameter to **Second Floor Plan**.

- Repeat this process for each view listed in Table 4.2.

- Drag each of the views to the appropriate Sheets as indicated in Table 4.2.

None of these views requires that the Crop Region be enabled. Simply edit the "Title on Sheet" parameter and then drag them to the appropriate sheet.

**Table 4.2** *Titles on Sheets for Plan Views*

| View Name | Title on Sheet | Drag to Sheet |
|---|---|---|
| Floor Plan Views | | |
| Level 2 | **Second Floor Plan** | A102 – Floor Plans |
| Level 3 | **Third Floor Plan** | A103 – Floor Plans |
| Level 4 | **Fourth Floor Plan** | A104 – Floor Plans |
| Roof | **Roof Plan** | A105 – Floor Plans |
| Site Plan | **Site Plan** | A100 – Site Plan |
| Ceiling Plan Views | | |
| Level 2 | **First Floor Reflected Ceiling Plan** | A106 – Reflected Ceiling Plans |
| Level 2 | **Second Floor Reflected Ceiling Plan** | A107 – Reflected Ceiling Plans |
| Level 3 | **Third Floor Reflected Ceiling Plan** | A108 – Reflected Ceiling Plans |
| Level 4 | **Fourth Floor Reflected Ceiling Plan** | A109 – Reflected Ceiling Plans |

Earlier when you added Levels to the project, if you ended up with a *Roof* ceiling plan, you can simply delete this view now. It will not be necessary.

8.  On the Project Browser, beneath *Ceiling Plans*, right-click the *Roof* view and choose **Delete**.

9.  Save the project.

## CREATE ELEVATION AND SECTION SHEET VIEWS

We will need a couple of Sheet views for our building elevations and building sections.

10. Repeating the process outlined above, create the sheet views indicated in Table 4.3.

- Repeating the process outlined above, edit the "Title on Sheet" parameter of each of the elevation and section views as indicated in Table 4.3.

- Drag each of the views to the appropriate Sheets as indicated in Table 4.3.

 **Note:** You are only creating three additional sheets here and that there are two views being placed on each of these sheets.

 **Tip:** If your elevation or section views are too wide for the sheet, you can use the crop region handles to resize them. The crop region is already on by default for section views. For elevation views, you must repeat the process outlined above for the *Level 1* view.

**Table 4.3**  *Titles on Sheets for Plan View*

| View Name | Title on Sheet | Drag to Sheet |
|---|---|---|
| Elevation Views | | |
| East | **East Building Elevation** | A202 – Building Elevations |
| North | **North Building Elevation** | A201 – Building Elevations |
| South | **South Building Elevation** | A201 – Building Elevations |
| West | **West Building Elevation** | A202 – Building Elevations |
| Section Views | | |
| Longitudinal | **Longitudinal Building Section** | A301 – Building Sections |
| Transverse | **Transverse Building Section** | A301 – Building Sections |

11. Create one more sheet numbered **A601** and titled **Schedules**.

- Drag each of the schedule views and drop them on this sheet.

If you zoom in on the schedules on the sheet, in some cases you will notice that the data in columns might wrap to a second line. You can leave it like this, or use the small triangular control handles to drag and resize the columns. The choice is yours.

When you have completed these Sheets, your Project Browser should resemble Figure 4.50. Again take note of the elevation and section tags in the various views. All of them now properly indicate the Sheet references and drawing numbers. All annotations were automatically enumerated in the order in which you dragged them to the Sheets.

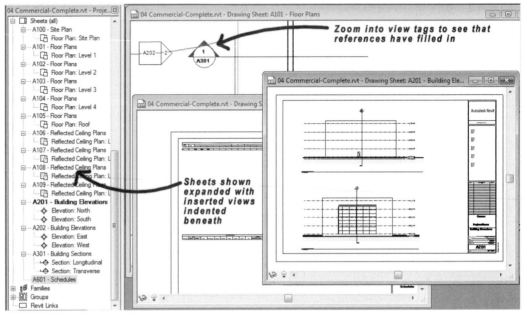

**Figure 4.50**  *Project Browser showing all sheets and the views they contain*

At this point, every sheet that we have created is likely still open. You can see which views and sheets are open on the Window menu. In a project this size, it is probably OK to have several windows open at once, however, as a matter of best practice, it is a good idea to periodically close unused windows. The easiest way to do this is with the **Close Hidden Windows** command on the Window menu. Views must be maximized first. If your screen shows view windows cascaded or tiled, please maximize before using the Close Hidden Windows command.

12. From the Window menu, choose **Close Hidden Windows**.

13. Save the project.

**BIM** *Manager Note:*   In some cases, when you increase the length of the title on the sheet parameter, the titlebar under the viewport will wrap the title onto two lines. If you prefer to have it not wrap, you can edit the View Title Family. To do this, expand *Families* on the Project Browser and then expand *Annotation Symbols*. Right-click on **View Title [M_View Title]** and then choose **Edit**. In the dialog that appears, click Yes. Click on the View Name label on screen and use the Drag handle on the right to make it wider. On the Design Bar, click the Load into Projects button. If a dialog appears showing multiple projects, check the box next to your *MRAC Commercial* project and then click OK. In the Reload Family dialog, place a checkmark in the "Override parameter values of existing types" box and then click Yes. Any sheets that previously wrapped the title to two lines should now be displaying it on a single line.

## CREATE A CUSTOM TITLEBLOCK FAMILY

The only Sheet we still need at this stage is a cover sheet. To create it, we will build a custom titleblock Family. Many times, firms have titleblocks already drawn in other CAD programs that they wish to import into Revit. You can do this if you wish. The exact steps will not be covered here, but you can create a new titleblock Family using the procedure covered here, and then import a CAD file into it using the procedures covered at the start of this chapter. The approach we will use here will be to create a very simple titleblock Family from scratch using Revit tools.

 **Note:** Note: If you prefer to skip this exercise, you can instead load the custom titleblock Family file provided with the files from the Mastering Revit Architecture CDROM. To do so, skip to the "Create a Cover Sheet" topic below.

14. From the File menu, choose **New > Titleblock**.

• In the "New" dialog, select the *E1 - 42 × 30.rft [A1 metric.rft]* template and then click Open.

A very simple set of tools will display on the Design Bar.

15. On the Design Bar, click the *Lines* tool.

• On the Options Bar, choose Wide Lines from the Type Selector, click the rectangle icon and type **-3/4" [-18]** in the Offset field.

• Draw a rectangle by snapping to the corners of the one already in the file.

The new rectangle will appear -3/4 [-18] smaller than the original one due to the offset.

16. Copy the bottom line up **6" [150]**.

17. Draw some smaller vertical lines within this space to separate it into a series of boxes (see Figure 4.51).

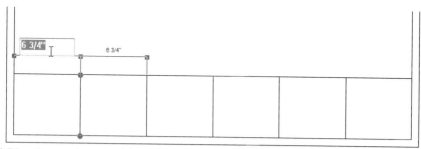

**Figure 4.51** *Draw a horizontal bar at the bottom and divide it into boxes*

18. On the Design Bar, click the **Text** tool.

• On the Options Bar, click the Properties button.

• In the Element Properties dialog, click the Edit / New button and then click the Duplicate button.

• Name the new Text Type: **1/2" Arial [12mm Arial]** and then click OK.

- For the Text Size type: **1/2" [12]**.

- Select the Bold checkbox and then click OK twice.

19. In the first box that you drew above, click a point in the upper left corner to place the text.

- Type the following:

**Consultant Name**

**Address 1**

**Address 2**

**Phone**

20. Copy this block of text to a few of the other boxes along the bottom of the titleblock.

- Leave a couple of boxes empty for a company logo and an Architect's seal.

- You can add a new text label to one of them that simply says: **Seal**.

You can either leave the logo box empty for now, or you can use the **File > Import / Link > Image** command to import your company logo. The file must be saved in JPG, JPEG, BMP, or PNG image file formats. Once you place the image, use the handle at the corner to resize it.

21. On the Design Bar, click the *Label* tool.

- Use the same procedure as above to create a new Type called **1" Arial Titles [25mm Arial Titles]**. Make the text size **1" [25]** and Bold.

- Click OK twice after creating the Type and then choose the Center icon on the Options Bar before placing the label.

22. Click a point near the top middle of the titleblock.

23. In the "Select Parameter" dialog, choose Project Name and then click OK.

- Drag the handles to make the title wider.

- Create another text Type with smaller text height and add additional centered labels beneath the title such as Project Address and Issue Date (see Figure 4.52).

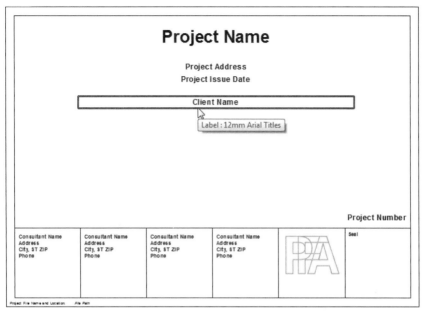

**Figure 4.52** *Draw a horizontal bar at the bottom and divide it into boxes*

Feel free to make any other embellishments you wish. You can draw additional lines, add more labels and text. Use labels when you wish to link the text to project data. When this titleblock is used in the project, the Project Name and date will automatically fill in. The consultants on the other hand will need to be edited manually in the titleblock later. While it is possible to make the consultant items link to labels as well, the process to do so is beyond the scope of this short tutorial. Consult the online help for more information on creating titleblocks if you wish to learn more.

24. Save the titleblock Family to the *Chapter04* folder.

- Name it: **MRAC EI 30 × 42 Cover Sheet.rfa** [**MRAC AI Cover Sheet.rfa**].

25. On the Design Bar, click the Load into Projects button.

## CREATE A COVER SHEET

Now that you have built your own custom titleblock Family for the cover sheet, we can use it in the project and create the cover sheet.

26. On the Project Browser, right-click *Sheets* and choose **New Sheet**.

- In the "Select a Titleblock" dialog, choose your new cover sheet titleblock and then click OK.

**Note:** If you decided to skip the titleblock Family creation exercise above, click the Load button, browse to the *Chapter04* folder with the other files for this chapter. Select the *MRAC EI 30 × 42 Cover Sheet.rfa* [*MRAC AI Cover Sheet.rfa*] file and then click Open.

- Right-click on the new sheet (currently *A602 – Unnamed*) and choose **Rename**.

- Re-number it to **G100** and rename it to **Cover Sheet**.

Why don't we add a 3D axonometric of the project to our Cover Sheet? We can use the default *{3D}* view, but it might be better to create a copy of it first.

27. On the Project Browser, expand *3D Views*, right-click the *{3D}* view and choose **Duplicate View > Duplicate**.

- Rename the view **Cover Axonometric**.

- Use the Dynamically Modify view tools to adjust the view to a pleasing vantage point.

 **Tip:** If you have a wheel mouse, hold down the SHIFT key and drag with the wheel to quickly spin the model.

28. Drag the *Cover Axonometric* view onto the Cover Sheet and position it.

When you drop the view on the Cover Sheet, it is probably too large for the Sheet. We can easily adjust the scale of a view after it has been placed.

29. On the Project Browser, right-click the *Cover Axonometric* view and choose **Properties**.

- In the Element Properties dialog, from the view Scale list choose **1/16"=1'-0"** [1:200].

We also do not want a title under this viewport on the cover. To do this, we need to duplicate the viewport Type.

30. Click the Edit / New button and then click the Duplicate button.

- Name the new Type: **Viewport w/o Title** and then click OK.

- For Show Title, choose **No** and then click OK twice.

 **Note:** If you are using the Metric files, you can use the *No Title* Type already in the project instead of creating a new one.

- Reposition the view as necessary.

 **Tip:** If the scale is still too large for the Sheet, you can choose **Custom** from the view Scale list and type in a value. In the Metric files, try a scale 1:250 for example.

## EDIT PROJECT INFORMATION

When we built the titleblock Family, we noted that the labels would automatically fill in with the appropriate project data. You may have noticed that most fields have not yet changed. The information that these labels reference is global to the entire project and is stored in a common System Family named "Project Information." We can edit its values at any time and the new values will immediately populate *all* Sheets in the project. Let's take a look.

31. On the Settings menu, choose **Project Information**.

- In the Element Properties dialog, input values for the various fields (see Figure 4.53).

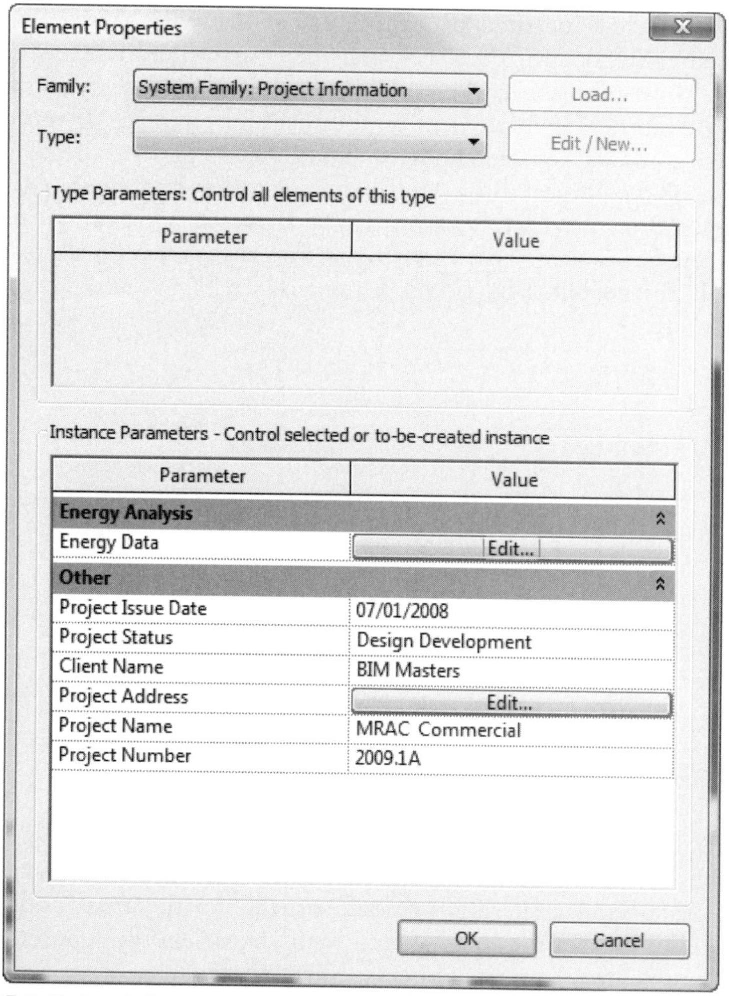

**Figure 4.53** *Edit Project Information*

32. Open any Sheet and view the Titleblock fields.

Notice the change to the values. You will also notice that some of the values such as Drawn By did not change. These are sheet specific parameters that must be edited on a sheet by sheet basis as is appropriate for this type of information. To make the edits, right-click a sheet and choose Properties, or refer to the "Create a Drawing List" topic below.

33. Save the project.

## PROJECT AND SHEET BROWSER ORGANIZATION

As the quantity of sheets in your projects increases, you may find it useful to explore other ways to organize your Project Browser. First, as noted above, you can open a view directly from the sheet entry in the browser by expanding the plus sign next to the sheet and then double-clicking the view indented beneath. You can see what this looks like above in Figure 4.50. You can also open the sheet, and then right-click the viewport and choose **Activate View** as we did above. This allows you to work directly in the view while maintaining the context of its placement on the sheet for reference. If you take advantage of either of these methods, you can choose to hide all views that are already on sheets from the *Views* node of the Project Browser. To do this, select the *Views (All)* node of the Browser, and then from the Type Selector, choose **Browser – Views: not on sheets**. This will filter the view list to show only those views that have not yet been placed on sheets. In our case, only the default *{3D}* view will remain (see Figure 4.54).

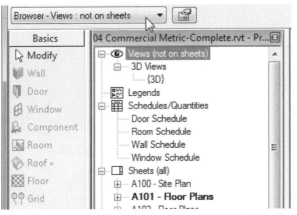

**Figure 4.54** *Change the Browser to display only the views not already on sheets*

There are other Browser organization types as well. One sorts by discipline, another by phase, and so on. The not on sheets type gives the most dramatic difference at the current stage of our project. Similar filtering types are available for the *Sheets* node. To see them, click on *Sheets (All)* in the Browser tree, and then choose another option from the Type Selector. For example, try **Browser – Sheets: Sheet Prefix**. This organization will give us two groupings, one for the "A" sheets and another for the "G" sheets. Often architects have sheet numbering schemes where the first two or three characters are codes for drawing types. With the numbering suggested above, the first character is the discipline code and the number immediately following it is a drawing type code where 1 equals plans, 2 equals elevations, 3 equals sections and so on. In this case, it might be useful to filter the list by the first two characters rather than just the first.

34. Select *Sheets (All)* from the Browser.

- From the Type Selector, choose **Browser – Sheets: Sheet Prefix**.

- With *Sheets (Sheet Prefix)* selected, click the Properties icon.

- Click the Folders button.

- Change the number of leading characters to **2** (see Figure 4.55).

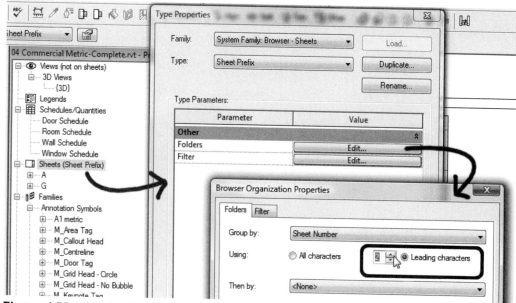

**Figure 4.55** *Edit the Sheet Prefix organization to group by 2 characters*

- Click OK twice.

Explore the results in the Project Browser. It should look like Figure 4.56.

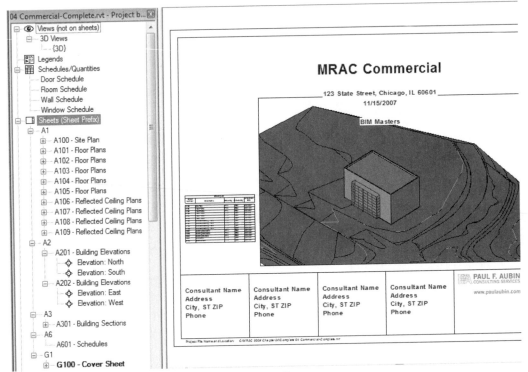

**Figure 4.56** *View the results of the 2 character prefix*

## CREATE A DRAWING LIST

Now that we have all of our sheets, we can compile a list of all the sheets. This list will resemble the other schedules that we already have and it can even be placed on a sheet as a sheet index.

35. From the View menu choose **New > Drawing List**.

- In the Drawing List Properties dialog, select Sheet Number and then click the Add -> button.

- Add additional fields like Sheet Name, Designed by, Drawn by, Checked by and Sheet Issue Date.

- Click the Sorting/Grouping tab.

- For Sort by, choose Sheet Number and then click OK (see Figure 4.57).

Earlier it was noted that the drawn by, checked by, and other similar fields had to be edited individually on each sheet. One of the benefits of the Drawing List schedule is that you can edit those values directly in the table.

| Drawing List | | | | | |
|---|---|---|---|---|---|
| Sheet Numb | Sheet Name | Designed By | Drawn By | Checked By | Sheet Issue Date |
| A100 | Site Plan | Designer | Author | PFA | 08/18/07 |
| A101 | Floor Plans | Designer | Author | Checker | 08/14/07 |
| A102 | Floor Plans | Designer | Author | Checker | 08/18/07 |
| A103 | Floor Plans | Designer | Author | Checker | 08/18/07 |
| A104 | Floor Plans | Designer | Author | Checker | 08/18/07 |
| A105 | Floor Plans | Designer | Author | Checker | 08/18/07 |
| A106 | Reflected Ceiling | Designer | Author | Checker | 08/18/07 |
| A107 | Reflected Ceiling | Designer | Author | Checker | 08/18/07 |
| A108 | Reflected Ceiling | Designer | Author | Checker | 08/18/07 |
| A109 | Reflected Ceiling | Designer | Author | Checker | 08/18/07 |
| A201 | Building Elevation | Designer | Author | Checker | 08/18/07 |
| A202 | Building Elevation | Designer | Author | Checker | 08/18/07 |
| A301 | Building Sections | Designer | Author | Checker | 08/18/07 |
| A601 | Schedules | Designer | Author | Checker | 08/18/07 |
| G100 | Cover Sheet | Designer | Author | Checker | 08/18/07 |

**Figure 4.57** *Create a Drawing List and optionally edit it directly*

You can use Sheet Issue Date column to help you plan the set's issuance over the life of the project. Like other schedules, if you wish you can drag this view onto one of your sheets such as the cover sheet. To do this, simply open the Cover Sheet, and then drag the drawing list and drop it on the sheet.

36. Save the project.

## DRAFTING VIEWS

One other type of view bears mention in this discussion: the Drafting view. A Drafting view is unlike other views in that it is not linked to anything in the model. The purpose of a Drafting view is to create details, diagrams, or sketches that are not easily modeled or that you and your project team have deemed not *worth* modeling. Remember that Building Information Modeling is as much about information as it is about graphics. In some cases, a simple diagram or some sketched embellishment on a detail is all that is required to convey design intent. In these cases, a Drafting view is often appropriate. Drafting views provide a blank sheet of paper suitable for importing legacy CAD details, image files or natively drawn sketches.

While we are not creating any Drafting views in this chapter, an entire chapter is devoted to the subject of detailing later in this book. We will have the opportunity to add Drafting views to our project at that time.

## PRINTING A DIGITAL CARTOON SET

Going though the upfront process of creating sheet views gives you a few benefits. First, this task can be handled by a single individual early in the project without the pressure of a looming deadline. If one person sets up all the necessary sheet, there is a greater chance for consistency from sheet to sheet in terms of naming and drawing placement. Following this process also gives you a digital cartoon set. Just like the traditional cartoon set, the digital version will help make good decisions about project documentation requirements and the impact on budget and personnel considerations. One extra advantage of the digital cartoon set over the traditional one is that it is the real building model and document set! Don't be concerned with the finality that this seems to imply. The model remains completely flexible and editable (as we have seen).

## PUBLISH A 2D DWF

As the final task in this chapter, go ahead and plot out the project sheets. You could print to a physical plotter, but to save paper, let's create a digital plot instead using the Publish to DWF function.

1. From the File menu, choose **Publish DWF > 2D DWF**.

   - In the "Publish dialog, beneath "Range," choose "Selected views/sheets."

2. Click the Select button.

   - In the "View/Sheet Set" dialog, clear the Views checkbox at the bottom.

   - Click the Check All button and then click OK.

   - In the message that appears, click No.

3. In the File name box, type: **Cartoon Set** and then click Save.

Revit will begin processing your request. A multi-sheet DWF file will be generated that you can open in Autodesk Design Review. This is a free application that you can download from the Autodesk web site. It allows you to view and reline DWF files that you share with your consultants and clients. DWF files are very small and read-only while maintaining full vector-based quality of the original model. In this way, you can share your designs with your extended team without fear of file size limitations or unauthorized editing.

After the publish is complete, browse to the folder where you saved it and double-click the file. Autodesk Design Review will open the file and you will see thumbnails of each sheet on the left side. You can view sheets, add comments, relines, and save them for later coordination back in Revit. If you CTRL click on the section and elevation tags and level heads, you will be taken to the corresponding view instantly. You can email the DWF file to a recipient who does not need to have Revit to open, view and print the DWF. All they need is the free Design Review application.

**BIM Manager Note:** Now that we have thoroughly explored Templates and views and have created a project that contains several views, we might be looking for ways to save time on the next project we begin. You can take any Revit Architecture project and save it as a template. This is very easy to do and allows you to reuse your work on similar projects in the future. To save the current project as a template, simply choose Save As from the File menu. In the Save As dialog, browse to a location to save the template file. The best location is a central location on the network server where all users have ready access. Give the new template file a name, change the "Save as type" to Template Files (*.rte) and then click Save. The template file will be saved to the location you specified and be ready for use.

## SUMMARY

- Template files provide a means to start new projects with consistent content and setup.

- All new projects should be created from an agreed upon office template file.

- Levels establish horizontal datums measured vertically in the Z direction for use as floor levels and other reference points.

- Walls heights can be constrained to Levels and will thus change height automatically if a Level changes.

- Site data can be imported from popular CAD programs and used to generate toposurface elements in Revit Architecture.

- Views are used to study, create, and manipulate model data.

- Views can be graphical like plans, sections and elevations or non-graphical like schedules and drawing lists.

- Edits made in one view are immediately seen in all appropriate views.

- Sheets are used to create drawing sets for printing.

- You can create your own custom titleblock to use in your Revit projects.

- Generate a Sheet List to plan your set or place on your cover page.

- Before printing to a paper plotter consider generating a digital plot instead—DWF files are small, high quality "digital plots" that you can use to share design data among the extended project team.

- Anyone with a free copy of Autodesk Design Review software can open, view, reline, and print a DWF file.

# Column Grids and Structural Layout

## INTRODUCTION

In this chapter, we will explore the layout of basic structural components. We will begin with the layout of the column grid lines for the commercial project started in the last chapter. We will also add columns and framing members to this project and begin to explore strategies related to separating and sharing data among the various disciplines involved in a project team. Finally, we will revisit the residential project to create a foundation plan.

## OBJECTIVES

In this chapter, we will begin by adding column Grid lines and Columns. The goal will be to understand the features and behaviors of Grids in Revit. We will then add Beams and Joists to complete the framing. After completing this chapter you will know how to:

- Add and modify column Grids
- Add and modify Architectural and Structural Columns
- Load Steel Shape Families
- Create Structural Framing elements
- Copy elements to Levels
- Create foundation Wall footings

## WORKING WITH GRIDS

In Chapter 1, Figure 1.1 we explored the hierarchy of Revit Architecture elements. In the annotation objects branch of the hierarchy, we find a category of datum elements that include Levels, Grids, and Reference Planes. In the last chapter, we worked with Levels. These are horizontal planes (expressed graphically as lines in views that show them) that typically define floor levels or stories of the building but that can also be used to define other horizontal datums such as "top of footing" or "bottom of steel." Grids in are very similar to Levels in almost every way except that they define vertical planes through the building. The most common use of Grids is to define the locations of structural columns in the building. Grids are typically added in two directions and their intersections are used to label unique locations for each column. Like Levels and section lines, Grids exhibit intelligent annotation characteristics and will appear automatically in all appropriate views such as plans, elevations, and sections at the proper scale and location. The third type of datum, the Reference Plane, also shares many characteristics of Levels and Grids, but is more generic in nature and does not have any annotation automatically associated with it. Reference Planes will be explored in later chapters.

## INSTALL THE CD FILES AND OPEN A PROJECT

The lessons that follow require the dataset included on the Mastering Revit Architecture CD ROM. If you have already installed all of the files from the CD, simply skip down to step 3 below to open the project. If you need to install the CD files, start at step 1.

1. If you have not already done so, install the dataset files located on the Mastering Revit Architecture CD ROM.

   Refer to "Files Included on the CD ROM" in the Preface for instructions on installing the dataset files included on the CD.

2. Launch Revit Architecture from the icon on your desktop or from the *Autodesk > Revit Architecture* group in *All Programs* on the Windows Start menu.

**Tip:** In Windows Vista, you can click the Start button and then begin typing **Revit** in the "Start Search" field. After a couple letters, Revit Architecture should appear near the top of the list. Click it to launch to program.

- If the New Features Workshop dialog appears, choose "Maybe later" and then click OK.

3. On the Standard toolbar, click the Open icon.

**Tip:** The keyboard shortcut for Open is CTRL + O. **Open** is also located on the File menu.

- In the "Open" dialog box, click the *My Documents* icon on the left side.
- Double-click on the *MRAC* folder, and then the *Chapter05* folder.

   If you installed the dataset files to a different location than the one listed here, use the "Look in" drop down list to browse to that location instead.

4. Double-click *05 Commercial.rvt* if you wish to work in Imperial units. Double-click *05 Commercial Metric.rvt* if you wish to work in Metric units

   You can also select it and then click the Open button.

The project will open with the last saved view visible on screen.

**Note:** The dataset for this chapter uses a simpler site plan than the one created in the previous chapter. In the next chapter, the site is returned to the state it was in at the end of the last chapter. The difference in site configurations will have no bearing on your education experience in this chapter.

## ADD GRID LINES

Let's begin laying out a column grid in this project. A Grid is a vertical datum element used mostly to represent column locations in the building. Grids help with locating objects in plans and elevations.

5. On the Project Browser, double-click to open the *Level 1* floor plan.

This view shows the first floor plan complete with an underlay of the site plan contour map. The contour map was imported in the previous chapter from a DWG file. For now, we will hide these contours. There are a few ways to accomplish this; here we will use the Temporary Hide/Isolate feature.

6. In the view window, click to select the contour map.

   • On the View Control Bar, click the Temporary Hide/Isolate pop-up icon and choose **Hide Element** (see Figure 5.1).

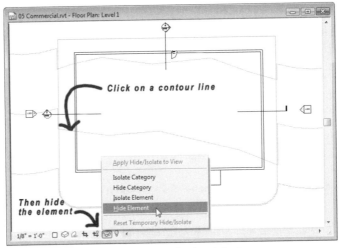

**Figure 5.1** *Temporarily hide the site contours in the Level 1 view*

The Hide/Isolate controls give you a quick way to temporarily hide objects on screen from display. When you choose the hide commands, the selected elements in the model are temporarily hidden in the current view. When you choose the isolate options, the selected element(s) remain visible and all other elements in the view are hidden. The Temporary Hide/Isolate commands provide an on-screen effect in the current work session only. In other words, they do not hide and isolate the elements when printed or exported. When a project is closed and reopened, all hidden or isolated objects are reset to their original state.

7. On the Design Bar, click the Basics tab and then click the Grid tool.

 **Tip:** The keyboard shortcut for Grid is **GR**. **Grid** is also located on the Drafting menu.

Like many elements we have already seen, some common tools appear on the Options Bar. In particular, there are two modes: Draw and Pick Lines. Draw is the default and allows you to click any two points to specify the extent of the grid line. Grid lines can be drawn at any angle: they can even be curved. You can also specify the grid lines based upon existing elements in the model. To do this, use the Pick Lines mode.

- Click any two points within the building footprint to add a Grid line.
- On the Design Bar, click the Modify tool or press the ESC key twice.
- Click on the Grid line just created.

Notice that many of the same handles and controls appear on the Grid line that appeared on the Level lines in the last chapter (see Figure 4.14 in the previous chapter). Place your pointer over each control to see a screen tip indicating its function (see Figure 5.2). Zoom in as required.

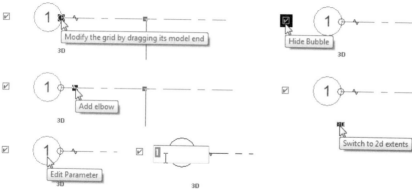

**Figure 5.2**  *Grids have many control handles and drag points*

Moving left to right in Figure 5.2, the following briefly describes each control.

- **Modify the grid by dragging its model end**—This round handle is used to drag the extent of the Grid. If the length or alignment constraint parameter is also active, dragging one Grid will affect the extent of the other constrained Grid lines as well.

- **Hide/Show Bubble**—Use this control to hide or show the Grid bubble at either end of the Grid line.

- **Add Elbow**—This small handle creates an elbow in the Grid line. This is useful when the annotation of two Grid lines overlap one another in a particular view. Click this handle to create the elbow, and then drag the resultant drag handles to your liking to make the annotation legible.

- **2D/3D Extents Toggle**—When 3D Extents are enabled, editing the extent of the Grid line in one view affects all views in which that Grid appears and is also set to 3D Extents. If you toggle this to 2D Extents, dragging the extent of the Grid line affects only the current view.

- **Edit Parameter**—Click the blue text to re-number the Grid bubble.

- **Length and Alignment Constraint** (not shown in the figure)—A padlock icon used to constrain the length and extents of one Grid line to the others nearby. This is useful to keep all of your Grid lines lined up with one another.

8. Undo the Grid line.

 **Note:** You can also delete the Grid, but the automatic numbering of the next Grid you add will start at number 2 since it is the next number in the sequence. If you undo the Grid instead, it returns the starting number to 1.

Sometimes it is easier to use existing geometry to assist in the creation of the Grid lines. This can speed up the creation process and ensure that design relationships are maintained.

9. On the Design Bar, click the Grid tool.

- On the Options Bar, click the Pick Lines icon.

- In the Offset field, type **4"** [**100**] and then press ENTER.

- Position the Modify tool over the inside edge of the bottom horizontal Wall.

- When the green dashed line appears on the inside of the building, click to add the Grid line (see Figure 5.3).

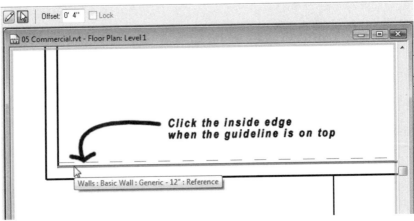

**Figure 5.3** *Create a Grid line based on a picked Wall edge*

Depending on whether you deleted the first Grid line or used Undo, the number parameter of this Grid line may be "1" or it may be "2." If it is 2, we can edit it.

- If necessary, click on the Grid bubble text parameter (blue text) and then change the value to 1.

- Make sure the Grid bubble shows on the left. If it is on the right, use the Show/Hide Bubble controls on each side to switch it.

You should now have Grid line 1 offset slightly to the inside edge of the bottom horizontal Wall with a bubble on the left. The extent of the Grid line matches the length of the Wall since we used the Wall to create it. Let's stretch it longer.

10. Click on Grid line 1 and using the model end handles (see above), stretch each end longer away from the outside edges of the model (see Figure 5.4).

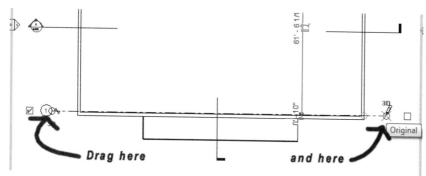

**Figure 5.4** *Lengthen the Grid lines by dragging the model end handles*

It is a good idea to get the extent of the first Grid line to your liking before you continue, as the additional Grid lines that you add will snap to this length automatically. However, new Grid lines will automatically constrain to their neighbors, so if you do not stretch it longer first, you can stretch any Grid line later, and they will all lengthen together.

11. On the Design Bar, click the Grid tool.

- On the Options Bar, verify that the Draw icon is selected.

  If the 4" [100] is still active, change it to **0**.

- Move the pointer to the right side, just above the previous Grid line.

A guideline will appear aligned with the previous Grid line and a temporary dimension will also appear.

- When the temporary dimension reads approximately 20'-0" [6000] click to set the first point—make sure it is still aligned.

- Move the pointer to the left and when the alignment guidelines appear, click to place the new bubble aligned to the first (see Figure 5.5).

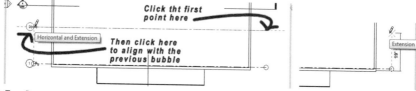

**Figure 5.5** *Draw a second Grid line aligned to the first one*

Notice the closed padlock icon that appears connected to the new Grid line. This indicates that the new Grid line is constrained to the previous one. Also take a careful look at the temporary dimensions. The witness lines are actually attached to the Walls, *not* the previous Grid line. We can adjust this and then edit the value to set the desired spacing between the two Grid lines.

12. With the Grid line command still active, move the pointer over the lower "Move Witness Line" handle (see the left panel in Figure 5.6).

- Drag the Witness Line up until it highlights Grid line 1 and then release (see the middle panel in Figure 5.6).

- Edit the temporary dimension value to **20'-0"** [**6000**] (see the right panel in Figure 5.6).

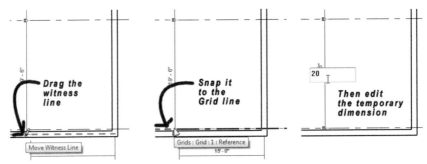

**Figure 5.6** *Move the Witness line and then edit the value of the temporary dimension*

At first it might be a bit tricky to do this operation with the command still active. If you prefer, you can press ESC twice to cancel out, select Grid line 2 with the Modify tool, and then perform the steps above. The result will be the same. The motion should be fluid. Move the pointer over the Witness Line handle, click and drag it to Grid line 1 and release. Grid line 1 will highlight as you drag it. When it highlights, you should release the mouse. At this point, we could continue adding Grid lines using the same process. Instead, let's test the alignment constraint, and then explore alternate techniques to create the additional Grid lines.

13. On the Design Bar, click the Modify tool or press the ESC key twice.

14. Select Grid line 2.

- Using the "Model end" drag handle, drag the bubble horizontally a bit.

Notice that both Grid bubble 1 and 2 move together.

- Position the bubbles where you want them.

15. Save the project.

## COPY AND ARRAY GRID LINES

We could draw each additional Grid line as noted above. We can also copy the existing ones using either the Copy tool or the Array tool. The Copy tool is basically a Move command that moves a copy instead of the original. The Array command can create several equally spaced copies in one action. It also has the option to "group and associate" the copies so that the parameters assigned to the array are maintained and remain editable. Let's explore a few options.

16. Select Grid line 2.

• On the Edit toolbar click the Copy tool.

**Tip:** The keyboard shortcut for Copy is **CO**. **Copy** is also located on the Edit menu (Note: choose **Copy**, not **Copy to Clipboard**.)

On the Options Bar, there are three checkboxes. The "Copy" box is already selected.

• On the Options Bar, place checkmarks in the "Constrain" and "Multiple" checkboxes.

The "Constrain" checkbox limits the movement of the copy in the vertical or horizontal direction and the "Multiple" option makes more than one copy in the same operation.

• For the start point, click at the end of the Grid line (snap to the endpoint).

• Move the pointer straight up.

• Using the temporary dimensions, place two copies above Grid line 2 spaced at **20'-0"** **[6000]** (see Figure 5.7).

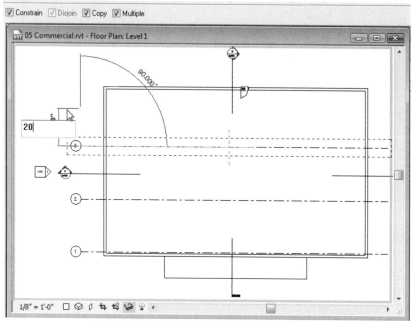

**Figure 5.7** *Make two copies of the Grid line above Grid line 2*

• On the Design Bar, click the Modify tool or press the ESC key twice.

Notice that the two new Grid lines numbered automatically as "3" and "4." Revit Architecture will always number the new Grid lines in sequence after the last one placed. If you were to delete one Grid line 4, and then add another Grid line, it would automatically number to "5." However, if you were to undo the placement of Grid line 4, then the next one would be "4" instead.

17. Click on either of the new Grid lines.

Notice that again, the temporary dimensions reference the Walls and not the other Grid lines. If you wish to move them with the temporary dimensions, you can first move the Witness Lines as we did above. You can also use the Move tool if you prefer. Let's move both Grid lines up slightly to make the middle bay larger.

18. Select Grid lines 3 and 4. (You can use the CTRL key or a Crossing selection to do this—be certain to select only the Grid lines.)

- On the Options Bar, click the Activate Dimensions button.

- Use the technique outlined above to move the Witness Line of the lower vertical dimension up to Grid line 2 (see the left side of Figure 5.8).

   When you have moved the Witness Line, the temporary dimension should read 20'-0" [6,000] (see the middle of Figure 5.8).

- Edit the value of the lower temporary dimension to **21'-2" [6,480]** (see the right side of Figure 5.8).

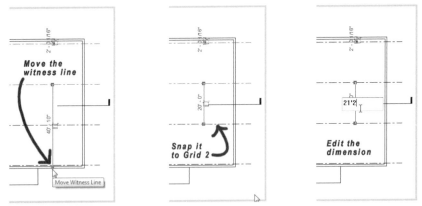

**Figure 5.8**  *Move Grid lines 3 and 4 using the temporary dimensions*

19. On the Design Bar, click the Grid tool.

- On the Options Bar, click the Pick Lines icon.

- In the Offset field, type **4" [100]** and then press ENTER.

- Position the pointer so that the green dashed line appears at the inside edge of the left vertical Wall and then click to add the Grid line.

• On the Design Bar, click the Modify tool or press the ESC key twice.

Notice that this Grid line became number 5. For the vertical bays, we want to use letters.

20. Click on the new vertical Grid line.

• Adjust its length so that it projects above and below the building footprint (like we did above) and then edit the bubble number to **A**.

21. Select Grid line A and then on the Edit toolbar click the Array tool.

 **Tip:** The keyboard shortcut for Array is **AR. Array** is also located on the Edit menu.

The Array command has many options. First, you can choose between a linear or radial array on the Options Bar. Next there is the "Group and Associate" checkbox. With this option enabled, all of the arrayed items (the original and the copies) will be grouped together. When thus grouped, you can select them later and edit the quantity of items in the array. The remaining items will re-space to match the new quantity either adding or deleting elements in the group. This can be very powerful and useful functionality that has many applications in building design. This being our first opportunity to explore this tool, we will test it out here with our Grid lines. Ultimately, we will want an ungrouped array because the final spacing of our column grid is not equal (as is typical in most buildings), however it will still be educational to explore briefly the "Group and Associate" option of array nonetheless.

• On the Options Bar, verify that the "Linear" icon is chosen and that "Group and Associate" is selected.

Array is much like the move and copy commands in that it requires you to pick two points on screen to indicate the spacing of the array. We have two options for what these two points can represent in the array: they can be the spacing between the each element, or the total spacing of the entire array. To set the spacing between each element, choose the "2nd" option. To set the total spacing, choose the "Last" option.

 **Tip:** In this case, the only bad thing about choosing the "Last" option is that the right-most Grid line will become Grid line B and then the four in the middle lettered sequentially left to right from C to F. You would then need to edit each Grid bubble parameter to fix this. Use the "2nd" option to avoid this problem.

• Choose the "2nd" option.

• For the "Number," type **6** and then place a checkmark in the "Constrain" box.

The number indicates the total quantity of elements in the final array. This quantity includes the original selection. The "Constrain" option works like the same option in the Copy command (see above) by limiting movement to horizontal and vertical only.

• Watch the Status Bar for prompts. Click the first point on the endpoint of the selected Grid line.

- Move the pointer to the right about 20'-0" [6000] and then click the second point (see Figure 5.9).

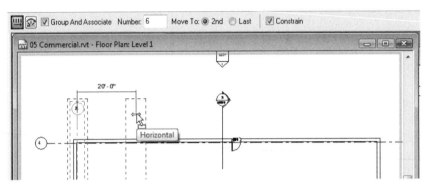

**Figure 5.9** *Array a total of 6 Grid lines horizontally*

The last element of the array is close to the inside edge of the right Wall. Using the temporary dimensions like we did above, we can adjust it to the exact location we need. Since the array is grouped, all items will adjust to maintain an equal spacing as we do this.

22. Click the last item (Grid line F on the right).

- On the Options Bar, click the Activate Dimensions button.

  The last Grid line (labeled F) needs to be 4" [100] from the inside face of the right Wall (like its counterpart on the left).

- Move Witness Lines and edit the temporary dimension value to **4" [100]**.

 **Tip:** Instead of dragging the Witness Line handle, click it. Each time you click it, it will shift to a new reference point in the Wall. It will cycle from inside to center to outside and then back again.

 **Note:** The Wall is 12" [300] thick. So you can place the Witness Lines in any convenient location and take this value into account to achieve the correct location.

Notice that all the Grid lines between A and F adjust with this change and remain equally spaced.

- Deselect Grid line F and then re-select it (click on it again).

A temporary dimension will appear on the arrayed items with a single text parameter that indicates the number of items in the array (see Figure 5.10).

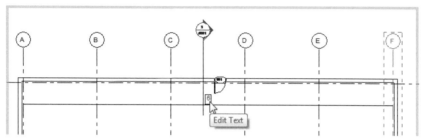

**Figure 5.10** *Edit the quantity of items in the array with the Edit Text parameter*

- Click to select this text parameter.
- Type **7** and then press ENTER.

Notice that a new Grid line is added and lettered automatically.

- Repeat the process typing **5** this time.

Notice that two of the Grid lines are deleted.

- Repeat the process one more time setting the value back to **6**. (You can also Undo twice if you prefer.)

This is the benefit of using the "Group and Associate" parameter when you create the array. However, as we mentioned above, the spacing of our commercial building's column grid is not actually equal. Therefore, we will actually ungroup this array. Before we do, feel free to experiment further with the array. For example, you can select any one of the Grid lines and move it. All of the others will move accordingly. This can be tricky, as the one you select actually stays stationary and all the others move relative to it. You can also choose the Edit Group button on the Options Bar with one of the elements selected. This will make an Edit Group mode active and a toolbar will appear. In this mode, you can manipulate the elements within the group. When you choose the Finish button, the change is applied to all group instances (all of the elements in the array in this case). We will explore Groups in more detail in Chapter 6. Be sure to undo any explorations you made returning to 6 Grid lines before continuing.

## DIMENSION THE GRID LINES

Let's ungroup the array and then adjust the Grid line spacing independently.

23. Select all vertical Grid lines—A through F (use the CTRL key or a crossing selection).

- On the Options Bar, click the Ungroup button.

The Grid lines will now move independently of one another. Try a few moves if you like, but be sure to undo before continuing. We are going to adjust several Grid lines now. This will be easier with some actual dimensions rather than the temporary dimensions.

24. On the Design Bar, click the Dimension tool.

- Starting with Grid line A, click successively on each lettered Grid line A–F (see Figure 5.11).

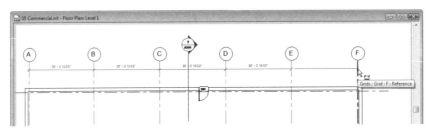

**Figure 5.11** *Add a dimension string to the lettered Grid lines*

- After you click F, move the mouse to a spot where you want to place the dimension string and then click to set the dimension string and end the command.

 **Tip:** Make sure that you click in an empty spot to place the dimension, if you click on another model element, it will add a Witness Line instead.

25. On the Design Bar, click the Modify tool or press the ESC key twice.

26. Select Grid line B

Notice that now that we have added a string of dimensions, in addition to the normal temporary dimensions we also can edit the dimension values of the string we just added.

 **Note:** If the temporary dimensions do not appear on the Options Bar, you can always click the "Activate Dimensions" button.

- Click on the blue dimension text between Grid line A and B, type **22'-2"** [**6740**] and then press ENTER.

Grid line B will move to the right slightly.

- Select Grid line C next and edit the distance between it and B to **14'-4"** [**4400**].

This time, Grid C will move. Remember, always click the element that you wish to move before editing dimension text.

- Repeat moving left to right for the remaining vertical Grid lines as shown in Figure 5.12.

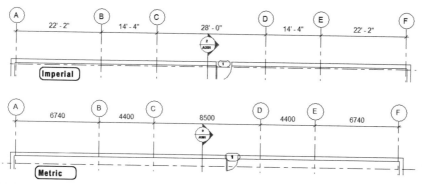

**Figure 5.12** *Edit dimension values to move Grid lines*

Our feature façade on the front of the building (we created a Mass for this in the last chapter) will require some structural support. We can add a few additional Grid lines in the front for these.

27. On the Design Bar, click the Grid tool.

- On the Options Bar, verify that the Draw icon is selected.

- Move the pointer to the inside of the building above the lower Wall and click between Grid lines C and D.

- Move down past the building and click again to place the bubble.

The bubble will automatically enumerate as "G."

- Click in the text parameter of the new bubble and change the value to **C.3**.

- Using the temporary dimensions, move the Grid line so that it is **9'-0"** [**2700**] from Grid line C (see Figure 5.13).

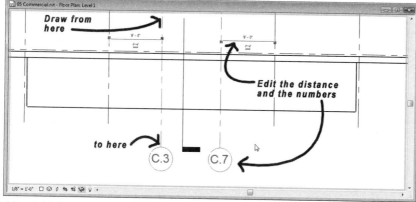

**Figure 5.13** *Add Grid lines at the bottom of the plan for support of the front façade*

- Add Grid line C.7 to the other side of the C-D bay (see Figure 5.13).

28. Add a vertical dimension string to the numbered bubble on the left.

29. Save the project.

## VIEWING GRID LINES IN OTHER VIEWS

All of the work we have done on Grid lines so far has been in the *Level 1* floor plan view. However, like other datums, grids will automatically appear in all orthographic views.

30. On the Project Browser, double-click to open the *West* elevation view.

Notice how the numbered Grid bubbles appear here running vertically on the elevation.

31. On the Project Browser, double-click to open the *South* elevation view.

This time only the lettered bubbles appear. If you find the bubbles in the middle to be too close together, you can use the drag handles to give them an elbow.

32. Click the "Add elbow" handle to make an elbow in Grid lines C.3 and C.7. Fine tune the shape as appropriate with the round handles (see Figure 5.14).

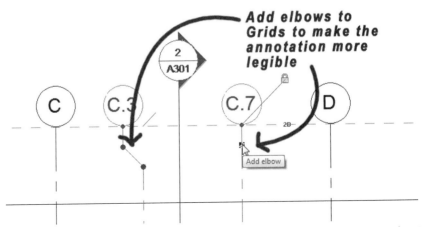

**Figure 5.14** *Create elbows in the Grid lines as needed to keep the bubbles from overlapping one another*

33. On the Project Browser, double-click to open the *Longitudinal* section view.

In this view, Grid lines C.3 and C.7 do not show since the section is looking toward the north. If the section were looking south, they would appear.

34. On the Project Browser, double-click to open the *Level 1* floor plan view.

- On the Window menu, choose **Close Hidden Windows**.

- Save the project.

## WORKING WITH COLUMNS

A collection of structural tools are included in Revit Architecture. A larger collection of structural tools appears in the Revit Structure application. Users of Revit Architecture can seamlessly share models back and forth between structural consultants using Revit Structure. If you are not collaborating with a Structural Engineer who uses Revit Structure, or if it is early in the project and you have not yet selected an engineer or received designs from the consultant, you can use the structural tools provided with Revit Architecture to begin the layout of Columns and if desired, Beams and Braces as well. Like all Revit elements, structural elements remain parametric and editable after they have been added to the model. Therefore, when the structural analysis comes back from the engineer, you can swap in the correct types and sizes to meet the requirements of the design.

### Disclaimer:

No structural analysis of any kind has been performed on the designs in this book. The shapes used in this book serve illustration purposes and are not to be construed as a recommendation of structural integrity or a design solution.

### ADD COLUMNS

Revit Architecture includes two types of columns: Architectural Columns (or just Columns) and Structural Columns. An Architectural Column provides the location and finished dimensions (including column wrap, furring, finishes, etc.) of a column in an architectural space. The Structural Column is the actual structural material responsible for supporting building loads without any enclosure or finish. Examples include steel or concrete columns. Let's take a look at both kinds.

Continue in the *Level 1* view opened above.

1. On the Design Bar, click the Modeling tab and then click the Column tool.

    This is the "Architectural" Column tool.

    - From the Type Selector, choose Rectangular Column : 24" × 24" [M_Rectangular Column : 610 × 610mm].

    - Move the pointer to the intersection of Grid lines 4 and A.

    The pointer will snap to the intersection automatically (see the left side of Figure 5.15).

 **Note:** Remember at the start of the chapter in the "Add Grid Lines" heading, we temporarily hide the site contours. If you closed Revit between lessons, the temporary hide mode resets itself. You can simply repeat the process above to re-hide the contours if necessary.

    - Click at the intersection to place the Column (see the right side of Figure 5.15).

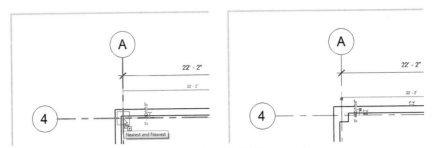

**Figure 5.15** *Add an Architectural Column to the A4 Grid intersection*

Notice that the Architectural Column automatically interacts with the Walls at the intersection and graphically displays as an integrated column.

    2. Continue placing Columns at all of the intersections around the perimeter of the building (see Figure 5.16).

> **Tip:** You can simply click at each intersection or you can use the Copy tool. If you use Copy, be sure to select "Multiple" on the Option Bar, and use the TAB key to select a good start point for the copy.

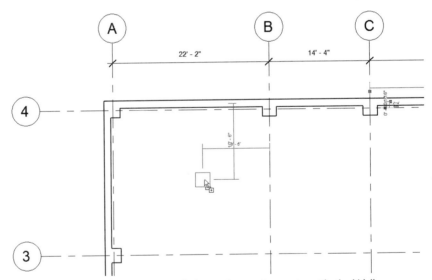

**Figure 5.16** *Architectural Columns around the perimeter integrate with the Walls*

    • On the Design Bar, click the Modify tool or press the ESC key twice.

Let's try some Structural Columns next.

    3. On the Design Bar, click the Structural tab and then click the Structural Column tool.

**Note:** If you do not see the Structural tab of the Design Bar, right-click on the Design Bar to display it.

- Open the Type Selector.

Notice that there are only two Structural Column types loaded in the current project (there is only one in the metric project). Structural Columns, like other Families in Revit Architecture can be loaded as needed from external libraries.

4. On the Options Bar, click the Load button.

If you are working in Imperial units, the *Imperial Library* folder should open automatically. If you are using Metric units, the *Metric Library* folder should open automatically. If the appropriate location fails to open automatically, browse there now. The easiest way to do this is to click the appropriate shortcut button on the left side of the dialog.

- Double-click the *Structural* folder, then double-click the *Columns* folder and finally double-click the *Steel* folder.

- Double-click the *W-Wide Flange-Column.rfa [M_W-Wide Flange-Column.rfa]* Family file.

The "Specify Types" dialog will appear on screen. This shows a list of industry standard steel-shape sizes. Scroll through the list to see all of the sizes. You can select one size or multiple items using the SHIFT and CTRL keys.

- Scroll in the list, locate W12x87 [W310x97] and select it (see Figure 5.17).

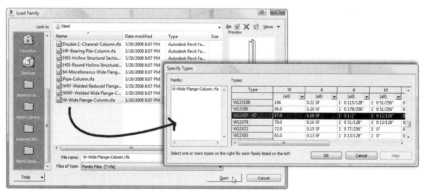

**Figure 5.17** *Common Industry Steel Shapes are available to load from the Family file*

- Click OK to load the Family.

  If a "Reload Family" dialog appears, simply click Yes.

- From the Type Selector, choose the newly loaded W12x87 [W310x97] type.

Structural Columns have many of the same parameters as Architectural Columns—and more. If you move the pointer around the screen before you click, you will note that like the Architectural Column, the Structural Column will pre-highlight Grid lines and Walls as possible hosts. (However if you insert a Structural Column at a Wall, it will not merge with the Wall the same way that the Architectural one did.)

In addition to behaviors shared with Architectural Columns, there are several additional options on the Options Bar. For example, two "Place By" options appear. The "Grid Intersection" button allows you to place several Structural Columns at once on a selection of Grid lines and the "Architectural Column" button which allows you to place a Structural Column at the location of a selection of Architectural Columns. Let's give these a try.

5.  On the Options Bar, click the "Grid Intersection" button.

6.  Click the pointer above Grid line 3 and to the right of Grid line E and then drag to the left of Grid line B and below Grid line 2 (see Figure 5.18).

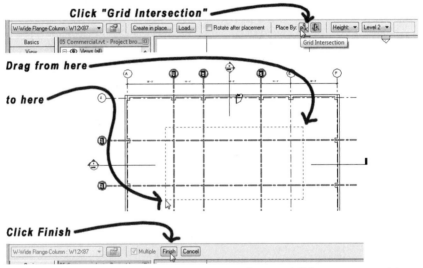

**Figure 5.18**  *Use the Grid Intersection option to create Structural Columns at several intersections*

The Grid lines will highlight in red and several ghosted columns will appear. If you are satisfied with this selection, you click the Finish button on the Options Bar. Otherwise, you simply select again until you are satisfied. If you decide not to add the Columns, you can click the Cancel button on the Options Bar.

•  On the Options Bar, click the Finish button.

You will now have a steel column at each Grid intersection. Let's try the "Architectural Column" option next. In addition, we will also adjust the height of the column as we add it.

7.  On the Options Bar, from the button that currently reads "Level 2," choose **Roof** from the pop-up.

This will create a single continuous Structural Column from Level 1 (the current Level) up to the Roof.

•  On the Options Bar, click the Architectural Column button.

- Using a crossing or inside selection (click and drag a selection window—either left to right or right to left will work in this case) surround the entire building to select all Architectural Columns.

- On the Options Bar, click Finish.

- On the Design Bar, click the Modify tool or press the ESC key twice.

## EDIT COLUMNS

We now have Structural Columns at all Grid intersections including the ones at the perimeter Walls. However, the ones in the center (not surrounded by Architectural Columns) span from Level 1 to Level 2, while the ones at the perimeter span the complete height of the building. We cannot see this in the current plan view. Let's take a look at the section.

8. On the Project Browser, double-click to open the *Transverse* section view.

When you switch to this view, you will likely not notice much detail with regard to the Structural Columns. This is because elements in Revit Architecture display at various levels of detail. This allows for simplified graphics in smaller scale drawings, and more detail in larger scale drawings. You can easily modify the default settings for this behavior, or simply adjust the level of detail display for an individual view. For example, Figure 5.19 shows the variation in the three detail levels when viewing Structural elements such as the Columns we have in our model. As you can see, the Coarse detail level shows simple linework for the steel shape in plan and a single line for it in section (and elevation). This is the currently active display detail level in both the floor plan and section views in our project. If you switch to Medium or Fine detail level, the graphics for the steel shape will get progressively more detailed in plan. In elevation and section, display is more detailed than Coarse, but the same graphics display for both higher levels of detail (see Figure 5.19).

 **Note:** Lineweight display adjusts automatically when you vary the scale of the view. Note the change in scale in each panel of the figure and experiment on your own screen to make the medium and high displays more legible.

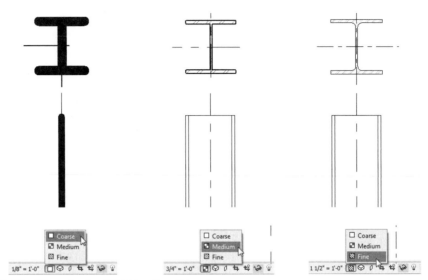

**Figure 5.19** *Three levels of Display Detail Level*

9. On the View Control Bar, change the Detail Level to **Medium**.

The Columns will now display with the correct dimensional thickness in the section rather than the diagrammatic single-line display. Zoom around the section as needed to see clearly that the Architectural Columns stop at Level 2 while the Structural ones go all the way to the Roof. Naturally it is likely that these Columns would actually be constructed in sections and not be a single continuous four-story tall Column. However, for design purposes, they function in the building as a single continuous column. Therefore, it is up to you decide in your own projects what the "Height" of the Column represents and build it accordingly. You can make the decision to create the Columns the actual height of the material that will be shipped to the site (two-stories tall for instance); this would be important if you use Revit Architecture to assist you in deriving construction quantities or when we send the model to our Structural Engineer for structural analysis and design. If you wish to create shorter Columns, simply choose a different Level to end the Columns at, or even choose **Unconnected** and type in an explicit dimensional height. Then open the plan view for the next Level and copy/paste aligned the Columns there. In this project, we will leave this decision to the Structural Engineers. For now, we will make our Structural Columns span the full height of the building and leave our Architectural Columns (which again represent the finish materials wrapping the Columns) spanning a single Level. Later we will copy these Architectural Columns to the other floors.

10. Zoom in on Grid line 2 or 3 at Level 1.

   • Move your pointer over the Columns and note the screen tips that appear (see Figure 5.20).

**Tip:** If you have trouble highlighting the overlapping elements, use the TAB key to cycle through adjacent elements.

**Figure 5.20** *In section view, we see columns in the middle of the building at different heights than those beyond*

In our section view, several of the Columns are in the same place. Some of these are the shorter ones in the center of the building and the tall ones appear beyond at the perimeter of the building. We can still select the ones we need to edit in this view, or if you prefer, we can return to the Level 1 plan view and select there. It makes no difference in which view you select them. If you select in section, you can easily avoid the full height Columns with a crossing or inside selection, however, it will be difficult to avoid the Architectural Columns in behind them. Furthermore, some of the Columns would not be selected as they are cropped away in the section view. However, in this case we can make use of another tool to assist in this selection.

11. Select one of the steel Columns (any one, tall or short).

- Right-click and choose **Select All Instances**.

Be careful with this command. It literally selects all instances of a Family Type in the entire project—both seen and unseen. In this case, that is exactly what we want, but in many cases, it may not be. Keep this in mind before using this tool in your work. For example, if you choose to select all instances of a door while working in a plan view, realize that you are also selecting similar doors on other floors of the building!

- On the Options Bar, click the Properties icon.

- In the Element Properties dialog, for Base Level, choose **Street Level**.

- For Top Level, choose **Roof** and then click OK (see Figure 5.21).

| Instance Parameters - Control selected or to-be-created instance | |
| --- | --- |
| Parameter | Value |
| **Constraints** | |
| Base Level | Street Level |
| Base Offset | 0' 0" |
| Top Level | Roof |
| Top Offset | 0' 0" |
| Moves With Grids | ☑ |

**Figure 5.21** *Edit the Base and Top Level parameters of the selected Columns*

Notice the height of all of the steel Columns now spans from below the first floor to the roof.

12. On the View toolbar, click the Default 3D View icon.

- Click on the Roof and then on the View Control Bar, from the Hide/Isolate pop-up, choose **Hide Element**.

You will notice that all steel Structural Columns were impacted by this change, not just the ones we could see in the section view. This is because as noted above, Select All Instances selects across the entire model, not just the elements visible in the current view.

## COPY AND PASTE ALIGNED

Let's copy our Architectural Columns and add them to other floors.

13. On the Project Browser, double-click to open the *Level 3* floor plan view.

Notice that only the Structural Columns show in this view. The Structural Columns show because they span multiple levels and therefore intersect this view's cut plane. The Architectural Columns do not show because they occur on Level 1 and are a single story in height.

14. On the Project Browser, double-click to open the *Level 1* floor plan view.

We could use the same selection method that we just used to select all of the Architectural Columns, but let's explore another technique this time.

15. Make an inside selection (from left to right) surrounding the entire building (don't include the Grid lines). (See item 1 in Figure 5.22.)

This will select all visible elements in the building model.

- On the Options Bar, click the Filter Selection icon (see item 2 in Figure 5.22).

- In the Filter dialog, click the "Check None" button.

- Place a checkmark in the Columns checkbox and then click OK (see item 3 in Figure 5.22).

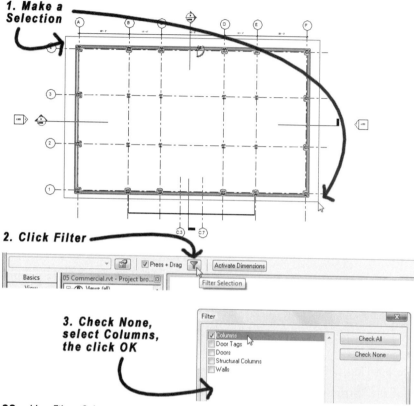

**Figure 5.22** *Use Filter Selection to modify a selection and remove unwanted items*

Only the Architectural Columns should now be highlighted in red. (Note that Architectural Columns are referred to in the Filter dialog simply as "Columns."

16. From the Edit menu, choose **Copy to Clipboard** or press CTRL + C.

• From the Edit menu, choose **Paste Aligned > Select Levels by Name**.

• In the "Select Levels" dialog, use the CTRL key and select Levels 2, 3, and 4 and then click OK.

17. On the Project Browser, double-click to open the *Level 3* floor plan view.

Notice that all of the Columns, both Structural and Architectural show in this view. The Architectural Columns are the copies that we just added. Check other views if you wish.

18. Save the project.

## ADDING CORE WALLS

Now that we have a column grid with associated Columns, it is a good time to sketch in the building core. To do this, we will use Structural Walls. Structural Walls are essentially the same as other Walls except that the "Structural Usage" parameter for them is automatically set to **Bearing**. We add Structural Walls in the same fashion as other Walls.

## ADD STRUCTURAL WALLS

1. On the Project Browser, double-click to open the *Level 1* floor plan view.

   - Zoom in on the two bays between column Grids C and E at the top of the plan (between Grid lines 3 and 4).

   **Tip:** Right-click and choose Zoom In Region or type **ZR. Zoom In Region** is also located on the View > Zoom menu.

2. On the Design Bar, click the Structural tab and then click the Structural Wall tool.

   - From the Type Selector, choose Basic Wall : Generic - 8" Masonry [Basic Wall : Generic - 225mm Masonry].

   - On the Options Bar, click the Properties icon.

   - In the Element Properties dialog, set the "Base Constraint" to **Street Level**.

   - Set the "Top Constraint" to **Up to level: Roof**.

These two settings instruct our Structural Wall to go the full height of the building from the lowest level to the highest. We will also need the core Walls to continue past the Roof to allow stair access to the roof. So let's also add a "Top Offset" parameter. Keep in mind that all of these parameters can be edited later as the project progresses and as design needs dictate.

   - In the "Top Offset" parameter field, type **12'-0" [3600]** (see Figure 5.23).

Take note of the "Structural Usage" parameter before you close the dialog—note that it is set to "Bearing" automatically.

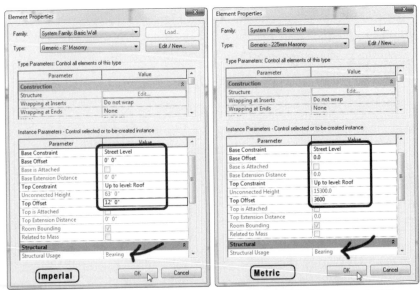

**Figure 5.23** *Configure the parameters for Structural Wall the full height of the building*

- Click OK to continue.

- On the Options Bar, click the Pick Lines icon.

- Make sure the Loc Line set to **Wall Centerline** and in the Offset field, type **1'-2"** **[350]** and then press ENTER.

With the Pick Lines option enabled, we will be able to select existing geometry from which to create the Walls. The Walls will be created at a distance of 1'-2" [350] from the selected geometry.

3. Highlight Grid line C and when the green dashed offset line appears to the right of it, click the mouse to create the Wall.

- Repeat for Grid line 3 (horizontal) when the offset line appears below the Grid line (see Figure 5.24).

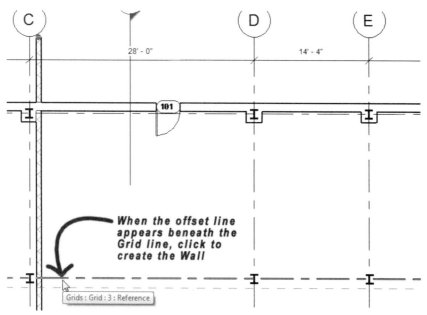

**Figure 5.24** *Create Walls from Grid lines using the Pick Lines option with an offset value*

- Change the Offset value on the Options Bar to **2'-8"** **[800]** and then create one more Wall to the right of Grid line E.

4. Using the Trim/Extend tool on the Tools toolbar, trim the three Structural Walls to appear as in Figure 5.25.

Use the Trim/Extend to Corner option for the two bottom intersections (remember to select the side of the Wall that you want to keep). Use the Trim/Extend Multiple Elements option for the upper two intersections to make them butt into the inside edge of the exterior horizontal Wall (remember to click the side of the Wall you want to keep). For more information on the Trim/Extend tool, refer to the tutorials in Chapter 3 or search the online Help.

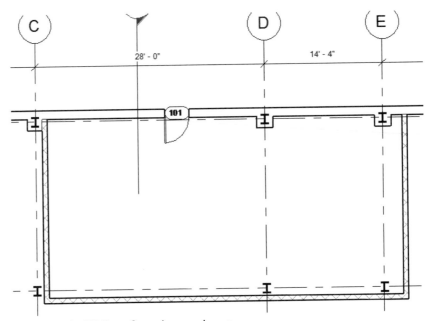

**Figure 5.25** *Trim the Walls to form the core layout*

- On the Design Bar, click the Modify tool or press the ESC key twice.

## CONVERT A WALL TO A STRUCTURAL WALL

Let's view the model in section to see what we have so far.

5. Double-click the blue Section Head between Grid lines C and D.

The *Transverse* building section view will open.

6. Click on the Crop Box to select it. Click and drag the triangular control handle at the top edge and drag it up enough to see the top of the core Walls (see Figure 5.26).

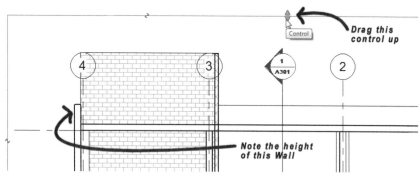

**Figure 5.26**  *Drag the Crop Box large enough to see the top of the new Walls*

Notice that the exterior Wall at the core only projects to the parapet height. In reality, this Wall would also need to be part of the core. Let's split this existing Wall and then change its parameters to match the rest of the core.

7. On the Project Browser, double-click to open the *Level 1* floor plan view.

8. On the Tools toolbar, click the Split tool.

 **Tip:** The keyboard shortcut for Split is **SL. Split Walls and Lines** is also located on the Tools menu.

Move the split pointer near the intersection between the vertical core Wall and the horizontal exterior Wall. (Be sure that "Delete Inner Segment" is not selected.)

• When the pointer snaps to the intersection, click the mouse to split the Wall.

• Repeat on the other intersection (see the top portion of Figure 5.27).

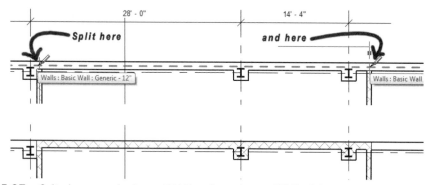

**Figure 5.27**  *Split the upper horizontal Wall to form the top Wall of the core*

• On the Design Bar, click the Modify tool or press the ESC key twice.

9. Select the middle horizontal Wall segment (created by the splits) and on the Options Bar, click the Properties icon.

- From the Type list, choose Generic - 12" Masonry [Generic - 300mm Masonry].
- Change the "Base Constraint" to **Street Level** and the "Base Offset" to **0** (zero).
- Change the "Top Constraint" to **Up to level: Roof** and the "Top Offset" to **12'-0"** [**3600**].
- Beneath the Structural grouping, change the "Structural Usage" parameter to **Bearing** and then click OK (see the bottom portion of Figure 5.27).

10. Double-click the blue Section Head between Grid lines C and D again.

Graphically, the Wall should now appear as the other core Walls do (the thickness varies).

 **Tip:** If you are unhappy with the way the Walls clean up at the corners, you can disjoin them using the drag handles at the ends. Drag the handles away from the intersection to disconnect the Walls from one another. Then use the Trim/Extend tool and the trim to a corner option to re-join the Walls.

11. Save the project.

## ADDING FLOORS

The most obvious omission that you might notice in our current sections is the lack of any floor slabs. Adding basic floor slabs is easy to accomplish with the Slab tool.

 **Note:** The Slab tool on the Structural tab of the Design Bar is functionally the same as the Floor tool on the Basics tab.

### CREATE A SLAB FROM WALLS

The easiest way to create a Floor slab and locate its boundaries is by using the existing Walls that bound it.

1. On the Project Browser, double-click to open the *Level 1* floor plan view.

2. On the Design Bar, click the Structural tab and then click the Slab tool.

   The Design Bar changes to sketch mode. Pick Walls should be active.

- If Pick Walls is not active, select it on the Design Bar now.
- Click on each of the exterior vertical Walls, the bottom horizontal one, and one of the top horizontal ones (see Figure 5.28).

 **Note:** Be sure to click on the inside edges. If you accidentally create sketch lines to the outside edge, use the Flip control at the middle of the sketch line to flip the sketch to the inside.

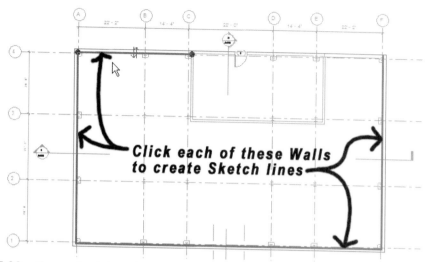

**Figure 5.28**    *Create sketch lines from Walls (Sketch lines enhanced for clarity)*

- Use the Trim/Extend tool to join the two open sketch lines.

    **Note:** The Trim/Extend tool will remember the last mode you used. Be sure to set it to Trim/Extend to Corner to complete this step.

- On the Design Bar, click the Floor Properties button.
- From the Type list, choose Generic - 12" [Generic 300mm] and then click OK.
- On the Design Bar, click the Finish Sketch button.
- In the dialog that appears, click Yes (see Figure 5.29).

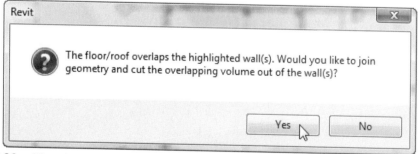

**Figure 5.29**    *Allow the new Slab to Join with the Walls*

You can always edit the automatic joins later, but in general, it is a good idea to allow Revit Architecture to apply joins when prompted. This connects the various building elements in logical ways and does so more efficiently than we can do manually.

## COPY FLOOR SLABS TO LEVELS

    3. Double-click the blue Section Head between Grid lines C and D again.

You can see the new Floor slab at Level 1.

    4. Select the new Floor slab, and then press CTRL + C.

    **Note:** If you prefer, choose **Copy to Clipboard** from the Edit menu.

- From the Edit menu, choose **Paste Aligned > Select Levels by Name**.
- In the Select Levels dialog, select Level 2, Level 3, and Level 4 and then click OK.

    **Tip:** Use the CTRL key to select more than one Level, or simply drag through the names.

## CREATE A SHAFT

You should now have a Floor slab at each Level and we already had a Roof at the top. However, these new floor slabs now interrupt our building core. We will need a shaft for the elevators and stairs that will come later. We will add a shaft now using approximate dimensions and then later, in Chapter 7, we will add the vertical circulation elements and fine tune the size of the shaft to suit the design as required.

    5. On the Project Browser, double-click to open the *Level 1* floor plan view.

    6. On the Design Bar, click the Modeling tab and then click the Opening tool.

- From the flyout that appears, choose **Shaft Opening** (see Figure 5.30).

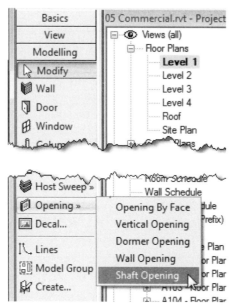

**Figure 5.30**  *Create a Shaft Opening from the Modeling tab*

The Design Bar changes to sketch mode. Lines should be the active tool. If it is not, click it now.

- On the Options Bar, click the Rectangle icon.

- Snap to the lower left outside endpoint of the core Walls.

- For the other corner, snap to the inside midpoint of the top exterior Wall (see Figure 5.31).

 **Note:** If you have difficulty snapping to the Midpoint, snap nearest to the location indicated in the figure instead.

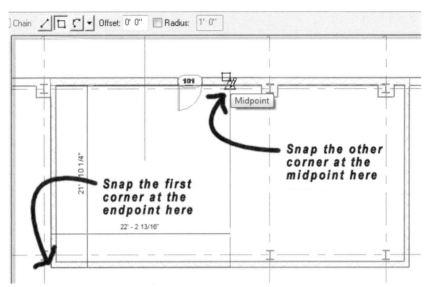

**Figure 5.31** *Sketch a rectangular Shaft Opening*

- On the Design Bar, click the Properties button.

- For the "Base Constraint" choose **Level I** and for the "Top Constraint" choose **Up to level: Roof** and then click OK.

- On the View toolbar, click the Default 3D View icon.

- Using the Spin controls (or hold down the shift key and drag with your wheel button) spin the model around so you can see down into the core.

- On the Design Bar, click Finish Sketch (see Figure 5.32).

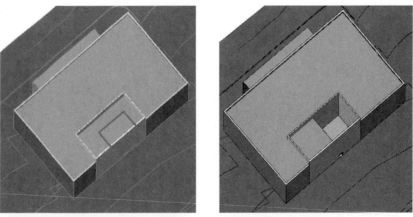

**Figure 5.32** *The completed Shaft cuts through all of the Floors*

7. On the Project Browser, double-click to open the *Level I* floor plan view.

8. On the Design Bar, click the Basics tab and then click the Section tool.

- Create a section (from left to right) through the top portion of the core.

- Click anywhere to deselect the section, and then double-click the new Section Head to open the view (see Figure 5.33).

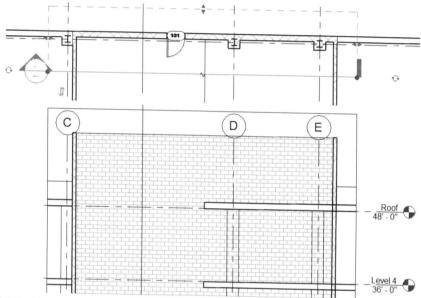

**Figure 5.33**  *View the Shaft in a new Section view*

Study the way that the shaft has cut the slabs in this and other views. When you are satisfied with the way that the Shaft has been created, save the project.

## CREATING STRUCTURAL FRAMING

So far we have explored Columns, Grids, and Slabs. In addition to these tools, Revit Architecture also includes tools to create Beams and Braces. If your primary job function is architectural, you may not be responsible for adding these elements to the model. This task might fall on the structural engineers on your team. However, like the Columns, even though they will ultimately be sized by the Structural Engineer, you can add the basic components to your model at any appropriate point in the design process and later re-size and reconfigure them as appropriate based upon your structural engineer's design and analysis. This is accomplished by simply selecting the elements and editing the Type applied in the properties dialog. In this section, we will take a brief look at the tools available. Feel free to explore beyond what is covered here.

### WORKING WITH BEAMS

You can add Beams to your model by sketching, or you can create them automatically based upon a column grid. In this exercise, we will use our grid to create Beams.

1. On the Project Browser, double-click to open the *Level 2* floor plan view.

2. On the Design Bar, click the Structural tab and then click the Beam tool.

Like the Columns above, we will first load a Family to use for the Beam shape.

- On the Options Bar, click the Load button.

- Double-click the *Structural* folder, then double-click the *Framing* folder and finally double-click the *Steel* folder.

- Double-click the *W-Wide Flange.rfa* [*M_W-Wide Flange.rfa*] Family file.

A list of industry standard steel-shape sizes will appear. Scroll through the list to see all of the sizes (similar to Figure 5.17 above).

- Scroll in the "Specify Types" dialog. Select W18x40 [W460x52] and then click OK.

   If a "Reload Family" dialog appears, click Yes to accept the reload.

- On the Options Bar, click the Grid button.

- Dragging from right to left, select all Grid lines.

- On the Options Bar, click the Finish button.

- On the Design Bar, click the Modify tool or press the ESC key twice.

After a short pause, it will appear as though the command is finished, but nothing will appear to have been created. This is because the Detail Level of our plan view is set to Coarse.

3. On the View Control Bar, change the Detail Level to **Medium**.

Beams will appear. They have been created at each major Grid line between Columns (see Figure 5.34).

**Figure 5.34** *Switch to Medium Detail Level to see the Beams in plan*

4. On the Project Browser, double-click to open the *Longitudinal* section view.

- On the View Control Bar, change the Detail Level to **Medium**.

- Zoom in on Level 2 and view one of the Beams.

Notice that the Beams appear at the same height as the top of the slab.

5. Select one of the Beams.

- Right-click and choose **Select All Instances**.

- On the Options Bar, click the Properties icon.

- Beneath the Constraints grouping, set the Start Level Offset and the End Level Offset parameters to **-4"** [**-100**] (negative values) and then click OK.

Adding this negative offset moves the Beams below the level enough to allow the topping slab to cover them. (Right now, we have a generic slab, but in later chapters, we can change the composition of the slab to show more detail. When we do, the Beams will already be at the correct elevation.)

There are other Beam creation methods and options. You can sketch Beams and create them specifically as Girders, Joists, Purlins, etc. When you use the Grid option as we have here, Revit determines the usage automatically. In this case we have Girders. You can view the Structural topic in the online help for more details on this subject.

Feel free to experiment further with the various Beam options before continuing. For example, you can use a Beam System tool. With this tool, you can use the girders to create sketch lines, or you can sketch manually. The shape that you sketch will be filled in with an automatically generated array of Beams. Try loading a bar joist Beam Family from the same folder as the girder above. Be sure to click the Structural Beam System Properties button on the Design Bar to edit the layout rules and spacing parameters and assign the bar joist to the system.

## WORKING WITH BRACES

To complete our preliminary structural layout of our commercial project, let's add a few cross braces at the building core.

6. On the Project Browser, double-click to open the *Level 1* floor plan view.

7. On the Design Bar, click the Structural tab and then click the Brace tool.

- On the Options Bar, click the Load button.

- Double-click the *Structural* folder, then double-click the *Framing* folder and finally double-click the *Steel* folder.

- Double-click the *L-Angle.rfa* [*M_L-Angle.rfa*] Family file.

Once again, a list of industry standard steel-shape sizes will appear. Scroll through the list to see all of the sizes (similar to Figure 5.17 above).

- Scroll in the list, select L6×6×3/8 [L152×152×9.5] and then click OK.

  If a "Reload Family" dialog appears, click Yes to accept the reload.

8. Verify that L6×6×3/8 [L152×152×9.5] is chosen from the Type Selector.

- On the Options Bar, from the "Start" list, choose **Level 1** from the drop down list and then type **2'-0"** [**600**] in the offset field next to it.

- From the "End" list, choose **Level 2** and type **-2'-0"** [**-600**] in the offset field.

- Snap the start point to the midpoint of the Column at Grid intersection 3C.

- Snap the end point to the midpoint of the Column at Grid intersection 4C.

- Repeat to create another Brace from Grid intersection 3D to 3E.

Feel free to add additional Braces or Beams as desired.

- On the Design Bar, click the Modify tool or press the ESC key twice.

As with the Beams, Braces will not appear in Coarse Detail Level. You can change this in the *Level 1* plan if you wish. To view either Brace in elevation, you can create additional Section views as we did above (see Figure 5.35).

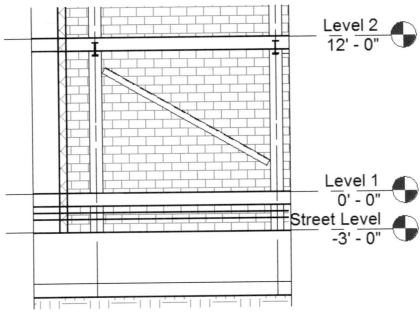

**Figure 5.35** *Create a new Section view to see the Braces*

## CREATE A NEW MODEL GROUP

The same Framing layout will occur on all four floors of the project. We could copy it to each Level as we did with the other elements, but let's take advantage of Groups to make it easier to edit the framing later. Anytime that you have a repetitive (or "typical") portion of your building design—such as a typical Stair, Toilet Room or Framing layout as in this case—you can use Groups to manage them. We saw Groups briefly at the start of the chapter with the Array command. In that instance, we actually ungrouped the array. In this case, we will leave the items grouped. The process is simple. Select the elements, Group them, and then insert instances in the model. Whenever you need to make a change, edit the Group. When you are finished editing, the change will apply to all instances of the Group across the entire project.

9. On the View toolbar, click the Default 3D View icon.

- Dragging from right to left, select the entire model.

- On the Options Bar, click the Filter icon.

- In the "Filter" dialog, click the Check None button.

- Place a checkmark only in the Structural Framing checkboxes (there should be two such items) and then click OK.

You may not be able to see the selected elements. It depends on the angle of your 3D view. It is not important to see the elements. Choosing "Structural Framing" selected all of the Beams and Braces. If you do wish to see the elements, use the Temporary Hide/Isolate icon and the Isolate Element option.

10. On the Tools toolbar, click the Group icon.

- In the "Create Model Group" dialog that appears, type **Typical Framing** and then click OK (see Figure 5.36).

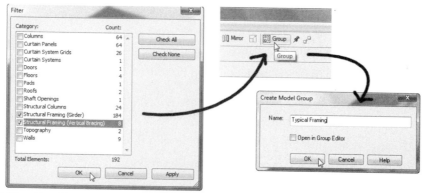

**Figure 5.36**   *Use Filter to help create and name a new Model Group*

11. On the Project Browser, double-click to open the *Level 1* floor plan view.

12. On the Project Browser, expand *Groups* (near the bottom of the list), then expand the *Model* branch.

Notice that "Typical Framing" is listed in the Model category.

- Right click on *Typical Framing* and choose **Select All Instances**.

Currently there is only one instance of this Group in the model, so this just selects that one instance. We could have used any other selection method here as well. Notice the small "L" shaped icon with grip handles and "X" and "Y" labels that appear at the center of the Group. This indicates the insertion point of the Group. Using the handle at the intersection of the two "L" shapes, we can relocate the insertion point. The other round handles allow you to rotate the axes. The "X" and "Y" labels allow you to flip the Group about either the X or Y-axis.

- Click and drag the insertion point handle (at the intersection of the "L") and snap it to the intersection of Grid lines A and 1.

13. On the Project Browser, double-click to open the *Level 2* plan.

14. Right-click the *Typical Framing* Group on the Project Browser and choose **Create Instance**.

   - Snap the new Group instance to the intersection of Grid lines A and 1.

   - Repeat for Level 3 and Level 4.

   - Open any section view to see the result.

## EDIT A GROUP

Now that we have a Group of our typical framing, if the design should change, we can edit the Group and the change will propagate across the project to all instances of the Group.

15. On the Project Browser, double-click to open the *Longitudinal* section view.

   - Zoom in on Level 1 and view one of the Beams.

   - Click on the Beam to select it.

This will select the entire Group. (A dashed box appears around a Group when you select it.)

   - On the Options Bar, click the Edit Group button.

     The Group edit toolbar appears.

   - Select the Beam again.

This time, since you are now in the Group edit mode, the Beam will highlight.

16. From the Type Selector, choose a different Beam Type.

   - On the Model Group: Typical Framing toolbar, click the Finish button.

Notice the change to the other instances of the Group (see Figure 5.37).

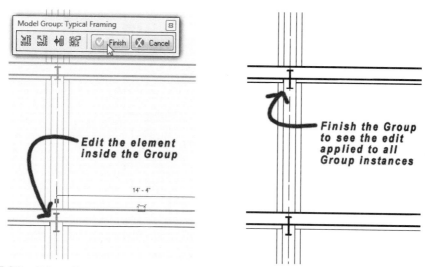

**Figure 5.37**   *Edit a Group to change all instances at once*

Continue to experiment with the Group if you wish. You can undo the changes when you are finished or you can leave them. It is up to you. This completes the work on the commercial project for this chapter.

As you experiment with the Group, it is important to note that editing the Group and ungrouping it do not yield the same results. When you ungroup a Group, you end up with individual elements that are no longer associated as a Group. Therefore, if you make changes to the ungrouped Elements, such changes will apply only to the Elements themselves and not be propagated back to the other instances of the Group. If you choose to Group the Elements again, you will be creating a new Group, not replacing the existing one. We will explore Groups in much more detail in Chapter 6.

   17. Save and close the project.

## ADDING FOOTINGS

To complete our brief exploration of structural tools in Revit Architecture we will have a look at the Footing tools provided. To do this, we will switch to the Residential project. Some new Footings are required for the addition on the back of the house.

### LOAD THE RESIDENTIAL PROJECT

Be sure that the Commercial Project has been closed and saved.

   1. On the Standard toolbar, click the Open icon.

 **Tip:** The keyboard shortcut for Open is CTRL + O. **Open** is also located on the File menu.

   • In the "Open" dialog box, click the *My Documents* icon on the left side.

- Double-click on the *MRAC* folder, and then the *Chapter05* folder.

  If you installed the dataset files to a different location than the one listed here, use the "Look in" drop down list to browse to that location instead.

2. Double-click *05 Residential.rvt* if you wish to work in Imperial units. Double-click *05 Residential Metric.rvt* if you wish to work in Metric units

  You can also select it and then click the Open button.

The project will open with the last opened view visible on screen.

## ADD CONTINUOUS FOOTINGS

You can create two types of Footings in Revit: Continuous and Isolated. A Continuous Footing is very easy to add and is associated with a Wall. An Isolated Footing is a Component Family that can be inserted anywhere a Footing is required, such as a pier footing, pile cap, etc. Footing tools are accessed from the Structural tab of the Design Bar.

3. On the Project Browser, double-click to open the *Basement* floor plan view.

4. On the Design Bar, click the Structural tab and then click the Foundation > Wall tool.

The Structural tab should be displayed already from the exercises above. If it is not, right-click on the Design Bar and choose **Structural** to display it.

- Choose Continuous Footing [Bearing Footing - 900 × 300] from the Type Selector.

  On the Status Bar, a "Select Wall(s)" prompt will appear.

- Click the vertical Wall on the left of the addition (top of the plan).

A "Warning" message will appear in the lower right corner of the screen. This is considered an "Ignorable" Warning message in Revit. It indicates a situation to which you may or may not be aware, but if you choose to ignore it, this condition does not prevent you from continuing your work. In this case, the Footing that we just added is not visible in the current view (see Figure 5.38).

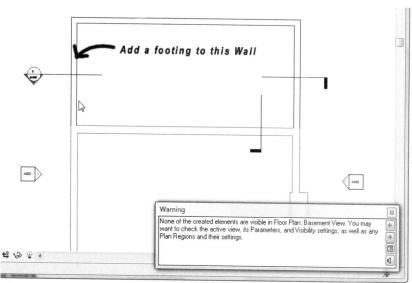

**Figure 5.38**   *Revit Architecture "Ignorable" Warning box*

We have to perform a few steps to correct this and display the Footings.

     5. Right-click in the *Basement* view window and choose **View Properties** (or just type VP).

     • Scroll to the bottom and click the Edit button next to "View Range."

     • In the "Primary Range" area, type **-2'-0"** **[-600]** in the Offset field next to Bottom.

     • In the "View Depth" area, type **-2'-0"** **[-600]** in the Offset field next to Level (see Figure 5.39).

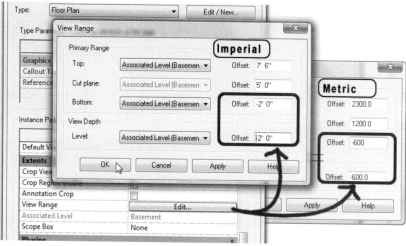

**Figure 5.39**   *Edit the View Range of the Basement Level Floor Plan*

- Click OK twice to see the results.

If you are working in the Imperial dataset, the Footing displays with a surface pattern. If you want to remove this, select the Footing, click the Properties icon on the Options Bar and then click the Edit/New button. This will edit the Type of the Footing. Next click on the Material, and then the little edit icon next to the Material name. This will load the "Materials" dialog. Finally, Click the edit button next to the Surface Pattern and click the No Pattern button. Click OK three times to return to the model.

If you are working in the Metric dataset, the lines will not automatically appear dashed. To edit this, choose **Object Styles** from the Settings menu. Next to the Structural Foundations item, choose Dash for the Line Style and then click OK (see Figure 5.40).

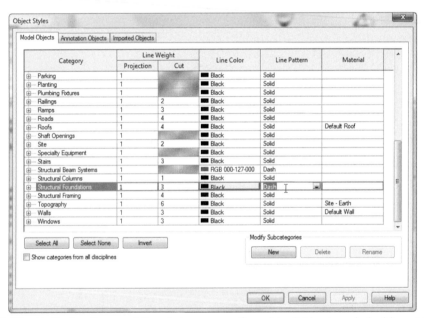

**Figure 5.40** *Verify that Structural Foundations are configured to display in a dashed line style in the Object Styles dialog*

6. Add two more Continuous Footings to the other two Walls of the addition.

## EDIT THE FOOTING

Like many building model elements in Revit Architecture, the Footings will interact intelligently with the Walls to which they are associated. For example, if we edit the Profile of the Wall to make a step at the bottom, the Footing will adjust accordingly.

7. On the Project Browser, double-click to open the *North* elevation view.

In this view, the terrain model is concealing the foundation Walls and Footings from view. For now, we will simply hide the Terrain so that we can work with the Footings and Foundations Walls.

8. Select the Toposurface, and then on the View Control Bar, choose **Hide Element** from the Temporary Hide/Isolate popup icon.

You should now be able to see the Footings and Walls. Remember that Temporary Hide Objects is limited to this work session only.

9. Select the Wall facing this elevation.

- On the Options Bar, click the Edit Profile button.

The Profile is a sketch that defines the elevation of the Wall. By default, this profile is a simple rectangular shape defined by the length and height of the Wall. We can edit the sketch of this profile to create a Wall with an alternate shape. When you choose this command from a plan view, Revit will prompt you select a more appropriate view from which to perform the edit.

- Using the Offset tool, offset the bottom edge down **1'-6"** [450].

- Draw a vertical sketch line (click the Lines tool on the Design Bar) at **11'-0"** [3300] from the right side.

- Using the Trim/Extend tool, complete the sketch as shown in Figure 5.41.

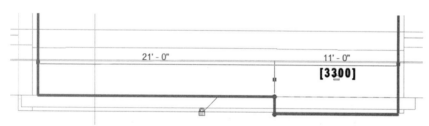

21' - 0"       11' - 0"
**[3300]**

**Figure 5.41** *Create a step in the Footing (sketch lines enhanced in the figure for clarity)*

10. On the Design Bar, click the Finish Sketch button.

The Footing will automatically adjust to match the new configuration of the Wall.

11. Perform similar Steps on the adjoining Wall (right side in the current elevation).

Study the model in various views. Remember that you will need to hide the Toposurface temporarily in other views to see the foundations (see Figure 5.42).

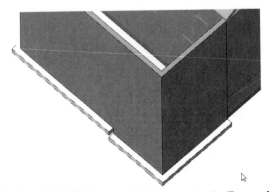

**Figure 5.42** *The completed foundation in the {3D} view (with the Toposurface hidden)*

There is plenty more that we could do to finish up the foundation of the residential project. We could add existing Footings (remember to work in that phase) or we could add an Isolated Footing to the chimney and the columns in the basement. We could also add isolated Footings to the Commercial project. When you choose the Isolated Footing tool, you will be prompted to load a Family for its use. This is left to the reader as an additional exercise. Feel free to experiment in both projects with these and other structural tools.

# SUMMARY

- Column Grid lines are datums that establish Column Grid intersections.

- Column Grid lines share many features with Levels and appear automatically in all orthographic views.

- Structural tools are accessible from the Structural tab of the Design Bar.

- Revit Architecture includes many Structural tools useful to Architects for basic structural layout.

- For complete Structural tools including interface with analysis packages, look into the Revit Structure application.

- Structural Columns represent the actual structural support members of the building

- Architectural Columns are typically used to represent the finished Column as it would be seen in the building. Use them for Column wraps, etc.

- Structural Walls are Walls that have their Structural Usage parameter set to Bearing.

- Floor Slabs can be created from bounding Walls.

- You can create Framing layouts and plans using Beams and Braces

- Group a collection of elements and reuse it elsewhere in the model. If the Group changes, all instances will update.

- A Continuous Footing is associated to Walls.

- An Isolated Footing is a Component Family that you place in your model.

# CHAPTER 6

# Groups and Links

## INTRODUCTION

In the previous chapter, we briefly covered the use of Groups. In this chapter, we will make a more detailed exploration of this feature. Groups provide a mechanism to standardize typical design elements throughout the project. A Group consists of a collection of elements that can be placed into the model as a single unit. You can edit a single instance of a Group and the changes will propagate to all instances throughout the model. Groups have many other features as well including the ability to have overrides applied to individual instances.

## OBJECTIVES

In this chapter, we will work with both Model and Detail Groups. We will explore how to create Groups, modify them and strategies for using them effectively in your projects. After completing this chapter, you will know how to:

- Create Groups
- Modify Groups
- Override elements in Group instances
- Create Attached Detail Groups
- Swap Groups with one another

## CREATING GROUPS

The dataset for this chapter will deviate from our Commercial and Residential projects to explore Groups in a dataset that will be more suitable to conveying the critical concepts. Groups can be created in any project. Creating them is as simple as making a selection of elements in your project and then clicking the Group icon. Groups appear on the Project Browser beneath the Families branch.

### INSTALL THE CD FILES AND OPEN A PROJECT

The lessons that follow require the dataset included on the Mastering Revit Architecture CD ROM. If you have already installed all of the files from the CD, simply skip down to step 3 below to open the project. If you need to install the CD files, start at step 1.

1. If you have not already done so, install the dataset files located on the Mastering Revit Architecture CD ROM.

   Refer to "Files Included on the CD ROM" in the Preface for instructions on installing the dataset files included on the CD.

2. Launch Revit Architecture from the icon on your desktop or from the *Autodesk > Revit Architecture* group in *All Programs* on the Windows Start menu.

   **Tip:** In Windows Vista, you can click the Start button, and then begin typing **Revit** in the "Start Search" field. After a couple letters, Revit Architecture should appear near the top of the list. Click it to launch to program.

   • If the New Features Workshop dialog appears, choose "Maybe later" and then click OK.
3. On the Standard toolbar, click the Open icon.

   **Tip:** The keyboard shortcut for Open is CTRL + O. **Open** is also located on the File menu.

   • In the "Open" dialog box, click the *My Documents* icon on the left side.

   • Double-click on the *MRAC* folder, and then the *Chapter06* folder.

   If you installed the dataset files to a different location than the one listed here, use the "Look in" drop down list to browse to that location instead.
4. Double-click *Understanding Groups.rvt*.

   You can also select it and then click the Open button.

The project will open with the last saved view visible on screen. Please note that for this chapter, units are immaterial to the lessons covered and as such only one dataset has been provided rather than the customary separate Imperial and Metric datasets of other chapters.

## EXPLORE THE DATASET

Groups are appropriate for nearly any repetitive (typical) design condition. Hotels, dormitories, apartment complexes, and condominiums give us plenty of opportunities to utilize Groups in very effective ways. In this example, we will work with a very simple hotel room layout.

5. On the Project Browser, double-click to open the *Architecture* floor plan.

This view shows the basic floor plan of a hotel guest room. Walls, Doors, Windows, and some basic fixtures are included.

6. Double-click to open the *Section 1* section view.

Here you will notice in addition to the items we can see in plan, there is also a multi-height ceiling plane in this model and some furniture items.

- Open other views to explore the dataset further if you wish before continuing (see Figure 6.1).

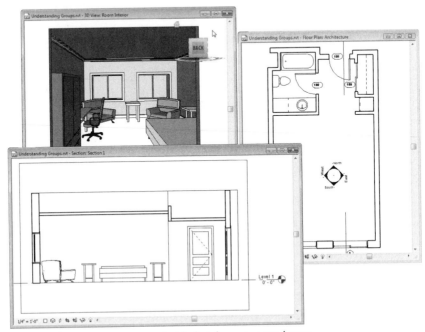

**Figure 6.1**   *The dataset represents a simple hotel guest room layout*

7. On the Project Browser, double-click to return to the *Architecture* floor plan.

- Maximize the view (if it is not already maximized).

- From the Window menu, choose **Close Hidden Windows**.

## CREATE A NEW GROUP

The first step to understanding Groups is to create one.

Continue in the Architecture floor plan.

8. Using a window selection, select all elements on screen. (Click above and to the left of the model and drag down to the right surrounding all elements).

- On the Edit toolbar, click the Group icon.

An error dialog will appear. When you made your selection marquee, the elevation view indicators and possibly the section view callout were included in the selection. View callouts cannot be included in a Group. Simply clicking OK in this warning will automatically exclude them from the selection set.

- Click OK in the warning dialog to dismiss it.

The "Create Model Group and Attached Detail Group" dialog will appear.

- In the Model Group Name field, type **Guest Room A**.
- Leave the Attached Detail Group name as **Group 1** and then click OK (see Figure 6.2).

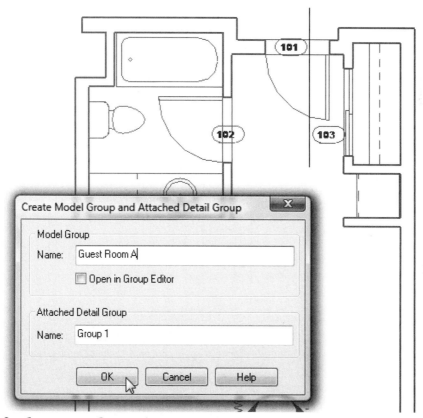

**Figure 6.2** *Create a new Group and give it a name*

Since we have selected both model and detail elements, Revit will actually create two groups. One will be a Model Group containing the Walls, Doors, Windows, and fixtures. The other will be a Detail Group that contains the door and window Tags. If is not possible for detail elements and model elements to be in the same group. The Detail Group will actually be an "Attached Detail Group." This means that this Detail Group is associated to its parent Model Group. Later, we can have instances of the Attached Detail Group automatically applied to instances of the Model Group. To see each Group, simply move your mouse over them on screen.

- Move your mouse near the edge of one of the Walls.

  You will see a dashed box appear around the Model Group with a screen tip indicating its name.

- Move your mouse over one of the Tags.

- You will see a dashed box appear around the Detail Group with a screen tip indicating its name (see Figure 6.3).

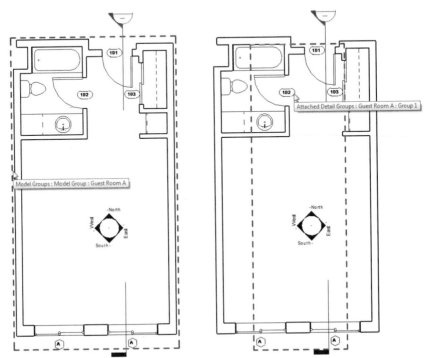

**Figure 6.3**  *Two Groups were created—A Model Group and an Attached Detail Group*

Groups that you create will appear in the Project Browser.

9. On the Project Browser, expand the *Groups* branch.

This reveals the Detail and Model branch.

- Expand the *Model* branch.

You will see the *Guest Room A* Group indented beneath the *Model* branch.

- Expand the *Guest Room A* entry to reveal the Attached Detail Group named *Group 1* (see Figure 6.4).

**Figure 6.4**  *Groups appear hierarchically on the Project Browser*

Each Model Group you create will appear beneath the *Groups > Model* node in the Project Browser. Attached Detail Groups will always appear beneath the Group to which they are attached. (If you create a Group from detail elements by themselves, without associated model geometry, they will appear beneath the *Groups > Detail* branch).

### CREATE A GROUP INSTANCE

Now that we have created a Group, we can easily add additional instances of the Group in our project. You can do this from the Modeling menu or the *Groups* branch of the Project Browser.

10. On the Project Browser, expand *Groups*, then *Model*, and then right-click *Guest Room A.*

- Choose **Create Instance**.

    A dashed rectangle will appear on screen with the mouse pointer in the center.

- Click on screen to place the new Group instance to the left side of the original.

- On the Options Bar, click the Finish button.

Another guest room will appear. (It does not include any annotation because as we saw above, the annotation is included in a separate Detail Group). When you create a Group, the geometric center of the Group becomes the insertion point by default. This is why when we added this instance; our mouse pointer was positioned in the center of the Group. You can move the origin to a more useful location simply by dragging it.

11. Select the Group instance that you just created.

In the center of the Group, a blue Group Origin icon will appear.

• Click and drag the round handle at the intersection of the two axes.

• Drop the icon on the Wall endpoint at the top left corner of the hotel room (see Figure 6.5).

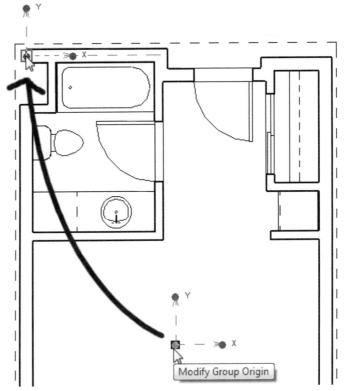

**Figure 6.5** *Move the Group Origin by dragging*

12. On the Project Browser, right-click *Guest Room A* again and choose **Create Instance**.

Notice the location of the mouse pointer relative to the Group outline this time.

• Click a point to the right of the original to place the new Group instance.

• On the Options Bar, click the Finish button.

If you select the original Group instance, you will notice that the insertion point for it is also at the upper left corner. Edits you make to one instance of a Group apply automatically to all instances.

## WORKING WITH ATTACHED DETAIL GROUPS

Take a look at the original guest room (the one that has Tags). Notice that there are three Doors, each with its own unique number. On the other hand, the Windows share the same designation of A. The default Revit Door Tags show the instance "Mark" parameter of Doors, which is unique for each Door while window Tags show the "Type Mark" for Windows which is the same for all instances of a given Type. Keep these observations in mind as we perform the next several steps.

To add tags to the other Group instances, we could manually tag each item in the Group. A faster method is to use an Attached Detail Group. An Attached Detail Group can automatically be applied to any instance of its parent Model Group.

13. Select one of the Groups without annotation.

   • On the Options Bar, click the Place Detail button.

   • In the dialog that appears, place a checkmark in the box next to Floor Plan:Group I and then click OK (see Figure 6.6).

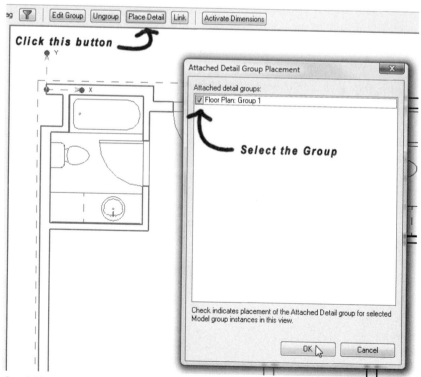

**Figure 6.6**   *Add an Attached Detail Group to the selected Model Group*

An instance of the Attached Detail Group will appear. Notice that each of the door Tags will have incremented sequentially to show unique numbers. The window Tags however will show the same designation that the originals did.

14. Repeat the process to add an Attached Detail Group to the other Model Group as well.

## EDITING GROUPS

You can edit a Group at any time. When you do, changes you make to the Group will be applied to all instances when the edit is complete. This is one of the most powerful benefits of using Groups. Furthermore, edits to a Model Group can also have an automatic impact on Attached Detail Groups.

### EDIT A GROUP

To understand the value and potential of editing a Group, we can start with a very simple modification.

15. Select one instance of the *Guest Room A* Group.

    • On the Options Bar, click the Edit Group button.

This enables the Group Edit mode. The background of the drawing area is tinted yellow and the Group Edit toolbar appears. The elements that are members of the Group remain bold, and all of the other elements on screen become grayed out. For this example, we will make a very simple edit.

16. Select one of the Windows.

    • On the Type Selector, choose **Slider with Trim:36"x48"**.

In the Group editor, the selected Window will immediately reduce in size and the associated Tag will change from A to E—even though this Tag resides in a separate Detail Group!

    • On the Group edit toolbar, click the Finish button (see Figure 6.7).

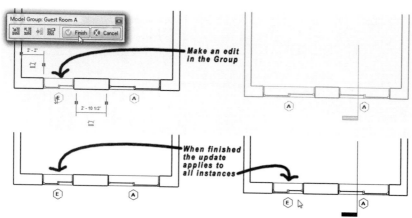

**Figure 6.7** *Make a change in the Group Editor*

When you have finished, the edit will be applied to all instances of the Group. Notice that the Attached Detail Groups update as well. Let's try another edit.

17. Select one instance of the *Guest Room A* Group.

- On the Options Bar, click the Edit Group button.

- Repeat the Window edit made above to the other Window.

- Create a new Window in the space between the two existing Windows. (Try a **Fixed:36"x48"**).

Notice that while the window Tag is also created, it comes in grayed out. This is because it is automatically excluded from the Group that we are currently editing.

- On the Group edit toolbar, click the Finish button.

18. Select the window Tag.

Notice that the Tag selects independently as a freestanding element in the project. It is not automatically added to the Attached Detail Group as you can see by examining the other two instances in this project (see the top half of Figure 6.8).

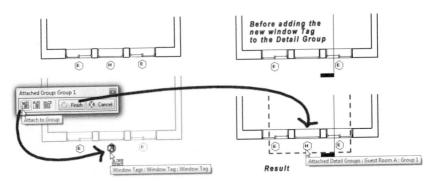

**Figure 6.8**  *Tags for newly added Group elements must be added manually to the Attached Detail Group*

19. Select one instance of the *Group 1* Attached Detail Group.

- On the Options Bar, click the Edit Group button.

- On the Group edit toolbar, click the Attach to Group icon.

- Select the window Tag (Tag H) and then click the Finish button (see the bottom half of Figure 6.8).

The Tag for Window Type H should now appear in all three instances of the Attached Detail Group. Again, since this Tag references a Type parameter (the Type Mark), the letter displays the same value in all instances of the Attached Detail Group. If you were to edit the Model Group again and delete one of the tagged elements (a Door or Window) then the tag in the Attached Detail Group would also be deleted automatically even though it is in a different Group—you would not have to separately edit the Detail Group.

## DUPLICATE AND EDIT A GROUP

Making a variation of a Group is simple to do. Once you have two or more variations, you can easily swap them out with one another.

20. Select one instance of the *Guest Room A* Group.

- Click the Properties icon (or right-click and choose **Element Properties**).

- At the top of the dialog, click the Edit/New button.

- Click the Duplicate button.

**Tip:** To do this quickly, press ALT + E and then immediately press ALT + D.

- In the "Name" dialog, type **Guest Room B** and then click OK three times (see the left side of Figure 6.9).

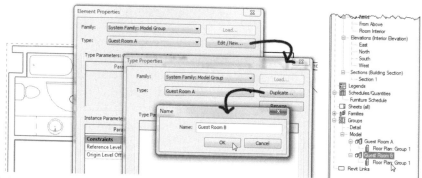

**Figure 6.9**   *Create a duplicate of the Model Group*

Notice the appearance of *Guest Room B* on the Project Browser. If you expand it, you will see that a copy of the *Group 1* Attached Detail Group has also been created and associated to the new Model Group (see the right side of Figure 6.9).

> 21. Select the same instance (now *Guest Room B*).
>
> • On the Options Bar, click the Edit Group button.
>
> • Delete the middle Window and make some other obvious change (such as enlarging the bathroom or flipping a Door).
>
> • Finish the Group (see Figure 6.10).

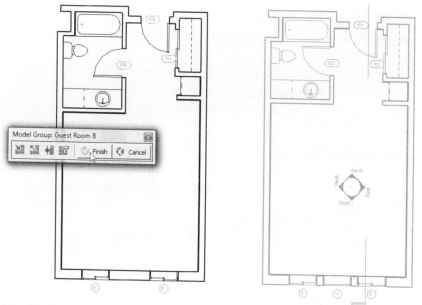

**Figure 6.10**   *Modify Model Group Guest Room B*

Not surprisingly, the change only affects the currently selected Group. This is because it is currently the only instance of *Guest Room B* in the project. At this point however, we can experiment with swapping Group instances and see the ease at which we can switch from one Group to another and also see one of the benefits of the Attached Detail Group functionality. Take notice of the names of the Attached Detail Groups on the Project Browser. For both Model Groups, the Attached Detail Groups have the same name—*Group 1*. This is important for the next experiment. If you use the same name for the Attached Detail Groups, they will automatically swap when the parent Model Groups swap. Let's take a look.

22. Select one instance of the *Guest Room A* Group.

- From the Type Selector, choose *Guest Room B*.

Since we tried to make the difference between the A and B Guest Rooms obvious, you should be able to spot the changes to the model right away. The most interesting change however is that the Attached Detail Group has changed automatically as well. Since the Attached Detail Groups for each Model Groups have the same name, Revit is able to swap them appropriately as well. This behavior works as long as the names of the Attached Detail Group is the same for each Model Group. In this example, we left the name "Group 1" but this name is not required. You can choose a more descriptive name if you wish.

23. On the Project Browser, right-click the Attached Detail Group (*Floor Plan: Group 1*) for *Guest Room A* and choose **Rename**.

- Change the name to **Tags** and then click OK.

- Repeat for the Attached Detail Group of *Guest Room B* and rename it to **Tags** as well.

- Repeat the process above to swap one of the *Guest Room B* instances in the project back to *Guest Room A*.

Notice that the Attached Detail Group continues to swap as well. With this in mind, you should try to pick useful and descriptive names for both your Model Groups and your Attached Detail Groups in your own projects. Careful naming often helps reduce errors by limiting the apparent complexity of projects.

## EDITING GROUP INSTANCES

As powerful as Groups are, situations will often arise in the course of a project where one instance of a Group needs to be slightly different from the other instances in the project. In this case, we could certainly repeat the process covered in the previous sequence and duplicate and edit another Group. However, doing so could begin to dilute the usefulness of Groups and make management of the multiple potential variations cumbersome and time consuming. In scenarios such as this, Revit offers us the ability to create overrides to individual Group instances. To illustrate the point, a simple example is appropriate.

## EXCLUDING GROUP MEMBERS

Continue from the previous Exercise.

1. Delete the two Groups copied above leaving only the original one and its Attached Detail Group.

Notice that the Attached Detail Groups are deleted automatically when their hosts are deleted.

2. Select the Model Group remaining on screen.

• Click the Mirror tool on the Edit toolbar.

• Click on the centerline of the vertical Wall on the right side of the Group as the mirror edge (see Figure 6.11).

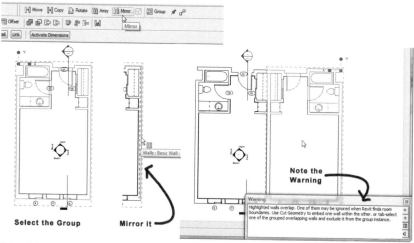

**Figure 6.11** *Mirror a copy of the Guest Room Group*

When you click the mirror Wall, a copy of the Group will appear, and a Warning dialog will also appear at the bottom right corner of the screen. This warning is not serious and can be ignored. All such "ignorable" warnings will appear in this location on screen. It is still a good idea the read the warning message as most of the time there is some useful information conveyed in them. In this case, Revit is alerting us that we now have two Walls overlapping in the same spot. While we can ignore this situation, the message further explains that Room boundaries may be affected:

"Highlighted walls overlap. One of them may be ignored when Revit finds room boundaries. Use Cut Geometry to embed one wall within the other. Or tab-select one of the grouped overlapping walls and exclude it from the group instance."

The right side of Figure 6.11 shows the warning message. If you wish, you can expand the warning dialog to get more detailed information. Do this with the small icon on the right side of the warning dialog or by choosing **Review Warnings** from the Settings menu. When you fully expand the error, you can click on each of the overlapping Walls to highlight them on screen (see Figure 6.12).

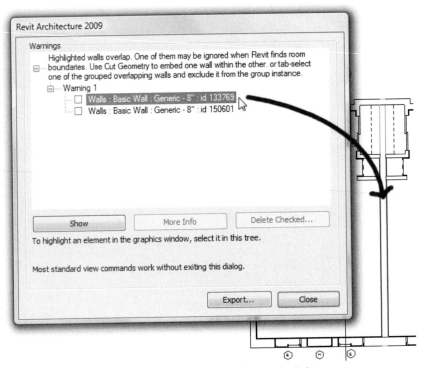

**Figure 6.12** *Click the Expand icon to see a more detailed error dialog*

If you wish to see the element highlighted in other views, click the Show button. When you are done with the warning, click the Close button to dismiss it.

In addition to potentially having an adverse effect on Rooms, you can also see that the overlapping Walls do not cleanup very nicely. The solution to both problems is simple: any element in any Group can be excluded from an individual instance of the Group. In this situation, we can exclude the Wall from one of the Groups.

3. Place your mouse over the double Wall.

Notice that the Group pre-highlights.

- Press the TAB key.

Notice that the other Group pre-highlights.

- Press the TAB key again.

This time, the Wall within one of the Groups pre-highlights.

- Click to select this Wall.

**Tip:** Notice the warning indicator on the Options Bar. This gives you a shortcut to the Review Warnings dialog, letting you know that there are warnings associated with the selected element.

- Click the Group Member icon to exclude this Wall from the Group (see Figure 6.13).

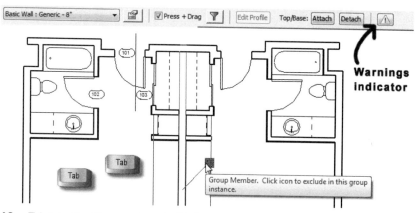

**Figure 6.13** *Tab into the Group, select the Wall and then click the icon to exclude it*

Notice that the extra Wall has been removed and the cleanup is now correct. It is important to realize that this change is not simply graphical override—Revit has actually removed one instance of one of the Walls. For example, were we to have counted the Walls before started and then re-count them now, there would be one Wall fewer in our model. Rather than count the Walls, which might prove tricky in this case, let's do another experiment using furniture.

    4. On the Project Browser, double-click to open the *Furniture* floor plan view.

Furniture elements will appear in the original guest room.

    5. Select all of the furniture elements.

**Tip:** use a window selection to select all the furniture and then click the filter icon on the Options Bar to remove any unwanted element categories.

- On the Edit toolbar, click the Group icon.

- In the Model Group Name field, type **King-01**.

- Using the technique covered in the "Create a Group Instance" topic above and move the origin point to the same location as the Guest Room Group.

    6. Mirror the furniture Group to the other guest room.

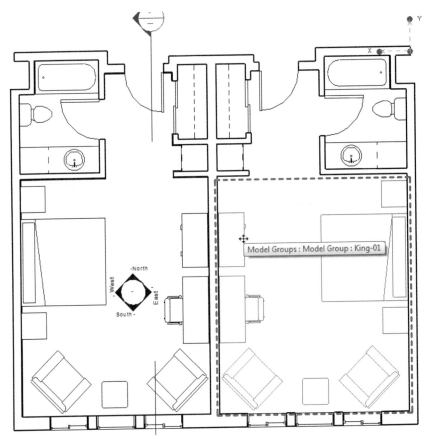

**Figure 6.14** *Group the furniture, move the origin point, and then mirror a copy*

7. On the Project Browser, beneath Schedules/Quantities, double-click to open the *Furniture Schedule* view.

Study the table and take note of the totals in the "Count" column. In particular, notice that we currently have 4 Side Chairs (Chair-Viper: Chair). This is impressive; Revit gives us an accurate count even when the items it is counting are inside Groups! Let's exclude one and see the impact on the Schedule.

8. Return to the *Furniture* floor plan.

- Using the TAB key, select one of the Side Chairs (near the Windows).

- Click the Group Member icon to exclude this chair from the Group (see Figure 6.15).

- Deselect the chair after excluding it to see it disappear.

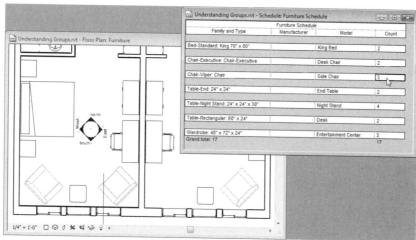

**Figure 6.15** *Excluding an item from a Group removes it from the schedule as well*

Now that is even more impressive; the schedule accurately reflects the quantity shown in the model. This example illustrates that you can use the exclude from Groups feature with confidence, as Revit will accurately reflect the exclusions throughout the model.

You should try a few more experiments to become comfortable with the full behavior of this feature. For example, move your mouse over the missing chair and it will pre-highlight as if it were there. In this way, you can TAB back into the Group and bring the element back (include it). Be careful when editing a Group that has overrides applied. You can edit either instance of the furniture Group on screen. However, if you edit the one with the excluded chair, you will not be able to edit the chair at all. If you instead edit the one without exclusions, you will have the ability to edit all its elements including both chairs. You can even move or otherwise edit the chair that is excluded in one of the other Group. The change will be visible in the Group (or Groups) that shows the chair and not visible in any that exclude it. Try some of these experiments now before continuing if you like.

    9. Save the project.

## ADDITIONAL GROUP DESIGN TECHNIQUES

As you refine your design using Groups, you will find some of the additional techniques covered here useful.

### CREATING ATTACHED DETAIL GROUPS FOR EXISTING MODEL GROUPS

The Attached Detail Groups that we made earlier were created at the same time as the host Model Groups. You can also create them later even after the Model Group has been created.

    1. On the Design Bar, click the Drafting tab.

    • Click the Tag tool and then choose By Category.

    • Tag each piece of furniture in one of the guest rooms only.

All of the furniture numbers have already been input in this dataset. To learn more about tagging and editing tag parameters, see Chapter 12.

- Select all of the furniture Tags, and then on the Edit toolbar, click the Group icon.

- In the Attached Detail Group Name field, type **Tags**.

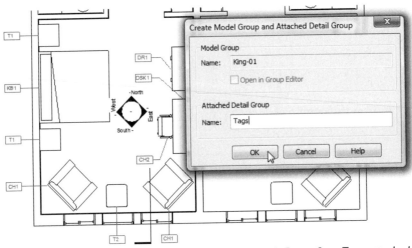

**Figure 6.16** *Revit will automatically create an Attached Detail Group from Tags attached to a Model Group*

As you can see, Revit will automatically recognize that the items being grouped are Tags associated with model elements contained in a Model Group. As a result, an Attached Detail Group is created automatically. Now that we have an Attached Detail Group for our furniture, we can add it to the other instance of our furniture Model Group.

## ADDING DETAIL GROUPS TO MIRRORED GROUPS

Attached Detail Groups can be very useful as we have seen. They do have limitations as well as we will see. If we return to the *Architecture* floor plan, we will notice that the mirrored Group has no Tags.

2. From the Project Browser, re-open the *Architecture* floor plan.

- Select the mirrored Group (on the right side).

- On the Options Bar, click the Place Detail button.

- In the Attached Detail Group Placement dialog, place a check in the box next to the *Tags* Detail Group and then click OK.

Notice that despite the Model Group's being mirrored, the Detail Group remains "right-reading." This is true if you mirror in any direction. Furthermore, you can mirror a selection of both Model and Detail Groups together in the same operation and the Detail Groups will remain right-reading as the Model Group mirrors.

3. Select both Model Groups and both Detail Groups.

   Be careful *not* to select the Section or Elevation tags.

   • On the Edit toolbar, click the Mirror tool.

   • On the Options Bar, click the Draw icon.

This allows you to sketch the mirror line rather than pick an object. In this case, we will mirror the selection up at a distance above the selection to allow room for a corridor between the rooms.

   • Using the temporary dimension as a guide, click the first point about 5'-0" above the top Walls to indicate the middle of the corridor.

   • Drag the mouse horizontally and then click again to complete the mirror (see Figure 6.17).

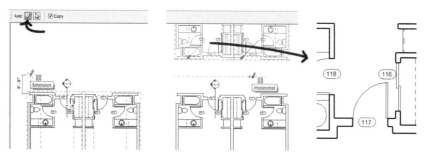

**Figure 6.17**   *Tags in mirrored Detail Groups remain right reading*

If you zoom in on the new rooms and study the tags, you will see that they are right-reading yet they each display a unique door number sequentially incremented from where the previous door numbering left off.

4. Select all Groups on screen (4 Model and 4 Attached Detail Groups).

   • Mirror the selection around the centerline of the rightmost vertical Wall.

As before, a warning will appear indicating that you have duplicate Walls again. Ignore this warning.

There are now eight total guest rooms each with its own Tags.

5. Select the four guest rooms in the middle.

   • From the Type Selector, choose *Guest Room B*.

As we saw earlier in the chapter, not only does the guest room geometry change, but the Attached Detail Group updates as well. The duplicate Walls warning will appear again.

   • Use the process covered in the "Excluding Group Members" topic above to exclude the extra Walls.

   • Save the project.

## DUPLICATE GROUPS ON PROJECT BROWSER

Returning to the Furniture view will reveal that while the Walls, Doors, and Windows contained in the duplicated Model Groups were copied to form additional guest rooms, the furniture was not. This is simply because the furniture is contained in a separate Model group. Let's duplicate our furniture Group and make an alternate for the other guest room type.

6. On the Project Browser, beneath Model Groups, right-click the *King-01* Group and choose **Duplicate**.

- Right-click the new copy and choose **Rename**. Call it: **Queen-01** (see Figure 6.18).

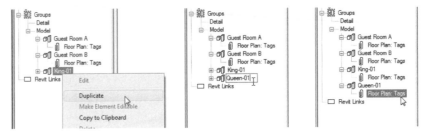

**Figure 6.18** *Duplicate and rename a Group on the Project Browser*

Notice that the *Queen-01* Group also has its own attached Detail Group named *Tags*. There are several commands on the right-click menu. Here is a brief description of each:

- **Duplicate**—Creates a copy of the Group and assigns it a default name.

- **Make Element Editable**—This command is only active in a project using Worksharing. Worksharing is a process enabling a team of people to work in the same Revit project. See Appendix B for more information.

- **Copy to Clipboard**—This copies the Group to the clipboard so you can paste it in other projects. This is a fast way to use the same Group in another project. If you paste it in the same project, it behaves like Duplicate.

- **Delete**—This deletes the Group definition from the project. You can only use this command if no instances of the Group are inserted in the project. Use with caution.

- **Rename**—Use to assign a new name to a Group definition.

- **Select All Instances**—Use this command to select all instances of the Group throughout the entire project. Be careful as this command selects all instances on all levels, even the ones that may not show in the current view.

- **Create Instance**—This adds an instance of the Group.

- **Match**—use this command to swap one Group with another on screen. You will be prompted to select the Group to use as the source and then the Groups to which to apply the source definition.

- **Edit**—This command will export the Group to a new Revit project and open it for editing. In this way, you can edit a Group independently of the current project and save it as its own file. To use the Group saved this way in a project, choose **Load from Library > Load File as Group** from the File menu and follow the prompts.

- **Save Group**—This command will also export the Group to a new Revit project but it will not automatically open it. You will simply be prompted for the file name and location in which to save it.

- **Reload**—This is a shortcut to the **Load from Library > Load File as Group** on the File menu. Use it to load an external file and replace the internal Group definition.

A similar list of commands appears when you right-click a Detail Group.

7. Select the furniture Group with the excluded chair.

- Right-click and choose **Restore All Excluded**.

   The previously excluded chair will be restored.

- With the Group still selected, choose *Queen-01* from the Type Selector.

8. On the Options Bar, click the Edit Group button.

- Select the bed and from the Type Selector, choose: ***Bed-Standard : Queen 60" x 79"***.

- Delete one of the lounge chairs and move the bed and nightstands down to fit the room better (see Figure 6.19).

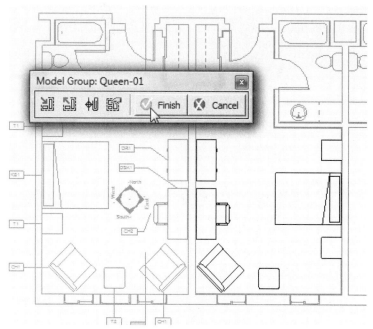

**Figure 6.19** *Swap in the Queen-01 Group and then modify it*

9. On the floating Model Group: Queen-01 toolbar, click the Finish button to complete the edit.

10. Select the *Queen-01* Group on screen.

- On the Options Bar, click the Place Detail button.

- In the Attached Detail Group Placement dialog, place a check in the box next to the *Tags* Detail Group and then click OK.

Notice that the Tags have automatically adjusted to the new furniture layout of the Group.

11. Using any of the techniques covered so far; add furniture and Model and Detail Groups to the remaining guest rooms (see Figure 6.20).

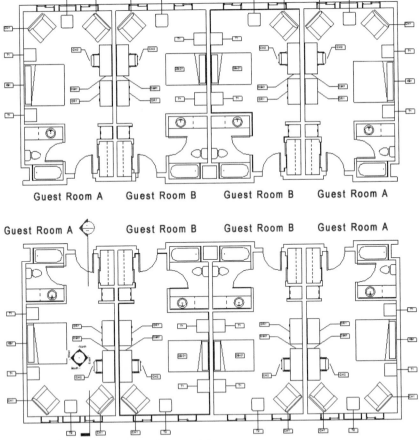

**Figure 6.20** *Add furniture groups to the remaining rooms*

## MAINTAINING GROUPS

As you can see, working with Groups so far has made it easier to compose our overall plan layout and quickly replicate a series of similarly configured spaces. After this initial design work, you may be tempted to ungroup your Groups to gain more direct access to the elements they contain. While it is certainly possible for you to do this, you may want to consider keeping your Groups active well into design development or even CDs. The reason for this is simply because despite our best efforts to minimize them, design changes continue to occur well into the construction document phase and even beyond. Groups can help you make such changes more efficiently.

## ADD MISSING ELEMENTS TO GROUPS

While design changes occur for any number of reasons, in this next example, we will consider a change resulting from an oversight during the design phase.

    1. On the View toolbar, click the Default 3D View icon.

Compare the original room that we started with to all of the copied versions and notice that the copies do not include the ceiling elements. Since we have been working exclusively in plan views, we did not notice that the ceiling were not included in the original selection set from which the Groups were created (see Figure 6.21).

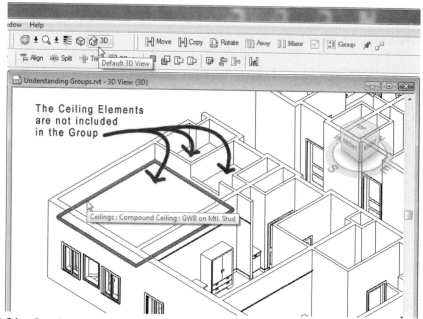

**Figure 6.21** *Switching to 3D view reveals elements missing from the Groups*

The fix for Guest Room A is simple. For Guest Room B there is an extra step. In the original Guest Room A Group (shown in the figure) the ceiling elements are positioned in the proper location. All we have to do is add those stray ceiling elements to the Group and they will appear in all instances of Guest Room A. For Guest Room B, we first need to copy the ceiling elements into position relative to one of our Guest Room B Groups and then add them to the Group. While the 3D view was useful to identify the problem, a ceiling plan view is the best choice for making the required edits.

2. On the Project Browser, double-click to open the *Level 1* Ceiling Plan.

   Zoom in on the original Guest Room A; in this case, it has an interior elevation tag within it, making it easy to locate. If you compare the elements in the original Guest Room A to the others, you will notice that there is a small Wall separating the main guest room from the entry foyer. We'll need to mirror this Wall from Guest Room A to Guest Room B to form the boundaries for the Ceiling elements.

3. Place your mouse over the Wall between the foyer space and the main room and then look at the Status Bar (at the bottom of the Revit screen).

   Press tab if necessary to highlight the Wall.

A message will appear such as "Walls : Basic Wall : Generic 5"." The format of this message is: *Category : Family : Type*. All Revit objects appear in this format when pre-highlighted on screen. (See the "Status Bar" heading in Chapter 2 for more information.)

   Depending on your settings, the same information may appear in a tooltip on screen (see Figure 6.22).

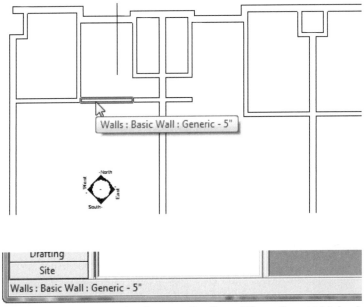

**Figure 6.22** *The Status Bar reports the Category, Family, and Type of pre-highlighted elements*

- When the Wall between the foyer space and the main room highlights, click it to select it.

4. On the Edit toolbar, click the Mirror tool.

- Using the center of the Wall between the two guest rooms, mirror the elements to the neighboring room (see Figure 6.23).

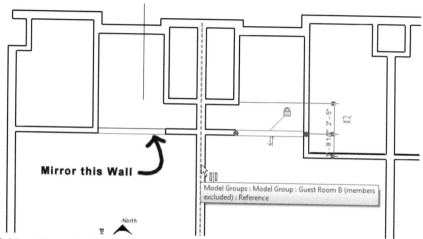

Model Groups : Model Group : Guest Room B (members excluded) : Reference

**Figure 6.23**    *Mirror the Wall required to enclose Ceilings*

While it would be possible to select and mirror the existing Ceiling objects to the other space as well, in this case it will be better to recreate them since the shapes of the rooms do not match.

5. Select the original instance of the Guest Room A Group.

- On the Options Bar, click the Edit Group button.

- On the Model Group: Guest Room A toolbar that appears, click the Add to Group icon.

This tool allows us to add items from the main model into the Group. When we are finished editing, these elements will appear in all instances of the Group.

6. Move your mouse near the edge of the toilet room in Guest Room A.

The Ceiling object will pre-highlight.

- Click the Ceiling to add it to the Group (see Figure 6.24).

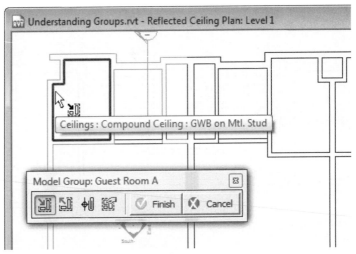

**Figure 6.24**  *Add Ceilings and the Wall to the Group*

- Repeat for the foyer Ceiling, the closet Ceiling, and the small Wall.

7. Finally, add the Ceiling in the main guest room space.

   Press the tab key if necessary to assist in adding any of the elements.

- Once you have added four Ceilings and the Wall to the Group, click the Finish button on the Model Group: Guest Room A toolbar.

If you return to the *{3D}* view, you should notice that the four Guest Room A Groups now have Ceilings. The easiest way to add Ceilings to the Guest Room B Groups, is to simply create new ones.

8. In the Ceiling Plan *Level 1*, select the instance of Guest Room B to which we mirrored the small Wall above.

- Click the Edit Group button.

- Add the small Wall to the Group.

9. On the Modeling tab of the Design Bar, click the **Ceiling** tool.

- From the Type Selector, choose **Compound Ceiling : GWB on Mtl. Stud**.

- Click inside each enclosed space to add Ceilings.

- Click the Finish button on the Group toolbar when done.

If you return to the *{3D}* view, you should now have Ceilings in all rooms.

## NESTING GROUPS

As we have seen, most model elements can be added to Model Groups. We can also make a Group that contains other Groups. So called "Nested Groups" can be useful, but can also present certain challenges as well. For example, in the dataset we have open, it might be useful to group all of the various guest rooms and their furniture into a single Group named something like "Typical Floor Layout." This approach may certainly prove valuable at the early stages of design where you stand to gain an advantage from the ease of being able to edit a Group instance and have the changes apply across the entire project. However, there are limitations. The most notable is that Attached Detail Groups only work one level deep. This means that you cannot make a Group containing both Model and Detail Groups as members. You will still be able to apply Attached Detail Groups to the nested instances of the Model Groups, but you will have to use the TAB key to select each instance before placement. With careful planning, you can certainly make a workable solution; the only caution is to plan your strategy carefully before execution.

To create a nested Group, you simply select objects (including other Groups) and then click the Group button as we have done already.

## CREATING A NESTED GROUP

Let's do a quick example of a nested Group in the project we have open. Suppose we wanted to explore adding some more Levels to this project and reusing the layout we have devised here on those Levels. We can certainly use Copy and Paste in that scenario, but creating a Group of the entire layout affords us the opportunity to make edits to the Group later and have those edits apply automatically to all Levels.

1. On the Project Browser, double-click to open the *Section 1* Building Section view.

2. On the Basics Design Bar, click the **Level** tool.

• Create a Level 12'-6" above the existing Level 1

• Rename it **Level 2** (see Figure 6.25).

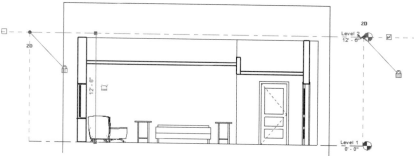

**Figure 6.25**   *Add and rename a new Level*

3. Return to the *Architecture* Floor Plan view.

4. Select all eight Model Groups on screen.

• Click the Group button on the Edit toolbar.

• Name the Group: **Typical Floor Layout** and then click OK.

Take notice of the Project Browser after you complete the Group. Typical Floor Layout will show Guest Room A and B indented beneath it. This indicates that these two Groups are nested within it (see Figure 6.26).

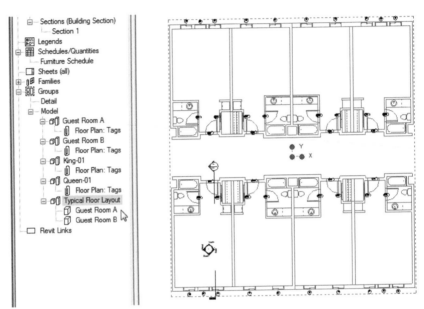

**Figure 6.26**  *Add and rename a new Level*

5. On the Project Browser, double-click to open the *Level 2* Floor Plan.

6. On the Project Browser, right-click Typical Floor Layout and choose **Create Instance**.

- Snap to the Group Origin of the previous Group.

- Ignore any errors that appear and then press ESC (see Figure 6.27).

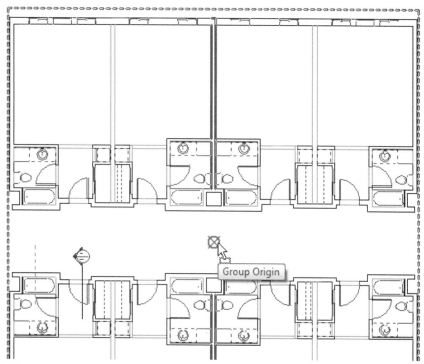

**Figure 6.27**   *Create a new instance of the Group at the Group Origin*

The error that appears is the same one we have seen before about overlapping Walls. It is safe to ignore in this case. We now have an identical layout on the second floor matching the one on the first floor.

## ADDING ATTACHED DETAIL GROUPS TO NESTED GROUPS

If you want to add the Tags Groups, you have to use the TAB key.

　　　7. Place your mouse over one of the Groups.

Notice that the entire floor layout Group pre-highlights.

- Press the TAB key.

- Click to select the nested Group instance.

- Click the Place Detail button, choose the Tags Group and then click OK (see Figure 6.28).

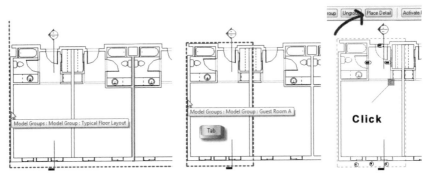

**Figure 6.28** *Add an Attached Detail Group to the nested Model Group (using TAB to select)*

You can repeat the process on the other Groups if you like. Following any of the procedures covered so far, you can also edit either the nested Groups or the overall Group and see the results throughout the model and in the Attached Detail Groups. Feel free to experiment further before continuing.

## GROUPS AND REVIT LINKS

We have explored many techniques and advantages of working with Groups so far in this chapter. You can begin to see the many advantages of including Groups in your workflow. While working with Groups directly in a project can prove a useful strategy for managing typical and repetitive design conditions, it is sometimes even more advantageous to export a Group to its own separate file. This can be achieved by saving the Group or converting it to a Linked file.

## SAVING A GROUP TO A FILE

A Group can be saved to a separate Revit file. This enables you to work on the Group independently of the project. This can be useful if the project is particularly large and/or if you want to have another colleague working on the Group at the same time as you or someone else is in the project file.

1. On the Project Browser, right-click the *King-01* Group and choose **Save Group**.

   - In the Save Group dialog, browse to the *Chapter06* folder, verify that the "Include attached detail groups as views" checkbox is selected, and then click OK (see Figure 6.29).

**Figure 6.29**  *Save a Group to a separate file*

The file name defaults to the same name as the Group: *King-01* in this case. Once the save is complete, you can open the file to study the result.

2. From the *Chapter06* folder, open the *King-01* file.

- On Project Browser, expand Views.

You should have two floor plan views: *Level 1* and *Tags*. *Tags* contains the annotation contained in the Attached Detail Group of the same name in the main project. The *Level 1* view contains the model geometry. You can make any edits here that you like. Upon saving those changes, we can reload them back into the main project.

3. In the *Level 1* floor plan view, make a change to the furniture layout.

- Save and close the *King-01* file.

- Back in the main project, right-click the *King-01* Group on Project Browser again and choose **Reload**.

- In the dialog that appears, browse to the *Chapter06* folder and select the *King-01* file to reload.

  If you are not in a view that shows the furniture, switch to one now to see the results.

## CONVERT A GROUP TO A LINKED FILE

Revit also provides the ability to embed other Revit files in your project as Revit links. A linked file provides many of the same advantages as Groups but remains a separate project file on your hard drive or server maintaining a live link for easy reloading. In this way, another individual can work in the linked file simultaneously. When the linked file is saved, you can capture the latest changes by reloading the linked file. The process is similar to the one just outlined, but the path to the link file is saved with the project so that we do not have to browse to each time we reload.

4. On the Project Browser, double-click to open the *Level 2* floor plan.

5. Select the *Typical Floor Layout* Group on screen.

- On the Options Bar, click the Link button.

- If a warning appears, click OK.

- In the "Convert Group to Link" dialog, click the Create New button (see Figure 6.30).

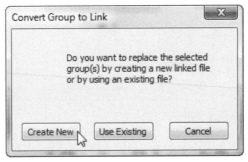

**Figure 6.30** *Save a Group to a separate file*

The first dialog was a warning about elements being deleted. This is because we had previously applied tags (via an Attached Detail Group) to the original model Group. When you convert a Group to a Link, the annotation cannot remain applied. The second message allows us to create a new file from the Group we are replacing, or to point to an existing file already on our hard drive or server to swap in its place. In this case, the Create New button was our obvious choice.

When the conversion is complete, you will see that the newly created file automatically appears beneath the Revit Links node of the Project Browser (see Figure 6.31). Here you can access features of the linked file via the right-click menu.

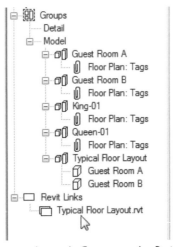

**Figure 6.31** *Linked Revit files appear beneath Groups on the Project Browser*

Feel free to open the linked file, make a few edits, then re-save and reload the file. When you try to open the linked file, Revit will warn you that it must be unloaded in the current project. Once edits are complete, you can right-click the linked file on the Project Browser and choose **Reload**.

## BINDING LINKED FILES (TO GROUPS)

The opposite of converting a Group to a linked file is "Bind," which converts a linked file into a Group.

1. Select the linked file.

- On the Options Bar, click the Bind button.

- In the dialog that appears, choose Attached Details and then click OK.

- Accept the default in the remaining dialog(s).

- Remove the link when prompted.

The Revit link should now be removed and in its place the Group that we started with should have been restored.

2. Save the project.

## WORKING WITH ROOMS IN GROUPS

When the time comes to add Room objects to our project we can choose to add them within the Groups or outside the Groups. If we add a Room to each Guest Room Group, they will appear in all instances like other objects. We can then tag them inside an Attached Detail Group or directly on the floor plan view.

Another approach is to simply add the Rooms outside of any Groups directly in the project. Since the Room object will conform automatically to the shape defined by the Walls, either approach is completely valid.

To compare methods, try both approaches in the current project.

3. Select one of the Guest Room Groups (use the TAB key to assist in selection).

- Click the Edit Group button on the Options Bar.

- Using the tool on the Design Bar, add a Room in the main space.

You will see the Room object conform to the shape of the main room plus the entry foyer. If you like you can repeat the process to add additional Rooms for the closet and toilet room. However, you may not want to have these Rooms separate and might instead prefer a Room that expands to include the closet and toilet rooms within it. To do this, we can select the Walls between the toilet room and main guest room area and make them "non room bounding."

4. Select the Walls that separate the toilet space and closet from the main space.

- Edit the properties of the selected Walls and turn of "Room Bounding."

- Re-edit the Group and add a new Room object. It should now fill the entire guest room layout (see Figure 6.32).

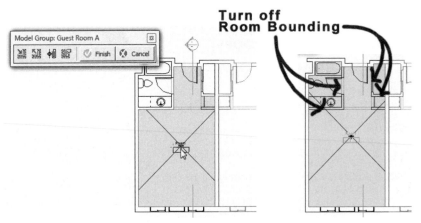

**Figure 6.32** *Adding Rooms to the Group and varying the Room Bounding behavior*

When you finish the Group the Room will be added to all instances. Again, the Room Tag will need to be attached separately as explained in "Edit a Group" above. If you wish to try the alternative method, simply exit the edit Group mode and add Rooms directly to the project. Tags can also be free-standing or grouped in Attached Detail Groups.

There are certainly plenty of other equally useful applications of Groups including typical toilet room layouts, typical stair tower, office furniture layouts, etc. For example, in the previous chapter we used a simple Group to create a typical floor framing condition that was copied to multiple floors in the building. There are almost limitless applications for Groups.

For your further experimentation, a larger and more complete dataset similar to the one utilized in this chapter has been provided. You will find two versions of "MRAC Hotel" in the *Chapter06* folder. One version named: *MRAC Hotel (With Rooms).rvt* has the Room objects embedded within the Guest Room Groups. The other version: *MRAC Hotel (Without Rooms).rvt* has the Rooms placed directly in the project (not in the Groups). You are encouraged to open each of these files and experiment further with all of the techniques covered in this chapter.

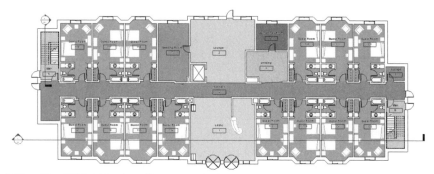

**Figure 6.33**  *The MRAC Hotel.rvt file is provide for your further experimentation*

## LINKED PROJECTS

Throughout the course of this chapter, we have worked in a separate dataset and not re-visited our commercial and residential projects. While theoretically any project can make use of Groups and Linked files, our residential project has no need for either. However, we have already added a Group to our commercial project at the end of Chapter 5 (for the structural framing) and we can also make use of Linked files for certain aspects of the project as well.

Many firms using Revit today utilize Linked files as a way of splitting up larger projects into more manageable pieces. It is common to see separations made along various disciplines (such as architectural, structural and mechanical) and sometimes between major functional areas of the project (like core, shell, and interiors). These are of course suggestions and each firm and in fact each project can and often will implement some variation of these. The one strategy that is common is to separate a large project into smaller files so that different team members can readily work in separate areas simultaneously.

**BIM** *Manager Note:*    Using Linked files is only one way in which Revit teams collaborate. The other method involves a series of tools called "Worksets." Worksets provide a means to separate a single building model into discrete portions for purposes of facilitating multi-user access to the same model. The process involves the creation of a "Central" model stored on a common network server and individual "Local" files on each team member's workstation. Revit keeps track of changes that each user makes by enabling object locking at the Workset level and the level of the individual element (referred to as "Borrowing" ). Worksets will be discussed more extensively in the Appendix B.

Now that we have explored most of the concepts of Groups and Links in the dataset files accompanying Chapter 6, let's return to our commercial project to see how some of these concepts might apply and allow the project to progress.

### LOAD THE COMMERCIAL PROJECT

Be sure that the *MRAC Hotel.rvt* and *Understanding Groups.rvt* Projects are saved and closed.

1. On the Standard toolbar, click the Open icon.

**Tip:** The keyboard shortcut for Open is CTRL + O. **Open** is also located on the File menu.

- In the "Open" dialog box, click the *My Documents* icon on the left side.
- Double-click on the *MRAC* folder, and then the *Chapter06* folder.

    If you installed the dataset files to a different location than the one listed here, use the "Look in" drop down list to browse to that location instead.

2. Double-click *06 Commercial.rvt* if you wish to work in Imperial units. Double-click *06 Commercial Metric.rvt* if you wish to work in Metric units.

    You can also select it and then click the Open button.

The project will open in Revit Architecture with the last opened view visible on screen.

## LINK THE SITE PROJECT

The project is in much the same state as we left it at the end of the previous chapter. However, some important changes have been made since we closed it there. For this reason, be certain that you use the new dataset provided for Chapter 6 and do not attempt to continue in your own files from the previous chapter. The building still looks the same, but the toposurface is no longer in the file. The geometry for the site was broken out into a separate Revit project. Using techniques covered in this chapter, we will link the Site project back into the project we have loaded.

3. On the View toolbar, click the Default 3D View icon.

    The view named {3D} will open showing an axonometric of the building.

4. On the Project Browser, right-click the *Revit Links* node and choose **New Link**.

- In the Add Link dialog, choose *06 Commercial-Site.rvt* [*06 Commercial-Site Metric.rvt*], accept the default Center to Center positioning and then click Open.

We just buried our building! When the site file comes in, it is centered around the building in plan properly, but it is inserted much too high. We can use the techniques we learned in Chapter 4 to move the link and save the coordinates for future use. This is best accomplished in an elevation view.

5. On the Project Browser, double-click to open the *South* elevation view.

6. Use the Align icon to align the Street Level in both projects.

- After clicking the Align icon, click the Street Level in the host project first. Then click the Street Level in the Site project next (see Figure 6.34).
- Do not click the lock icon.
- On the Design Bar, click the **Modify** tool.

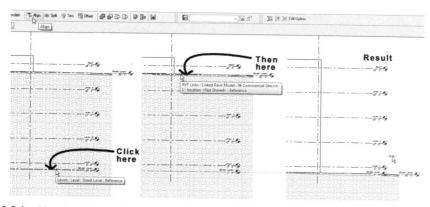

**Figure 6.34** *Use the Align tool to correct the mismatched height*

## ADD SHARED COORDINATES

The process of establishing shared coordinates was already covered in Chapter 4. This brief sequence will serve a quick refresher. Please refer to Chapter 4 for more information.

7. Select the linked site model.

- On the Options Bar, click the Properties icon (or right-click and choose **Element Properties**).

- Next to Shared Location, click the <Not Shared> button.

In Chapter 4, we used the Acquire option. Here we will stick with the default Publish option. This will keep the primary coordinates in the host file and share them with the site project. (Acquire would do the opposite.)

- A named location already exists in the site file called "Contours." We do not want to overwrite this, so click the Change button.

- Finally, click the Duplicate button and name the new Location: **Site** (see Figure 6.35.)

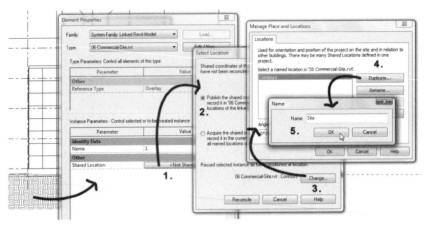

**Figure 6.35** *Set up shared coordinates*

- Click OK twice, then click Reconcile and finally click OK again.

8. Save the project.

- In the "Save Modified Linked Model" dialog, click Save and then OK.

This saves the new shared coordinates in both files. You may notice that the site uses a different contour file for its toposurface than the one we used in Chapter 4. This represents a simple refinement that simulates a later stage of the project where more detailed site information is now available.

## CONVERT A GROUP TO A LINK

Now that we have completed setup of our site model, we will next isolate the structural elements and create a separate linked file from them. This will leave only architectural elements (Walls, Doors, Windows, Column enclosures (Architectural Columns), Floors, and Roof) in the *06 Commercial.rvt [06 Commercial Metric.rvt]* project file. You may recall that in the previous chapter, we created Columns, Beams, Beam Systems, and Braces in our commercial project. These elements and the core Walls are the ones that we will separate out to their own structural model. However, some of these elements like the core Walls actually need to appear in both files. We'll look at a special way to achieve that as well.

The task of separating the structural elements into their own file can be accomplished in a few ways. We could create a separate file, and then select the required elements and copy and paste them from our project to the new one. We can also use the techniques already covered in this chapter and isolate the required objects using Groups. The process is as follows: select the desired objects, make a Group from them and then convert that Group to a Link. (We could also save the Group without converting it to a Link as well.) Once we have the Group saved as a separate file, we could open it directly, or import into another project. If you prefer to use copy and paste in the next sequence, feel free to do so. The steps that follow will highlight the approach using Groups.

9. On the View toolbar, click the Default 3D View icon.

The view named {3D} will re-open.

10. Using a Window selection, select all elements on screen.

• Click the filter icon, deselect everything except: Model Groups and Structural elements (see Figure 6.36).

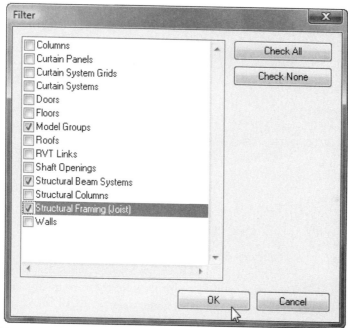

**Figure 6.36**   *Filter the selection to just framing elements*

At this point, we can go directly to the Group step, however, it is often good practice to use the Temporary Hide/Isolate icon (on the View Control Bar) to isolate the selected elements first to be sure you have the desired selection.

11. On the View Control Bar, click the Temporary Hide/Isolate icon and choose **Isolate Element** from the pop-up.

You should only have Beams, Braces, and Joists selected on screen.

• Click the Group icon on the Edit toolbar.

• In the "Create Model Group" dialog, type: **Structure** for the Name and then click OK (see Figure 6.37).

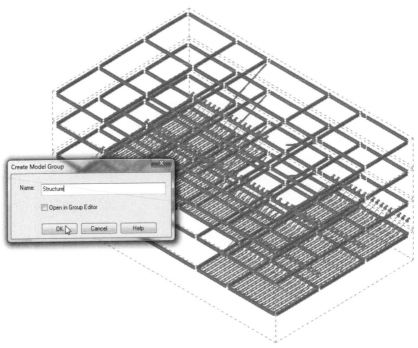

**Figure 6.37** *Group all the Structural framing*

At this point we have two options: we can simply save the Group to a file (as seen in the "Saving a Group to a File" topic above) or we can convert it to a Link (using the procedure covered in the "Convert a Group to a Linked file" topic above).

12. With the Structure Group still selected, click the Link button on the Options Bar.

- In the dialog that appears, choose Create New option and then click Save.

Using this method, we have quickly and efficiently gathered all of the structural framing and moved it to a separate linked Revit file. If you were to unload the Link now, none of the Structure would remain in the *06 Commercial.rvt* [*06 Commercial Metric.rvt*] project file. At this point we could open the file and continue the exercise in there. However, one limitation of either the save to Group or convert to Link method is that the resulting file is not based on the default Revit template. This means the resulting file has no levels, few annotations, and only the bare minimum of views. As a result, it will prove better in practice to create a new file using your preferred template first, and then insert the newly created file into it as a Group. In this way, we can ensure that the Structural file (or any file created this way) benefits from the office standards embedded in a template project.

## WORKING WITH COPY/MONITOR

To save a few steps, a file has already been created from the standard template and included with the other Chapter 6 files. We'll open this file, make a few preparations and then insert our Group into it. The preparations that we need involve copying the Levels and Grids from our main commercial project over to the structural file. While it is possible to simply copy these items to the structural file using Groups or copy and paste, a better approach is the Copy/Monitor tool which is specifically designed for this purpose. The Copy/ Monitor tool allows you to copy certain elements (Levels, Grids, Walls, Floors and Columns) from a linked file and keep them linked back to the originals. In this way, you can monitor changes as they occur and update the copies to match. This provides a very practical way for project teams to collaborate on shared elements even when they are not physically located in the same office.

13. Save and Close the *06 Commercial.rvt* [*06 Commercial Metric.rvt*] project file.

14. On the Standard toolbar, click the Open icon.

   **Tip:** The keyboard shortcut for Open is CTRL + O. **Open** is also located on the File menu.

- In the "Open" dialog box, click the *My Documents* icon on the left side.

- Double-click on the *MRAC* folder, and then the *Chapter06* folder.

   If you installed the dataset files to a different location than the one listed here, use the "Look in" drop-down list to browse to that location instead.

15. Double-click *06 Commercial-Structure.rvt* if you wish to work in Imperial units. Double-click *06 Commercial-Structure Metric.rvt* if you wish to work in Metric units.

   You can also select it and then click the Open button.

The project will open in with the *South* elevation view visible on screen. We had to close the commercial project above because Revit will not allow you have both projects (the host and the Link) open at the same time.

16. On the Project Browser, right-click the *Revit Links* node and choose **New Link**.

- In the Add Link dialog, choose *06 Commercial.rvt* [*06 Commercial Metric.rvt*].

- For Positioning, choose "By shared coordinates" and then click Open. In the warning that appears, click OK.

**BIM Manager Note:**   The warning simply informs us that the file has Links of its own that will not carry through to the current host. If we wanted them to, we could edit the Link type in the commercial project to be Attachment rather than Overlay. This is done in the Manage Links dialog. Overlaid reference file are direct links. Attachments can nest several levels deep.

17. On the Tools toolbar, click the Copy/Monitor icon and then choose the **Select Link** option from the pop-up.

- Click on the commercial project linked file on screen to select it.

   The Design Bar will change to the Copy/Monitor mode and all other tabs and tools will be hidden.

18. On the Design Bar, click the **Copy** tool.

- On the Options Bar, click the Multiple checkbox.

- Select Levels 1 through 4 and the Roof in the linked file.

- On the Options Bar, click the Finish button (see Figure 6.38).

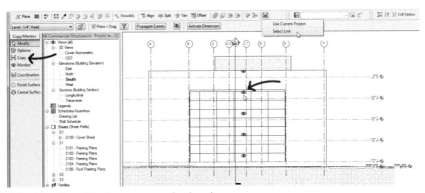

**Figure 6.38**  *Use Copy/Monitor to copy the Levels*

A small "eyeball" icon will appear next to each monitored item. If a Level is changed in the main commercial project, the structural engineer can re-enter the Copy/Monitor mode and use the **Coordination** tool on the Design Bar to synchronize the changes.

19. On the Design Bar, click the Finish Mode button.

## CREATE STRUCTURAL PLANS

20. From the View menu, choose **New > Floor Plan**.

- In the dialog that appears, select all of the Levels and then click OK.

21. On the Project Browser, expand Floor Plans, select *Level 1*, hold down the SHIFT key and then select *Roof*.

- From the View menu, choose **Apply View Template**.

- Select Structural Framing Plan and then click OK.

Having assigned this view template we now see only structural components. This will make selecting the remaining items for copy/monitor much easier.

## COPY/MONITOR GRIDS, WALLS AND COLUMNS

Now that we have copied the Levels and set up framing plans we are ready to copy and monitor the remaining structural items.

22. On the Project Browser, double-click to open the *Level 1* plan view.

If necessary, zoom to fit (type **ZF**).

- Use the Copy/Monitor steps above (with the multiple option) to copy all of the Grid, all of the structural (steel) columns, the floor slab and the four core Walls. (These are the only items currently showing on screen).

  The Grids, Walls, and Columns span all floors, but you will need to copy each floor slab separately per floor.

- When finished, click Finish Mode.

## INSERT A GROUP

To save a few steps, a file has already been created from the standard template and included with the other Chapter 6 files. We'll open this file and insert our Group into it now.

23. On the Project Browser, double-click to open the *South* elevation view.

- Use Temporary Hide/Isolate to hide the linked model.

24. From the File menu, choose **Load from Library > Load File as Group**.

  The "Load File as Group" dialog should open to the Chapter06 folder. (If you saved your Group file above to a different folder, browse there now).

- Select the *Structure.rvt* file and then click Open. (If a message regarding duplicate Types appears, click OK.

After a short pause, you will note that the Structure Group is now available on the Project Browser beneath the Model Groups branch.

25. On the Project Browser, beneath the Model Groups branch, right-click Structure and choose **Create Instance**.

- To insert it in the correct location, simply type **0** and then press ENTER.

- On the Options Bar, click the Finish button.

26. Save and Close the Structural model.

## USING RELOAD FROM TO SWAP A LINK WITH ANOTHER FILE

Now that have completed the setup of our structural model, we are ready to load it into our main commercial project. Since we already have a link to the Structural Group created above, the process will involve simply swapping file referenced by this Link. This is done in the Manage Links dialog.

27. Re-open the *06 Commercial.rvt* [*06 Commercial Metric.rvt*] file.

28. From the File menu, choose **Manage Links**.

- In the "Manage Links" dialog, click the RVT tab.

- Click on the *Structure* entry at the left.

- Click the Reload From button at the bottom (see Figure 6.39).

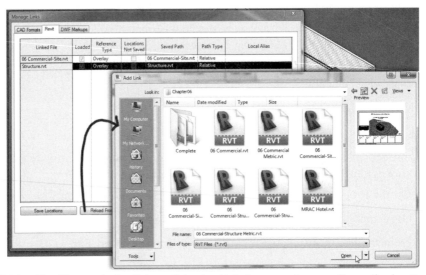

**Figure 6.39** *The Manage Links dialog—Reload a Link from an alternate file and/or location*

29. Select *06 Commercial-Structure.rvt* [*06 Commercial Metric-Structure.rvt*] and then click Open.

- If a dialog appears about other linked models, simply click OK.

- Click OK to close the Manage Links dialog.

In this example, we used "Reload From" because we wanted to point the linked file to a different project file than the one originally used. In normal circumstances, you will want to use the Reload button that simply loads the latest saved changes from the same RVT file. If you no longer want the Linked file, you can use the Remove button.

**BIM** *Manager Note:* If you try to open a project that is actively linked by the currently open project, Revit will display a warning indicating that the Linked file must first be unloaded before it can be opened. In other words, you cannot have both projects open at the same time in the same Revit session. This limitation does *not* prevent two different users from working simultaneously in each of the projects since each team member will be working on a different system. However, if both users are actively changing their respective models, you should save frequently and use the "Manage Links" dialog periodically to reload the linked project(s).

The best place to see the results of the work we have done here is in one of the section views. Open a section view and select the linked structural file. You will see it highlight on screen and be able to see clearly the elements that have been moved and copied to this linked file in comparison to those that remain in the host commercial project. In the next chapter, we will continue to refine the commercial project and its linked files. For now however, our work with Groups and Links in this chapter is complete.

30. Save and close all project files.

# SUMMARY

- Groups offer a powerful means to create and manage typical design conditions and keep all instances of them coordinated throughout the project.

- Any selection of objects can be grouped. Model element and Annotation elements however cannot occupy the same Group.

- When creating a Group from a selection of Model elements and attached Annotation elements (such as Tags) the Annotation becomes a separate Attached Detail Group.

- Attached Detail Groups contain annotation that is linked to the model elements in the corresponding Model Group.

- The insertion point of a Group can be adjusted by dragging the control on screen. Subsequent instances of the Group will insert relative to this location.

- When you edit any instance of a Group, the changes are applied to all instances.

- Elements can be excluded from individual instances of Groups making a unique condition. Changes to the Groups still apply to the other elements in the Group.

- Groups can contain other Groups creating so-called "nested" Groups.

- Groups can be saved to files thereby becoming independent Revit projects.

- Groups can be converted to linked to Revit projects.

- A separate Revit project can be linked to your current project. If the external project changes, the link will update to reflect the changes.

- Linked projects can be converted to Groups.

- Maintain proper positioning of linked projects using Shared Coordinates.

- Use Copy/Monitor to copy certain kinds of elements from a linked file and keep the copies synchronized with the original linked file.

- You can swap instances of Groups and Links with other Groups or Links respectively.

# CHAPTER 7

# Vertical Circulation

## INTRODUCTION

In this chapter, we will look at Stairs and Railings. We will explore these elements in both of our projects. The Residential Project contains existing Stairs on the interior and an existing exterior Stair at the front entrance. The core plan of the commercial building will include Stairs, Railings, and elevators. The exterior entrance plaza leading up to the commercial building calls for Stairs, a Ramp, and Railings.

## OBJECTIVES

We will add Stairs for the existing conditions of the residential project and lay out the core for the commercial project. Our exploration will include coverage of the Stairs, Railings, Ramps, and Elevators. After completing this chapter you will know how to do the following:

- Add and modify Stairs
- Add and modify Railings
- Add and modify Floors and Shafts
- Add and modify Ramps
- Add Elevators

## STAIRS AND RAILINGS

Revit Stairs are "sketch-based" objects. Sketch-based objects require that you sketch out their basic form with simple 2D sketch lines. This linework is then used to generate the 2D and/or 3D form of the final object. We have seen other examples of sketch-based objects such as the Roof and floor Slabs in the previous chapters. Stairs (like Walls) are System Families. This means that the Families are predefined in the software and any Types associated with them must be part of the original template from which our project was built. If we want to use a Stair Type that is not part of this original template, either we have to create a duplicate of one of the existing ones within the project, or we have to import a Type from another project (Transfer Project Standards or Copy and Paste can be used for this purpose). Stairs, like most Revit Architecture elements, have both instance and type parameters. Type parameters include riser and tread relationships, stringer settings, and basic display settings. Width and height parameters and clearances belong directly to the Stair object (instance parameters). Stairs automatically create Railings by default. For situations where the Railing is not required, we simply delete it after we build the Stair.

## INSTALL THE CD FILES AND OPEN A PROJECT

The lessons that follow require the dataset included on the Mastering Revit Architecture CD ROM. If you have already installed all of the files from the CD, simply skip down to step 3 to open the project. If you need to install the CD files, start at step 1.

1. If you have not already done so, install the dataset files located on the Mastering Revit Architecture CD ROM.

   Refer to "Files Included on the CD ROM" in the Preface for instructions on installing the dataset files included on the CD.

2. Launch Revit Architecture from the icon on your desktop or from the *Autodesk > Revit Architecture* group in *All Programs* on the Windows Start menu.

 **Tip:** In Windows Vista, you can click the Start button, and then begin typing **Revit** in the "Start Search field". After a couple letters, Revit Architecture should appear near the top of the list. Click it to launch to program.

   • If the New Features Workshop dialog appears, choose "Maybe later" and then click OK.
3. On the Standard toolbar, click the Open icon.

 **Tip:** The keyboard shortcut for Open is CTRL + O. **Open** is also located on the File menu.

   • In the "Open" dialog box, click the *My Documents* icon on the left side.

   • Double-click on the *MRAC* folder, and then the *Chapter07* folder.

   If you installed the dataset files to a different location than the one listed here, use the "Look in" drop-down list to browse to that location instead.
4. Double-click *07 Residential.rvt* if you wish to work in Imperial units. Double-click *07 Residential Metric.rvt* if you wish to work in Metric units

   You can also select it and then click the Open button.

The project will open with the last saved view visible on screen.

## ADD A STAIR TO THE RESIDENTIAL PLAN

This is the Residential project that was begun in Chapter 3. (We added some footings to it in Chapter 5, and the second floor existing conditions have also been laid out). In Chapter 3, you may recall that we completed the existing conditions without adding any Stairs. Our first exploration of Stairs will be to add the existing stairs to the Residential project now.

Adding Stairs is much like adding any other Revit Architecture sketch-based object. Use the *Stair* tool on the Design Bar to begin creating a Stair. Since it is a sketch-based object, the Design Bar immediately changes to Sketch mode. Sketch the required Stair components and then finish the sketch to create the Stair and its associated Railings.

1. On the Project Browser, double-click to open the *First Floor* plan view.

**BIM** *Manager Note:* An additional view named *First Floor (Chapter 3)* of the first floor also appears in the Project Browser. This view contains the dimensions that were used in Chapter 3 to set the locations of the Walls. While you can delete the dimensions without removing the associated constraints, it is more convenient to modify those constraints later if required if you do not delete them now. The recommended approach (used here) is to duplicate the view and apply dimensions to the copied view. Think of this as a "working view" as opposed to the original from which it was copied. The original is more likely to be presented and printed on Sheets; the "working" view simply provides a convenient place for us to work. Remember, in Revit Architecture views are simply live snapshots of the total model. Edits you make to model geometry in one view affect all other appropriate views automatically. Therefore, even though the dimensions appear in only one of the floor plan views, the constraints apply to the elements in the model in every view.

We'll start with a very simple Stair—the existing front entrance stairs.

2. Right-click and choose **Zoom In Region**.

 **Tip:** The keyboard shortcut for Zoom In Region is **ZR**. **Zoom In Region** is also located on the Views > Zoom menu.

• Click and drag a region around the front door at the bottom of the plan (see Figure 7.1).

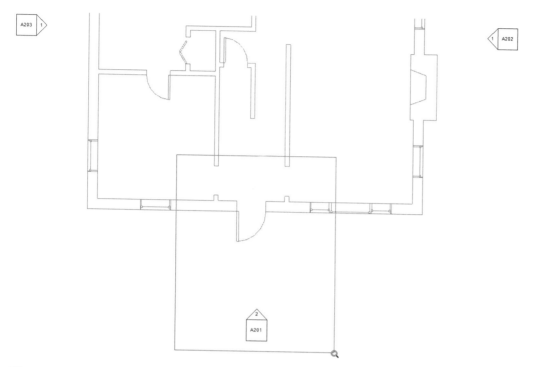

**Figure 7.1** *Zoom in on the front entrance*

3. On the Design Bar, click the Modeling tab and then click the **Stairs** tool.

The Design Bar changes to sketch mode and shows only the Stair tools.

The Design Bar will go into sketch mode showing only the tools available in that mode. The common "Modify," "Dimension," "Ref Plane," "Finish Sketch," and "Quit Sketch" tools will appear. In addition, Stairs have two sketch creation modes: "Boundary and Riser" and "Run." (Note that Boundary and Riser are two different buttons that are used together as parts of the same sketch.) You basically sketch out the plan view of the Stair in simple 2D linework and when you finish the sketch, Revit creates a Stair element from it that includes all of the correct parameters. A Stair sketch requires at least two boundary lines and several riser lines. If you use the "Run" option, you simply draw the path of the Stair, and Revit will automatically create the required Boundary and Riser lines. If you prefer, you can choose either the "Riser" or "Boundary" mode to draw these items manually. In general, you can create most common stair configurations including straight runs, U-shaped, L-shaped, etc. with the "Run" option. Use the other modes when you wish to modify the automatically created graphics to add a special feature to the Stair or when you want to build a sculptural Stair. We will explore these options later in the tutorial. For now, let's stick with the "Run" option.

Also like other sketch-based Revit Architecture objects, the Stair has a "Stair Properties" button. You use this button to access the instance and type parameters of the Stair that you are sketching. You cannot use the Properties icon on the Options Bar for this. This is because the Options Bar shows options for the current element. When you are in sketch mode, the current element is the Model Lines that you are sketching, *not* the object (Stair in this case) that the sketch lines will become. An additional button also appears for Stairs. This is the "Railing Type" button. When you create a Stair, Revit automatically adds a Railing, which is a separate object. Most stairs require railings, so Revit Architecture adds them automatically. Use this control to choose the Railing Type you wish it to add, or if you prefer, choose "None" to instruct the tool not to add a Railing (see Figure 7.2).

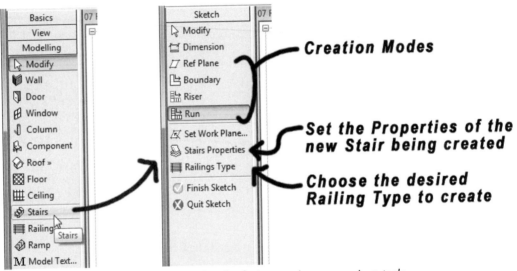

**Figure 7.2**   *The Sketch-mode Design Bar for Stairs contains some unique tools*

- On the Design Bar, verify that the "Run" option is selected and then click the Stair Properties button.

- In the Element Properties dialog, from the Type list, choose **Monolithic Stair**.

- Beneath the "Constraints" grouping, for the "Base Level" choose **Site**, and for the "Top Level" choose **First Floor**.

- Beneath the "Graphics" grouping, uncheck all the boxes.

These boxes control where and when arrows and annotation will appear indicating the direction of a Stair in plans. In this case, this is a simple existing entrance stair. It does not require any graphics in plan.

- Beneath the "Dimensions" grouping, set the "Width" to **4'-0"** [**1200**].

- Beneath the "Phasing" grouping, from the "Phase Created" list, choose **Existing** and then click OK (see Figure 7.3).

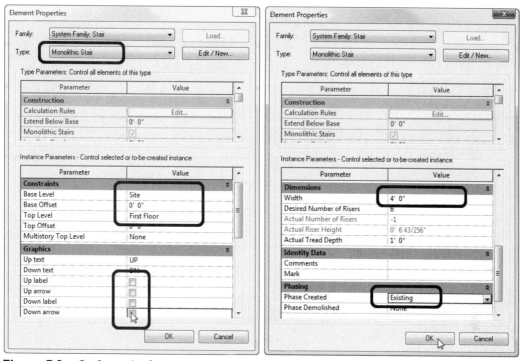

**Figure 7.3**  *Configure the Stair parameters*

4. Move your pointer to the middle of the front Door.

- When the guideline appears at the midpoint, click the mouse to set the first point of the Stair run (see the left side of Figure 7.4).

When you create the Stair, the first point is at the bottom of the run and the last point you click is at the top. If you click just two points, you get a straight run. As you drag the mouse (after the first pick) a note in grey text will appear indicating how many risers you have placed. If you click again before you use up all the risers, you create a landing. In this case, we only have 6 risers and want a straight run, so we will click only one more point. However, because of the direction required for this stair, we are going to click the second point "up" (inside the house) and then move the resultant Stair sketch to the proper location outside the house. As an alternative to this method, we could have created Reference Planes first to place the Stair in the precise location. We will try the Reference Plane approach below.

- Move the pointer straight up and when the number of risers reads "0 Remaining" click the second point (see the middle of Figure 7.4).

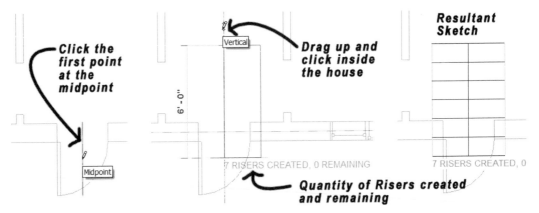

**Figure 7.4**  *Create a straight run of Stairs with two clicks*

A sketch of blue and green lines will appear indicating the Stair (see the right side of Figure 7.4). The green lines are the boundary lines of the Stair, and the blue lines are the risers. You can edit this sketch in any way before clicking Finish Sketch on the Design Bar. In this case, the only edit we need to do is move the sketch to the correct location.

5. Dragging from left to right, surround the entire Stair sketch.

  **Note:** Don't worry about selecting the neighboring elements. When you are in sketch mode, you can only select elements of the sketch.

All the sketch lines will highlight red.

6. On the Tools toolbar, click the **Move** tool.

- Snap to the Endpoint at the top of the sketch (see the left side of Figure 7.5).

- Snap to the Intersection of the sketch and the outside edge of the house (see the middle of Figure 7.5).

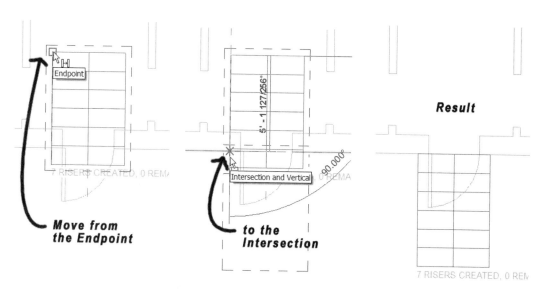

**Figure 7.5** *Move the Stair sketch and snap it to the outside of the house*

The new location of the sketch should match the image on the right side of Figure 7.5.

    7. On the Design Bar, click the Railings Type button.

      • In the Railings type dialog, choose **Handrail - Rectangular [900mm]** and then click OK.

    8. On the Design Bar, click the Finish Sketch button.

You will exit the sketch mode and the Stair will appear in the plan. You can view it in other views as well, to see that it was created properly. Try any section, elevation, or 3D view for this. To get the best look at it, we are going to view it in elevation perpendicular to the direction of travel.

## MODIFY THE RAILING

Sometimes when you create an element, it looks fine in plan view, but then you look at it in another view and notice that something is not correct. This is the case here.

    9. On the Project Browser, double-click to open the *East* elevation view.

      • Use Zoom In Region again, to zoom in on the Stair (see Figure 7.6).

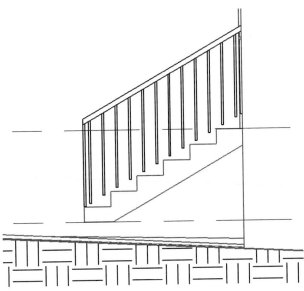

**Figure 7.6** *Study the Stair in an elevation view*

The Stair itself looks fine; notice that the balusters do not attach to the Stair treads. There is an easy fix for this.

10. On the Project Browser, double-click to open the *First Floor* plan view.

- Move your mouse over the Stair to pre-highlight it, and then over one of the Railings.

Notice that the Railing is actually next to the Stair rather than overlapping it. This is why the balusters terminate the way they do. If we flip the Railings to make them sit on top of the Stair, then the balusters will project down to the treads.

11. Select one of the Railings.

- Click the Flip Railing Direction control. Repeat for the other Railing.

- On the Project Browser, double-click to open the *East* elevation view (see Figure 7.7).

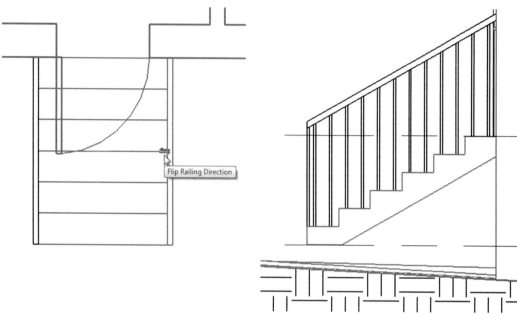

**Figure 7.7**  *Measure a distance with the **Tape Measure** tool*

    12. Select both Railings, edit their properties, verify that their Phase Created is set to Existing, and then click OK.

## MODIFY THE TOPOSURFACE

As we continue to view our Stair in the *East* elevation view, we discover that there is another perhaps even more obvious issue; the Stair does not actually sit on the grade. We could edit the Stair and adjust it accordingly, but upon further consideration, it is typically not desirable to have the terrain slope toward the building structure. Therefore, a better approach in this situation would be to adjust the Toposurface to provide better grading.

    13. On the Project Browser, double-click to open the *Site* plan view.

- Select the Topography: Surface object.

Be sure to select the topography and note the linked DWG file.

- On the Options Bar, click the Edit button.

    14. On the Design Bar click the **Point** tool.

- On the Options Bar, type **-3'-1"** [**-940**].

- Snap a point to each of the lower points of the building and to the two lower points of the Stair (see the left side of Figure 7.8).

- On the Options Bar, type **-2'-10"** [**-860**].

- Snap a point between the corner of the building and the Stair on each side of the Stair (see the right side of Figure 7.8).

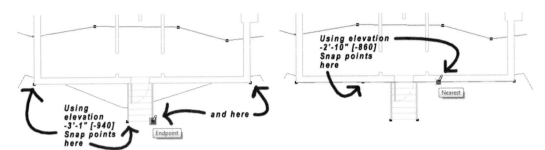

**Figure 7.8** *Add points along the front of the house and Stairs*

15. On the Design Bar, click the Finish Surface button.

16. On the Project Browser, double-click to open the *East* elevation view.

Zoom in again if necessary. Notice that the Stair is now touching the ground instead of floating in space and the terrain now has positive drainage at the front of the house (see Figure 7.9).

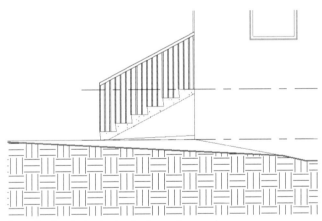

**Figure 7.9** *After the edit, the Stair sits properly on the grade*

17. Save the project.

## COPY AND DEMOLISH A STAIR

At the back door of the existing house is another Stair like the one in the front. However, this one will be demolished to make way for the new addition. The easiest way to create this Stair is to copy, rotate and modify it from the one we just created at the front.

18. On the Project Browser, double-click to open the *First Floor* plan view.

19. Select both the Stair and its Railings. (Use a marquee selection or the CTRL key.)

- On the Tools toolbar, click the **Copy** tool and make a copy of the selected Stair and Railings and place it near the back of the house within the space of the new addition (the exact location is not important yet).

- With the Stair and Railings still selected, click the **Rotate** tool on the Tools toolbar.

- On the Options Bar, type **180** in the Angle field and then press ENTER.

- Move the Stair and Railings so that it touches the house centered on the existing back Door (see Figure 7.10).

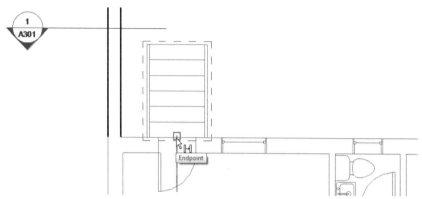

**Figure 7.10**  *Copy, Rotate, and then Move the Stair and Railings into place at the back Door of the existing house*

20. On the Design Bar, click the **Modify** tool or press the ESC key twice.

21. Double-click the blue vertical Section Head (cutting through the new addition) or open the *Transverse* section view from the Project Browser.

You will note, similarly to above, that the Stair floats above the terrain at the back of the existing house. This is because the grade slopes from the front of the house down toward the back. So by the time we reach the back of the house, the Stair requires a few additional risers.

22. On the Tools toolbar, click the **Tape Measure** tool.

- Snap to the bottom corner of the Stair for the first point.

- Click on the terrain directly below the Stair for the second point (see Figure 7.11).

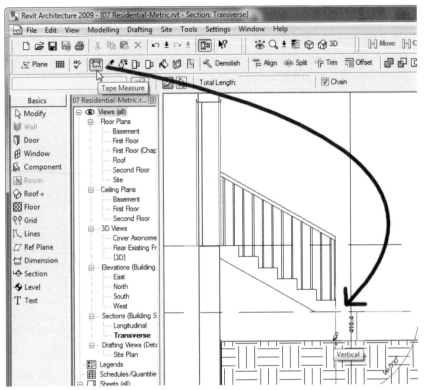

**Figure 7.11** *Measure the distance between the bottom of the Stair and the terrain*

The distance rounded off should be 1'-4" [410]. We will use this value to adjust the Stair height, which will add risers to it.

23. Select the Stair, and then on the Options Bar, click the Edit button.

A dialog will appear indicating that you need to choose a view better suited to the edit than the currently active section view. The sketch was created in a plan view and it would be difficult or impossible to edit it from section.

24. From the "Go To View" dialog, choose *Floor Plan: First Floor* and then click Open View.

- On the Design Bar, click the Stair Properties button.

- In the Base Offset field, type **-1'-4" [-410]**.

- Scroll down and change the "Desired Number of Risers" to **8**.

- Verify that the Phase Created is set to Existing and then click OK.

25. Select the two riser lines at the top of the sketch (the bottom two risers in the Stair) and on the Edit toolbar, click the Copy icon.

- Snap to the points indicated in Figure 7.12 to add two new risers to the sketch.

26. Use the Trim/Extend tool to extend the Boundary lines to the new risers as shown on the right side of Figure 7.12.

- On the Design Bar, click the Finish Sketch button.

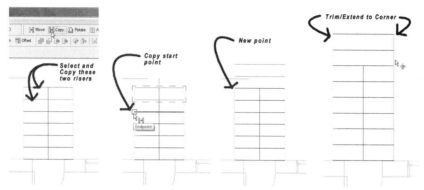

**Figure 7.12**   *Change the Phasing parameters of the Stair and Railings to demolish them*

27. On the Tools toolbar, click the Demolish tool.

- Click on the Stair and each of the Railings to mark them as demolished in the current phase.

- On the Design Bar, click the **Modify** tool or press the ESC key twice.

 **Note:** You can also select the Stair and Railings, edit their Properties, and then change the "Phase Demolished" setting to New Construction.

The Stair and Railings will turn dashed to indicate that they are to be demolished.

## CREATE AN EXISTING CONDITIONS VIEW

The next Stair that we will build is the main stair in the existing house. Before we do, let's explore a technique to make working with phases a bit simpler. In Chapter 3, we temporarily changed the current Phase of the *First Floor* plan view (named *Level 1* in that chapter) to "Existing" to add all of the existing construction. At the end of the lesson, we set the Phase back to "New Construction." We could repeat that process here. As an alternative approach, it is easier to maintain two first floor plan views—one set to the Existing phase, and the other set to the New Construction phase. In this way, you can simply open the view for the phase in which you wish to work.

28. On the Project Browser, right-click the *First Floor* plan view and choose **Duplicate View > Duplicate**.

- Right-click the new *Copy of First Floor* view and choose **Rename**.

- For the new Name type **Existing First Floor** and then click OK.

  The *Existing First Floor* view should have opened automatically (it should be bold on the Project Browser). If it did not, double-click to open it now.

- Right-click in the drawing window and choose **View Properties**.

- From the "Phase" list (scroll down), choose **Existing** and then click OK.

This will set the current Phase to Existing. This means that any elements we add while working in this view will be assigned to the Existing construction phase automatically. We did not *need* to create this view, but sometimes it is easier than remembering to set the Phase parameter of each object we create or changing the Phase of the view back and forth. With Existing set as the current Phase, notice that the rear Stair is no longer dashed. It is still set to be demolished, but since we are now viewing the plan as it looks during the Existing Phase, these Stairs are not yet demolished at this point in time. The Walls of the addition have also disappeared. Again, during the Existing Phase, those Walls do not yet exist.

29. Double-click to open the *First Floor* plan view.

Notice that this view, still set to New Construction Phase, continues to show all of the new construction Walls and the demolished Stair.

30. Double-click to open the *Existing First Floor* plan view.

## CREATE A STRAIGHT INTERIOR STAIR

The existing interior Stair is also a simple straight-run stair going from the first to the second floor. However, there are a few custom details on the first few treads and some Railing treatments that will make the Stair a bit more interesting.

31. Working in our newly created *Existing First Floor* plan view, Zoom and Pan to the middle of the existing house layout.

32. On the Design Bar, click the Modeling tab and then click the **Stairs** tool.

    The Design Bar again changes to sketch mode and shows only the Stair tools.

Sometimes it is difficult to set the precise location of the element that you are placing. In the case of Stairs, the Run option adds the Stairs from the midpoint at the bottom of the run. We do not currently have a convenient point to snap to at that location. We could simply draw the sketch at a random location and then move it into position relative to the Walls. In many cases, you will find this to be the easiest approach. An alternative approach that we will explore here is to use Reference Planes. Reference Planes are simple datum objects in Revit. You can add them anywhere that you need assistance in achieving alignments or reference points. Think of them as analogous to blue pencil lines in hand drafting.

33. On the Design Bar, click the **Ref Plane** tool.

- On the Options Bar, click the Pick Lines icon.

- In the Offset field, type **1'-6 ½"** [**471**]. (This is half the width that the Stair needs to be.)

- Highlight the right side of the vertical Wall on the left of the stair corridor. When the green guideline appears in the middle of the stair space, click to create the Reference Plane (see the left side of Figure 7.13).

- On the Options Bar, change the Offset parameter to **3'-6"** [**1100**].

- Offset a Reference Plane up from the inside face of the bottom exterior Wall (see the right side of Figure 7.13).

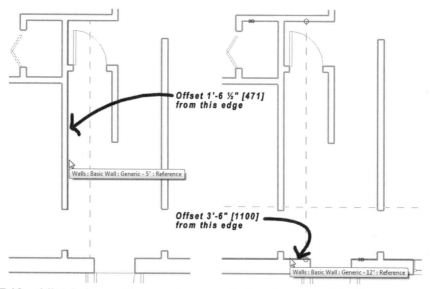

**Figure 7.13**   *Offset Reference Planes to the create the insertion point of the Stair*

34. On the Design Bar, click the **Modify** tool or press the ESC key twice.

**BIM Manager Note:**   The way in which Reference Planes are presented here is one acceptable use for them. However, they are also part of the constraint mechanism in Revit. This means that they are very similar to Levels & Structural Grids (see previous chapters). If an element somewhere else in the model is constrained to a Reference Plane, the location of the Reference Plane will drive the position and/or shape of the constrained geometry. This important feature can be exploited directly in the project, but is most often seen in the Family editor. Detailed coverage of the Family editor and Reference Plane usage within Families can be found in Chapter 10.

35. On the Design Bar, click Modeling, choose Stairs, and then click the **Run** tool.

- On the Design Bar, click the Stair Properties button.

Notice that Revit remembers the settings of the previously created Stair. Several settings must be adjusted to make this Stair match the interior Stair occurring in the existing house.

- In the Element Properties dialog, from the Type list, choose **7" max riser 11" tread [*190mm max riser 250mm going*]**.

- Verify that the "Base Level" is set to **First Floor** and the "Top Level" is **Second Floor**.

- Beneath the "Graphics" grouping, place checkmarks in only the "Up label" and "Up arrow" boxes; be sure the others are cleared.

- In the "Dimensions" grouping, change the Width to **3'-1"** [**942**] and then click OK (see Figure 7.14).

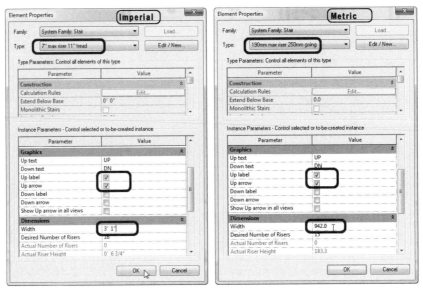

**Figure 7.14**   *Set up the parameters for the Stair*

36. Click the first point of the Stair run at the intersection of the two Reference Planes.

- Move the pointer straight up far enough to place all of the risers in a single run and then click.

The sketch created will be too long for the space. The parameters configured above were all "instance" parameters. Instance parameters apply directly to each individual object in the model. To adjust our Stair to fit the space, we need to modify the "type" parameters. Type parameters apply to all objects sharing the same type. We can edit them directly, but any changes would affect all Stairs in the model with type: *7" max riser 11" tread [190mm max riser 250mm going]*. A more prudent practice is to first create a new type and then adjust its parameters as appropriate. In this way, we preserve the existing type for use on other Stairs while creating a variation suited to the Stair we are currently creating.

## CREATE A NEW STAIR TYPE

In this sequence, we will edit the parameters of the Stair Type and make some minor modifications to help it conform to the existing conditions. This will involve applying less stringent tread and riser rules on our Stair. Since the stairs existing in this house were built before current code requirements were in place, an alternate Stair Type needs to be created allowing more freedom.

37. On the Edit menu, choose **Undo Run** (or press CTRL + Z).

This will remove the sketch so that we can adjust the parameters and re-create it.

38. On the Design Bar, click the Stair Properties button.

• At the top of the dialog, click the Edit/New button.

• Click the Duplicate button.

 **Tip:** To do this quickly, press ALT + E and then immediately press ALT + D.

• In the "Name" dialog, type **Existing House Stair** and then click OK.

Here you can enter a range of values for minimum and maximum tread and riser. You can also enter building code values for your jurisdiction in a rule-based calculator (click the Edit button next to "Calculation Rules" to do this). Values assigned to these rules will constrain the parameters of the Stair as it is being sketched. Again, because the Stair we are building is an existing Stair, we will set the limits very broadly so that we can enter actual field values without restriction.

• Beneath the "Treads" grouping, type **8"** [**200**] for the "Minimum Tread Depth."

• From the "Nosing Profile" list, choose **Default**.

• Beneath the "Risers" grouping, **12"** [**300**] for the "Maximum Riser Height" (see Figure 7.15).

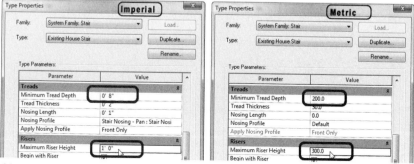

**Figure 7.15** *Edit the Type parameters of the new Stair Type*

39. Click OK to accept the values and return to the "Element Properties" dialog.

• Beneath the "Dimensions" grouping, set the "Actual Tread Depth" value to **8"** [**250**].

• Set the "Desired Number of Risers" parameter to **14** and then click OK (see Figure 7.16).

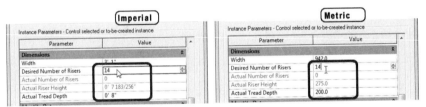

**Figure 7.16** *Set the actual Tread and Riser settings of the existing house Stair*

40. On the Design Bar, click the **Run** tool.

- Click the first point of the Stair run at the intersection of the two Reference Planes.

- Move the pointer straight up far enough to place all of the risers in a single run and then click.

Notice that the Stair sketch now fits the available space (see Figure 7.17).

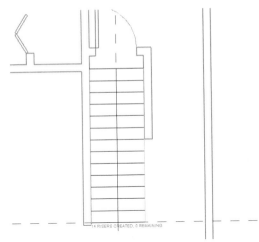

**Figure 7.17** *The Stair sketch based on the newly created Stair Type properly fits the space*

41. On the Design Bar, click the Finish Sketch button.

The Stair object will appear with its railings. Notice that a cut line appears automatically. Also, since we asked only for "up" annotation, we only get an arrow and label pointing up, not up and down. We will add the Stair going down to the basement later.

**EDIT THE STAIR TYPE**

When you look at the Stair we created, it appears to be too wide—part of it overlaps the neighboring Walls. This overlap is actually the stringer of the Stair and the Railing that sits on top of the stringer. We will need to make some adjustments to the Stair Type to address this.

42. Select the Stair.

- On the Options Bar, click the Properties icon. (You can also right-click and choose **Element Properties**.)

- Click the Edit/New button (or press ALT + E).

- Beneath the "Stringers" grouping, choose **Open** from both the "Left Stringer" and the "Right Stringer" lists.

- For the "Trim Stringers at Top" option, choose **Match Level**.

An "Open" stringer is notched in the shape of the treads and risers and supports the treads from underneath. The "Closed" stringer occurs at the edges of the Stair with the treads and risers spanning in between. A typical wooden stair would use an open stringer, while a steel pan stair typically would be closed. The tops of the stringers can be cut to match the Level or they can be left uncut. Illustrations of several of stringer settings are shown in Figure 7.18.

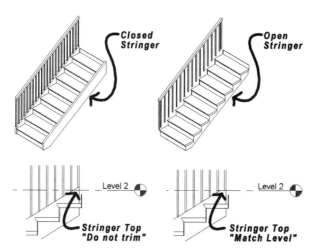

**Figure 7.18** *Exploring the various Stringer settings*

In addition to these settings, we can also edit the dimensions of the stringer material such as the height and thickness. In this case, we will leave those settings as they are.

43. Click OK twice to apply the changes and return to the model.

It appears as though nothing has changed in our plan view. However, if you slowly pass your mouse over the Stair and its edge, you will note that when the Stair pre-highlights, it has indeed reduced in width. The overlapping portion that we still see is actually the Railing as on the other Stair we worked with earlier.

44. Click on one of the Railings.

- Click the small "Flip Railing Direction" control to flip the Railing to the other side of the stringer.

- Repeat on the other side. (This is similar to the situation pictured above in Figure 7.7.)

## CREATE THE "DOWN" STAIR

In addition to the Stair we have added here, the house also has an existing Stair going down to the basement. The simplest way to create this one is with Copy and Paste.

45. On the Design Bar, click the Basics tab and then click the **Section** tool.

   - Create a section line running vertically through the Stair.

   - Deselect the Section Line (click away from it) and then double-click the Section Head to open the associated view.

In this view, you can see very clearly that we have a single Stair spanning from first to second floor.

46. Select the Stair.

   - From the Edit menu, choose **Copy to Clipboard** (or press CTRL + C).

   - From the Edit menu, choose **Paste Aligned > Select Levels by Name**.

   - In the "Select Levels" dialog, choose Basement and then click OK (see Figure 7.19).

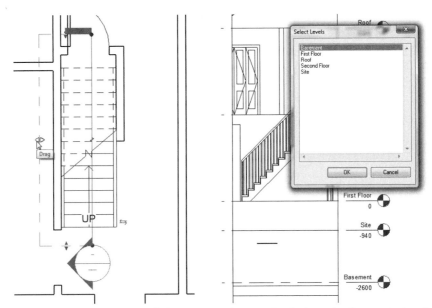

**Figure 7.19**  *Add a section view and then copy and paste the Stair to the Basement Level*

47. With the new Stair still selected, on the Options Bar, click the Properties icon.

   - Verify that the "Base Level" is Basement and that the "Top Level" is First Floor.

   - Remove the "Top Offset" by typing a value of **0** (zero) and then click OK (see Figure 7.20).

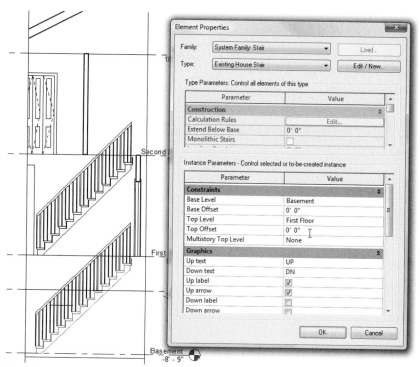

**Figure 7.20** *Copy and Paste the Stair to the Basement*

48. On the Project Browser, double-click to open the *Existing First Floor* plan view.

49. Select the basement Stair (the one just pasted—you may need to use the TAB key to select the right one).

   • On the Options Bar, click the Properties icon.

   • Beneath the "Graphics" grouping, clear the "Up arrow" and "Up label" checkboxes.

   • Place checkmarks in the "Down arrow" and "Down label" checkboxes and then click OK.

50. On the Project Browser, double-click to open the *Basement* floor plan view.

Notice only the basement Stairs appear here. If you open the Second Floor plan view as well, you will note that only the upper Stairs appear. However, since we have not yet added any Floor elements to our residential model, they will show even beneath the closet at the front of the house. We'll adjust this in later chapters.

## EDIT THE STAIR SKETCH

So far we have used the "Run" option to automatically create very simple Stair sketches based upon our riser and tread dimensions. However, you can edit these simple sketches to add architectural detailing to your Stairs. You can also completely create your own custom sketch to create very complex Stair configurations. In this example, we will widen the lower portion of the Stair and add a bull nose to the bottom two treads.

51. On the Project Browser, double-click to open the *Existing First Floor* plan view.

- From the Window menu, choose **Close Hidden Windows**.

52. Select the "Up" Stair. (Use the TAB key if necessary to assist in selection.)

- On the Options Bar, click the Edit button.

This returns you to sketch mode and reveals the sketch lines created when we added the Stair. By adjusting these sketch lines, we can change the shape of the Stair. The Stair sketch uses three colors on screen to indicate the function of each line. Green indicates a boundary line. The Stair must have two boundary edges. The blue line at the center indicates the Stair run. Editing this line has an effect on the overall Stair. It behaves much like the Run sketch option used to create the Stair. The black lines are the individual risers. A gray text note will also appear near the sketch indicating how many risers are created and required (see Figure 7.21).

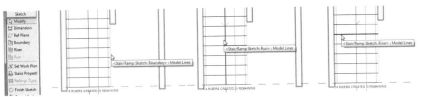

**Figure 7.21**    *Stairs include three types of sketch lines*

**BIM** **Manager Note:**    If you position your pointer over one of the black riser lines or green boundary lines to pre-highlight it, you will notice in the screen tip that appears that the sketch lines of the Stair are actually "Model Lines." There are two types of line elements in Revit: they are Model Lines and Drafting Lines. A Model line, while two-dimensional, is a part of the building model. Therefore, like other elements of the model, it will appear in *all* appropriate views. Imagine it as if the line was actually painted on the floor or wall much like the stripping on roads. A Drafting line on the other hand appears *only* in the view in which it is drawn. Model Lines are used to represent real things, while Drafting Lines are like annotation, used to embellish a particular view to convey intent. Model Lines are used in the Stair sketch because they indicate the actual form of the Stair in the model.

When you are working on a Stair sketch, there are three sketching methods on the Design Bar: Boundary, Riser, and Run. We have seen Run already in each of the Stairs we built earlier. If you have placed all of the required risers, the Run option will be unavailable. To edit the existing run, you would edit the blue sketch line. When you wish to customize an existing Stair, or when you wish to draw a Stair free-form, you use the Riser and Boundary options. In this case, we'll edit the boundary.

53. On the Design Bar, click the **Boundary** tool.

- On the Options Bar, select the Draw and Line icons and place a checkmark in the "Chain" checkbox.

- Click the first point of the new boundary Line at the intersection of the right-hand Boundary Line and the bottom of the existing Wall (see "Point 1" in Figure 7.22).

- Click the next point at the other side of the Wall's width (see "Point 2" in Figure 7.22).

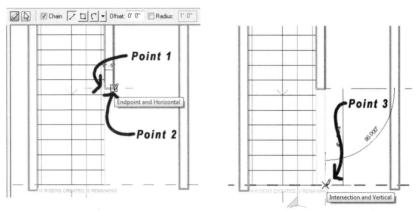

**Figure 7.22**   *Sketch two new Boundary Lines snapping to the existing Wall*

- Place the last point aligned with the original boundary Line's bottom end and the Wall's right face (see "Point 3" in Figure 7.22).

54. On the Design Bar, click the **Modify** tool or press the ESC key twice.

- Use the **Trim/Extend** tool (with the "Trim/Extend to Corner" option) to remove the unneeded segment (see Figure 7.23).

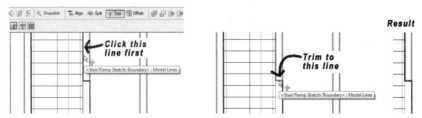

**Figure 7.23**   *Trim the existing Boundary Line to join it to the new sketch*

- With Trim/Extend still active, on the Options Bar, choose the "Trim/Extend Multiple Elements" icon.

- For the Trim/Extend boundary, click the newly drawn boundary Line (vertical one).

- Click each of the riser Lines to extend them to this boundary (see Figure 7.24).

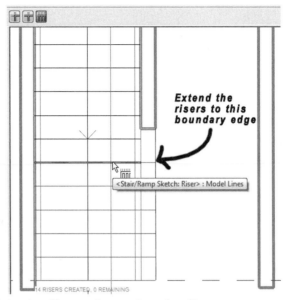

**Figure 7.24** *Extend the riser Lines to the new boundary Line*

- On the Design Bar, click the **Modify** tool or press the ESC key twice.

On the lower portion of the Stair, the risers will now come out flush with the Wall. Before we add the bull nose to the lower two steps we need to stretch the Boundary Line back a bit. Otherwise, the stringer will follow the shape of the bull nose.

55. Click the right Boundary Line.

- Using the drag handle at the bottom, drag the endpoint up and snap it to the end of the third Riser Line.

- When the error dialog appears, click the Unjoin Elements button (see Figure 7.25).

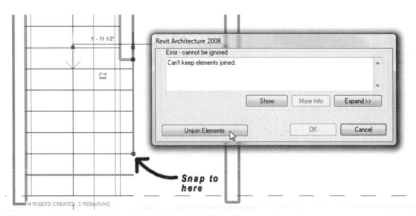

**Figure 7.25** *Reduce the length of the boundary Line by two treads*

The boundary line is joined to each of the riser lines. This is required to create a valid Stair. When we stretch the boundary line as indicated above, it breaks this relationship with the lower risers. Next we will edit the shape of the riser lines to make the bull nose condition. Before Revit will allow us to finish the sketch, we must join them back to the boundary.

## ADD A BULL NOSE RISER

To add the bull nose riser, we simply sketch them in using the riser sketch tool on the Design Bar.

56. On the Design Bar, click the **Riser** tool.

- On the Options Bar, clear the "Chain" checkbox.

- Click the "Arc passing through three points" icon.

- Place the first point of the arc at the right endpoint of the bottom Riser Line (see "Point 1" in Figure 7.26).

- Place the second point at the endpoint of the third Riser line from the bottom (see "Point 2" in Figure 7.26).

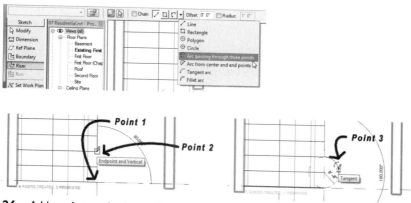

**Figure 7.26**   *Add an Arc to the lower Riser*

- Move the pointer to the right and when the shape snaps to a half-circle, click to place the last point (see "Point 3" in Figure 7.26).

- Repeat the process to create a smaller Arc (one tread deep) between the second and third risers (see Figure 7.27).

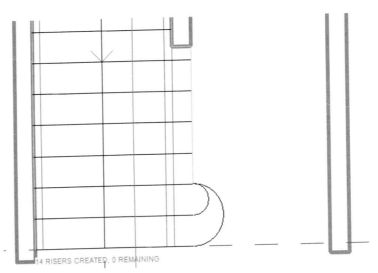

**Figure 7.27** *Add an additional Riser Arc*

57. On the Design Bar, click the Finish Sketch button.

## VIEW THE STAIR IN 3D

At this point, it might be easiest to understand what we have created if we view the edited Stair three-dimensionally. We have used the "Default 3D View" icon several times already. However, this tool will show the entire model in 3D. With all of the exterior Walls visible, it will be impossible to see the Stairs inside the building from such a vantage point. There are a few tricks we can use to quickly view just the Stairs in 3D.

58. On the Window menu, choose **Close Hidden Windows**.

 **Note:** If this command is not available, then you already have only the current view window open.

- On the View toolbar, click the Default 3D View icon.

- From the Window menu, choose **Tile**.

 **Tip:** The keyboard shortcut for Tile is **WT**.

59. Click in the *First Floor* view window to make it active.

- Select the Stair, Railings, and their immediately adjacent Walls (see the left side of Figure 7.28).

 **Tip:** Use the CTRL key to click on each object that you wish to add to the selection. Try to avoid selecting the basement Stair.

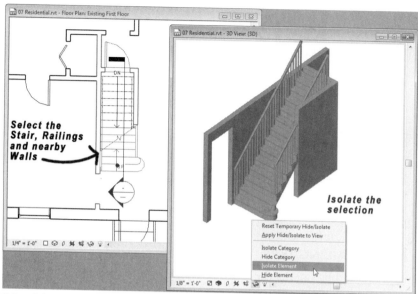

**Figure 7.28**  *Select elements in the plan view, isolate them in the 3D view*

  **Tip:** The keyboard shortcut for Isolate Object is **HI**. **Isolate Object** is also located on the View > Temporary Hide\Isolate menu.

- From the View Control Bar in the 3D view window, choose **Isolate Element** from the Temporary Hide/Isolate pop-up icon (see the right side of Figure 7.28).

Feel free to use the "Dynamically Modify View" controls to spin the isolated selection around to an alternative viewing angle. Notice the way the two treads at the bottom of the Stair have the bull nose applied to them. Notice also the way that the stringer stops before these two treads. These are direct effects of the way that we sketched the Stair above.

## ADJUST A RAILING SKETCH

Having completed the edit of the Stair above, it is now apparent that the Railing could use a bit of adjustment as well. You edit Railings the same way as Stairs—edit the sketch.

60. Zoom in on the Railing on the right side.

As you can see, this Railing has matched the shape of the Boundary Line that we sketched above and makes a slight jog as it goes up the run of Stairs. Let's eliminate the portion of the Railing above the jog (adjacent to the Wall).

61. Select the Railing on the right side (use TAB if necessary to select it. Select in either view).

- On the Options Bar, click the Edit button.

- Delete the top two sketch lines, (the vertical one and the short horizontal one that are adjacent to the Wall).

- On the Design Bar, click Finish Sketch (see Figure 7.29).

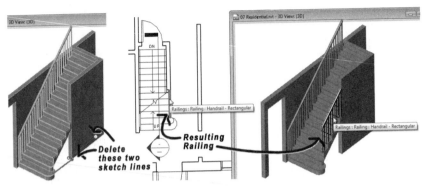

**Figure 7.29**  *Edit the Railing sketch*

## CREATE A NEW RAILING TYPE

The Railing on the other side of the Stair does not need balusters or posts. It is attached to the Wall. To represent this correctly, let's duplicate the existing Railing Type and then modify it.

62. Select the long Railing on the left side of the Stair.

- On the Options Bar, click the Properties icon.

- At the top of the dialog, click the Edit/New button.

- Click the Duplicate button.

**Tip:** To do this quickly, press ALT + E and then immediately press ALT + D.

- In the "Name" dialog, type **Existing House Railing – No Balusters** and then click OK.

- In the Type Properties dialog, beneath the "Construction" grouping, click the Edit button next to "Baluster Placement."

- In the "Edit Baluster Placement" dialog, in the "Main Pattern" area (at the top), choose **None** from the Baluster Family list for element 2 (see the top of Figure 7.30).

- In the "Posts" area (at the bottom), choose **None** for all three components.

- Clear the "Use Baluster Per Tread On Stairs" checkbox (see the bottom of Figure 7.30).

- Click OK three times to return to the model window.

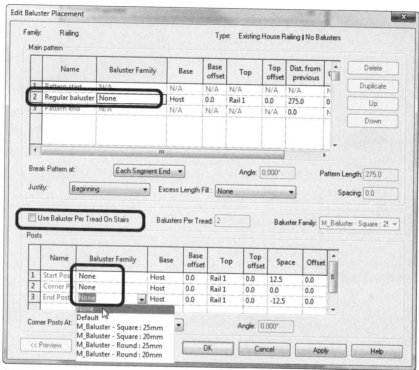

**Figure 7.30**   *Remove balusters and posts from the new Railing Type*

## IMPORT A NEW RAILING TYPE

Railings (like Stairs, Walls, Floors, and Roofs) are System Families. System Families are so called because they are "built in" to the system. This means that the parameters, behavior and overall geometric characteristics of System Families are predefined in the software and not editable by the user. In the case of Railings, there is a single Railing System Family. Each System Family can, however, have multiple Types. A Type is a collection of parameters that are applied automatically to all elements assigned to the Type. If a Type parameter is modified, it affects all elements of that Type in the entire project.

The item we just created: ***Existing House Railing – No Balusters***, was a Type. In some cases, you will build a Type like this and later wish to use it in another project. However, unlike Component Families (Doors, Windows, Furniture, etc.) we cannot save and load Railings (or other System Families) in individual RFA files. You can however copy and paste them from other projects or use the Transfer Project Standards command (on the File menu) to copy Types between various projects. To illustrate this approach, in the Autodesk Web Library, there is a project file which contains several Railing examples.

63. From the Window menu, choose **I Recent Files**.

- On the right side, click the Revit Web Content Library link (see Figure 7.31).

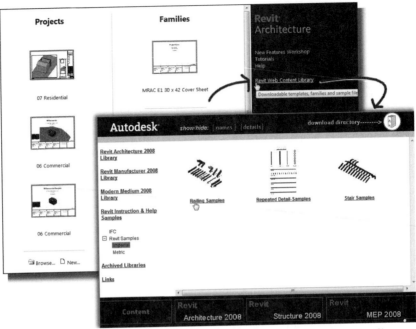

**Figure 7.31**  *Access the Revit Web Content Library and open the Railing Samples file*

Your web browser will open and the Web Library page will load.

- Click the *Revit Instruction & Help Samples* link

- Click the *Revit Samples* link then click either the *Imperial* or *Metric* link

- Finally, click the *Railing Samples* [*Railings*] item to open the file.

- In the dialog that appears, click the Open button to download the file and open it in your Revit session.

 **Note:** If you do not have access to the web or for some reason cannot locate the files noted here, copies have been included with the files installed from the Mastering Revit Architecture CD ROM. You will find them in the \MRAC\Library\Stair Samples folder wherever you installed the dataset files. If you are using Imperial units, be sure to open the *Railing Samples* file and open the *Railings* file if you are using Metric.

When the file opens on screen, you will see a three-dimensional view that has several Railing samples placed in it. Each has a text label next to it indicating the name of the Railing Type. To use one of these Railings in your project, simply copy it from this project and then paste it into your project.

Zoom and pan around the file and locate the ***Stair Handrail – Residential*** Type.

64. Select one of the ***Stair Handrail – Residential*** Railings and then press CTRL + C.

- From the Window menu, choose the *Existing First Floor* plan view.

- Press CTRL + v and then in the dialog that appears, click OK.
- Click a point anywhere on screen to place the pasted Railing.
- On the Options bar, click the Finish button.
- Select the newly pasted Railing and then press the DELETE key.

You must actually place the pasted element to import the Type into the project. Once you have done this however, you can safely delete the extraneous Railing without fear of deleting the newly imported Type. The **Stair Handrail – Residential** Type is now part of the Residential project and we can assign it to our existing Railing.

65. On the Project Browser, double-click to open the {3D} view.

66. Select the short Railing (the one with balusters).

- On the Options Bar, choose **Stair Handrail – Residential** from the Type Selector.
- Following the process outlined above, select the Railing, edit its properties, and then edit the Type.
- Edit the Baluster Placement and change the End Post to None.
- Click OK three times to return to the model and view the results (see Figure 7.32).

**Figure 7.32**  *After pasting in the new Railing Type, apply it to the Railing and remove the top post*

Examine the Railing in both 3D and plan after placement. It may be necessary to flip it to seat the balusters on the treads properly.

## ADJUST WALL PROFILE

Although our Stair looks fine, the 3D view reveals that the Wall on the right does not continue under the Stair and the Basement Door is too tall and intersects the Stair.

67. In the 3D view window, on the View Control Bar, choose **Hidden Line** from the Model Graphics popup.

**Tip:** The keyboard shortcut for Hidden Line is **HL**. **Hidden Line** is also located on the View menu.

- From the View menu, choose **Orient > East**.

The options on the View > Orient menu allow you to match the 3D view to several preset vantage points or to match to another existing view, such as a section. If you wish, you can also save the view as a new view in the project.

68. Select the Wall to the right of the Stair.

   It is to the right in plan, it is in front of the Stairs in the 3D view.

- On the Options Bar, click the Edit Profile button.

- Edit the sketch lines as indicated in the left two panels of Figure 7.33.

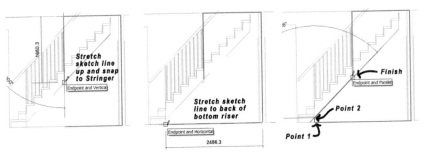

**Figure 7.33**   *Edit the sketch of the Wall profile*

- On the Design Bar, click the **Lines** tool.

- On the options Bar, place a checkmark in the "Chain" checkbox.

- Draw two sketch lines to close the Wall profile shape (see the right panel of Figure 7.32).

69. On the Design Bar, click the Finish Sketch button.

- From the View menu, choose **Orient > Southeast** (see Figure 7.34).

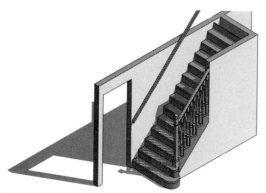

**Figure 7.34** *The completed residential Stair hallway (shown with shadows on in the figure)*

When you close the project, the Temporary Hide/Isolate will be reset automatically. It cannot be saved between sessions. If you would like to preserve the isolated 3D view of the Stairs as we see it here, you can create a custom view for this. To do so, select all of the items in the 3D view. Then choose the Reset Temporary Hide/Isolate from the pop-up icon. Next, right-click on screen and choose **Select Previous**. Using Temporary Hide/ Isolate again, hide those elements (instead of isolate this time). Next, zoom out and select everything on screen. Right-click in the view and choose **Hide in View > Elements**. Finally, reset the Temporary Hide/Isolate again, and the Stair and Railings should reappear. This time however, the effect will be permanent. Therefore, right-click the *{3D}* view and rename it to something like **3D Stair**. There are other ways to achieve a similar effect. We will explore an alternative in the coming chapters using the "Orient to other view functionality."

70. Save and Close the Residential project.

## COMMERCIAL CORE LAYOUT

While we have done much with the Stair tools in the residential project, there is still more we can explore. To continue our exploration of vertical circulation, we will switch to the Commercial Project. In this project, we will add a multi-story Stair and some elevators in the building core and we will add ramps to the front entrance of the building.

### LOAD THE COMMERCIAL-STRUCTURE PROJECT

Be sure that the Residential Project is saved and closed.

1. On the Standard toolbar, click the Open icon.

**Tip:** The keyboard shortcut for Open is CTRL + O. **Open** is also located on the File menu.

- In the "Open" dialog box, click the *My Documents* icon on the left side.
- Double-click on the *MRAC* folder, and then the *Chapter07* folder.

If you installed the dataset files to a different location than the one listed here, use the "Look in" drop-down list to browse to that location instead.

2. Double-click *07 Commercial-Structure.rvt* if you wish to work in Imperial units. Double-click *07 Commercial-Structure Metric.rvt* if you wish to work in Metric units.

You can also select it and then click the Open button.

The project will open in Revit Architecture with the last opened view visible on screen.

 **Note:** Please start with the files provided for Chapter 7 rather than attempting to continue from your own saved version from the previous chapter. A few changes have been included here that were not detailed in the previous chapter.

3. On the Project Browser, double-click to open the *Level 1* floor plan view.

• Right-click in the workspace and choose **Zoom In Region**.

 **Tip:** The keyboard shortcut for Zoom In Region is **ZR. Zoom In Region** is also located on the View > Zoom menu.

• Drag a box around the core to zoom in on it (see Figure 7.35).

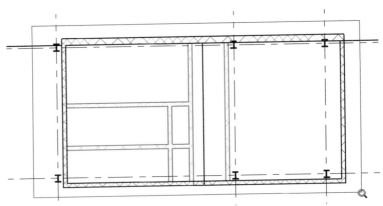

**Figure 7.35** *Zoom in on the core and note the new Walls and Doors added since the previous chapter*

## PERFORM A COORDINATION REVIEW

In the previous chapter, we established links between the structural model and the commercial project. We used Copy/Monitor to create monitored copies of the geometry from the commercial project that is pertinent to the structural file. As you can see, some additional Walls, Doors, and Openings have been added to the building core area in the linked commercial project since we established the Copy/Monitor link. In addition, there may be other less obvious changes that require coordination. Returning to the Copy/Monitor tool covered in the previous chapter, we can perform a coordination review and reconcile the differences between the two files.

4. On the Tools toolbar, click the Copy/Monitor icon and then choose the **Select Link** option from the pop-up.

- Click on the commercial project linked file on screen to select it (it is easiest to click on one of the core Walls shaded in halftone).

The Design Bar will change to the Copy/Monitor mode and all other tabs and tools will be hidden.

5. On the Design Bar, click the **Coordination** tool.

- Expand "Shaft Openings" then "Monitor Opening Sketches."

- Next to the "Sketches are different" item in the Action column, choose **Copy Sketch to "Opening Cut"** and then click Apply (see Figure 7.36).

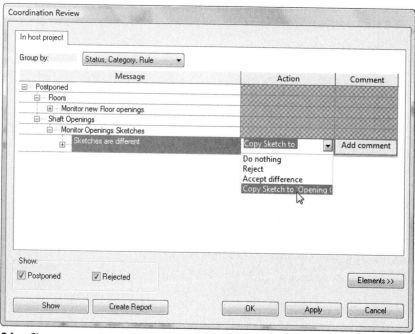

**Figure 7.36**  *Choose an action for the difference in coordination review*

6. Click OK to dismiss the "Coordination Review" dialog.

There may be other items listed for review, but for now, the shaft opening sketch is the only one that requires immediate attention. As you can see, this tool allows you to synchronize changes between the two linked files for elements that are monitored. Before leaving this file, let's also copy the new core Walls to the structural file using the procedure we followed in the previous chapter.

7. On the Design Bar, click the **Copy** tool.

- On the Options Bar, click the Multiple checkbox.

- With the CTRL key held down, select each of the internal (shaded in halftone) core Walls (see Figure 7.37).

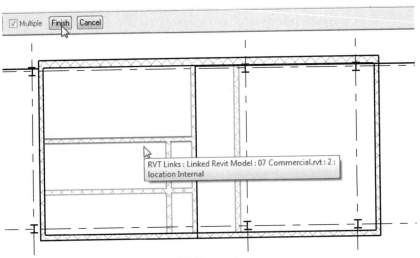

**Figure 7.37** *Select all of the internal core Walls to copy*

- On the Options Bar, click the Finish button.
- On the Design Bar, click the Finish Mode button.

8. Save and close the *07 Commercial-Structure.rvt* [*07 Commercial-Structure Metric.rvt*] file.

## ADD A NEW STAIR TO THE COMMERCIAL PLAN CORE

Now that we have resolved any coordination issues in the core area, let's open the commercial project and begin adding some Stairs. We will add a U-shaped egress Stair to the space in the upper left corner of the core.

9. On the Standard toolbar, click the Open icon.

 **Tip:** The keyboard shortcut for Open is CTRL + O. **Open** is also located on the File menu.

- In the "Open" dialog box, click the *My Documents* icon on the left side.
- Double-click on the *MRAC* folder, and then the *Chapter07* folder.

    If you installed the dataset files to a different location than the one listed here, use the "Look in" drop-down list to browse to that location instead.

10. Double-click *07 Commercial.rvt* if you wish to work in Imperial units. Double-click *07 Commercial Metric.rvt* if you wish to work in Metric units.

    You can also select it and then click the Open button.

The project will open in Revit Architecture with the last opened view visible on screen.

11. On the Project Browser, double-click to open the *Level 1* floor plan view.

• Zoom in to the same area of the plan as above.

12. On the Design Bar, click the Modeling tab and then click the **Stairs** tool.

The Design Bar changes to sketch mode and shows only the Stairs tools.

Previously, in the "Create a Straight Interior Stair" heading, we used the **Run** tool to create a straight Stair for the residential project. To create a Stair with a landing, you start the same way but create more than one run within the same Stair sketch. To do this, simply click short of the complete run. This leaves some of the treads unplaced. You then start another run nearby. When you do this, a landing will be created in your sketch automatically. We will use this approach to create a U-shaped Stair.

• On the Design Bar, click the Stair Properties button.

• Verify that the Type is **7" max riser 11" tread [190mm max riser 250mm going]**.

• In the "Instance Parameters" area, beneath the "Dimensions" grouping, change the Width to **3'-8" [1100]**.

• Set the "Desired Number of Risers" to **21** and then click OK.

13. Locate the first point on the inside edge of the right vertical Wall of the Stair core.

• Drag the mouse horizontally to the left. When the message reads "11 Risers Created, 10 Remaining" click the mouse to set the next point (see Figure 7.38).

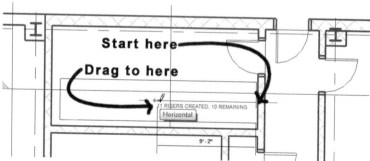

**Figure 7.38** *Start the Stair sketch and place half the treads*

• Move the pointer above the point just clicked, and then click again (see Figure 7.39).

The point is approximately 2'-0" [600] above the previous one, but the exact location is not important right now as we will adjust it below.

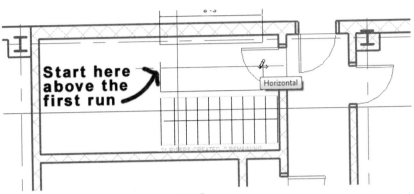

**Figure 7.39** *Sketch the second run above the first*

- Drag horizontally to the right and click to place the remaining risers.

A sketch of the U-shaped Stair will appear. The size is not exactly correct and it is not in the right spot. We can adjust both of these issues easily.

14. Select the green vertical sketch line on the left.

This is part of the Stair Boundary.

- Click the **Move tool on the Edit toolbar.**

- Move the Boundary line to the left **1'-0" [300]**.

Notice that the other Boundary lines remain attached to the moved line.

15. Using a window selection, select the entire Stair sketch.

**Note:** Don't worry about selecting the neighboring elements. When you are in sketch mode, you can only select elements of the sketch.

- Using the **Move** tool again, snap the start point of the move to the lower left endpoint.

- Snap the end point of the move to the endpoint of the Wall at the lower left inside corner of the room (see Figure 7.40).

**Tip:** If you have trouble getting Revit to automatically present the endpoint, press the TAB key to cycle to it. You can also type **SE** to enable snap to endpoint.

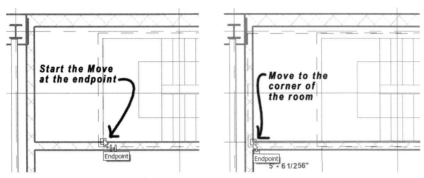

**Figure 7.40** *Move the entire sketch to the corner of the room*

This positions the Stair in the right location. However, the Stair Type that we are using here uses Closed stringers instead of open ones like the Stair above. Therefore, we need to take the thickness of our stringers into account when establishing our final position.

16. Using Move again, move the entire sketch up vertically **2"** **[50]** and right horizontally **2"** **[50]** (see Figure 7.41).

 **Tip:** Tip: Simply start the Move, click any start point, move the mouse horizontally or vertically in the desired direction and then type the move value (**2"** **[50]** in this case) on the keyboard followed by an ENTER.

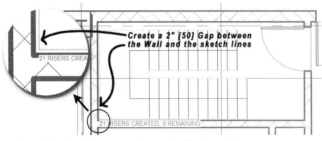

**Figure 7.41** *Move the entire sketch to create a gap for the stringer*

17. Using a window selection, surround just the top run of the Stair sketch (see Figure 7.42).

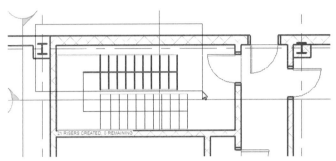

**Figure 7.42** *Select the top run of the Stair sketch*

- Using the same techniques, move it from the top left endpoint of the selection, to the inside horizontal edge of the room.
- With the same sketch lines still selected, move them down **2"** **[50]** (see Figure 7.43).

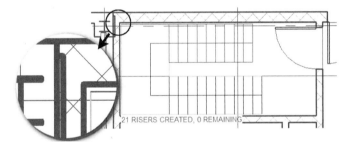

**Figure 7.43** *Move the top run of the Stair sketch and create a gap for the stringer*

18. On the Design Bar, click the Railing Type button.
   - In the "Railing Type" dialog, choose **Handrail – Pipe** [**900mm Pipe**] and then click OK.

When you create a Stair in Revit, the Railing is created automatically for convenience. The Railing is actually a separate element with its own sketch. You can of course edit this later, but choosing a Railing Type now saves a step.

   - On the Design Bar, click Finish Sketch.

19. Save the project.

## CREATE A MULTI-STORY STAIR

Let's take a look at our Stair in a section view.

20. A section line runs horizontally through the Stair. Double-click the blue section head to go to that view.

As you can see, the Stair that we built only occupies the first floor. We can edit the properties of the Stair to make it apply to multiple stories.

21. Select the Stair object in the current Section view.

Be sure to select the Stair and not the Railing for this step.

- On the Options Bar, click the Properties icon.

- In the "Instance Parameters" area, beneath the "Constraints" grouping, choose **Roof** from the "Multistory Top Level" list.

- Click OK to apply the change (see Figure 7.44).

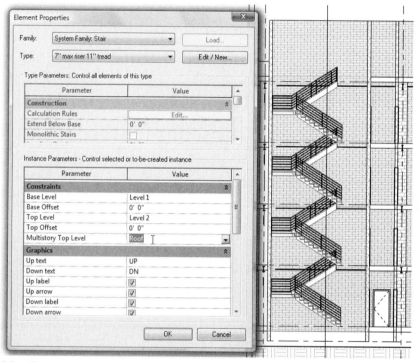

**Figure 7.44** *Change the Stair to a multistory Stair*

22. On the Project Browser, double-click to open the *Level 1* floor plan view.

Notice that only the UP Stair annotation displays on the first floor and that the Stair above the cut plane is shown dashed.

23. On the Project Browser, double-click to open the *Level 3* floor plan view.

Notice that in this view (and in *Level 2* and *Level 4* as well) that both the UP and DOWN Stair annotations display.

24. On the Project Browser, double-click to open the *Roof* plan view (see Figure 7.45).

Notice that here, only the DOWN annotation shows.

**Tip:** The UP and DOWN annotations can be moved. First, select the Stair. A blue control dot appears next to the text. Click and drag the text to a better location.

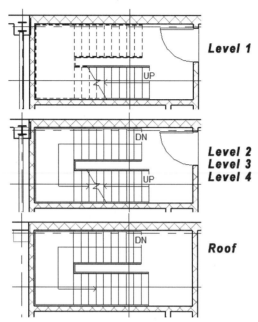

**Figure 7.45** *Stairs automatically display directional annotations appropriate to the level*

    25. Save the project.

## FLOORS, LANDINGS AND SHAFTS

In the previous chapter, we created some simple Floors and a Roof for the project. We used a Shaft element to cut these horizontal surfaces through all floors at the stair and elevator shafts. Another glance at the section in our current model will reveal that there are no landings on the entrance side (right) of the Stairs. We have a single continuous Floor slab across the entire floor plate with a single Shaft cutting out both the Stair and Elevator spaces.

At this point, to accommodate the Stair elements that we have added, and add landings on the entrance side to meet the Stairs, we need to adjust our existing Shaft. We can edit the shape of the Shaft to reveal more of the slabs in the shape of these landings, or we can remove the Shaft, edit the Slab sketches, and/or build the landings as separate Floor elements. The exact approach that we take is really a matter of personal preference. If the shape of the void is the same on all levels, the shaft approach is easier. If the shape of each landing varies, it is better to edit the sketch of each Floor.

### EDIT THE SHAFT

Regardless of the approach we choose, the Shaft will require editing. Editing the Shaft is easy. We simply select it and then edit the sketch.

    1. On the Project Browser, double-click to open the *Level 2* floor plan view.

    2. Move the pointer near the outside edge of the Stair or Elevator spaces.

- When the screen tip reads "Shaft Opening: Opening Cut" and the Shaft pre-highlights, click the mouse to select it.

  The Shaft will turn opaque and block the other elements while it is selected.

- On the Options Bar, click the Edit button.

- Click the top horizontal sketch line and move it down to the outside of the top elevator Wall.

- Move the bottom line up and the right side over to the left to match the shape of the elevator core (see Figure 7.46).

  You can use the Move tool for this or the Align tool.

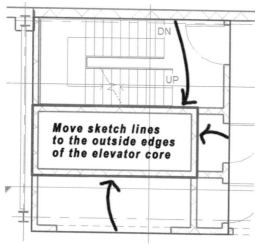

**Figure 7.46**   *Move sketch lines of the Shaft to fit the elevator core (sketch lines in the figure enhanced for clarity)*

   **Tip:** You can make your sketch lines appear bolder on screen as you work by editing their lineweight in the Line Styles dialog (Settings menu).

3. On the Design Bar, click the Finish Sketch button.

The most obvious effect of this change is that the Stair is now covered up by the Floor slabs. The change will not be very apparent in plan views for the elevator shaft. If you wish to see the elevator shaft more clearly, you can go to the default {3D} view, hold your SHIFT key down and drag with the wheel on your mouse to spin the model around and see the effect. If you don't have a wheel, or if you prefer, you can also use the Dynamically Modify View (F8) icon instead. Spinning the model in this way will reveal that the Shaft now only cuts the Floor slabs at the elevator shaft.

## EDIT THE FLOORS

If you spun the model in 3D above, you noticed that the previous edit has now restored our Floor slabs in the Stair tower. This will make it difficult for us to climb these stairs. As was noted above, a Shaft provides a quick and convenient way to cut several Floors at once, however the alternative to a Shaft is to edit the sketch of the Floors themselves to model them in their true shape rather than carve a hole from it later using the Shaft element. It should be noted again that we are showing both approaches here for the education value. In your own projects, either approach can be employed based upon your team's preferences. You can also use a combination of techniques as appropriate.

4. On the Project Browser, double-click to open the *Section at Stair* view.

- Select the Floor element at Level 2.

- On the Project Browser, double-click to open the *Level 2* view.

The Floor should still be selected when you switch from section to plan, although it may not appear so. The easiest way to tell is that the Edit button should appear on the Options Bar. Otherwise, repeat the process and try again.

5. On the Options Bar, click the Edit button.

- On the Design Bar, click the **Lines** tool and on the Options Bar, click the Pick Lines icon.

6. Click the edges indicted in Figure 7.47.

- Use the **Split** tool (with the "Delete Inner Segment" option chosen) to remove the middle segment of the existing line at the top of the Stair.

- Use the **Trim** tool as required to complete the cleanup (see Figure 7.47).

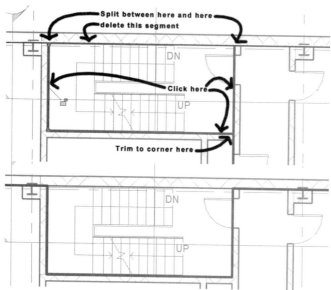

**Figure 7.47**   *Add sketch lines around the Stair tower space (sketch lines in the figure enhanced for clarity)*

- On the Design Bar, click the Finish Sketch button.

  If an error appears regarding overlapping lines, return to the sketch and use the **Trim** tool to clean up the corners.

- In the Alert dialog that appears, click Yes to accept the joining of geometry.

7. Repeat the edit Floor sketch on the *Level 3, Level 4,* and *Roof* views.

**Tip:** A handy shortcut to selecting the Floor is to return to the section view, select the next Floor, and then immediately click the Edit button on the Options Bar. You will be prompted for an appropriate floor plan view to open.

8. Check the model in the *{3D}* and/or one of the *Section at Stairs* views.

**Tip:** If you open the *{3D}* view, you can display it in hidden line, wireframe, or shaded. The Shaded with Edges option often offers the best contrast and readability.

9. Save the project.

## ADD A LANDING (FLOOR)

Our commercial Stair Tower is nearly complete. We have completed the work we needed on the Floors and Roof. We are now ready to add some new Floor elements for the Stair Landings. The first floor is fine as is, so let's start with the second floor.

10. On the Project Browser, double-click to open the *Level 2* floor plan view.

11. On the Design Bar, click the Modeling tab and then click the **Floor** tool.

   The Design Bar changes to sketch mode and shows only the Floor tools.

   - On the Design Bar, click the **Lines** tool.

   - On the Options Bar, click the Pick Lines icon.

   - Click the inside edge of the right vertical Wall and each of the inside edges of the two horizontal Walls (see Figure 7.48).

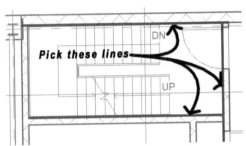

**Figure 7.48** *Use the Pick Lines option to create sketch lines from existing Wall edges (sketch lines in the figure enhanced for clarity)*

   - Continuing with Pick Lines, click each of the tread lines of the Stair on the right side (see the left side of Figure 7.49).

   - Use the **Trim/Extend** tool with the "Trim/Extend to Corner" option to finish the sketch (see the right side of Figure 7.49).

 **Tip:** Remember to pick the side of the lines that you wish to keep.

12. On the Design Bar, click the **Floor Properties** tool.

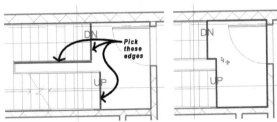

**Figure 7.49** *Pick the remaining lines and then Trim the corners (sketch lines in the figure enhanced for clarity)*

   - From the Type list, choose **LW Concrete on Metal Deck** [**Floor : Insitu Concrete 225mm**] and then click OK.

   - On the Design Bar, click the Finish Sketch button.

## COPY THE LANDING TO LEVELS

13. Double-click the blue section head running horizontally though the Stair to go to that view. (This is the *Section at Stair* view.)

You should be able to see the new Floor landing at the second floor. (It is not as thick as the main Floor object.) The other floors will require landings as well, so let's copy and paste this one to the other levels.

14. Select the Floor landing, and from the edit menu choose **Copy to Clipboard** (or press CTRL + C).

- From the Edit menu, choose **Paste Aligned > Select Levels by Name**.

- In the "Select Levels" dialog, select *Level 3*, hold down the CTRL key and then select *Level 4* and *Roof*.

- Click OK to complete the operation.

Landings will appear at the other levels.

15. Zoom in on the top of the Stair tower at the Roof Level.

Notice that the landing is lower than the roof. The vertical position of the Roof element is different from that of the Floor slabs. Notice that the Roof sits on top of the level line, while the Floors sit below them. The construction of both the Floors and the Roof includes steel framing joists, metal deck, and concrete topping. The Roof will have a membrane roofing layer on top as well. At this stage of the project, we have only a simple generic structure in place to represent the overall thickness of this structure. Later in the project, we can edit the Floor and Roof Types to add the appropriate layers. For now, just getting the Roof positioned properly relative to the level is all that is necessary.

16. Select the Roof element.

- On the Options Bar, click the Properties icon.

- Beneath the "Constraints" grouping, edit the "Base Offset From Level" to **-1'-0"** [**-400**] and then click OK (see Figure 7.50).

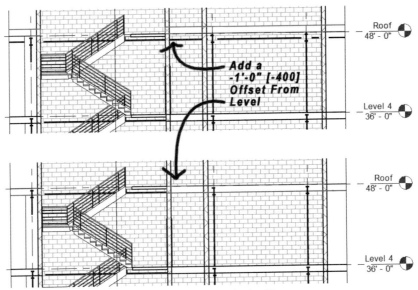

**Figure 7.50** *Adjust the Roof offset from level to match the Floors*

## EDIT A MATERIAL (IMPERIAL ONLY)

If you are working in the Imperial dataset, the Type chosen for the landing Floor includes a stipple pattern in the plan views. This pattern is not necessary for our plan views. Removing it is easy. If you are working in the metric dataset, the following steps are not required, but feel free to follow along to understand the process if you wish.

17. On the Project Browser, double-click to open the *Level 2* floor plan view.

18. Select the Landing.

- On the Options Bar, click the Properties icon.

- Next to the Type list, click the Edit/New button (or press ALT + E).

- In the "Type Parameters" area, beneath the "Construction" grouping, click the Edit button next to Structure.

The "Edit Assembly" dialog will appear revealing the structure of the selected Floor type. We are editing at the type level right now, so any change we make will automatically apply to all elements that use this Type in the entire project. This means that *all* of our landings will receive the change when we are finished editing.

- Next to Layer 2, click in the Material cell.

A small expand (down pointing arrow) icon will appear in the right side of the field. This icon is used to open a browse window to choose a material for the layer (see Figure 7.51).

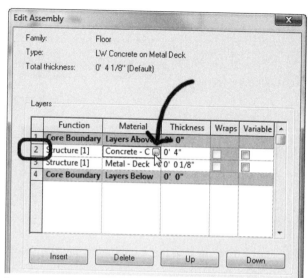

**Figure 7.51** *Each Layer of the Floor's Structure is assigned a Material*

- Click the browse icon to open the "Materials" dialog.

Notice that the currently selected Material is highlighted in a long list of available Materials.

19. At the bottom of the dialog, click the Duplicate button.

**BIM** *Manager Note:* We could edit the existing Material directly, but it is good practice to get in the habit of renaming (or duplicating) any Families, Types, or Styles that you edit. This way, you can always quickly locate the elements that you have modified. In an office environment, establishing a good logical naming scheme for such elements is also highly recommended. For example, using the company name or initials as a prefix or suffix to the name is a standard best practice recommendation.

- Name the new Material **MRAC Landing Concrete Topping** and then click OK.

**Tip:** If you want your custom elements to sort automatically to the top of the list, add an underscore _ to the front of the name. **_MRB Landing Concrete Topping** in this case.

- On the right side of the "Materials" dialog, in the "Surface Pattern" area, click the drop-down menu icon.

- From the list that appears, choose <none> (see Figure 7.52).

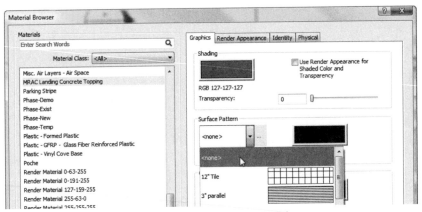

**Figure 7.52** *Remove the Surface Pattern from the new Material*

20. Click OK four times to return to the model and view the results.

21. Save the project.

## EDIT RAILINGS

One final change remains for our Stair tower. The Railings should wrap around at the landings. We can accomplish this by editing the Railing sketch. When you create a Stair, the Railings are added automatically. By default, they match the same shape as the Stair. However, like the Stairs, Railings have a sketch that we can edit.

22. On the Project Browser, double-click to open the *Level 2* floor plan view.

23. Select the inside Railing.

Make sure you select the Railing and not the Stair; use the TAB key if necessary.

- On the Options Bar, click the Edit button.

The Design Bar changes to sketch mode and shows only the Railing tools.

- On the Design Bar, click the **Lines** tool.

- On the Options Bar, place a checkmark in the "Chain" checkbox.

- Draw the two segments indicated in Figure 7.53.

**Figure 7.53**  *Draw two segments in the Railing sketch*

24. On the Design Bar, click the Finish Sketch button.

The best way to view the result is in a 3D view. If we use the default 3D view, we will have to isolate the Stair (as we did above in the residential project) in order to see the Railing. Another option is to create a new 3D view of just the stairs. In this case, creating a 3D view from the *Section at Stair* view would be the best choice. It will give use a three-dimensional section view in which to study the Stairs and Railings.

25. On the Project Browser, right-click the {3D} view and choose **Duplicate View > Duplicate**.

   • Right-click the *Copy of {3D}* view and choose **Rename**.

   • In the "Rename View" dialog change the name to **3D Stair Section** and then click OK.

26. Click anywhere in the 3D window to shift focus from the Project Browser to the model window.

   • From the View menu, choose **Orient > To Other View**.

   • In the "Orient To Other View" dialog, select the *Section at Stair* view and then click OK (see Figure 7.54).

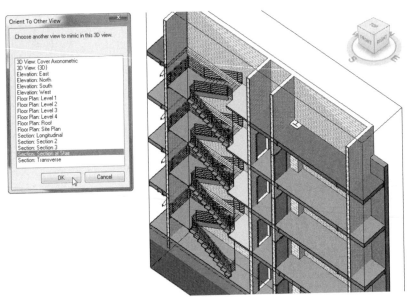

**Figure 7.54** *Orient the view to the Section at Stair Tower then spin it in 3D*

27. Hold your SHIFT key down and drag with the wheel on your mouse to spin the model around and see the effect.

 **Note:** If you don't have a wheel, or if you prefer, you can also use the Dynamically Modify View (F8) icon instead.

28. Zoom in on the second floor at the point where we edited the Railing.

- Pan to other floors as well.

Notice that the Railing now wraps around the gap between the two runs and this change has occurred on all Levels (see Figure 7.55).

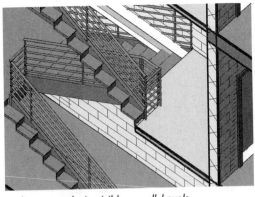

**Figure 7.55** *The Railing edit we made is visible on all Levels*

For the top and the bottom of the Stair, you can add additional Railings if you like to complete the layout. The easiest way to do this is to select the Railing, right-click, and choose **Create Similar**. Sketch the new Railing and use the ***Set Host*** tool on the Design Bar to attach it to the Landing.

 **Tip:** If you witness a strange artifact for the balusters in the 3D view, you can edit your sketch and pull the vertical sketch line back slightly from the horizontal segment. Leave the vertical segment about an inch [few millimeters] short.

### ADJUST OUTSIDE RAILING

Previously in the "Create a New Railing Type" heading of the residential project, we created a new Railing Type that did not have any balusters. A Railing Type similar to this has been provided in the dataset that we can use for the Railing along the Stair core Walls.

29. Select the outside Railing (the one adjacent to the Wall).

 • From the Type Selector on the Options Bar, choose **MRAC Handrail Only** for the Type.

30. Save the project.

## RAMPS AND ELEVATORS

In this section, we will add some ramps and elevators. Ramps and elevators are grouped here to round out our look at vertical circulation, but the approach to working with each one is quite different. Creating a ramp in Revit is much like adding a Stair. Elevators on the other hand are simply Families that we insert in our models like other components.

### ADD A RAMP

The ***Ramp*** tool is provided on the Modeling tab of the Design Bar. Ramps are very similar to Stairs. To save a bit of time and effort, the dataset includes a floor slab at the front of the building as an entrance patio. Some Reference Planes have been included here as well. These Reference Planes are provided to make it easy for us to the sketch the ramp in the following example. When building your own Ramps and Stairs, you can also use Reference Planes to assist you, or any other technique you find helpful in laying out the sketch.

 1. On the Project Browser, double-click to open the *Entrance Plan* plan view.

This is an enlarged floor plan view showing just the front portion of the building.

 • Zoom in on the lower right corner at the front of the building.

You will notice four reference planes in this location. We will use these to help us sketch the ramp.

 2. On the Design Bar, click the Modeling tab and then click the ***Ramp*** tool.

Notice that the Design Bar changes to a series of tools that are identical to the Stair Design Bar we saw previously.

 • On the Design Bar, click the ***Ramp Properties*** tool.

- In the "Instance Parameters" area, beneath the "Dimensions" grouping set the Width to **3'-0"** [**900**].

- Verify that the "Base Level" is set to **Street Level** and the "Top Level" is set to **Level I** and then click OK.

- Starting at the lower left endpoint of the inclined Reference Plane, click on each Reference Plane endpoint moving first left to right, then up, then right to left (see Figure 7.56).

 **Note:** Like when we created the U-shaped Stair previously, you only click the start and end of the runs. The landings fill in automatically.

Also, note in the figure, that the first click in the run will show a tooltip for the Endpoint, but not highlight the point with the customary square snap indicator. The square snap indicator does appear on the second click of the run as shown near point 8 in the figure. This anomaly should not prove detrimental to completing the exercise.

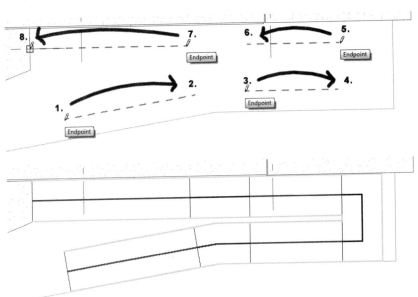

**Figure 7.56**  *Sketch the ramp using the provided Reference Planes*

If you wish, you can edit the shape of the green outline on the right side at the landing to make the ramp match the building Pad in the linked Site model. Try moving the vertical green boundary line to the right about 1'-0" [300] to widen the landing, or even change the shape to angle or curve it if you like.

**Note:** While we have provided the Reference Planes in this exercise to make the process of sketching the ramp go smoothly, you can use any of the sketching techniques that we covered above on Stairs when adding ramps in your own projects. Remember that you can sketch the Run, or edit the Risers or Boundaries directly with the appropriate tools on the Design Bar.

3. On the Design Bar, click the Finish Sketch button.

• View the ramp in various views to see the results (see Figure 7.57).

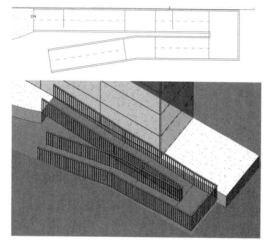

**Figure 7.57**  *The completed ramp*

4. Return to the *Entrance Plan* view and then Mirror the ramp and its Railings to the other side.

   Use any selection technique, but be sure to select only the Ramp and Railings.

• Use the Draw option to pick a midpoint about which to mirror.

   Be sure that "Copy" is selected so that you maintain the original when mirroring.

5. Save the project.

## ADD A CUSTOM SHAPED STAIR

We can add a few steps up to the patio leading to the main building entrance. To do this, we simply add another Stair. However, we'll sketch it a bit differently than the ones above.

6. On the Design Bar, click the Modeling tab and then click the **Stairs** tool.

   The Design Bar changes to sketch mode and shows only the Stairs tools.

• On the Design Bar, click the Stair Properties button.

• Set the Type to: **Monolithic Stair.**

- Verify that the Base Level is Street Level and the Top Level is Level 1. If they are not, correct them and then click OK.

7. On the Design Bar, click the **Boundary** tool.

- On the Options Bar, select the Draw icon and then click the Line icon.

- Set the Offset to **6'-0"** [**1800**].

- Click the first point at the midpoint of the curved edge of the patio (see Figure 7.58).

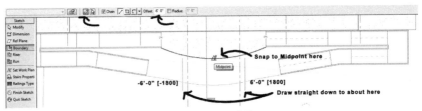

**Figure 7.58**  *Add two Boundary lines drawn with the Offset option*

- Draw straight down and click below the bottom edge of the Ramps as shown in the figure.

- Press ESC when done.

- Repeat the process changing the Offset to **-6'-0"** [**-1800**].

  You should have two vertical green lines centered on the patio and spaced 12'-0" [3600] apart. These are the Boundaries of the Stair (this is a monolithic Stair, but in other Stairs, these lines would become the stringers).

8. On the Design Bar, click the **Riser** tool.

- On the Options Bar, click the Pick Lines tool.

- Set the Offset to **5'-0"** [**1500**].

9. Highlight the curved edge of the patio. Move the mouse slightly to make green dashed line appear below, and then click to create the Riser line (see top left panel of Figure 7.59).

- Change the Offset to **1'-0"** [**300**].

- Highlight the Riser line you just added, and offset a new line up from it. Repeat until you have created 6 total (see top right panel of Figure 7.59).

  The last one will be right on top of the patio edge.

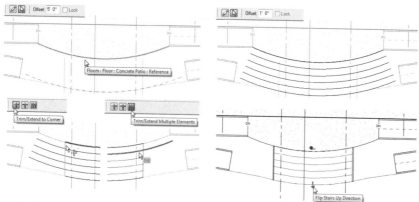

**Figure 7.59** *Create the Riser lines by offsetting from the curved patio edge and then Trim to complete sketch*

10. Use the Trim tool to clean up the sketch.

   • Use the Trim/Extend to Corner option to trim and extend the green Boundary lines to the curved black Riser lines.

   • Use the Trim/Extend Multiple Elements to trim the curved lines back to the green Boundary lines (see bottom right panel of Figure 7.59).

   Remember to click the green Boundary line as the trimming edge first, and then pick the side of the curve you wish to keep when trimming.

11. On the Design Bar, click the Finish Sketch button.

An ignorable warning will likely appear. (Warnings that appear in the lower right corner of the screen can be ignored.) You should always read the warning however, and address them when you can. In this case, the warning tells us that we do not have the desired number of risers. Normally you would include the landing at the top of the stair flight as part of the Stair and therefore include one more Riser. In this case, we have the patio slab instead and will be fine ignoring this message for now.

One problem not indicated by the warning or obvious from the plan view is that the stair direction is wrong. The highest tread is at the bottom of the screen and lowest by the patio. This is easy to fix.

12. Select the Stair and click the Flip Stairs Up Direction control as shown in the bottom right panel of Figure 7.59.

 **Tip:** If your Stair disappears after clicking the flip control, undo and then edit the Stair sketch. Zoom in closely on the Stair. Select the green Boundary on one side of the Stair and nudge it slightly. (Press the left or right arrow keys on the keyboard one time.) Finish the sketch and try again.

- Flip both Railings as indicated in the "Modify the Railing" topic earlier in this chapter.
- Study the Stair in section or 3D views before continuing.

## ADD RAILINGS TO THE ENTRY STAIR

If you want to add railing extensions to the Railings, just edit the sketch.

13. Return to the *Entrance Plan* view.

14. Select one of the Stair Railings and on the Options Bar, click the Edit button.

A single sketch line appears.

- On the Design Bar, click the **Lines** tool.
- At the bottom endpoint of the line, draw a new line **1'-0"** [**300**] long.
- At the top, draw a **2"** [**50**] long vertical line and then use the pick option to offset a curved line **2"** [**50**] from the patio edge.
- Select one of the new lines and on the Options Bar, choose Flat for the Slope (see Figure 7.60).

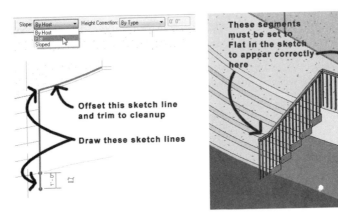

**Figure 7.60** *Add lines to the Railing Sketch*

- Repeat for the other two lines added in this sequence.
- On the Design Bar, choose Finish Sketch.

15. Return to the *Entrance Plan* view.

- Edit the other Railing instead and repeat the process.

If you wish, you can add a center Railing on the Stairs. To do this, click the Railing tool on the Design Bar. Sketch a line down the center of the Stairs (snapping to midpoints). Add an additional line at the top and another at the bottom each 1'-0" [300] long. You will have three vertical lines running end to end. Set the two short ones to Flat on Slope option. Finally, click the Set Host tool on the Design Bar and then select the Stair to attach the Railing to the Stair. Finish the sketch.

## ADD RAILINGS TO THE PATIO

Let's add guardrails to the Patio.

Remain in the *Entrance Plan* floor plan view.

16. On the Design Bar, click the Modeling tab and then click the **Railing** tool.

    • On the Design Bar, be sure that the **Lines** tool is chosen.

    • On the Options Bar, click the Pick Lines icon.

    • Set the "Offset" value to **4"** [**100**].

17. Using Figure 7.61 as a guide, place Railing sketch lines around the patio edges.

    • On the Design Bar, click the **Set Host** tool and then click the patio slab.

    • Finish the sketch.

 **Note:** The figure shows all the required sketch lines, however, Revit does not allow disconnected Railing segments. Therefore, you will need to create two separate Railing elements.

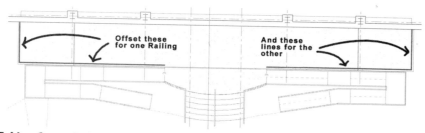

**Figure 7.61**    *Create Railings from the patio slab (sketch lines in the figure enhanced for clarity)*

18. Save the project when finished.

If you open the default 3D view, your model should look something like Figure 7.62. As we study this view, it now becomes obvious that the curtain wall at the entrance needs revision. That task will be completed in a later chapter. For now, we are finished with the commercial building's front entrance.

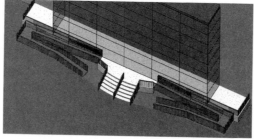

**Figure 7.62**    *The completed Ramps, entry Stairs and Railings*

## ADD ELEVATORS

Unlike Ramps and Stairs, we do not have a dedicated tool or element category for elevators in Revit. Elevators are like other component Families and we use the *Component* tool to add them in the same way that we added plumbing fixtures in Chapter 3. Elevator Families can be created from scratch or derived from manufacturers drawing files. Some sample elevator Families are provided in the Autodesk web library that we accessed previously in the "Import a New Railing Type" topic. As noted in that topic, if you do not have Internet access, the Family files have been provided with the files installed from the Mastering Revit Architecture CD ROM in the *\MRAC\Library\* folder. Feel free to access the following content from that location instead of directly from the Web library as indicated in the following steps.

19. On the Project Browser, double-click to open the *Level 1* floor plan view.

20. From the Window menu, choose **1 Recent Files**.

- Open the Revit Web Content Library.

The Web Library will open in your default web browser. If it does not appear as the active window, look down on your Windows Task Bar for it. Click the icon on the Task Bar to bring it to the front. This is a typical web page with navigation links on the left side and a main preview pane on the right. Autodesk provides a library for Revit Architecture, Revit Structure, and Revit MEP accessed by the buttons at the bottom of the page.

For this exercise, we will go directly to the component we need for our project, but please feel free to spend some time exploring in this web site and download any of the components you wish and add them to the current project. A new version of the dataset will be provided in the next chapter, so you do not have to worry about straying from the exercise. Play time is highly encouraged and is often a great way to learn more about Revit Architecture!

- Browse to *Revit Architecture Library* > *US Library* > *Families* > *Specialty Equipment* > *Conveying Systems* [*UK Library* > *Families* > *Specialty Equipment* > *Elevators and Lifts*].

- On the right, click on *Elevator-Hydraulic* [*Lift Assembly*].

- In the dialog that appears, click Open. (If a further warning appears, click Allow.)

When you open the file, it will open directly in the Revit Architecture workspace in "Family Edit" mode. Family Edit mode is a special mode for editing Revit Architecture Families. The tools on the Design Bar will be unique to this mode, but otherwise most of the interface will remain the same. Here you could edit the Family file before adding it to your project. We will not edit it, but simply save the file and add it to our project. The Family Editor is covered in detail in Chapter 10.

21. On the Design Bar, click the Load into Projects button (see Figure 7.63).

**Note:** If you have more than one Revit Architecture project open, you will be prompted to select the project in which you wish to load the Family. Choose the *07 Commercial [07 Commercial Metric]* project and then click OK.

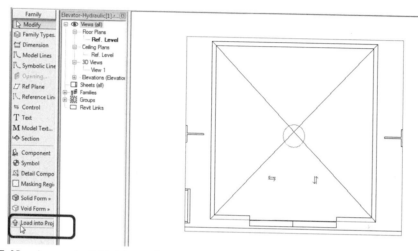

**Figure 7.63**   *Access the Web Library from any "Open" dialog*

The elevator Family is now loaded into your project file and ready to be placed in the model.

 **Note:** If you do not have Internet access, or are having trouble locating the files in the web library, open the file from the *Library* folder of the dataset installed from the Mastering Revit Architecture CD ROM instead.

To load from the dataset folder instead, choose **Load from Library > Load Family** from the File menu and then browse to the *Library\Imperial Library\Specialty Equipment\Conveying Systems [Library\Metric Library\Specialty Equipment\Elevators and Lifts]* folder in the location where you installed the CD ROM files.

Once you have loaded the Family using either method described here, you can continue to the next steps.

22. On the Design Bar, click the Basics tab and then click the **Component** tool.

The Elevator should now be loaded and already selected on the Type Selector.

• If necessary, from the Type Selector, choose **Elevator-Hydraulic [Lift Assembly]**.

• Place two elevators in the elevator core space.

Notice that they will snap automatically to the nearby Walls. Snap them to the lower Wall of the core space (see Figure 7.64).

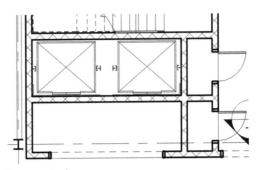

**Figure 7.64** *Place two elevators in the core*

- On the Design Bar, click the **Modify** tool or press the ESC key twice.

- Open each of the other floor plan views to see the elevators.

- In a section view, you can select the elevator and adjust the height using the control handle.

23. Save and close the project.

## SUMMARY

- Stairs and Railings offer flexible configuration with Type-based parameters and Element-based variations.

- Stairs are "sketch-based" objects—create them with a 2D footprint sketch that includes risers and boundary edges.

- Like other Revit Architecture elements, Stairs and Railings can be demolished or added to any phase.

- Edit a Stair Type to apply changes to all Stair elements sharing that Type.

- Edit the Stair Sketch to make customizations such as bull nosed treads.

- Edit the Railing Sketch to modify the extent of the Railing or add railing extensions.

- Adjust Wall Profiles to cut away the space where they intersect Stairs.

- Create Stairs with integral landings by drawing short runs in the sketch. Revit automatically fills in the landing in between runs.

- Stairs can be made multiple stories tall by assigning the height constraints.

- You can use Floor elements to model landings where necessary.

- Ramps are similar to Stairs and are also sketch-based.

- Custom shaped Stairs can be sketched using the Boundary and Riser tools instead of the Run tool.

- Railings are added automatically to newly drawn Stairs. However, you can use the Set Host tool to sketch a custom Railing and apply it to an existing Stair, Slab, or Ramp.

- Pre-made elevator component families can be loaded into the project and placed in the model.

# CHAPTER 8

# Floors and Roofs

## INTRODUCTION

In this chapter we will focus on Floors and Roofs in both our residential and commercial projects. Roofs and Floors in Revit are "sketch based" objects. This means that, as we saw with Stairs in the previous chapter, a simple two-dimensional sketch is utilized to indicate the shape and form of the Floor or Roof you create. We will begin in the residential project with gable Roofs. We will understand how to make Walls attach to Roofs, edit Roof structure, and how to apply edge conditions and gutters. The commercial project will give us the opportunity to explore flat Roofs and Floors.

## OBJECTIVES

In Revit Architecture you can construct Roofs in a variety of ways. Roofs can interact with the Walls of the building and we can apply custom treatment to their edges. After completing this chapter you will know how to do the following:

- Build Roofs
- Create a Custom Roof Type
- Add and modify Floors
- Work with Roof Edges and Gutters
- Understand and utilize Slope Arrows
- Attach Walls and Join Roofs

## CREATING ROOFS

Since we are working concurrently on two different projects in this book, we will get an opportunity to look at both traditional sloped residential roofs and "flat" commercial roofs. The only real difference between the two in Revit is the slope parameters that we assign and the way that we treat the edges. To get started, we'll begin with the residential project.

### INSTALL THE CD FILES AND OPEN A PROJECT

The lessons that follow require the dataset included on the Mastering Revit Architecture CD ROM. If you have already installed all of the files from the CD, simply skip down to step 3 below to open the project. If you need to install the CD files, start at step 1.

1. If you have not already done so, install the dataset files located on the Mastering Revit Architecture CD ROM.

   Refer to "Files Included on the CD ROM" in the Preface for instructions on installing the dataset files included on the CD.

2. Launch Revit Architecture from the icon on your desktop or from the *Autodesk > Revit Architecture* group in *All Programs* on the Windows Start menu.

 **Tip:** In Windows Vista, you can click the Start button, and then begin typing **Revit** in the "Start Search" field. After a couple letters, Revit Architecture should appear near the top of the list. Click it to launch to program.

   - If the New Features Workshop dialog appears, choose "Maybe later" and then click OK.
3. On the Standard toolbar, click the Open icon.

 **Tip:** The keyboard shortcut for Open is CTRL + O. **Open** is also located on the File menu.

   - In the "Open" dialog box, click the *My Documents* icon on the left side.
   - Double-click on the *MRAC* folder, and then the *Chapter08* folder.

   If you installed the dataset files to a different location than the one listed here, use the "Look in" drop-down list to browse to that location instead.

4. Double-click *08 Residential.rvt* if you wish to work in Imperial units. Double-click *08 Residential Metric.rvt* if you wish to work in Metric units.

   You can also select it and then click the Open button.

The project will open with the last saved view visible on screen.

## CREATE AN EXISTING ROOF PLAN VIEW

In the previous chapter, we introduced the concept of creating a separate view for the existing construction and setting its parameters accordingly. We did this while adding Stairs in the first floor. Let's use the same technique now to create an "Existing Conditions Roof Plan" view.

5. On the Project Browser, right-click the *Roof* plan view and choose **Duplicate View > Duplicate**.

   - Right-click on *Copy of Roof* and choose **Rename**.
   - Name the new view **Existing Roof** and then click OK.
   - On the Project Browser, right-click on *Existing Roof* and choose **Properties**.

- In the Element Properties dialog, beneath "Instance Parameters" choose **Existing** for the "Phase" and then click OK.

The new construction will disappear and only the existing second floor will show in gray as an underlay to the current view. Recall that in Chapter 3 we assigned the Second Floor as an underlay to the *Roof.* Copying the view here also copies this underlay parameter to our *Existing Roof* view.

## ADD THE EXISTING ROOF

You may note that the temporary Roof provided in previous chapters has not been included for this chapter project. There is valuable experience to be gained in creating the Roof from scratch, as we will do in the tutorial that follows. We will begin with a simple gable Roof on the existing house and then add a slightly more complex double gable on the new addition.

6. On the Design Bar, click the Basics tab and then click the **Roof** tool. Then from the flyout that appears, choose **Roof by Footprint**.

The Design Bar changes to sketch mode and shows only the Roof tools.

- On the Design Bar, verify that the **Pick Walls** tool is active.

- On the Options Bar, verify that the "Defines slope" checkbox is selected.

- In the Overhang field, type **6″ [150]**.

- Verify that "Extend to wall core" is *not* selected.

- Move the pointer over the topmost horizontal Wall and when the green dashed line appears above and to the outside, click the mouse (see Figure 8.1).

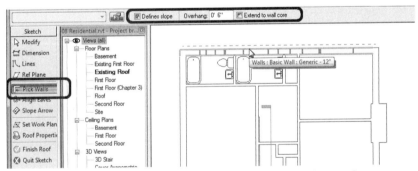

**Figure 8.1**   *Set the parameters for the first edge of the Roof on the Options Bar*

- Repeat this on the lower horizontal Wall to create another sketch line below and to the outside of that Wall.

  Near both of these sketch lines, a triangular slope icon will appear.

7. On the Options Bar, clear the "Defines slope" checkbox and then click the outside vertical Wall on the left.

- Also click to the outside of vertical Walls as shown in Figure 8.2.

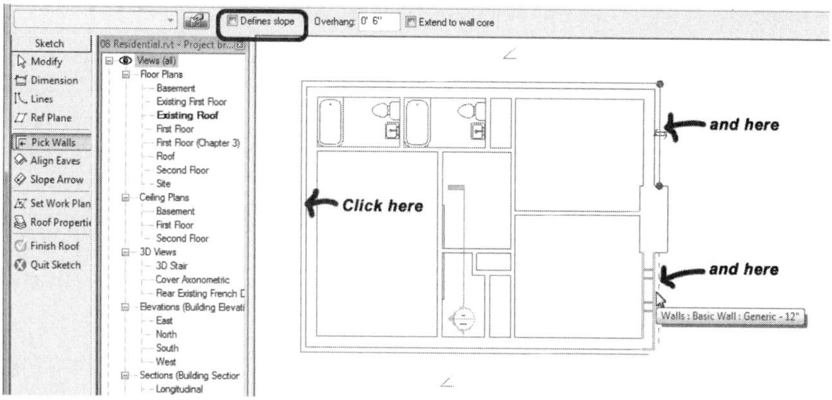

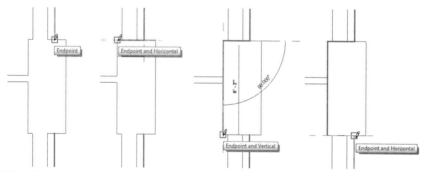

**Figure 8.2**   *Create the gable ends by clearing "Defines slope"*

Since we have the chimney on the right side, we need to make the Roof cut around it. We can accomplish this by drawing the remaining sketch lines relative to the chimney.

8. On the Design Bar, click the **Lines** tool.

- On the Options Bar, place a checkmark in the "Chain" checkbox.

- Using Figure 8.3 as a guide, sketch the remaining three sketch lines.

**Figure 8.3**   *Sketch around the chimney*

- On the Design Bar, click the **Modify** tool or press the esc key twice.

Before we complete the sketch, let's adjust the slope and properties of the roof.

9. Click on the top horizontal sketch line, hold down the ctrl key and then click the bottom horizontal line as well.

- On the Options Bar, click the Properties icon.

- In the "Element Properties" dialog, beneath the "Dimensions" grouping, change the Roof slope. If you are working in Imperial units, type **6″** for the Rise/12″ parameter. If you are working in Metric, set the Slope Angle to **26.57**.

- Click OK to dismiss the "Element Properties" dialog.

 **Tip:** As an alternative, you can select the sketch line and then edit the slope directly with the temporary dimension that appears next to the slope indicator of the slope defining line.

10. On the Design Bar, click the Finish Roof button.

11. On the View toolbar, click the Default 3D View icon (or open the {3D} view from the Project Browser).

Notice that the two Roof edges that we designated as "Defines slope" have a pitch sloping up to a single gable ridge down the middle of the existing house. Feel free to use the "Dynamically Modify View" tools (or hold down SHIFT and drag with the mouse wheel) to spin the model around and see it from different angles.

## ADJUST THE CHIMNEY

Notice also that the Roof has a cutout for the space of the chimney. (However, currently the chimney is too short). We can adjust the height of the chimney by simply dragging the control handles. To do this with a bit of accuracy, let's first lay down a reference plane.

12. On the Project Browser, double-click to open the *East* elevation view.

13. On the Design Bar, click the Basics tab and then click the **Ref Plane** tool.

- Click the first point above the roof to the left of the chimney.

- Click the other point in a horizontal line above the top of the Roof to the right of the chimney.

Depending on how high above the roof you drew your Reference Plane, an "ignorable warning message" will most likely appear when you place the second point of the Reference Plane (see Figure 8.4). An ignorable warning message is simply an informational message that alerts you of a situation that may not otherwise be apparent. This type of dialog appears in the lower right corner of the screen and is tinted in color that can be customized in the Options dialog. An ignorable warning requires no critical or immediate user action and can be ignored. Simply continuing to draw or picking other tools closes this warning dialog. You do not need to click the close button to close it.

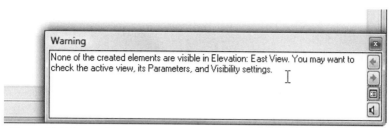

**Figure 8.4** *"Ignorable Warning" dialogs report information that may be useful, but can be safely ignored*

This message appeared because the elevation view has a crop boundary enabled. This crop region is visible by default in the section views, but invisible in elevation views. We could turn on the visibility of Crop Region in this view so that the Reference Plane would become visible. However, it is likely that we sketched it too high above the building anyway. Let's use the temporary dimension to move it first and see if it comes into view before we take any action on the Crop Region. Incidentally, the display of the Crop Region is easily toggled on or off from the icon on the View Control Bar.

14. With the Reference Plane (the one you just drew) still active, click in the blue text of the temporary dimension.

- Type a value of **8'-6″ [2600]** and then press ENTER (see Figure 8.5).

 **Note:** Be certain that the temporary dimension measures from the Roof Level to the Reference Plane. If it does not, adjust the witness lines first.

- On the Design Bar, click the **Modify** tool or press the ESC key twice.

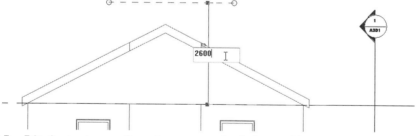

**Figure 8.5** *Edit the temporary dimension to move the Reference Plane*

If you deselected the Reference Plane already, you can click the "Do Not Crop View" icon on the View Control Bar to toggle off the Crop Region. This will disable cropping in this view and the Reference Plane will appear above the Roof. Click to select the Reference Plane and then adjust the temporary dimension as noted above. Toggle the Crop Region on again when finished.

15. Click on the chimney to select it.

You may recall that the chimney was constructed in Chapter 3 from an In-Place Family.

- Drag the control handle at the top and snap it to the Reference Plane.

- Click the small padlock icon (close the lock) to constrain the top edge of the chimney to the Reference Plane (see Figure 8.6).

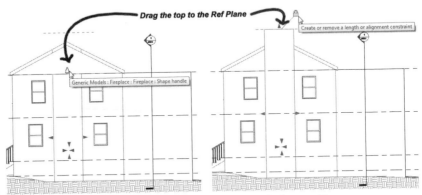

**Figure 8.6** *Drag the top of the chimney and lock it to the Reference Plane*

Constraining the geometry to the Reference Plane is not required, but it makes it easier to adjust the height of the chimney later if required. Simply move the Reference Plane and the top of the chimney will follow. Try it now if you like.

16. Save the project.

## ATTACH WALLS TO THE ROOF

It may not be apparent from the elevation view, but the Walls do not project all the way to the Roof. In some cases when you draw a Roof, you will be prompted to automatically attach the Walls to the Roof. Usually for this to occur, the Walls must intersect the Roof. In this case, the Walls stopped beneath the Roof. However, we can still manually attach them to the Roof. It is a good idea to get in the habit of checking for this condition.

17. From the Window menu, choose **Close Hidden Windows**.

- On the View toolbar, click the Default 3D View icon (or open the {3D} view from the Project Browser).

- From the Window menu, choose **Window Tile**.

 **Tip:** The keyboard shortcut for Tile is **WT**.

18. Pre-highlight one of the exterior Walls of the existing house.

- Press the TAB key to highlight a chain of connected Walls.

- With all of the existing house Walls pre-highlighted, click the mouse to select them (see Figure 8.7).

**Note:** If you have trouble chain-selecting the existing Walls, you can use other selection methods such as the CTRL key, or you can even duplicate the {3D} View and make an "Existing Conditions 3D" view from which to perform these steps like we did previously for the plan.

19. On the Options Bar, click the Attach button.

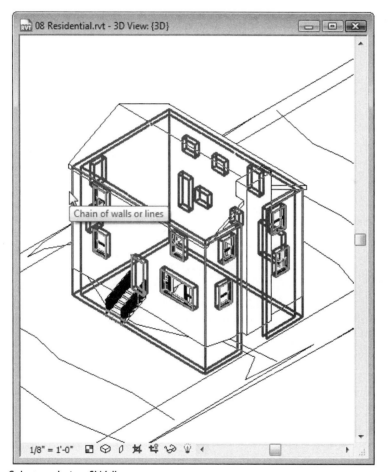

**Figure 8.7**  *Select a chain of Walls*

- On the Options Bar, verify that "Top" is selected and then click the Roof (see Figure 8.8).

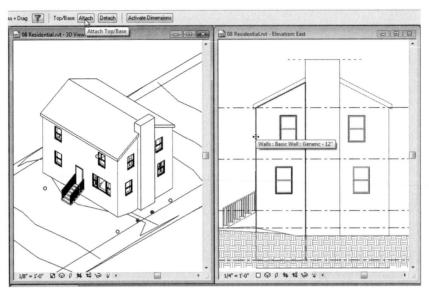

**Figure 8.8** *Attach the Walls to the Roof*

- Deselect the Walls.

## EDIT THE SECOND FLOOR WALL LAYOUT

There are many other Roof options to explore. The best place to do this in this project is on the new construction in the addition rather than the existing construction. The second floor of the addition will extend the two existing bathrooms and have an outdoor patio on the left side of the plan. On the north and west sides of the patio, we will have low height parapet-type Walls. The Roof will cover the other portions of the addition, but the patio will be uncovered. Before we add the Roof, we need to modify the Walls a bit to reflect these design features.

20. On the Project Browser, double-click to open the *Second Floor* plan view.

21. Using techniques covered in Chapter 3, add two Walls as shown in Figure 8.9. Use the same Type as the other three Walls in the addition.

- Set the Location Line to **Finish Face:Exterior**.

- Set the "Base Constraint" to **Second Floor** and the "Top Constraint" to **Up to level: Roof**.

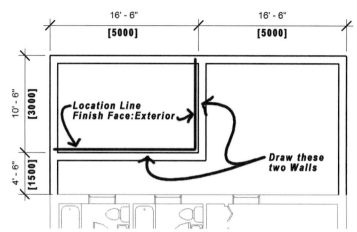

**Figure 8.9** *Add two new Walls to the Second Floor*

The space that we have just described in the top left corner is the outdoor patio. As noted above, it will have low height Walls on two sides. Now let's edit the original exterior Walls to reflect this condition. To do this, we are going to edit the Profile of the Walls.

22. Select the top horizontal Wall.

- On the Options Bar, click the Edit Profile button.

    The "Go To View" dialog will appear.

- In the "Go To View" dialog, choose *Elevation:North* and then click Open View (see Figure 8.10).

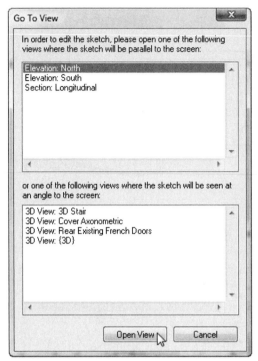

**Figure 8.10** *Certain edits prompt for a more appropriate view in which to work*

When you begin an operation that cannot easily be performed in the current view, Revit will prompt you to open a more appropriate view. The "Go To View" dialog suggests all appropriate views in the current project in which to perform the operation. You should now be looking at the back of the house with the selected Wall in sketch mode. In this mode we can edit the shape of the Wall to "sculpt" it to meet the needs of the design.

23. On the Design Bar, click the **Lines** tool.

- On the Options Bar, click the Pick Lines icon.

- Click on the right vertical edge of the intersecting Wall in the middle of the second floor (see Figure 8.11).

 **Note:** This is the outside edge of the Wall we drew above.

- On the Options Bar, type **3'-6″ [1050]** in the "Offset" field.

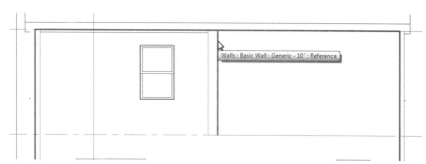

**Figure 8.11** *Add a sketch line to the Wall Profile Sketch relative to the intersecting Wall*

- Highlight the Second Floor Level line.
- When the green line appears above the Level line, click to create the sketch line (see Figure 8.12).

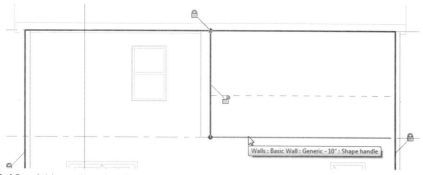

**Figure 8.12** *Add another sketch line relative to the Second Floor Level Line*

24. Use the Trim/Extend tool to cleanup the sketch and close all of the corners (see Figure 8.13).

- Use the Trim/Extend to Corner option on the Options Bar.

**Note:** Remember to click the side of the line that you wish to keep.

25. On the Design Bar, click the Finish Sketch button.

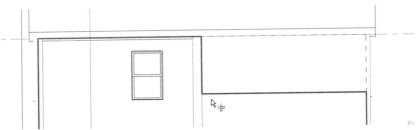

**Figure 8.13** *Complete the Wall Profile Sketch (sketch lines enhanced for clarity)*

26. Select the other Wall (the vertical one on the right in the current view—on the left in the plan view) and repeat the process.

- When prompted, open the *Elevation:West* view.

- Again use the outside edge of the Wall we drew above to create the vertical sketch line and then offset a line up from the Second Floor Level Line as before.

- Trim/Extend to complete the sketch (see Figure 8.14).

- On the Design Bar, click the Finish Sketch button.

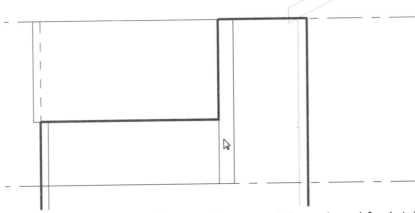

**Figure 8.14** *Edit the Profile Sketch of the opposite Wall (sketch lines enhanced for clarity)*

The new addition will be brick veneer on a stud wall backup. We can apply a Wall Type that represents this type of construction to our new construction Walls.

27. Select all of the Walls in the addition (the three original ones and the two new ones we just added).

- From the Type Selector, choose **Basic Wall : Exterior - Brick on Mtl. Stud.**

28. On the View toolbar, click the Default 3D View icon (or open the {3D} view from the Project Browser).

- Hold down the SHIFT key and drag the wheel on your mouse to spin the model around and see the edits to the Walls (see Figure 8.15).

 **Note:** If you do not have a wheel mouse, use the "Dynamically Modify View" tool on the View toolbar instead.

 **Note:** If the {3D} view does not show the new construction, right-click the view and edit its Properties. Change the Phase to New Construction.

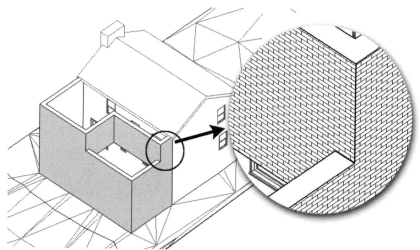

**Figure 8.15**   *View the model in 3D to see the completed Wall edits*

29. On the Project Browser, right-click the {3D} view and choose **Duplicate View > Duplicate**.

- Name the new view **New Addition Axon**.

30. Save the project.

## ADD THE NEW ROOF

Now that we have prepared the second floor Wall layout, we are ready to begin roofing the addition.

31. On the Project Browser, double-click to open the *Roof* plan view.

**Note:** Be sure to open "*Roof*" this time, not "*Existing Roof.*"

32. On the Design Bar, click the Basics tab, click the **Roof** tool, and then from the flyout choose **Roof by Footprint**.

- On the Options Bar, place a checkmark in the "Defines slope" checkbox.

- In the Overhang field, type **6″ [150]**.

- Place the two sketch lines indicated in Figure 8.16.

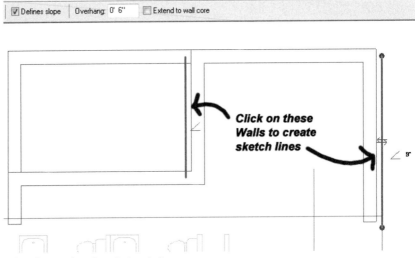

**Figure 8.16**  *Create the sloped sketch lines*

33. On the Design Bar, click the **Modify** tool or press the ESC key twice.

34. Select both of these sketch lines.

- On the Options Bar, click the Properties icon.

- In the "Element Properties" dialog, beneath the "Dimensions" grouping, change the Roof slope. If you are working in Imperial units, type **6″** for the Rise/12″ parameter. If you are working in Metric, set the Slope Angle to **26.57**.

- Click OK to dismiss the "Element Properties" dialog and deselect the sketch lines.

 **Tip:** As an alternative, you can select the sketch line and then edit the slope directly with the temporary dimension that appears next to the slope indicator.

35. On the Design Bar, click the **Lines** tool.

- Clear both the "Defines slope" and "Chain" checkboxes.

- Draw a horizontal line aligned to the edge of the existing house Roof and the width of the two sketch lines we already have.

- Use the Pick Lines option and a **6″ [150]** offset to create the top sketch line (see the left side of Figure 8.17).

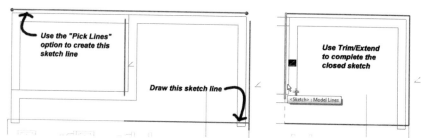

**Figure 8.17**   *Create two horizontal sketch lines that do not define slope*

36. Using the Trim/Extend tool, cleanup the rectangular sketch shape (see the right side of Figure 8.17).

- On the Design Bar, click the Finish Roof button.

- On the Project Browser, double-click to open the *New Addition Axon* 3D view.

- Spin the model to see the interaction of the two Roofs clearly.

## JOIN ROOFS

As you can see in the 3D view, the new Roof does not intersect with the existing one. This is easily corrected.

37. On the Tools toolbar, click the Join Roofs icon.

   **Tip: Join/Unjoin Roof** is also located on the Tools menu.

Take a look at the Status Bar (lower left corner of the screen) and notice the message. You are prompted to: "Select an edge at the end of the roof that you wish to join or unjoin." We want to select one of the edges of the new construction Roof. The next prompt will ask us to select a face to which to join. In that case, we will select the face of the existing Roof.

- Click on the edge of the new construction Roof as indicated on the left side of Figure 8.18.

- Click on the face of the existing construction Roof as indicated on the right side of Figure 8.18.

   **Note:** To pick a Roof face of the existing Roof, move the Modify tool over any of the edges of the existing Roof. The perimeter of the existing Roof will pre-highlight. Use the TAB key to cycle the selection, and click when the desired Roof face is pre-highlighted to select it.

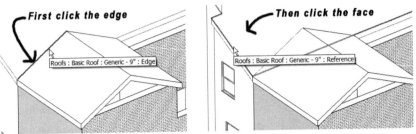

**Figure 8.18** *Select an edge of the Roof to join to the face of the other Roof*

The new construction Roof will now extend over the existing Roof and form nicely mitered intersections. The same tool can be used in reverse if you ever need to unjoin a Roof.

38. Repeat the entire process to create the shorter new Roof on the other side of the addition (see Figure 8.19).

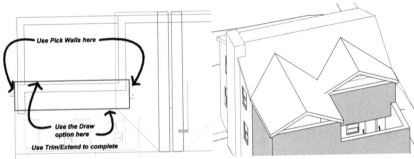

**Figure 8.19** *Create the Roof on the other side of the addition*

Join Roofs works when the two Roofs meet perpendicular to one another. We will use a different technique to join the two new Roofs together at the small valley between them. For that we will use the more generic *Join Geometry* tool.

39. On the Tools toolbar, click the Join Geometry icon.

 **Tip: Join Geometry** is also located on the Tools menu.

Again, watch the Status Bar prompts.

- Click on the first new construction Roof, then click the other to join them.

- On the Design Bar, click the **Modify** tool or press the ESC key twice.

40. Using the technique covered above, Attach the Walls to the new Roofs (see Figure 8.20).

**Tip:** If you have trouble attaching the tall Wall on the East side of the house to the new Roof, attach it first to the existing construction Roof and then attach it to the new construction Roof.

41. Save the project.

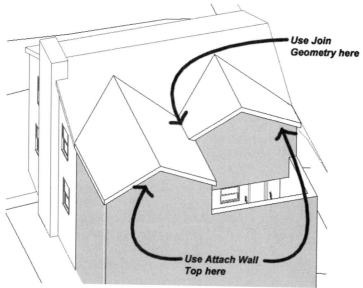

**Figure 8.20** *Join the two new construction Roofs with Join Geometry*

## EDITING ROOFS

Now that we have created the Roofs for the Residential project, let's turn our attention to editing Roofs that already exist in a Revit Architecture model. In this case, they will be the ones that we have just added. In your own projects, you will work with Roofs in much the same way as other Revit Architecture elements—start by laying them out with simple generic parameters and then over the course of the project, layer in additional details and parameters as design and project needs dictate. In this sequence, we will explore some of the possibilities available for editing Roof elements.

### MODIFY THE ROOF PLAN VIEW

You may have noticed that the *Roof* plan view looks a bit odd. The Roof elements are being cut in plan which shows us only part of the sloped surface. While appropriate for an attic space, this kind of display is not how we would typically represent an actual Roof Plan. To show the roof correctly, we want to see the entire Roof element uncut. To achieve this, we need to modify the view range of the *Roof* plan view.

Continuing in the Residential Project.

   1. On the Project Browser, double-click to open the *Roof* plan view.

   • Right-click in the workspace and choose **View Properties**.

- In the "Element Properties" dialog, beneath the "Extents" grouping, click the Edit button next to "View Range."
- In the "Primary Range" area, change the "Top" Offset to **15'-0"** [**4500**].
- Change the "Cut Plane" Offset to **10'-0"** [**3000**] and then click OK (see Figure 8.21).

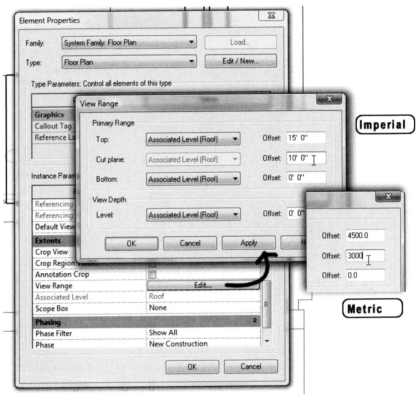

**Figure 8.21** *Adjust the Cut Range of the Roof Plan view*

2. Click OK twice to return to the "Element Properties" dialog.

- Beneath the "Graphics" grouping, choose **None** for the "Underlay."
- Click OK to return to the *Roof* plan view and see the results (see Figure 8.22).

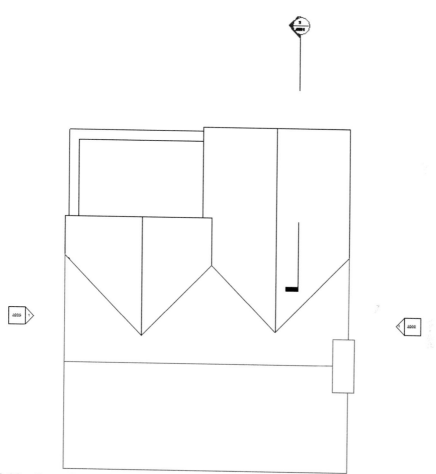

**Figure 8.22**   *Turn off the Underlay of the Second Floor and view the Roof Plan*

## UNDERSTAND ROOF OPTIONS

The Roof element has several options that we have not yet explored. Let's take a look at some of them now.

3. On the Project Browser, double-click to open the *Longitudinal* section view.

- Zoom in Region around the eave at the left side of the section (see Figure 8.23).

Notice the way the Roof intersects the attached Wall.

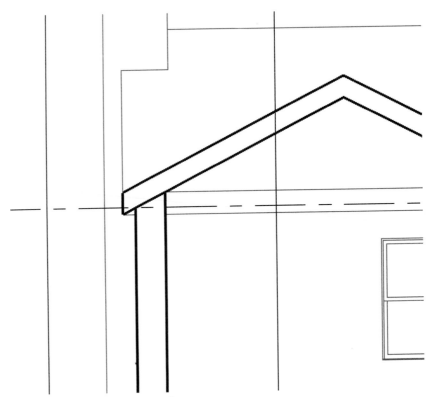

**Figure 8.23** *Zoom in on the eave condition*

Above we swapped in a more detailed Wall Type for these Walls. However, since we are currently viewing the model in "Coarse" display mode, we do not see any difference in the Wall structure. To see the more detailed Wall structure, we need to adjust the detail display level of the current view.

4. On the View Control Bar, click the Detail Level icon and choose **Medium** (see Figure 8.24).

**Figure 8.24** *Change the Detail Level to Medium*

The Layers that makeup the Wall Type's structure will appear including brick, a stud layer and air gap. This Detail Level will make it a little easier to understand the various Roof options that we are about to explore.

  5. Select the Roof.

  • On the Options Bar, click the Properties icon.

  • Beneath the "Construction" grouping, change the "Rafter Cut" to **Two Cut – Plumb** and then click OK.

The results will probably not be satisfactory.

  • Return to the Roof Properties and set the "Fascia Depth" to **6″ [150]**.

  • Repeat the process and choose **Two Cut – Square** (see Figure 8.25).

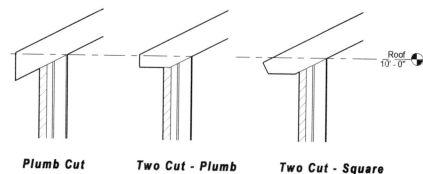

| *Plumb Cut* | *Two Cut - Plumb* | *Two Cut - Square* |

**Figure 8.25**  *Comparing the Rafter Cut options (Illustrations simplified for clarity)*

The section shown in the figure has been simplified to remove any of the beyond components (the Far Clip Offset of the section was adjusted to achieve this). However, the cut components of your model should closely resemble the figure. The panel on the left shows the original condition—which is the default. The middle and right panels show the conditions that we just tried.

  6. Return to the Roof "Element Properties" dialog once more and change the "Rafter Cut" to **Two Cut – Plumb** and then click OK.

When we were adding the Roof above, you may have noticed the "Extend to wall core" checkbox on the Options Bar. (We did not use this option when creating the Roof.) When you choose this option, the Overhang setting will be measured relative to face of the core layer rather than the finish face of the Wall. Figure 8.26 shows this option used with and without attaching the Walls to the Roof. (The Core layer of the Wall has been shaded gray in the figure for clarity). If you wish, you can experiment with these options. Remember, if you choose to edit the Roof while the section view is open, you will be prompted to open a more appropriate view—choose *Floor Plan:Roof.* Cancel or undo any edits made if you do experiment with this setting.

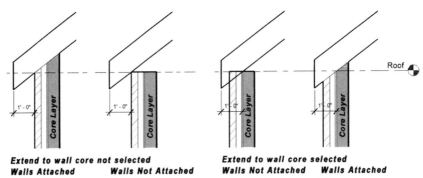

**Figure 8.26** *Understanding the effect of the "extend to wall core" and Attach options*

There is one other setting worth exploring at this point. When you create your Roof, Revit offers two modes of construction: Rafter or Truss. The difference between these two settings is simply the point that is used as the spring point for the Roof.

7. Select the Roof.

- On the Options Bar, click the Properties icon.

- Beneath the "Construction" grouping, change the "Rafter or Truss" to **Truss** and then click OK.

Notice how the entire Roof appears to move up. Rafter measures the plate of the Roof from the inside edge of the Wall. Truss measures from the outside edge. This option is only available for Roofs created using the "Pick Walls" option.

8. Return to the Roof "Element Properties," change the "Rafter or Truss" to **Rafter** and then click OK.

9. Select the other two Roofs (one new construction, one existing) and then edit the Properties.

- Change the "Rafter Cut" to **Two Cut – Plumb**.

- Set the "Fascia Depth" to **6″ [150]** and then click OK.

- Set "Rafter or Truss" to **Rafter**.

10. Save the project.

## CREATE A COMPLEX ROOF TYPE

Much like Walls, Roofs can be composed of several Layers of structure. You can create your own Roof Type that contains the structure you need, or transfer an appropriate Type from another project using the Transfer Project Standards feature. In this example, we will build a new Roof Type from scratch.

11. Select all of the Roofs.

- On the Options Bar, click the Properties icon.

12. Next to the Type list, click the Edit/New button.

**Tip:** A shortcut for this is to press ALT + E.

The Type Properties dialog will appear.

- Next to the Type list, click the Duplicate button.

**Tip:** A shortcut for this is to press ALT + D.

A new Name dialog will appear. By default "(2)" has been appended to the existing name.

- Change the name to **MRAC Wood Rafters with Asphalt Shingles** and then click OK (see Figure 3.40 in Chapter 3 for an example).

13. In the "Type Properties" dialog, next to the "Structure" click the edit button.

In the "Edit Assembly" dialog, you can see that the Roof Type currently contains only a single generic Layer. This is because we originally duplicated it from the generic Roof Type. You can add, edit, and delete Layers to the Roof structure in this dialog. We will keep the structure of our custom Roof Type simple. We need a structure layer, which will be wood rafters, a plywood substrate, and asphalt shingles for the finish layer. When you build a new Type in Revit Architecture, there are several things to consider. We can add as much or as little detail to the structure of the Roof Type as we wish. In some cases, it will prove valuable to represent each piece of the Roof's actual construction. However, also consider the potential negative impact that highly detailed Types can have on drawing legibility and computer performance. As a general rule of thumb, you should seek to build your models as accurately as possible while remembering that any architectural drawing includes a certain degree of abstraction as a matter of industry convention and the facilitation of clarity. All of these points hold true in other areas of Revit as well, such as creating and editing Wall Types. We will see more on this in coming chapters. With these issues in mind, we will abstract our Roof construction to just the three Layers noted previously, thereby excluding building paper, insulation, or interior finish. These items can be added to the Roof Type later (which will automatically apply to all Roof elements that reference the Type) or we can apply these items graphically as drafting embellishment in a Detail view (see Chapter 10 for more information on Detail views).

## EDIT ROOF STRUCTURE

The "Edit Assembly" dialog lists each Layer of the Roof Type in a list with a numeric index number next to each one. There are four columns next to each item.

- **Function**—Click in this field for a list of pre-defined functions. The functions include "Structure," "Substrate," "Finish," and others. The number next to the function name indicates the priority of the Layer with regard to material "wrapping" joins. In this way, the Structure Layer of one Roof will attempt to join with the Structure Layer of another. One [1] is the highest priority, while five [5] is the lowest. "Membrane Layers" have zero thickness and thus do not have priority nor do they join.

- **Material**—Materials determine the graphical characteristics of each Layer. Material properties include patterns, render material, and shading.

- **Thickness**—This is the dimensional thickness of the Layer.

- **Wraps**—Controls if the Layer wraps around corners at the ends or at openings. If this is not selected, the Layer simply cuts perpendicular at the ends and openings.

- **Variable**—When the Type is applied to a flat Roof, one component on the list can be given a variable thickness. Editing tools on the Options Bar provide a means to add high and low points to the surface of the Roof element. When the variable box is selected, such height variations are applied only to the variable component. (This is used to represent tapered insulation on flat roofs for example—see below.) When no layers use the variable option, thickness variations are applied to the entire Roof essentially warping the surface.

14. In the "Edit Assembly" dialog click in the Material cell for the Structure Layer (Layer number 2).

    A small expand (downward pointing arrow) icon will appear at the right side of the cell.

- Click the small browse icon to open the "Materials" dialog.

- From the "Name" list, select **Structure - Wood Joist/Rafter Layer [Structure - Timber Joist/Rafter Layer]** and then click OK.

- In the Thickness field, type **7 1/4" [190]** and then press ENTER (see Figure 8.27).

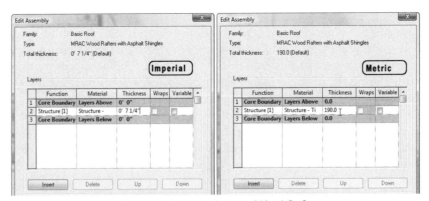

**Figure 8.27**  *Edit the existing Structure Layer to become Wood Rafters*

15. Beneath the list of Layers, click the Insert button.

    A new zero thickness Structure Layer will appear.

Notice that the new Layer appears within the "Core Boundary." Roofs and Walls have a Core that contains the structural Layers. You can have additional Layers on either side or both sides of the Core. If a new Layer that you insert does not appear in the desired location, select it and then use the Up and Down buttons to adjust its position in the structure of the Roof.

- With the new Layer highlighted, click the Up button to place it above the Core Boundary.
- Change the Function of the new Layer to **Substrate [2]**.
- Change the Material to **Wood - Sheathing – plywood**.
- Set the Thickness to **5/8"** [**16**].

16. Click the Insert button again.

- Change the Function of the new Layer to **Finish [4]**.
- Change the Material to **Roofing – Asphalt** [**Roofing – Asphalt Shingle**].
- Set the Thickness to ¼" [**6**].

If you wish to see how the Type looks so far, click the Preview button at the bottom left corner of the dialog (see Figure 8.28).

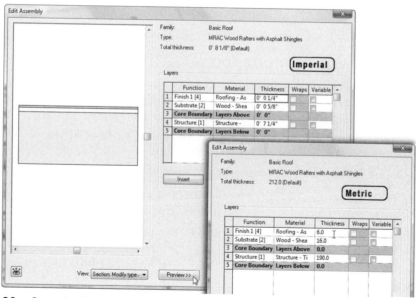

**Figure 8.28**   *Open the Preview window to see the completed structure graphically*

17. Click OK three times to return to the workspace window.

Notice that the new Roof Layers appear in the section view—if you switched to a different view, please return to the *Longitudinal* section view now. Also note that the new Layers will only appear if you left the section view in the Medium Display mode from above. If you set it back to Course, the graphics will simplify to show the outer edge of the Roof only. If you have difficulty seeing the sheathing and asphalt shingle Layers, try toggle the "Thin

Lines" setting. You can find an icon for this on the View toolbar. Thin Lines is also available from the View menu (see Figure 8.29).

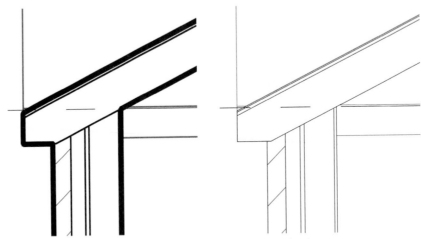

**Figure 8.29** *Toggle Thin Lines to see the Roof Structure more clearly*

Toggle Thin Lines back off after you are satisfied with the Roof structure.

### APPLY A HOST SWEEP TO ROOF EDGES

Creating the new Type and editing its structure provides a satisfactory representation of the overall Roof construction. However, the edges of the Roof could use some further embellishment. A Host Sweep allows us to apply a detailed profile condition the edges of a Roof or along the length of a Wall. In this example, we will explore the use of Host Sweeps to apply fascia boards and gutters to our Roofs.

18. On the Project Browser, double-click to open the *New Addition Axon* view.

19. On the Design Bar, click the Modeling tab and then click the **Host Sweep** tool. Then choose **Roof Fascia** from the flyout menu that appears.

There are some Fascia Types already resident in the project. However, as with the Roof Type above, we will create our own. A Fascia is a simple element whose primary parameter is the assignment of a Profile shape. The Roof here has a 6″ [150] roof edge as defined above so we will want a Profile close to this depth.

20. On the Options Bar, click the Properties icon.

- Next to the Type list, click the Edit/New button.

- Next to the Type list, click the Duplicate button.

- Change the name to **MRAC Simple Fascia** and then click OK.

- From the "Profile" list, choose **Fascia-Flat : 1 x 6 [M_Fascia-Flat : 19 x 140mm]** and then click OK twice.

A glance at the Status Bar will reveal the following prompt: "Click on edge of Roof, Soffit, Fascia, or Model Line to add. Click again to remove." With the *New Addition Axon* view open, it is easy to accomplish this.

21. Move the pointer over the various edges of the Roof elements on screen.

Notice that both the top and bottom edges of any given edge will pre-highlight (see Figure 8.30).

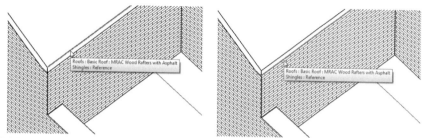

**Figure 8.30** *You can Pre-Highlight either the top or bottom edge*

- Click on a top edge to apply the Fascia.
- Place Fascia boards on the top edges of all of the horizontal (fascia) edges of the new Roof and all edges of the existing Roof as shown in Figure 8.31.

 **Note:** Be sure to click the top edge of each Roof edge, not the bottom.

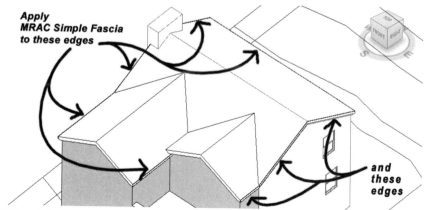

**Figure 8.31** *Adjust Gutter Horizontal Offset*

- On the Design Bar, click the **Modify** tool or press the ESC key twice.

We did not apply a Fascia to the sloped (rake) edges of the new construction Roof yet because we are going to use a different Type for these. However, before we create the Type, we need a more complex Profile Family from the library.

22. On the File menu, choose **Load from Library > Load Family**.

Either the *Imperial Library* or *Metric Library* folder should open automatically. If this has not occurred, you can use the shortcut icons on the left side of the dialog to jump to those locations. If you do not have these icons, the appropriate files have been provided with the files installed from the Mastering Revit Architecture CD ROM in the *Library* folder. You can navigate manually to that location and access the Families from there if necessary.

- Double-click on the *Profiles* folder and then double-click the *Roofs* folder.

- Select the *Fascia-Built-Up.rfa* file and then click Open.

This loads the new Profile Family into the current project. We now need to repeat the steps above to create a new Fascia Type using this Profile.

23. On the Design Bar, click the Modeling tab and then click the **Host Sweep > Roof Fascia** tool.

- On the Options Bar, click the Properties icon.

- Next to the Type list, click the Edit/New button and then the Duplicate button.

- Change the name to **MRAC Built-up Fascia** and then click OK.

- From the "Profile" list, choose *Fascia-Built-Up : 1 x 8 w 1 x 6 [M_Fascia-Built-Up : 38 x 140mm x 38 x 89mm]* and then click OK twice.

24. Click each of the remaining Roof edges (remember to click the top edge).

- On the Design Bar, click the **Modify** tool or press the ESC key twice.

Zoom in on one of the intersections between the Rake and Fascia conditions. If you select one of the Fascia boards, you will notice a drag control at the ends. You can stretch this control to modify the way the two Fascia boards intersect. Also, with one of the Fascia or Rake boards selected, you will notice a "Change Mitering Option" button on the Options Bar. The options are "Horizontal," "Vertical," and "Perpendicular." Try them out if you wish to see how each option behaves.

25. Save the project.

## ADD GUTTERS

Another type of Host Sweep available to Roofs is a Gutter. These are conceptually the same as Fascia boards. They use a Profile to determine the cross-section shape and sweep it along the path of the Roof edge(s).

26. Zoom out to see the whole Roof.

27. On the Design Bar, click the Modeling tab and then click the **Host Sweep** tool. Then choose **Roof Gutter**.

 • Click one of the horizontal edges of the new construction Roof.

If the gutter is not visible, it needs to be flipped. There are flip controls on the selected Gutter.

 • Click the Flip control to flip the gutter to the outside.

 • Before clicking the next segment, click the Start next button on the options Bar.

When you create Host Sweeps, multiple segments can be added to a single Sweep. If you later click to select it, you will notice that they all select as one. In this case, we will have more flexibility to select and flip our gutters if we start each segment as a new Sweep.

28. Add Gutters (and flip as required) to the remaining horizontal new construction Roof edges (see Figure 8.32).

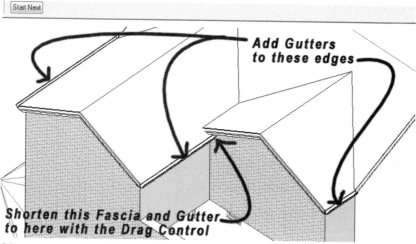

**Figure 8.32**   *Add Gutters to the new construction Roof*

After you place the Gutters, pre-highlight them and notice the one in the valley between the two new Roofs is too long. You can use the Drag control on the Fascia board (and the Gutter if necessary) to shorten it.

29. On the Project Browser, double-click to open the *Longitudinal* section view.

30. Zoom in on one of the Gutters (turn on Thin Lines if necessary).

Depending on which line you clicked (the Roof edge or the Fascia edge) when adding the gutters, you may notice that the Gutter overlaps the Fascia board. We can delete the gutter and re-add it, or we can apply an offset to the Gutter equal to the thickness of the Fascia profile to compensate for this.

31. Select any of the Gutters that require adjustment.

- On the Options Bar, click the Properties icon.

- In the "Element Properties" dialog, change the "Horizontal Profile Offset" to ¾" [19] and then click OK (see Figure 8.33).

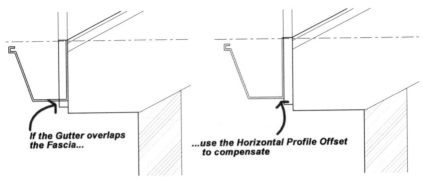

**Figure 8.33** *Adjust Gutter Horizontal Offset*

## SKYLIGHTS

As a finishing touch to the residential Roof, let's add a skylight in one of the new Roofs. To do this, we must load another Family into the project.

32. On the Project Browser, double-click to open the *Roof* plan view.

33. On the File menu, choose **Load from Library > Load Family**.

- Double-click on the *Windows* folder.

- Select the *Skylight.rfa* [*M_Skylight.rfa*] file and then click Open.

34. On the Design Bar, click the Basics tab and then click the **Window** tool.

- From the Type Selector, choose **Skylight : 28" x 38"** [**M_Skylight : 0711 x 0965mm**]

35. Place it approximately as indicated in Figure 8.34.

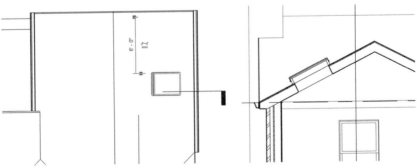

**Figure 8.34** *Place a skylight in the Roof*

A vertical temporary dimension will remain after you place the skylight. Its witness lines should be between the outside horizontal Wall and the top edge of the skylight.

- Adjust the witness lines if necessary and edit the value of the temporary dimension to **5'-0"** [**1500**].

36. On the Project Browser, double-click to open the *Longitudinal* section view.

The skylight should appear cutting through the Roof.

37. Save the project.

## CREATING FLOORS

Use Floor elements to model the floor planes or slabs in your projects. Most often, the Floor element is a simple flat structure that you create via a closed sketch. In some cases, the Floor may slope like in parking garages or theaters. In this topic, we will add and modify some Floor elements.

### ADD FLOORS

We have already worked with Floors a little in our commercial project. However, you have probably noticed that our residential project currently has no Floors.

1. On the Project Browser, double-click to open the *First Floor* plan view.

2. On the Design Bar, click the Basics tab and then click the **Floor** tool.

- On the Design Bar, click the **Pick Walls** tool.

- On the Options Bar, place a checkmark in the "Extend into wall (to core)" checkbox.

3. Click on the vertical Wall of the addition on the left.

- Click on each of the other two exterior Walls of the new addition.

- Click the existing Wall between the house and the addition.

Sketch lines will appear at all four Walls. The one for the existing Wall is currently on the inside edge, but we need it to be on the exterior side adjacent to the new construction.

- On the Design Bar, click the **Modify** tool or press the ESC key twice.

4. Drag the horizontal sketch line between the existing house and the addition to the other side of the existing Wall (see Figure 8.35).

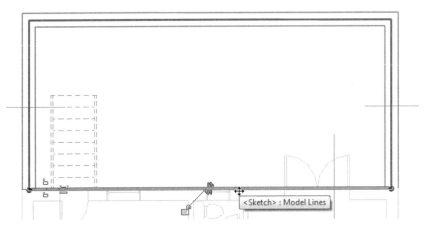

**Figure 8.35**  *Modify the sketch lines*

    5. On the Design Bar, click the Floor Properties button.

- From the Type list, choose **Wood Joist 10″ - Wood Finish [Standard Timber-Wood Finish]** and then click OK.

The structure of the Floor is similar to the structure of the Roof that we explored above. Feel free to edit the Type Parameters to study the structure but be sure to not make any changes at this time.

    6. On the Design Bar, click the Finish Sketch button.

- In the dialog that appears, click Yes to join the geometry (see Figure 8.36).

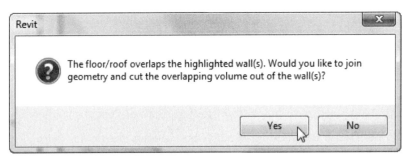

**Figure 8.36**  *Allow Revit Architecture to join the Floor to the neighboring Walls*

    7. Repeat the entire process on the second floor.

Create the sketch the same way initially, however, modify the sketch to conform to the "L" shape (excluding the outdoor patio) of the interior space of the addition. You can use an additional "Pick Walls" action to do this (see Figure 8.37). Remember to edit the Floor Properties and choose the same Type as the first floor.

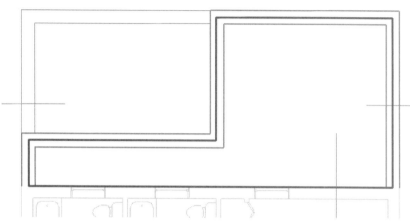

**Figure 8.37**  *Create the sketch for the second floor*

## CREATE A FLOOR TYPE

The last Floor that we need to make is the one for the patio space on the second floor. The process is similar to the above.

8. Click the **Floor** tool and use the same options.

9. Create sketch lines for each of the four Walls that make up the patio and use the Trim/Extend tool to complete the shape (see Figure 8.38).

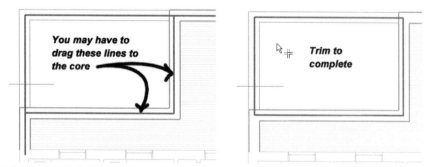

**Figure 8.38**  *Create the sketch for the patio floor*

10. On the Design Bar, click the Floor Properties button.

- From the Type list, choose **Wood Joist 10″ - Wood Finish** [***Standard Timber-Wood Finish***].

- Next to the Type list, click the Edit/New button and then the Duplicate button.

- Change the name to **MRAC Patio Floor** and then click OK.

The patio will have a wood deck built up on top of sleepers to provide drainage below. Therefore, we can use nearly the same component makeup as the other Floor system and simply insert the sleeper Layer. While this Floor Type would have a membrane Layer and flashing, we will not include those in the model but rather show those components later in details (see Chapter 11 for more information on detailing).

11. Click the Edit button next to Structure, select Layer 1, and then click the Insert button to add a new Layer.

- Move it to down to beneath the Finish Layer (but above the Core Boundary).

- Set the new Layer Function to **Substrate [2]**.

- Change the thickness of the new Layer to **1 ½" [40]** and change the Material to **Wood - Stud Layer**.

- Click OK three times to return to the sketch.

- On the Design Bar, click the Finish Sketch button.

- In the dialog that appears, click Yes to accept joining with the Walls.

## ADJUST FLOOR POSITION AND JOINS

A couple of issues remain with the two Floors on the second floor. First, the Walls that highlighted automatically for join did not include the two Walls we added at the start of the chapter. Second, the patio Floor is too low relative to the other one. We can see these issues best in section.

12. On the Project Browser, double-click to open the *Longitudinal* section view.

Examine the section to see both conditions noted here.

13. Select the patio Floor (it is on the right in the section).

- On the Options Bar, click the Properties icon.

- For the "Height Offset From Level" value, type **1 ½" [40]** and then click OK.

This is the same amount as the thickness of the sleepers.

14. On the Tools toolbar, click the Join Geometry icon.

- For the first pick, click the patio Floor.

- For the second pick, click the vertical Wall (see Figure 8.39).

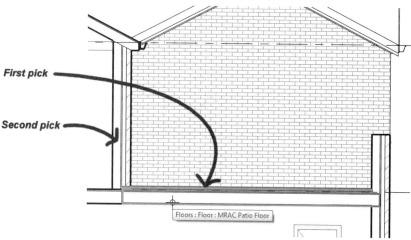

Figure 8.39 *Join the Floor to the Wall*

The Join command will remain active. You can perform several Join operations in a row.

15. Join the same Wall to the other Floor.

16. Join the two Floors to one another (see Figure 8.40).

 **Tip:** Note the "Multiple Join" option on the Options Bar. Place a checkmark in this box to Join several elements in one operation.

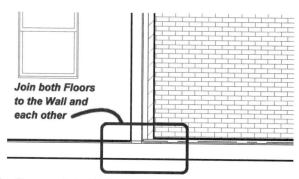

Figure 8.40 *Join the Floors and the Wall*

17. On the Project Browser, double-click to open the *Second Floor* plan view.

   • Create a section cutting vertically (parallel to *Transverse*) through the patio (see Figure 8.41, left side).

   • Rename it **Patio Section**.

   • Open this section, change the Detail Level to **Medium** and then repeat the steps here to join the Floors and Wall (see Figure 8.41, right side).

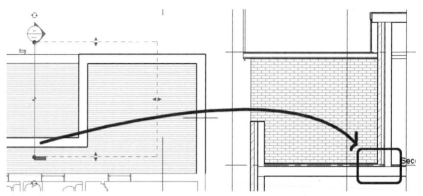

**Figure 8.41**  *Cut a section to assist in joining the remaining Floors and the Wall*

18. On the Design Bar, click the **Modify** tool or press the ESC key twice.

Previously we mentioned the priorities of the Layers within the Roof (also Wall and Floor) structure. In these two sections, you can see this interaction very clearly. Notice the way that the structural Layer of the Floor cuts into the Core of the Walls.

19. Save and close the project.

## COMMERCIAL PROJECT ROOF

Our commercial project already has a Roof. However, it is currently just a flat slab. Also, the Stair tower does not yet have a Roof. Our aim in this section will be to refine the Roof element already in the commercial project and to add additional required Roof elements. We will also begin work on our commercial project's Roof plan view.

### LOAD THE COMMERCIAL PROJECT

Be sure that the Residential Project has been saved and closed.

1. On the Standard toolbar, click the Open icon.

 **Tip:** The keyboard shortcut for Open is CTRL + O. **Open** is also located on the File menu.

- In the "Open" dialog box, click the *My Documents* icon on the left side.

- Double-click on the *MRAC* folder, and then the *Chapter08* folder.

    If you installed the dataset files to a different location than the one listed here, use the "Look in" drop-down list to browse to that location instead.

2. Double-click *08 Commercial.rvt* if you wish to work in Imperial units. Double-click *08 Commercial Metric.rvt* if you wish to work in Metric units

    You can also select it and then click the Open button.

**Note:** Some minor modifications have been made to the model since the last chapter. The layout of Walls on the Roof level is slightly different than it was in the previous chapter. As such, the core Walls were shortened to the level below and a new set of Walls added on the Roof. For this reason, please be sure to use the dataset provided with this chapter rather than attempting to continue in your previous files.

The project will open in Revit Architecture with the last opened view visible on screen.

## CREATE A ROOF BY EXTRUSION

Until now we have created our Roofs with the footprint option. It is also possible to create a Roof by sketching the profile of it in section and extruding this profile to form the Roof. In the Quick Start chapter we looked briefly at creating a Roof by Extrusion. Let's look at that process in more detail for the Roof at the top of our Stair tower.

    3. On the Project Browser, double-click to open the *South* elevation view.

To assist us in placing the Roof, we will add some new Levels. You may recall from Chapter 4 that a Level can be created with automatically associated plan views, or it can be created without them and simply used for reference. We do not need a separate plan for the Roof of the Stair Tower, so the Levels we add here will not have associated plan views.

    4. Select the Roof Level line.

- On the Tools toolbar, click the Copy tool.

- Click anywhere to set the start point.

- Move the mouse straight up, type **8'-0"** [**2400**] and then press ENTER.

- Repeat the process and create another copy **4'-0"** [**1200**] above the previous copy (or **12'-0"** [**3600**] above the Roof Level).

Notice that both copies have black-colored Level heads. A Level head will be blue if it has an associated floor plan view, and black if it does not. Take a look at the Floor Plans on the Project Browser to confirm that no new floor plan views have been created. (If you wish to create a Level using the *Level* tool instead, it defaults to adding plan views: however, you can turn this option off on the Options Bar before drawing the Level.)

    5. Click on the blue text of the new Level heads and rename the lower one to **Stair Roof Low** and the upper one to **Stair Roof High** (see Figure 8.42).

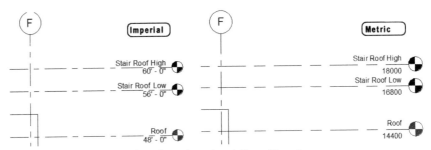

**Figure 8.42** *Copy two new Levels without Associated Floor Plan views*

6. On the Design Bar, click the Basics tab and then click the **Roof** tool and then from the flyout, choose **Roof by Extrusion**.

A "Work Plane" dialog will appear. The Work Plane is the plane in which we will sketch. In this case, because we are making a Roof by extrusion, an effective Work Plane will be perpendicular to the Roof Levels. Any of our numbered Grid Lines can serve this purpose—the numbered Grid Lines form planes parallel to the screen in the current elevation view. Once we have chosen a plane, we will be able to sketch the shape of our Roof. When we finish the sketch, it will extrude perpendicular to the selected plane.

- In the Roof "Work Plane" dialog, choose Grid 4 (from the "Name" list) and then click OK (see Figure 8.43).

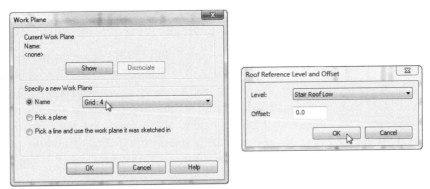

**Figure 8.43** *Choose a Column Grid Line as the Work Plane*

As you can see in the figure, it is also possible to pick the face of some geometry such as a Wall in the model to set the Work Plane. In this case the named Plane associated with the Grid line works best.

- In the "Roof Reference Level and Offset" dialog that appears, select **Stair Roof Low** and then click OK.

The Roof must be associated with a Level to determine its height. As you can see in the dialog, you can choose to create the Roof at any Level in the project and add an offset above or below the Level if appropriate. In this case, we specifically created Stair Roof Low for the task at hand, so no offset is required.

The Design Bar changes to sketch mode and shows only the Roof tools.

7. On the Options Bar, click the "Arc passing through three points" icon.

- For the Arc start point, click the intersection of the Stair Roof Low Level and the left edge of the core Wall (see the top panel of Figure 8.44).

- For the Arc end point, click the intersection of the Stair Roof High Level and the right edge of the core Wall (see the middle panel of Figure 8.44).

- For the Arc intermediate point, click somewhere near the middle of the top edge of the core Wall (see the bottom panel of Figure 8.44).

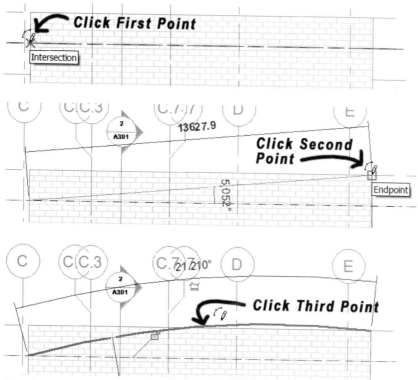

**Figure 8.44** *Sketch an Arc Profile for the Extruded Roof*

You can click on the sketch line and use its drag controls to fine tune the shape to your liking. When you draw a Roof by extrusion, you create an open shape, *not* a closed shape. The thickness of the Roof material will be determined by the Roof Properties and the Type assigned to it just like the other Roofs; therefore, we don't need to sketch the thickness of the roof.

- On the Design Bar, click the **Modify** tool or press the ESC key twice.

8. On the Design Bar, click the Properties button.

- In the "Element Properties" dialog, from the Type list, choose **Steel Truss - Insulation on Metal Deck – EPDM [Steel Bar Joist - Steel Deck - EPDM Membrane]**.

In the "Instance Parameters" area, notice that the Work Plane parameter is set to Grid 4. (This is unavailable for edit—to change the Work Plane after creation, Select the Roof, then click the "Edit Work Plane" button on the Options Bar.).

- For the "Extrusion Start" type **-2'-0" [-600]**.

- For the "Extrusion End" type **22'-0" [6600]**.

- Click OK to return to the sketch.

You probably noticed that there were no overhang parameters in the "Element Properties" dialog or on the Options Bar. To create an overhang, you simply edit the sketch line, or add additional segments.

9. On the Design Bar, verify that **Lines** is still selected.

- On the Options Bar, click the Line icon.

- Add a 2'-0" [600] long horizontal line at each end of the Arc (see Figure 8.45).

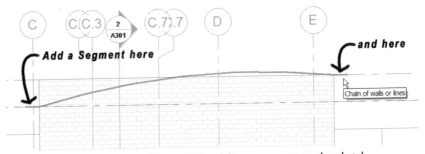

**Figure 8.45**  *Sketch overhangs by adding additional Line segments to the sketch*

10. On the Design Bar, click the Finish Sketch button.

The Roof should appear with its thickness determined by the Type that we chose.

11. On the Project Browser, double-click to open the {3D} 3D view (see Figure 8.46).

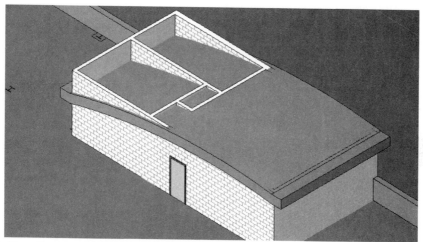

**Figure 8.46**    *The new Roof intersects the core Walls*

It appears as though our Walls could use some adjustment.

    12. On the Project Browser, double-click to open the *Roof* plan view.

- Dragging from left to right, surround the entire core.

Walls, Stairs, and other objects will be selected by this action.

- On the Options Bar, click the Filter Selection icon.

- In the Filter dialog, click the Check None button, click the checkbox next to Walls, and then click OK.

This will remove all other elements from the selection leaving only the core Walls on the Roof level selected.

    13. On the Project Browser, double-click to open the *South* elevation view.

If you prefer, you can tile the two windows with the command on the Window menu instead. The selection of Walls will remain active.

- On the Options Bar, click the Attach button.

- Verify that Top is selected on the Options Bar and then click on the Roof (see Figure 8.47).

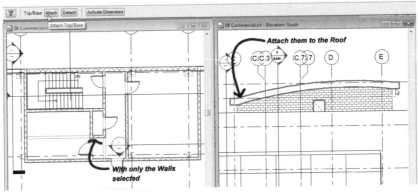

**Figure 8.47** *Attach the core Walls to the imported Roof*

You can study this change in other views as well, if you like, such as the {3D} view. You may recall that in previous chapters we copied the Walls to the Structural model using the Copy/Monitor tool. To simplify this exercise and future exercises, the copied Walls have been removed from the Structural model. As such, you will not be required to synchronize the changes at this time. You can choose to return to the structural model later, run the Copy/Monitor tool, and perform a coordination review to see if any changes require synchronization. Refer to the previous chapter for more details on the process. Since the model was simplified and the task is not required at this time, this task is left to the reader as an additional optional exercise.

## WORKING WITH ROOF SUB-ELEMENTS

The final exercise in this chapter will be to add drainage sloping to the commercial project flat roof. In the previous edition of this book, this task was accomplished with Slope Arrows. The newer Roof sub-element editing tools (introduced in Revit 2008) are now considered a better way to represent the slope on flat Roofs and slabs. However, since Slope Arrows can still be considered useful, the original tutorial is included in PDF format in the *Chapter08* folder with the dataset files installed from the CD ROM. If you wish to perform that exercise, please perform a Save As (from the File menu) to create a copy of your project in its current state. You can perform the steps here in one copy, then re-open the saved copy and perform the steps in the PDF. If you do not wish to do the tasks in the PDF, you can skip the Save As step.

## USING SUB-ELEMENT EDITING TOOLS

When you have a flat Roof or Floor element in your model (no edges sloped), you will see the sub-element editing tools on the Options Bar when the Roof or Floor is selected. If even one edge of the Roof or Floor is set to slope defining, the tools will not appear (see Figure 8.48).

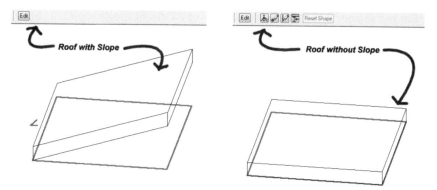

**Figure 8.48**  *Roofs or Floors with no sloping edges in their sketch have access to the sub-element editing tools (sketch lines superimposed and enhanced for clarity)*

1. On the Project Browser, double-click to open the *Longitudinal* section view.

**Tip:** While it is possible to select a Roof or Floor in plan views using the TAB key to cycle to it, you will often find it easier to select in section and then change views if necessary to edit.

- Select the Roof element.

  Notice the collection of sub-element tools shown on the right side of Figure 8.48 appears on the Options Bar.

Moving left to right, the tools are as follows:

- Modify Sub-Elements—This tool can be used to adjust the height of any points or edges drawn with the Draw Points and Draw Split Lines tools.

- Draw Points—Use this tool to add points to the surface of the Roof or Floor. Each point has a height that you can adjust to either a negative or positive offset from the Roof or Floor level.

- Draw Split Lines—This tool adds elevation changes using lines instead of points. Like the Draw Points tool, each line has a height that you can adjust to either a negative or positive offset from the Roof or Floor level.

- Pick Supports—If you have structural supports set at accurate levels, you can use them to indicate the level changes of the Roof or Floor.

- Reset Shape—The button (grayed out in the figure) is used to remove all edits and return the shape of the Roof or Floor to flat with no slopes.

2. On the Project Browser, double-click to open the *Roof* plan view.

   The sub-element tools should still be visible on the Options Bar—changing views should not deselect objects.

3. On the Options Bar, click the Draw Split Lines icon.

   - Click a point on the outside edge of the core Wall just to the left of the Door.

   - Drag straight down and click a point on the inside edge of the parapet Wall (see Figure 8.49).

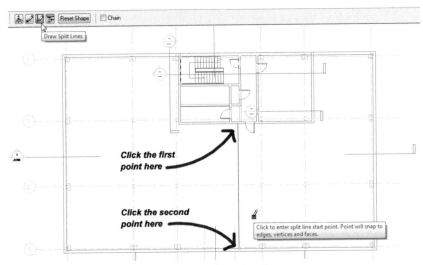

**Figure 8.49**  *Add a Split Line down the middle of the Roof*

4. Draw another vertical Split Line along column grid line B.

   - Start just above intersection 3B and end just below intersection 2B.

5. Draw a final vertical Split Line on the right side of the plan parallel to column grid line E.

   - Start at the corner of the core Walls and draw straight down to just below and to the right of intersection 2E (see Figure 8.50).

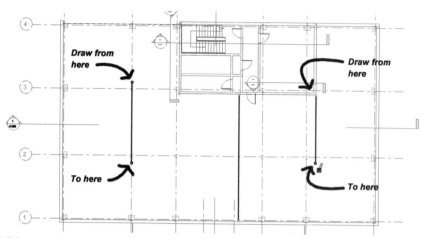

**Figure 8.50**  *Add two additional Split Lines (sketch lines enhanced for clarity)*

6. On the Options Bar, click the Modify Sub-Elements icon.

• Select the first Split Line drawn (the one in the middle).

   An elevation label will appear in the familiar blue temporary dimension color.

• Click on the label (currently reading 0) and change it to **4″** [**100**].

7. Select one of the other two Split Lines drawn and change the elevation to **-4″** [**-90**].

• Repeat for the remaining Split Line (see Figure 8.51).

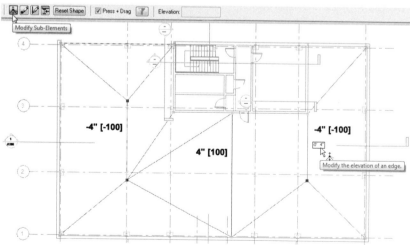

**Figure 8.51**  *Edit the elevation of the Split Line edges*

8. On the Project Browser, double-click to open the {*3D*} 3D view.

Changing views will terminate the sub-element editing mode. You can also click the *Modify* tool first if you prefer. Notice that the Roof now displays the edges of the ridges and valleys for the sloping planes (see Figure 8.52).

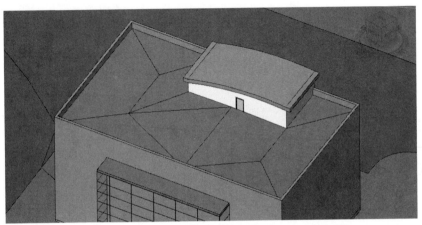

**Figure 8.52** *Add Slope Arrows to the other Roof and study the model in 3D*

The default behavior in the sub-element editing mode is for the entire Roof slab to be affected by the slope. If you prefer, you can edit the structure of the Roof Type applied to the Roof element and make its thickness variable. When doing so, the bottom edge of the Roof will remain flat, while the top edge slopes according to the Split Lines and elevation points added above. The best way to see these sometimes subtle variations is in a section view.

9. On the Project Browser, double-click to open the *Longitudinal* section view.

Take a close look at the Roof element. Notice that both the top and bottom edges are tapered (maintaining a uniform thickness). We can assign one of the layers in the Roof Type structure to a variable thickness. When doing so, the bottom layers will remain flat, the variable layer will have a flat bottom and sloping top edge, and any layers on top of the variable one will continue to slope (with uniform thickness).

10. Select the Roof.

- Right-click and choose **Element Properties**.

- Click the Edit/New button and then the Edit button next to Structure.

- Place a checkmark in the "Variable" column next to layer 2 and then click OK three times (see the top half of Figure 8.53).

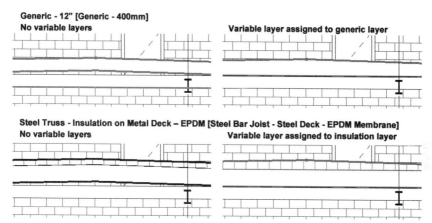

**Figure 8.53** *Comparing the Roof with and without a variable thickness layer*

Currently we are using a Generic Type with only a single layer. To appreciate fully the effect of a variable component, we should to assign a complex Roof Type.

11. Select the Roof and from the Type Selector, choose **Steel Truss - Insulation on Metal Deck – EPDM [Steel Bar Joist - Steel Deck - EPDM Membrane]**.

- Select the Roof again.

- Right-click and choose **Element Properties**.

- Click the Edit/New button and then the Edit button next to Structure.

- Place a checkmark in the "Variable" column next to layer 2 (the insulation this time) and then click OK three times (see the bottom half of Figure 8.53).

Now that we have a more detailed structure in place you can see that we have effectively represented tapered insulation and the support structure beneath it remains level. However, this Roof Type's lowest layer is meant to represent the steel bar joists. This means that the entire roof structure sits too high in the model.

12. Select the Roof, right-click and choose **Element Properties**.

- In the "Base Offset From Level" field, type **-1'-6″ [-450]** and then click OK.

Finally, the Door from the stair tower to the roof is still too low. We can move it up using nearly the same procedure.

13. Select the Door, right-click and choose **Element Properties**.

- In the "Sill Height" field, type **1'-0″ [300]** and then click OK.

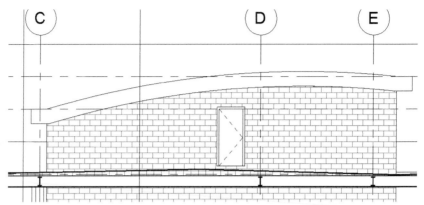

**Figure 8.54** *Move the Roof slab down with a negative offset from the level and move the Door up with a positive sill height offset*

If you wish, you can follow procedures covered in the previous chapter to add a few steps up to the Door inside the building on the *Roof* plan. Feel free to study the model in other views and experiment further with the Roofs and Floor slabs.

14. Save and Close the commercial project file.

# SUMMARY

- Roofs are sketch-based elements that can be generated from existing Walls or manually drawn sketch lines.

- Walls can be attached to Roofs and remain attached as the model is modified.

- Join Roofs together to resolve the intersection of complex Roof planes.

- Roofs have many options for their construction and how they interact with neighboring elements.

- You can apply edge conditions to each Roof Fascia and Rake.

- Gutters and other sweep profiles and Host Sweeps can be applied to Roof edges.

- Skylights interact with and cut holes in Roofs in the same way Windows interact with Walls.

- Create complex Roof Types that include layers of structure that share many features with complex Wall Types.

- Floors are added and modified via sketch mode similar to Roofs.

- Floors may also have complex structure like Roofs and Walls.

- Edit a Roof shape at any time by editing the sketch of the Roof.

- Roofs can be created with a plan sketch like Floors or via an extruded profile shape.

- Use Slope Lines to create complex Roof sloping patterns.

- Flat Roofs can accurately represent subtle drainage slope using the sub-element editing tools.

- Making Roof Type layers Variable allows the sub-element sloping to be applied to a single Roof component such as tapered insulation.

# CHAPTER 9

# Developing the Exterior Skin

## INTRODUCTION

In this chapter, we will enclose our commercial project with a building skin. The skin will be comprised of a masonry enclosure on three sides, with various curtain wall elements on the front and sides of the building. The front façade curtain wall begins on the second floor and spans the height of the third and fourth floors. In Chapter 4, we created a massing element to suggest this design element and applied a temporary Curtain System to it. We will now replace this with a more refined Curtain Wall.

## OBJECTIVES

In order to complete the shell of the commercial project we will apply a detailed Wall type to the skin Walls already in the project. We will also build Curtain Walls and Curtain Systems for the front and side façades. Upon completion of this chapter, you will be able to:

- Create and swap Wall types
- Add Curtain Walls to the model
- Modify a Curtain Wall
- Build a Curtain System type
- Build and add a Stacked Wall

## CREATING THE MASONRY SHELL

With the exception of the large portion of the front façade that was cut away in previous chapters, the majority of the skin of the commercial building is comprised of masonry Walls. We will perforate portions of this masonry skin with Curtain Systems later in the chapter, but we will begin by swapping out the simple generic Wall types used in early chapters with a more detailed Wall type appropriate to the design at this stage.

### INSTALL THE CD FILES AND OPEN A PROJECT

The lessons that follow require the dataset included on the Mastering Revit Architecture CD ROM. If you have already installed all of the files from the CD, simply skip down to step 3 below to open the project. If you need to install the CD files, start at step 1.

1. If you have not already done so, install the dataset files located on the Mastering Revit Architecture CD ROM.

   Refer to "Files Included on the CD ROM" in the Preface for instructions on installing the dataset files included on the CD.

2. Launch Revit Architecture from the icon on your desktop or from the *Autodesk >* *Revit Architecture* group in *All Programs* on the Windows Start menu.

> **Tip:** In Windows Vista, you can click the Start button, and then begin typing **Revit** in the "Start Search field. After a couple letters, Revit Architecture should appear near the top of the list. Click it to launch to program.

   • If the New Features Workshop dialog appears, choose "Maybe later" and then click OK.

   • From the File menu, choose **Close**.

This closes the empty new project that Revit automatically creates upon launch.

3. On the Standard toolbar, click the Open icon.

> **Tip:** The keyboard shortcut for Open is CTRL + O. **Open** is also located on the File menu.

   • In the "Open" dialog box, click the *My Documents* icon on the left side.

   • Double-click on the *MRAC* folder, and then the *Chapter09* folder.

   If you installed the dataset files to a different location than the one listed here, use the "Look in" drop down list to browse to that location instead.

4. Double-click *09 Commercial.rvt* if you wish to work in Imperial units. Double-click *09 Commercial Metric.rvt* if you wish to work in Metric units

   You can also select it and then click the Open button.

The project will open with the last saved view visible on screen.

You may find the Structural model distracting as we work through this chapter. Therefore, choose Manage Links from the File menu, click the Revit Links tab, and Unload the Structural file.

5. From the File menu, choose Manage Links.

   • Click the Revit tab, select the *09 Commercial-Structure* file and then click the Unload button.

   • When asked to confirm, click Yes and then click OK.

## CREATING A WALL TYPE

As you can see, the project is largely unchanged from the previous chapter. We still have the very simple generic Wall type used for the building skin. The first thing we'll do is refine that a bit.

6. On the Project Browser, double-click to open the *Level 1* floor plan view.

7. Pre-highlight one of the exterior Walls (try the vertical one on the left).

• Press the TAB key once to pre-highlight a chain of Walls.

All of the exterior Walls should pre-highlight including the one at the core.

• Click the mouse to select the chain of Walls.

• Hold down the SHIFT key and then click to remove the masonry Wall(s) at the core.

All of the exterior Walls except the masonry Wall at the core (the horizontal one at the top middle of the plan) should now be selected.

**Note:** The Type Selector should read *Generic - 12"*[*Generic - 300mm*]. If it does not, then you still have the core Wall selected. Use SHIFT click to remove them.

8. On the Options Bar, click the Properties icon.

• In the Element Properties dialog, click the Edit/New button.

• Click the Duplicate button.

**Tip:** If you prefer, you can hold down the ALT key and the press E followed by D.

• For the name type **MRAC Exterior Shell** and then press OK (see Figure 9.1).

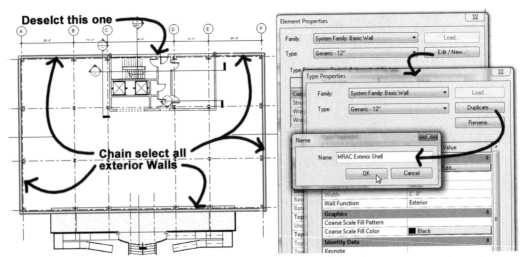

**Figure 9.1**  *Select a chain of Walls, remove the core Wall, and then use Properties to duplicate the type*

## EDIT WALL TYPE STRUCTURE

9. In the Type Parameters dialog next to "Structure," click the Edit button.

If you completed the previous chapter and worked through the roof tutorials, you saw a very similar dialog there. The structure of Wall types and Roof types have much in common. Wall types, however, have more parameters available.

10. Highlight Layer 2 (Structure [1]).

- In the Thickness column, type **7 5/8"** [**190**].

- In the Material column, click the browse icon to choose a Material.

- In the Materials dialog, choose **Masonry - Concrete Masonry Units** [**Masonry - Concrete Blocks**] and then click OK.

11. Beneath the Layers list, click the Insert button.

- Click the Up button to move the new Layer up above the Core Boundary.

- From the Function list, choose **Thermal/Air Layer [3]**.

Functions determine the way that Walls join with one another. A pre-defined list of Functions is built into Revit Architecture. Structure layers will clean up with other Structure layers and interrupt layers of other functions. Substrate layers have the next highest priority and will clean up with other Substrate layers, be interrupted by Structure layers, and interrupt layers of all lower priorities and so on. For more information, refer to the online Help.

- Change the Material to **Air Barrier - Air Infiltration Barrier**.

- Change the Thickness to **2"** [**50**].

12. Insert one more Layer above the air gap Layer.

- Set its Function to **Finish 1 [4]**.

- Set its Material to **Masonry – Brick**.

- And set its Thickness to **3 5/8"** [**90**].

13. At the bottom of the Edit Assembly dialog, click the << Preview button (see Figure 9.2).

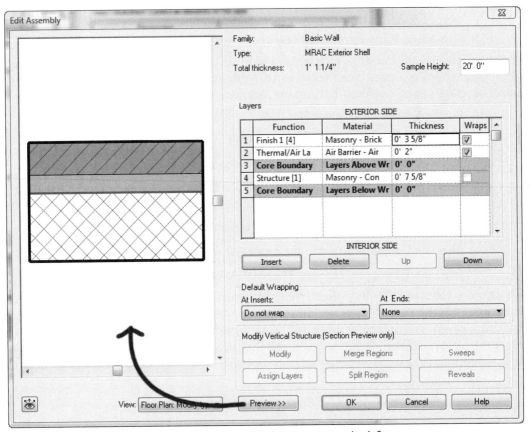

**Figure 9.2** *Edit the Structure of the Wall type and preview it on the left*

14. Click OK three times to return to the workspace window.

15. On the View Control Bar, change the Detail Level to Medium.

- Study the results in the floor plan view.

- Open the {3D} view and study the results there as well.

When studying the 3D view, you will note that we still have the simple "stand-in" curtain wall feature on the front of the building. (This was created in Chapter 4.) We are going to re-design this façade below. However, it should be clear from looking at the current state of the façade that we need to modify the Wall at the front of the building to at least allow for an entry lobby to the building. To do this, we'll simply split the front Wall and a few wing Walls to add some depth to the main façade.

16. On the Project Browser, double-click to open the *Entrance Plan* floor plan view.

17. On the Tools toolbar, click the **Split** tool.

- On the Options Bar, place a checkmark in the "Delete Inner Segment" checkbox.

We'll split from one edge of the patio to the other.

- Click the first point of the split on the left side of the plan (use the edge of the patio as a guide).

- Click the second point of the split on the right side of the plan (again using the edge of the patio as a guide).

- On the Design Bar, click the **Modify** tool or press the ESC key twice.

Recall that the exterior Walls use a Finish Face: Exterior Location Line. If you click to select one of the exterior Walls (there are now two after the split), you will notice that the control handle (small round blue dot) is on the outside edge of the Wall.

18. Select the Wall on the left side of the plan.

- Drag the control handle to snap at the intersection formed in the notch of the patio.

- Right-click the Wall and choose **Create Similar**.

- Draw a new Wall as shown in Figure 9.3.

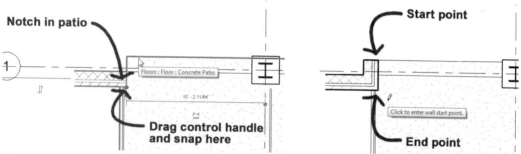

**Figure 9.3**   *Edit the end of the split Wall and add a new short segment*

19. Repeat the process on the other side of the plan.

    When drawing the second Wall, reverse the order of the first and second clicks. Click from bottom to top, otherwise you will have to reverse the Wall after drawing it.

20. Open the {3D} view and study the results there as well.

Make sure that the height of the new Walls matches the height of the original Walls. If they do not, select the two new (short perpendicular) Walls and click the Properties icon (or right-click and choose **Element Properties**) and change the Base Constraint to Street Level and the Top Constraint to Up to level: Roof with a 4'-0" [1200] Top Offset.

21. On the Project Browser, double-click to open the *Entrance Plan* floor plan view again.

Notice the way that the Wall layers terminate at the freestanding end of the Wall. The air gap is exposed on the end of the Wall. We can wrap the ends of the materials to close such gaps. We have the ability to do this both at the ends of Wall and at openings created by Doors and Windows (inserts).

22. Select one of the exterior Walls.

    You do not need to select them all. Type-level edits will apply to all Walls of that type.

    - Click the Properties icon (or right-click and choose **Element Properties**).

    - Click the Edit/New button (or press ALT + E).

23. Next to Structure, click the Edit button.

    - Below the list of layers, in the "Default Wrapping" area, Choose Exterior from the "At Ends" drop-down.

This does not fully solve the problem. As you can see in the preview, the brick now wraps over the end of the Wall, but so too does the air gap.

    - In the Layers list at the top, remove the checkmark from the "Wraps" column next to layer 2 (Thermal/Air Layer [3]).

You should now see that only the brick wraps and the air gap is covered over by the wrapping brick layer (see Figure 9.4).

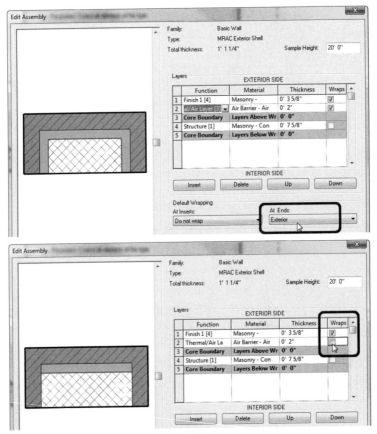

**Figure 9.4** *Place the first point of the Curtain Wall*

24. Click OK three times to return to the workspace window and study the results.

25. Save the Project.

## WORKING WITH CURTAIN WALLS

Curtain Walls in Revit Architecture, like their real-life counterparts, are panelized wall systems. They come in two varieties: Curtain Walls and Curtain Systems. Curtain Walls are drawn the same way as Walls and using the same tool. To create a Curtain Wall, you simply choose the appropriate type from the Type Selector as you draw the Wall. A Curtain System is typically created from the faces of other geometry such as a Mass.

A Curtain Wall type sets up a panel modulation along the length and/or height of the Wall. Each panel can be assigned specific Family elements such as glass or stone panels or even other Wall types. The edges between the panels are mullions. You can decide which edges should receive mullion elements and which type of mullion each edge should use. You can even control the way that the mullions intersect with one another.

### DRAWING CURTAIN WALLS

The simplest ways to create a Curtain Wall is to draw it using the Wall tool. The Wall tool allows you to create three kinds of Wall: Basic Wall, Curtain Wall, and Stacked Wall (these are Wall Families). All of the Walls that we have created so far are Basic Walls. Basic Walls can have one or more Layers in their type's Structure (as seen in the previous exercise). "Basic" does not necessarily mean simple or generic. A Basic Wall is defined by having a single continuous set of Layers both horizontally and vertically. The "Generic" type Walls we used in the early chapters are among the most "basic" Wall types available, but the Wall type we just created above with three Layers is still a Basic Wall. This is because all of the Layers run the full length and height of the Wall. A Curtain Wall, as mentioned above, defines a panel system along the length and/or height of the Wall. A Stacked Wall actually "stacks" two or more Basic Wall types on top of one another. We will look at Stacked Walls briefly later in this chapter. The major focus of this chapter will be on Curtain Walls.

### DRAW A CURTAIN WALL

Using the Wall tool, let's create our first Curtain Wall at the front entrance to the building on the first floor. We'll draw it in the place where the Wall we split out used to be.

1. On the Project Browser, double-click to open the *Entrance Plan* floor plan view.

2. On the Design Bar, click the Basics tab and then click the **Wall** tool.

   • From the Type Selector, choose **Curtain Wall: Curtain Wall 1 [*Curtain Wall*]**.

   • On the Options Bar, verify that Draw is selected.

   • From the "Height" list, choose **Level 2**.

   • Clear the "Chain" checkbox.

3. Click the first point of the Curtain Wall at the midpoint of the short vertical masonry Wall at the left (see the left side of Figure 9.5).

4. Click the end point of the Curtain Wall at the midpoint of the short vertical masonry Wall on the other side (see right side of Figure 9.5).

 **Tip:** While the command is active, hold down the wheel on your mouse and drag to pan to the other side, or use the scroll bars.

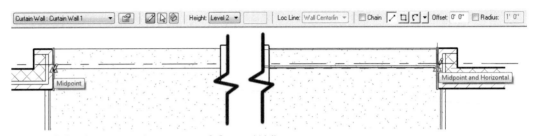

**Figure 9.5** *Draw a single segment of Curtain Wall*

The new Curtain Wall will remain selected with a Temporary Dimension showing.

5. Using the "Move Witness Line" drag control, move the witness line to the outside edge of the horizontal masonry Wall.

Start dragging the control and move over the outside edge of the Wall. Use the TAB key to highlight the outside edge and then release.

• Change the value of the temporary dimension to **6" [150]** (see Figure 9.6).

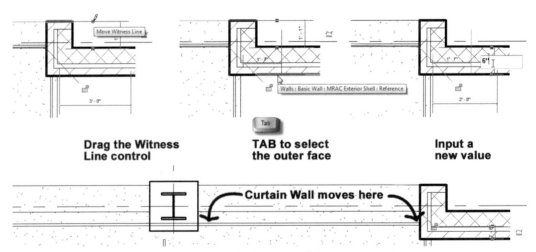

**Drag the Witness Line control**

**TAB to select the outer face**

**Input a new value**

**Figure 9.6** *Adjust the Witness Line and edit the value of the temporary dimension*

• On the Design Bar, click the **Modify** tool or press the ESC key twice.

## EDIT THE CURTAIN WALL

Let's take a look at what we have and make some adjustments.

6. From the Project Browser, open the {3D} view.

   If necessary, spin the model around to the front so that you can see the new Curtain Wall.

Two things are immediately noticeable: the new Curtain Wall is too tall even though we set the Top Constraint to Level 2 and the stand-in Curtain Wall created in Chapter 4 is obscuring any clear view of the new one. First we will hide the Curtain Wall (actually a Curtain System) from Chapter 4 to make the view easier to read. With the new Curtain Wall height, it turns out that a height offset was applied to the Curtain Wall by default as we were drawing it. This was because the other Walls had a height offset and this Wall simply inherited it. Let's remove it now.

7. Select the Curtain System (click near the top just beneath the small roof section).

- Right-click and choose **Hide in View > Elements**.

   The Curtain Wall drawn above should now be clearly visible and easy to select.

8. Select the newly drawn Curtain Wall and on the Options Bar, click the properties icon (see Figure 9.7).

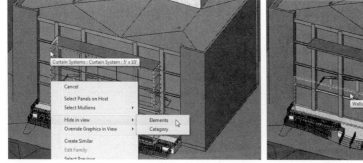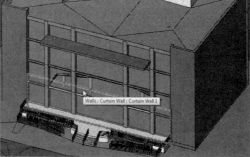

**Figure 9.7**   *Study the model and the new Curtain Wall in the {3D} view*

Notice that the Base Constraint is set to Street Level and that there is a 4'-0" [1200] Top Offset.

- In the "Element Properties" dialog, beneath the "Constraints" grouping, change the Base Constraint to **Level 1** and the "Top Offset" to **0** (zero) and then click OK.

The Curtain Wall should now match the height of the first floor only.

A Curtain Wall is comprised of a Grid Pattern in both the horizontal and vertical directions. Within each of the cells defined by these grids, is a panel. Panel Families can be made to represent anything from glass to stone panels to louvers to Doors. The number of grid divisions in the pattern is either a parameter of the Curtain Wall type or can be defined on the Curtain Wall instance directly. You can select the grid lines and panels independently. By default, the Curtain Wall itself pre-highlights first. To select the internal elements like grid edges, mullions, and panels, you can use the TAB key. Each time you TAB, a different portion of the Curtain Wall will pre-highlight. As is consistent throughout Revit Architecture, you can then click the mouse to select that element. With Curtain Walls, you can also select a single element and then right-click to get additional selection options such as selecting all panels of mullions along a continuous line horizontally or vertically. We'll try all of these techniques below.

## CREATE A WORKING VIEW

For this particular Curtain Wall, we have one large panel of glazing. Therefore tabbing will not yield very interesting results yet. Naturally, it would be unlikely to have a single continuous panel of glass across the entire front of the building. We will learn how to sub-divide the Curtain Wall next. We can sub-divide the Curtain Wall in any view. However, it will be easiest to create a new view to see and work on the Curtain Wall isolated from the rest of the model.

9. On the Project Browser, double-click to open the *Level 1* floor plan view.

• On the Design Bar, click the Basics tab and then click the **Section** tool.

• Draw a Section line in front of (below in plan) the Curtain Wall just a little wider both left and right.

• Using the Control Handle, drag the "Far Clip Offset" back to just the inside of the Curtain Wall (see Figure 9.8).

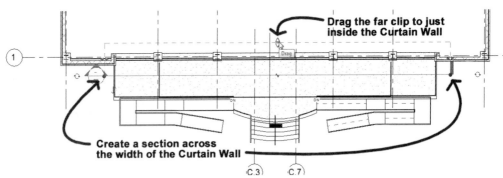

**Figure 9.8** *Create a Section around just the Curtain Wall*

10. Deselect the Section line and then double-click the blue section head to open the section view.

• Drag the top of the crop box down to just above the top edge of the Curtain Wall (slightly above Level 2).

The Curtain System hidden above will be visible here again.

• Select the Curtain System (you may need to move you mouse around near column line B or E to find it) and then right-click and choose **Hide in View > Elements** again.

11. On the Project Browser, right-click *Section I* and choose **Rename**.

- Change the name to: **Entry Curtain Wall Section** and then click OK.

12. On the Project Browser, right-click the *{3D}* view and choose **Duplicate View > Duplicate**.

- Right-click *Copy of {3D}* and choose **Rename**.

- Change the name to: **Entry Curtain Wall 3D** and then click OK.

As you can see, this view currently shows the entire extent of the model. We can orient and simultaneously crop the view to match any other view on the Project Browser list, such as the section view we just created.

13. Click anywhere in the view window to make it active.

- On the View menu, choose **Orient > To Other View**.

 **Tip:** If this command is grayed out, remember to click in the *Entry Curtain Wall 3D* view window to make it active first.

- In the "Orient To Other View" dialog, choose *Section: Entry Curtain Wall Section* and then click OK.

14. On the View Control Bar, change the Model Graphics Style to **Hidden Line** (see Figure 9.9).

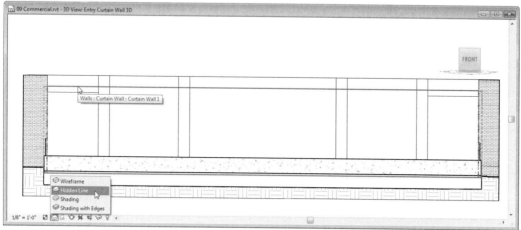

**Figure 9.9**   *Orient the copied 3D view to the Section at Curtain Wall and view it as Hidden Line*

We now have a section and a 3D view that are cropped to show only the Curtain Wall across the front entrance. In the 3D view, if you hold down the SHIFT key and drag the model to spin the view, you will notice that the view is cropped to match the section in all directions; including the depth. To return to the head-on (elevation) viewpoint, choose **Orient > To Other View** from the View menu again. This view will make it easier to work on the Curtain

Wall. The 3D view is good for checking your progress, but you will get some additional editing functionality in the section view. We will use both below as we work.

## ADD CURTAIN GRIDS

Now that we can see the Curtain Wall clearly without the rest of the model cluttering our view, let's sub-divide the Curtain Wall with Curtain Grids.

15. On the Project Browser, double-click to open the *Entry Curtain Wall Section* view.

16. Select the Curtain System (you will see it running vertically along Column lines B and E).

- Right-click and choose **Hide in View > Elements**.

17. On the Design Bar, click the Modeling tab and then click the **Curtain Grid** tool.

- Move the pointer near the top edge of the Curtain Wall.

A Grid line and temporary dimensions will appear.

 **Tip:** So you don't get confused about where the top of the Curtain Wall is in the hidden line display, click the Hide Crop Region toggle icon on the View Control Bar.

- Click to create a grid line in the middle bay near column C.

- Click to create another in the same bay near column D (see Figure 9.10).

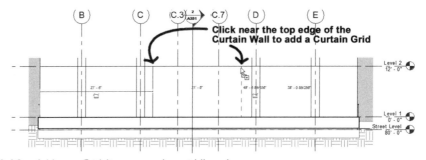

**Figure 9.10** *Add two Grid lines near the middle columns*

- On the Design Bar, click the **Modify** tool or press the ESC key twice.

18. Select the new Grid Line on the left (you should be able to just click it, but use the TAB key if necessary).

Beneath the temporary dimension a small dimension icon will appear.

- Click on the dimension icon beneath the temporary dimension on the left (see Figure 9.11).

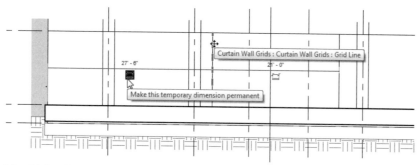

**Figure 9.11** *Make the temporary dimension permanent for the selected Grid Line*

- Repeat by selecting the Grid line on the right and then making the temporary dimension on the right permanent as well.

  Do not make the temporary dimension in the middle permanent.

19. Click away from the selected element to deselect it (or on the Design Bar, click the **Modify** tool or press the ESC key twice).

- Drag the leftmost witness line of the dimension on the left to the right face of the column.

- Drag the rightmost witness line of the dimension on the right to the left face of the column (see Figure 9.12).

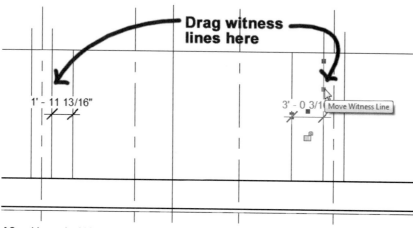

**Figure 9.12** *Move the Witness Lines of the Dimensions to the faces of the Columns*

20. Select the new Grid Line on the left (you should be able to just click it, but use the TAB key if necessary).

Notice how the temporary dimensions now include the dimension that we just created and modified. We can now edit the value of this dimension to place the Grid Lines where we would like relative to the Columns.

- Click on the temporary dimension between the Grid Line and the Column (the one edited above).

- Input a value of **1'-0"** [**345**].

- Repeat on the other side.

You should now have a middle bay that has a Grid Line 1'-0" [345] away from the Columns on each side. The distance between the Grid Lines should be 24'-0" [7200]. If this is not the case, make any required adjustments (see Figure 9.13).

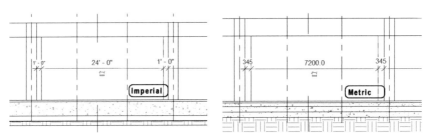

**Figure 9.13** *Move both Grid Lines relative the neighboring Columns*

We can add additional Curtain Grid Lines by sketching them like the ones above, or we can copy them from the existing ones. Let's copy the ones we have to create three wide bays and two small bays for the entrance to our building.

21. Select the Grid Line on the left.

22. On the Tools toolbar, click the Copy tool.

- On the Options Bar, place a checkmark in the "Multiple" checkbox.

- Pick any start point and then drag to the right.

- Type **6'-0"** [**1800**] and then press ENTER to create the first copy.

- For the next copy, type **3'-0"** [**900**] and then press ENTER.

- Create another one at an offset of **6'-0"** [**1800**] and then one more at **3'-0"** [**900**].

- On the Design Bar, click the **Modify** tool or press the ESC key twice (see Figure 9.14).

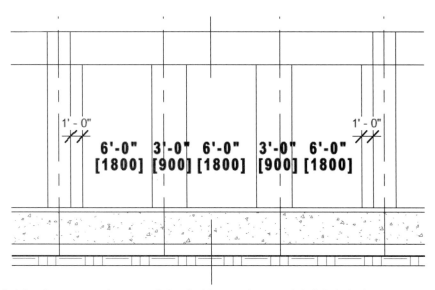

**Figure 9.14** *Create several copies of the Grid Line making an A B A B A rhythm*

23. Zoom out so you can see the entire Curtain Wall.

24. On the Design Bar, click the Modeling tab and then click the **Curtain Grid** tool.

- Move the pointer near the left edge of the Curtain Wall.

  A Grid line and temporary dimensions will appear.

- Click to create a horizontal Grid Line 4'-0" [1200] from the top (use the temporary dimensions to position it if necessary).

At this point we have created several Grid Lines in our Curtain Wall. We can now use the tabbing technique mentioned above to see that this procedure has also cut the one large Curtain Panel into several smaller ones at each division we created. We can select any or all of these panels and assign them to other types. Some of the bays could be made solid construction like stone or brick while some could remain glass.

25. Save your project.

## ASSIGN CURTAIN PANEL TYPES

Now that we have several Curtain Grid Lines dividing our Curtain Wall into individual panels, let's assign some panel types to these bays. Our building needs an entrance. The three bays that we have roughed out in the front need some Doors. A Curtain Wall Door is actually a Panel Family that looks like and schedules like a Door.

26. Zoom in on the bay between column line C and D.

- Place your Modify tool (mouse pointer) over the edge of one of the wider bays created above.

- Press TAB until the panel in this bay pre-highlights—when it does, click to select it (see Figure 9.15).

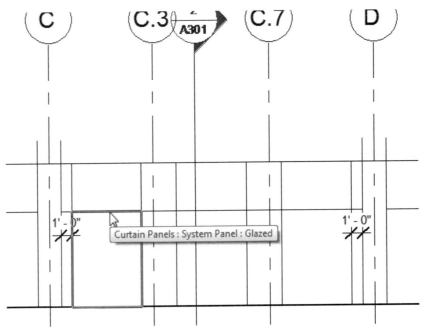

**Figure 9.15** *Select the larger bays at the entrance*

- Pre-highlight the next wide panel using the same technique.

- Hold down the CTRL key and click to select it.

You should have two wide panels selected after this action.

 **Tip:** You can temporarily switch the view to Shaded with Edges instead of Hidden Lines to make the selected elements show better.

- Repeat once more to select the remaining wide panel.

27. With the three panels selected, click the Properties icon on the Options Bar.

- In the "Element Properties" dialog, click the Load button.

The Open dialog should default to the location of your installed Revit Architecture libraries. By default, the US version opens to the *Imperial Library* folder. If you installed both Imperial and Metric content, you should also have Imperial Library and Metric Library icons on the icon bar at the left. If you do not have these icons, the Families that are referenced below have been provided with the files installed from the Mastering Revit Architecture CD ROM. Feel free to access the required content from that location instead.

- In the Open dialog, double-click the *Doors* folder.

If you do not see a *Doors* folder, browse first to the location where you installed the Mastering Revit Architecture CDROM files and then double-click the *Imperial Library* [*Metric Library*] folder, then the *Doors* folder.

- Select the *Curtain Wall-Store Front-Dbl.rfa* [*M_Curtain Wall-Store Front-Dbl.rfa*] file and then click Open.

**BIM** *Manager Note:* Please note that even though the files referenced above are stored in a *Doors* folder, they are actually Curtain Panel Families and not Door Families. Normal Doors cannot be placed in a Curtain Wall. So if you create your own custom Curtain Wall doors, be sure to choose the dedicated Family template for that purpose. Family templates and Family creation techniques are covered in the next chapter.

28. In the "Element Properties" dialog, click OK to see the results (see Figure 9.16).

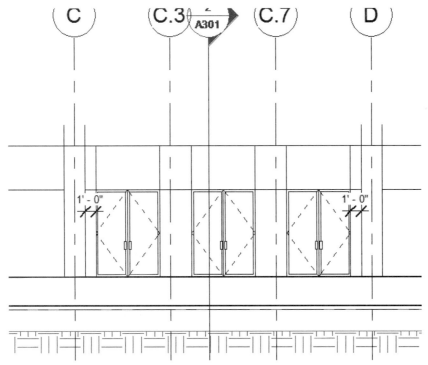

**Figure 9.16** *Swap in Storefront Doors for the three wide bays*

## ASSIGN MULLIONS

Our next task is to apply some mullions to the Curtain Grid Lines.

29. On the Design Bar, click the Modeling tab and then click the **Mullion** tool.

- From the Type Selector, verify that **Rectangular Mullion : 2.5" × 5" rectangular** [**Rectangular Mullion : 50 × 150mm**] is selected.

On the Options Bar, there are three methods to create Mullions: a single Grid Line Segment, an Entire Grid Line, or All Empty Segments. Entire Grid Line is the default choice.

- With the "Entire Grid Line" option selected, click on the horizontal Grid Line (the one 4'-0" [1200] from the top).

If you are zoomed in close enough, you should notice that the Doors adjust in size to accommodate the mullion.

- Try the "Grid Line Segment" option next on any Grid segment.

- Use the "All Empty Segments" option to complete the process of adding mullions (see Figure 9.17).

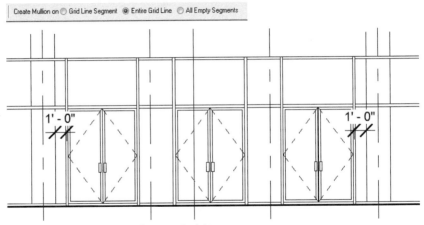

**Figure 9.17**  *Add Mullions to the Curtain Grid Lines*

## EDIT MULLION JOINS

You will notice that the vertical mullions have been given priority over the horizontal ones. If you prefer to have the horizontal mullions continuous and have the verticals stop and start at each intersection, you can toggle the Mullion Join behavior for each join.

30. Select one of the vertical mullions at the Doors.

Notice the Join icons that appear at each end of the Mullion and on the right-click menu (see Figure 9.18).

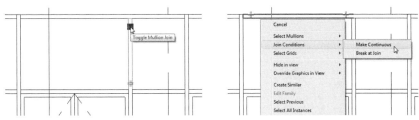

**Figure 9.18**  *Edit Mullion Joins using the control*

- Click the Toggle Mullion Joins icon to model the join to your liking.

In addition to the control icons, you can right-click to access the same options. To use the right-click options effectively, select a single mullion, right-click, and choose **Select Mullions > On Gridline**. Once the entire gridline is selected, right-click again and choose **Join Conditions > Make Continuous**.

## CREATE ADDITIONAL GRID LINES AND MULLIONS

The middle entrance bay is complete, but the other bays could use some more embellishment.

31. Using the techniques covered above, create a new Grid Line **1'-0"** [**345**] from the edges of the two middle columns as shown in Figure 9.19.

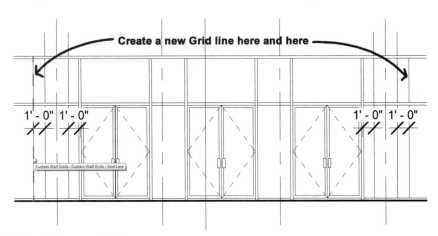

**Figure 9.19**  *Add more Grid Lines*

32. Add Mullions to each of these new Grid Lines.

33. Repeat the entire process to create Grid Lines and Mullions on each side of the remaining Columns (see Figure 9.20).

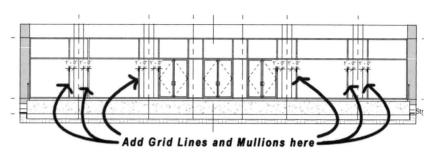

**Figure 9.20** *Add more Grid Lines and Mullions at the remaining Columns*

## USING A WALL TYPE FOR A PANEL

At each of the Columns, we want to remove the horizontal Mullions and swap out the glazing for a solid Wall panel.

> 34. Zoom in on one of the Columns.
>
> - Using the CTRL key, select all three of the horizontal mullions crossing through the Column.

 **Tip:** If you have trouble with the selection, remember your TAB key.

You only want the piece of Mullion between the two vertical Mullions adjacent to each column.

> - Delete the three selected Mullions (press the DELETE key) (see the left panel of Figure 9.21).
> - Select the horizontal Grid Line.
> - On the Options Bar, click the Add or Remove Segments button.
> - Click the Grid Line in-between the two vertical Mullions (see the middle panel of Figure 9.21).

The Grid Line will turn dashed between these Mullions to indicate that a portion of it has been removed.

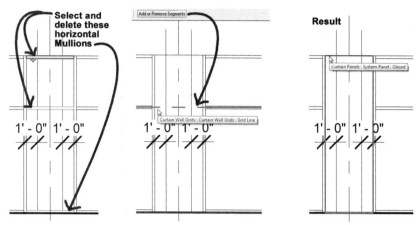

**Figure 9.21**   *Delete Mullions and Remove a Grid Line*

You will now have a continuous vertical panel at the Column as shown in the right panel of Figure 9.21. You may need to use the TAB key to pre-highlight it to see this.

35. Using the TAB key, pre-highlight and then select the new full height panel at the Column.

- From the Type Selector, choose **Basic Wall : Generic - 8"** [**Basic Wall : Generic - 200mm**].

36. Repeat the entire process for the other three Columns on the front façade.

- On the Project Browser, double-click to open the *Entrance Plan* floor plan view.

- Zoom in on the Columns at the front of the façade to see the result.

- Select each Door (use the TAB key), and using the flip controls, flip the swing outside (see Figure 9.22).

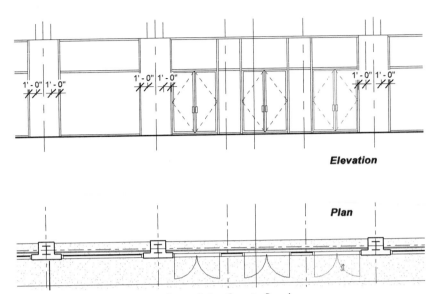

**Figure 9.22** *Swap in a Wall type in place of the Curtain Panels*

37. On the Project Browser, double-click to return to the *Entry Curtain Wall Section* view.

38. Using the TAB and CTRL keys, select each of the four lower panels (two to the right of the entrance, two to the left).

- From the Type Selector, choose **Curtain Wall : Storefront** (see Figure 9.23).

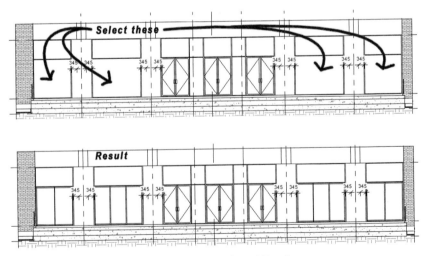

**Figure 9.23** *Swap in a Curtain Wall type for the selected Panels*

There are two interesting points worth mention on this step: first, we are able to use one Curtain Wall type as a Panel on another. Second, the ***Storefront*** Curtain Wall type has built-in subdivisions of Grid lines and Mullions pre-assigned to it. We will explore this further in the remainder of the chapter.

39. Make one last substitution by selecting the two square panels between and above the Doors.

- From the Type Selector, choose **System Panel: Solid**.

40. On the Project Browser, double-click to open the *Entry Curtain Wall 3D* view.

- Hold down the SHIFT key and then drag the mouse to spin the model and study the results (see Figure 9.24).

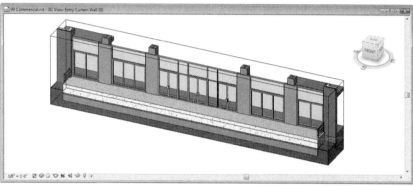

**Figure 9.24** *Spin the model to study the results*

If you wish to see the Curtain Wall in the context of the rest of the building, open the *{3D}* view and study it there.

41. Save the project.

## CREATING HOSTED CURTAIN WALLS

The Curtain Wall we created above filled a space that had no previous enclosure. We drew it just like any other Wall and then edited it. You can also create Curtain Walls from existing Walls or even have embedded within existing Walls. In this way the Wall will "host" the Curtain Wall. In this sequence we will use one of the types provided in the current file and draw a Curtain Wall hosted in the exterior shell Walls of the building.

### DRAW A CURTAIN WALL

The first part of this exercise is similar to the previous one. We will draw a Curtain Wall using a type already in the file.

1. On the Project Browser, double-click to open the *Level 2* floor plan view.

2. On the Design Bar, click the Basics tab and then click the **Wall** tool.

- From the Type Selector, choose **Curtain Wall : Storefront**.

- On the Options Bar, verify that Draw is selected.

- From the "Height" list, choose **Level 4**.

Recall from above that a Top Offset was automatically applied to the Curtain Wall as it was drawn. In this case (as we did above), we will remove this offset.

- On the Options Bar, click the Properties icon.

- In the "Element Properties" dialog set the "Top Offset" to **0** (zero) and then click OK.

- On the Options Bar, clear the "Chain" checkbox.

   Add the Curtain Wall to the vertical Wall on the left side of the plan.

3. Click the first point of the Curtain Wall halfway between column lines I and 2 on the inside brick face of the vertical masonry Wall (see Figure 9.25).

   Use the TAB as necessary to assist you in selection of the proper point.

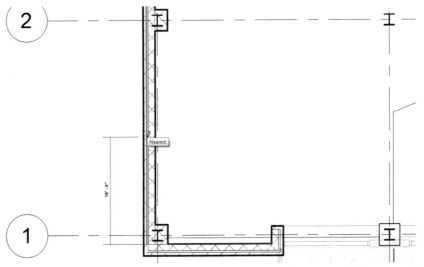

**Figure 9.25**  *Draw a new Curtain Wall on the second floor directly on top of the existing Wall on the face of the brick*

Draw the Curtain Wall **45'-0"** [**13500**] long (a little more than half way between Column lines 3 and 4).

The easiest way to do this is to move the mouse vertically, type the desired length, and then press ENTER (see Figure 9.26).

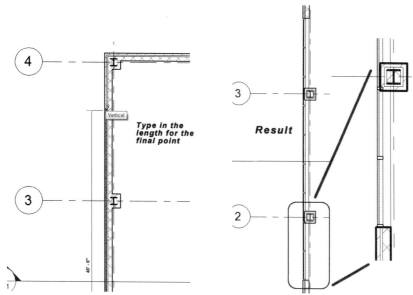

**Figure 9.26**   *Draw an embedded Curtain Wall*

On the Design Bar, click the **Modify** tool or press the ESC key twice.

- The Curtain Wall should cut the host Wall automatically. In some cases it will not (for example when using the **Curtain Wall I** Type). If it does not cut automatically, we can use the Cut tool on the Tools toolbar to cut the Wall manually with the Curtain Wall.

4. Zoom in on the column at column line 3.

Because we chose the **Curtain Wall : Storefront** Curtain Wall type above, this Curtain Wall already has Grid lines and mullions assigned to the Grid Lines. However, the Curtain Wall is too close and the mullions intersect the Columns.

5. Select the Curtain Wall.

The glazing should appear on the outside of the Curtain Wall. If it does not, click the "Flip Wall Orientation" control.

- On the Options Bar, click the "Activate Dimensions" button.

- Edit the temporary dimension in the thickness of the Wall (it may be up near the top) to move the Curtain Wall out (away from the interior).

- Experiment to find the right value. Try about **7"** **[175]** (see Figure 9.27).

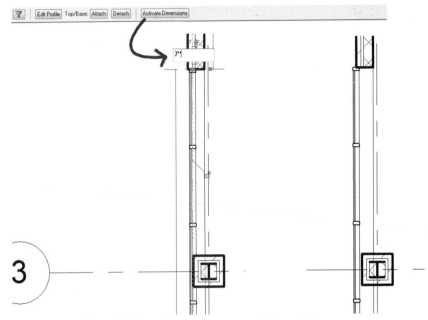

**Figure 9.27**  *Flip the Curtain Walls and move them toward the outside*

6. On the Project Browser, double-click to open the {3D} view.

- Spin the model as needed to gain a clear view of the Curtain Wall.

## ADJUST THE CURTAIN GRID SPACING

As you can see, this Curtain Wall type defines a large vertical spacing of grid bays. In addition, the floor element between the second and third floors is clearly visible. Let's make a few adjustments to address both of these issues.

7. Select the Curtain Wall.

- Right-click and choose **Element Properties**.

- In the "Element Properties" dialog, click the Edit / New button and then click the Duplicate button.

- Name the new type: **MRAC Storefront**.

Below, in the "Create a Custom Curtain Wall/System Type" topic, we will explore many of the settings in this dialog in detail. For now, we'll focus on just a few settings required by the current task.

- Beneath the "Horizontal Grid Pattern" grouping, change the Spacing to **4'-0"** [**1200**] and then click OK twice.

  Keep the Curtain Wall selected.

8. In the center of the Curtain Wall, click the Configure Grid Layout control.

Two lines appear through the middle of the grid: one vertical the other horizontal. To shift the horizontal grid lines up or down, edit the dimension on the horizontal origin line.

- On the left side of the Curtain Wall, click the Horizontal Curtain Grid Origin temporary dimension.

- Type a new value of **2'-0"** [**600**] and then press ENTER (see Figure 9.28).

 **Tip:** There are two temporary dimensions, both the horizontal and vertical origin lines. One moves the grid, the other rotates it. Be sure to use the tooltip to find the correct one before editing. If you want to experiment with the rotation as well, feel free to do so, but undo before continuing.

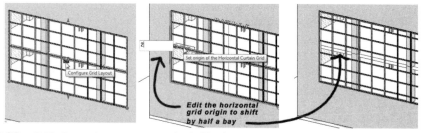

**Figure 9.28**    *Shift the grid pattern origin by half a bay*

We now have a short bay at the top and the bottom of the overall Curtain Wall and a full grid centered at the floor line between second and third floors. Let's replace those with spandrel glass.

9. Select one of the panels in the middle of the Curtain Wall height (one that occurs at the floor line).

 **Tip:** Place your mouse near the edge of the panel you want to select and then press TAB until it pre-highlights, then click.

- Right-click and choose **Select Panels > Along Horizontal Grid**.

Notice the small pushpin icons attached to each Panel—you may need to zoom to see them all. Since the Curtain Wall type that we used here assigned the Grids, Panels, and Mullions automatically, these Panel definitions are pinned to the type. If you look at the Type Selector, you will notice that you cannot change the type (it is grayed out). To change a Panel, you must unpin it first. You can simply click the pin icon to unpin it, but since there are many selected, this would not be the most efficient approach.

10. From the Edit menu choose **Unpin Position**.

 **Tip:** The keyboard shortcut for Unpin Position is UP.

- Keeping the same selection of Panels, from the Type Selector choose **System Panel: Solid**.

The entire row of Panels should now be solid (see Figure 9.29).

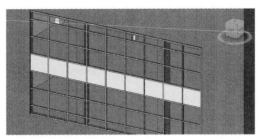

**Figure 9.29** *Change the middle row of Panels to solid*

### MIRROR THE CURTAIN WALL

11. On the Project Browser, double-click to open the *Level 2* floor plan view.

12. Select the Curtain Wall.

Keep in mind that the Curtain Wall segment is made up of many subparts. The whole curtain wall is represented by a green dashed that will appear when the Modify tool is passed over the top of the wall.

13. On the Tools toolbar, click the Mirror tool.

 **Tip:** The keyboard shortcut for Mirror is **MM**. **Mirror** is also located on the Edit menu.

- On the Options Bar, verify that the "Copy" checkbox is selected.

- Click the Draw option icon.

- Snap to the midpoint of the Floor slab curve (in the center of the plan) and then drag straight up and click again (see Figure 9.30).

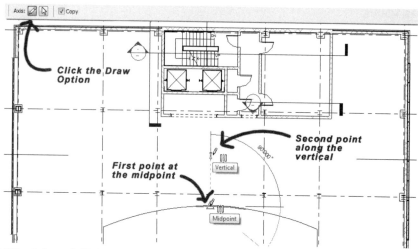

**Figure 9.30**   *Select all Mullions on the corner gridline*

The Curtain Wall will mirror over to the other side of the plan. If it automatically cuts the Wall, then skip the next steps; otherwise continue.

## USING CUT GEOMETRY

You can perform the next action in either a plan or 3D view.

14. On the Tools toolbar, click the Cut Geometry icon. (You can also find it on the Tools menu.)

• At the "First Pick" prompt, click the Wall.

• At the "Second Pick" prompt, click the Curtain Wall.

15. On the Project Browser, double-click to open the {3D} view.

• Orbit the model and study the results (see Figure 9.31).

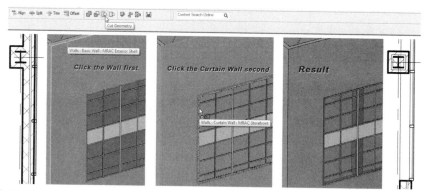

**Figure 9.31**   *Four Corner Mullion conditions are included in Revit*

482

## ADD WINDOWS

Some basic punched Windows will help complete the masonry portions of the façade.

16. Add ordinary punched Windows to the masonry Walls on the first and forth floors above and below the Curtain Walls.

- Add Windows on all floors in the north and south Walls.

Simply open the *Level 1* floor plan, add some Windows to one side, mirror them to the other side, and then copy them to the *Level 4* plan using paste aligned.

17. Study the results in the {3D} view (see Figure 9.32).

**Figure 9.32** *Add Windows to the other floors*

18. Save the project.

## CREATE A CUSTOM CURTAIN WALL/SYSTEM TYPE

We have now worked with a simple Curtain Wall with no divisions (in which we added all of the Curtain Grid Lines and Mullions manually) and a Curtain Wall type that included Grid spacing and Mullions pre-assigned within its type. In this sequence, we will combine techniques starting with the creation of our own Curtain Wall type. The goal is to try to include as many of the divisions and Mullions within the type so that they occur automatically. We can then finish the design using the manual Grid line and Mullion techniques already covered.

## EXPLORE CURTAIN WALL TYPE PROPERTIES

Before we build our own Curtain Wall type, let's take a closer look at the one we used in the previous sequence. There are three Curtain Wall types provided in the out-of-the-box templates. The simplest is named **Curtain Wall 1 [Curtain Wall]**. It has no built-in divisions and defaults to a single continuous panel of glass. We used this one in the first exercise at the start of the chapter. The next type is slightly more complex including a default grid spacing in both the horizontal and vertical dimensions. This type is named **Exterior Glazing**. Finally we have the type used in the previous sequence: **Storefront**. As we have seen, not only does this one include pre-defined grid spacing but also adds mullions on all edges automatically (see Figure 9.33).

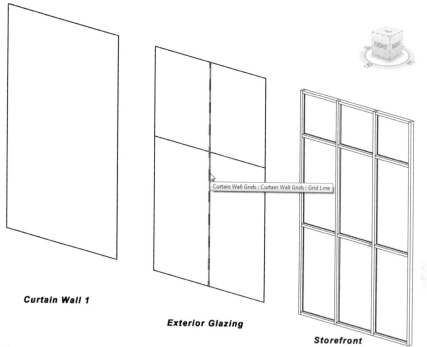

**Curtain Wall 1**

**Exterior Glazing**

**Storefront**

**Figure 9.33**   *Three Curtain Wall types are included in the out-of-the-box template files*

I. On the Project Browser, locate the *Families* node and click the plus (+) sign icon next to it to expand it.

• Click to expand *Walls* next, then *Curtain Wall* (see Figure 9.34).

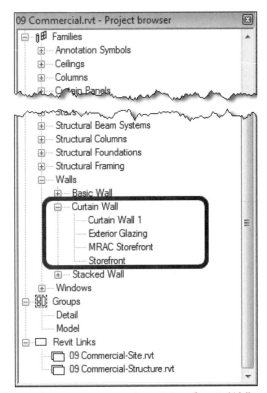

**Figure 9.34** *Use the Project Browser to access the existing Curtain Wall types in the current project*

Each of the types currently resident in the existing project will appear in the tree listing. Notice that the three shown in the previous figure are included as well as the one we created in the previous exercise. You can duplicate and modify new types directly from the Project Browser, or using the steps that we followed in the "Adjust the Curtain Grid Spacing" topic above. You can also add an instance of a particular type to the model directly from here as well.

2. Right-click on *Storefront* and choose **Properties**.

This takes you directly to the "Type Properties" dialog for this type. This dialog is similar to any other "Type Properties" dialog. Following is a brief description of some of the important parameters seen here.

Under the "Construction" grouping are the following parameters:

- **Wall Function**—Since Curtain Walls are a type of Wall, they have the same list of "Functions" as other Walls. We can create Curtain Wall types that fulfill a variety of functions such as "Exterior" and "Interior." (Some of the other Functions like "Foundation" or "Retaining Wall" would not make sense for Curtain Walls.)

- **Automatically Embed**—When this checkbox is selected, the Curtain Wall will attempt to embed itself automatically within another Wall like the one that we created in the previous exercise. When this function is off you can still use the Cut tool to embed a Curtain Wall in a Wall manually.

- **Curtain Panel**—This parameter gives a list of all possible panel types that are loaded into the project. Choose the default type for the Curtain Wall here. All panels will default to this type and they will appear pinned (as we saw above) when selected. You can always unpin them and override the Panel type used for one or more Panels in the specific instance of the Curtain Wall element in the model (we also did this above).

- **Join Condition**—Early in this chapter in the "Edit Mullion Joins" topic, we edited the default join condition used by Mullions. You assign the default condition to a Curtain Wall type here. Several options appear for the default way that Mullions will join to one another. You can make the horizontal or the vertical continuous, make the border continuous or choose not to assign an option.

Beneath the "Vertical Grid Pattern" and the "Horizontal Grid Pattern" groupings are three parameters each:

- **Layout** and **Spacing**—Layout choices include "None," "Fixed Distance," "Fixed Number" and "Maximum Distance." When None is chosen the Curtain Wall will include a single panel across the entire horizontal or vertical dimension as appropriate. Fixed Distance will create Curtain Grid Lines at the spacing indicated in the "Spacing" parameter. There may be panel space left over when using this option. Maximum Distance is similar in that it specifies the *largest* that the spacing can become but will size all bays equally up to the size indicated. In this case, there will not be any left over. The Spacing parameter is used for both of these settings. Choosing the "Fixed Number" option will disable the "Spacing" parameter. The number of bays is an instance parameter. This means that when you choose the "Fixed Number" option, you must then assign the number of bays individually to each instance of the Curtain Wall. You cannot do it globally in the type properties dialog.

- **Adjust for Mullion Size**—This parameter adjusts the location of Curtain Grid lines to maintain equal Panels even when Mullion sizes vary.

Beneath the "Vertical Mullions" and "Horizontal Mullions" groupings, you can choose from a list of available Mullion Families to use for the Borders and the Interior Mullions. These Mullions will be created automatically and will be pinned to the Curtain Wall as it is created (see above). You can unpin any Mullion, as we did in the exercise above, and assign an alternate type as design needs dictate.

The remaining parameters are the standard identity data properties that we have seen in other object types. These are mostly used when doing quantities, schedules, and takeoffs.

3. On the Project Browser, beneath the *Families* node, expand *Curtain Systems*.

   A single Curtain System Family appears named simply: **Curtain System**.

4. Expand *Curtain System*.

   A single type appears named: **5' × 10' [1500 × 3000mm]**.

Notice that the parameters here are nearly identical to the Curtain Wall. The only differences are that a Curtain System does not have the "Wall Function" or "Embed" options and the two sets of grid parameters are named more generically: "Grid 1" and "Grid 2." You are also limited to Panel Families for Panels and cannot use Wall types as you can with Curtain Walls. In all other ways, Curtain Systems share the functionality and behavior of Curtain Walls. Curtain Systems however allow for more "freeform" designs than Curtain Walls. At the start of this book, in Chapter 4, we created a Mass element to represent the front façade of the building. To that Mass, we applied a simple generic Curtain System. At the start of this chapter, we hide that Curtain System to make it easier to work on the other portions of the façade. At this point, we are ready to redisplay this element and begin refining its design. To get started, we'll unhide it and then build a custom Curtain System type to finish up the front façade of the building.

5. On the Design Bar, click the **Modify** tool or press the ESC key twice.

## EDITING MASSING ELEMENTS

Most of the techniques that follow can be performed on either Curtain Systems or Curtain Walls in your own projects. Choose the object that best suits your building design, preferences, and needs. Since we created a Mass and attached Curtain System earlier in the book, we will continue with that approach here.

You can create a Curtain System two ways: by lines or by faces. When you create it by lines, you pick two edges in your existing model, or sketch two lines. The lines or edges chosen should define opposite edges of the surface you wish the Curtain System to occupy. Feel free to experiment with this option in another file. In this situation, we used the "faces" option. (Review the "Add a Mass" topic in Chapter 4 for a refresher).

6. On the Project Browser, double-click to open the {3D} view.

7. On the View Control Bar, click the Reveal Hidden Elements icon (small light bulb).

A dark maroon border will appear around the view window and the hidden elements (the Curtain System in this case) will reappear in dark maroon as well.

- Right-click the Curtain System and choose **Unhide in view > Elements** (see Figure 9.35).

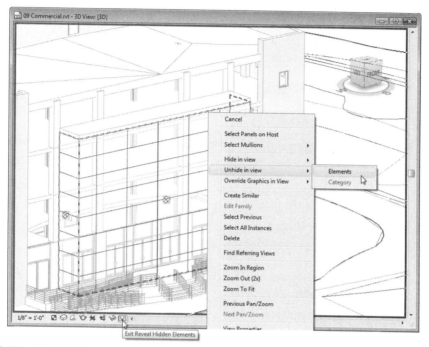

**Figure 9.35**   *Unhide the Curtain System in the {3D} view*

8. On the View Control Bar, click the Exit Reveal Hidden Elements icon.

9. On the View toolbar, click the Show Mass icon (next to the Default 3D View icon).

The Mass will reappear on screen. It will be a little difficult to see since Masses default to a transparent purple material that is very similar in appearance to the glass material. However, the show/hide Mass command is the only display command that affects all views, not just the active one. Therefore, if you switch to another view, the Mass will be visible there as well.

10. On the Project Browser, double-click to open the *Entrance Plan* view.

• Using the TAB key, highlight and then select the Mass at the front of the building.

• On the Options Bar, click the Edit button (see Figure 9.36).

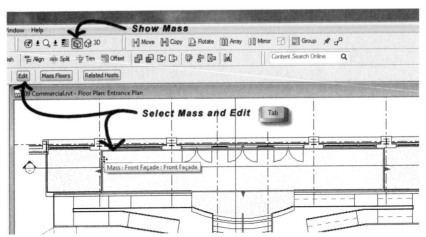

**Figure 9.36** *Select the Mass using the tab key and the click the Edit button*

> The drawing window grays out and the Design Bar changes to Mass mode and shows only the Massing tools.

Since we already have a Massing object, we'll simply edit it.

> 11. Click to select the Form: Extrusion object.

You should not need to TAB this time since this is the only element within this Mass object. Each Mass element can contain one or more "Form" objects that together describe the overall shape of the Mass.

> • On the Options Bar, click the Edit button again to edit the sketch of the Extrusion.

> 12. Make edits to the sketch as shown in Figure 9.41.

> • Use the Align tool to align the left edge of the sketch to the left edge of the column.

> • Repeat for the right side.

> • Align the top edge of the sketch to the outside face of the Wall.

> • Use the control handles to change the shape of the sketch as shown. (Use temporary dimensions or guidelines to help.)

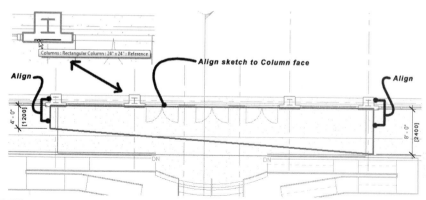

**Figure 9.37**   *Edit the shape of the Extrusion sketch (Dimensions added for clarity)*

13. On the Design Bar, click the Extrusion Properties button.

  • In the "Element Properties" dialog, change the Extrusion Start to: **12'-0"** [**3600**].

  • Change the Extrusion End to: **48'-0"** [**14400**] and then click OK.

  • On the Design Bar, click the Finish Sketch button.

Since we moved the bottom edge up to 12'-0" [3600], the Extrusion will no longer be visible in the plan view. We'll be able to see it in the 3D view however. We are also still editing the Mass, so let's finish that first.

14. On the Design Bar, click the Finish Mass button.

15. On the Project Browser, double-click to open the {3D} view.

Your newly shaped Mass will appear as we edited it, but the existing Curtain System and Roof elements made from the Mass back in Chapter 4 no longer match its shape. This is easily fixed.

16. Select the small Roof at the top of the façade element.

  • On the Options Bar, click the Remake button.

  • Repeat for the Curtain System (see Figure 9.42).

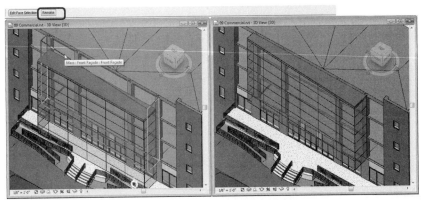

**Figure 9.38** *Use the Remake button to update the Roof and Curtain System to match the edited Mass*

17. On the View toolbar, click the Show Mass icon.

This will toggle the display of Masses off again.

### CREATE A NEW CURTAIN SYSTEM TYPE

Now that we have modified the shape of the Mass element, we are ready to create a new Curtain System type and apply it to the front façade.

18. Select the Curtain System on screen, right-click and choose **Element Properties**.

- Click the Edit / New button and then the Duplicate button.

- Name the new type: **MRAC Front Façade**.

By now you have likely noticed that we sometimes select the object and edit its properties, and other times we use the Families branch of the Project Browser. Ultimately, it is a matter of personal preference and both methods lead to the same result.

19. Beneath the "Construction" grouping, choose **System Panel: Glazed** for the Curtain Panel.

- Change the "Join Condition" to **Grid 2 Continuous**.

20. Change the parameters of the "Grid 1 Pattern."

- Change the "Layout" type to **Fixed Distance**.

- Set the "Spacing" to **11'-0"** [**3300**].

21. Change the parameters of the "Grid 2 Pattern."

- Verify that the "Layout" type is set to **Fixed Distance**.

- Set the "Spacing" to **12'-0"** [**3600**] (see Figure 9.39).

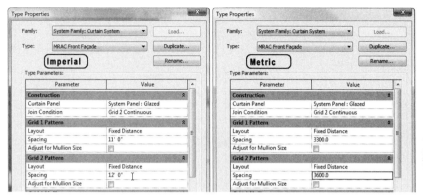

**Figure 9.39** *Configure the Grid Pattern parameters for the new Curtain System*

22. Click OK twice to accept and apply the changes.

23. On the Project Browser, double-click to open the *South* elevation view.

As we can see, the new spacing has been applied to the Curtain Wall in both directions. It may be easier to see the pattern clearly if we hide the column grid lines.

- Select one Column Grid line, right-click and choose **Hide in view > Category**.

You should now be able to see the pattern clearly in the elevation view. Note that the spacing begins from one end of the Curtain Wall façade and is not centered. In some designs, this may be the desired result, but in this one, we want the pattern centered. This is controlled by the Justification parameter. Unlike the spacing parameters themselves (which were type parameters), the Justification of the pattern is an instance parameter.

24. Select the Curtain System, right-click, and choose **Element Properties**.

Be sure to select the Curtain System; press TAB as necessary.

- Beneath the "Grid 1 Pattern" grouping, set the "Justification" to **Center** and then click OK.

- Note the change to the justification (see Figure 9.40).

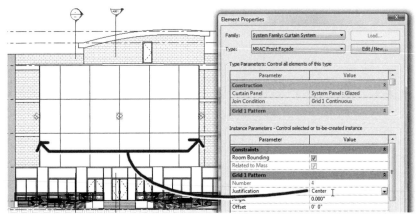

**Figure 9.40** *Center the grid pattern*

**Note:** The "Angle" and "Offset" parameters here are the same settings that we edited above in the "Adjust the Curtain Grid Spacing" topic using the on screen temporary dimensions.

25. Save the project.

## EDIT THE HORIZONTAL MULLIONS

We now have the overall horizontal and vertical spacing established. Next we'll address the shape and size of the mullions. In this sequence, we will edit the horizontal mullions to represent spandrels and the vertical Mullions to represent piers. We need to define a few new Mullion types and then apply them to the horizontal and vertical orientations of the Curtain System. We can define these types on the Project Browser.

26. On the Project Browser, locate the *Families* node again. It should still be expanded from above—if not, click the plus (+) sign icon next to it to expand it.

- Click to expand *Curtain Wall Mullions*, then *Rectangular Mullion*.

27. Right-click on **2.5" × 5" rectangular [50 × 150mm]** and choose **Duplicate**.

- For the name, type **MRAC Spandrel** and then press ENTER.

- Right-click on **MRAC Spandrel** and then choose **Properties**.

There are several parameters in the "Type Parameters" dialog for Mullions. Let's take a look at several of them. Under the "Constraints" grouping are the following parameters:

- **Angle**—Use this constraint to rotate the Mullion relative to the Curtain Wall (see top left of Figure 9.41).

- **Offset**—Input a positive or negative value here to shift the position of the Mullion in or out relative to the Curtain Wall (see middle right of Figure 9.41).

Under the "Construction" grouping are the following parameters:

- **Profile**—By default there are two basic shapes for Mullion profiles: rectangular and circular. You can create custom profile Families that can be loaded into your project and then will appear in the list here (see middle left of Figure 9.41).

- **Position**—Two options exist for Position: Perpendicular to Face is the default condition. This orientation, as its name implies, sets the mullion profile relative to the Curtain Wall. Parallel to Ground rotates the mullion profile and is useful in conditions such as in a sloped glazing or a sloped curtain system (Not shown).

- **Corner Mullion**—This checkbox indicates if the Mullion defines a corner condition. This setting is read-only. You cannot make custom corner Mullions (Not shown).

- **Thickness**—This is the depth of the Mullion as measured along its axis perpendicular to the Curtain Wall (see top right of Figure 9.41).

Under the "Materials and Finishes" grouping, you can choose a Material for the Mullion from the list of Materials available in the project. Unlike the "Thickness" parameter listed above, width is divided into two separate width parameters under the "Dimensions" grouping (see top right of Figure 9.41):

- **Width on side 1** and **Width on side 2**—If you set both of these parameters to the same value, then your Mullion will be centered on the Curtain Grid line (see bottom left of Figure 9.41). If you specify different settings, you will essentially shift the Mullion relative to the Grid line (see bottom right of Figure 9.41). In either case, the overall width will be the total of both settings.

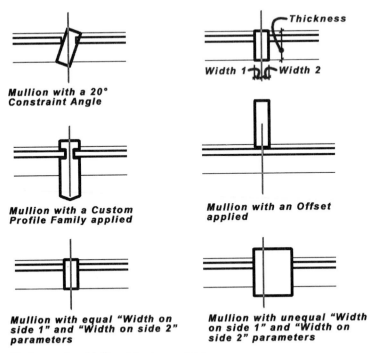

**Figure 9.41** *Understanding Mullion type parameters*

- In the "Type Properties" dialog, type **I'-0"** **[300]** for the "Thickness," "Width on side 1," and "Width on side 2."

  Remember that the total width is the sum of both the "Width on side 1" and "Width on side 2" parameters, so in this case, this makes the Mullion 2'-0" [600] wide.

- Click OK again dismiss the "Type Properties" dialog and return to the model.

The Mullion you just created will not appear in the model yet. All we did was create a new Mullion type. We have not used it in the actual model yet.

28. On the View toolbar, click the Default 3D view icon (or open the {*3D*} view from the Project Browser).

   If you are not looking at the front façade, spin the model to see it.

29. On the Project Browser, beneath the *Families > Curtain Systems* node, right-click on the **MRAC Front Façade** Curtain System type and choose **Properties**.

As an alternative you can select the Curtain System element (be sure to select the Curtain Wall and not the mullions or Grid lines—use the TAB if necessary). Right-click and choose **Element Properties**. Click the Edit/New button (or press ALT + E) to edit the type properties.

- Beneath the "Grid 2 Mullions" grouping, choose **Rectangular Mullion : Spandrel** for each of the Mullion conditions: "Interior Type, "Border 1 Type," "Border 2 Type."

- Click OK to see the result (see Figure 9.42).

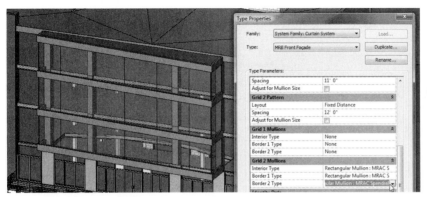

**Figure 9.42** *Assign the new Mullion type to the horizontal Mullions*

## EDIT THE VERTICAL MULLIONS

Repeating nearly the same process, we can edit the Vertical Mullions. We also need a new type here.

30. On the Project Browser, return to the *Families > Curtain Wall Mullions > Rectangular Mullion* node again.

31. Right-click on **2.5" × 5" rectangular [50 × 150mm]** and choose **Duplicate**.

- For the name type **MRAC Pier Mullion** and then press ENTER.

- Right-click on **MRAC Pier Mullion** and then choose **Properties**.

- Set the "Thickness" to **1'-0" [300]**.

- Set both the "Width" parameters to **6" [150]** and then click OK.

This will make a 1'-0" [300] square pier Mullion.

32. Return to the "Type Properties" dialog for the **MRAC Front Façade** type again.

- Beneath the "Grid 1 Mullions" grouping, for "Interior Type," choose **Rectangular Mullion : Pier Mullion**.

Leave "Border 1 Type" and "Border 2 Type" set to None.

- Click OK to accept the changes and close the dialog.

Notice that the spandrels in the middle of the Curtain System are continuous horizontally. This is because we configured Grid 2 as continuous in the type parameters. However, you can see that the top and bottom border Mullions are not continuous.

33. Return to the "Type Properties" dialog for the **MRAC Front Façade** type.

Remember, you can right-click on the Project Browser beneath Families, or select the Curtain System in the model, edit its Properties, and then click Edit/New to access the "Type Parameters."

- Change the "Join Condition" to **Border and Grid 2 Continuous**.

- Click OK to see the result (see Figure 9.43).

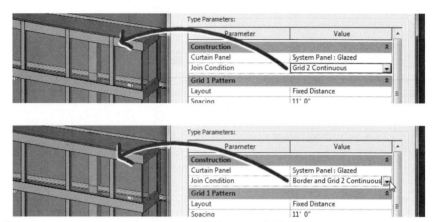

**Figure 9.43** *Assign the new Mullion type to the horizontal Mullions*

34. Save the model.

## ADJUST MULLION POSITION

Take a close look at the Curtain Wall in the South elevation view and you will note that the size of the horizontal bays is not equal.

35. On the Project Browser, double-click to open the *South* elevation view.

Notice that the Level lines for Level 3 and Level 4 pass through the middle of the Spandrel Mullions, but at Level 2 and the Roof they do not. Since our façade is a Curtain System, and since the Curtain System is based on a Mass, we'll have to edit the Mass once more.

36. On the View toolbar, click the Show Mass icon (next to the Default 3D View icon).

37. Place your mouse near the edge of the Curtain System, press tab.

- Continue pressing TAB until the **Mass: Front Façade: Front Façade** element pre-highlights and then click to select it.

- On the Options Bar, click the Edit button.

- Click in the same spot to select the Extrusion.

- Right-click and choose **Element Properties**.

- Change the Extrusion Start to: **11'-0" [3300]** and the Extrusion End to: **49'-0" [14700]** and then click OK.

38. On the Design Bar, click the Finish Mass button.

The Mass is now 1'-0" [300] taller at both the top and the bottom than the Curtain System (see the left panel of Figure 9.44).

- Select the Curtain System and then on the Options Bar, click the Remake button.

- Remake the Roof as well.

This adjusts the Curtain System to the correct height, but now we have an extra Mullion occurring at the top and the Mullions at the middle floors are not centered on the Level lines (see the middle panel of Figure 9.44).

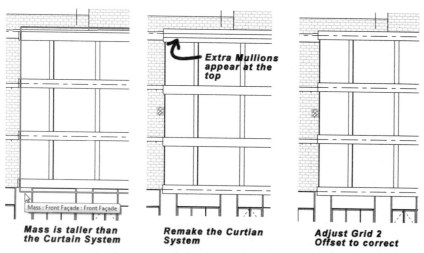

**Figure 9.44**   *Edit Mass, remake the Curtain System, and shift the start point of the bays*

To correct this and get each of the Spandrel Mullions centered on a Level line, we need to shift the grid offset as we did above in the "Adjust the Curtain Grid Spacing" topic. We can do this with the temporary dimensions on screen, or in the Element Properties dialog. In the above referenced topic, we used the onscreen temporary dimensions. This time let's try the Element Properties dialog.

39. Select the Curtain System, right-click and choose **Element Properties**.

- Beneath the "Grid 2 Pattern" grouping, change the Offset to **1'-0"** [**300**] and then click OK (see the right panel of Figure 9.44).

Looking at the right panel of Figure 9.44, it appears that this action solved both the centering of the middle levels and the extra Mullions at the top. However, if we were to investigate carefully (using the TAB key), we would discover that there are actually two Mullions at each location along the top and bottom edges. The reason for this is that the overall spacing of the horizontal bays (Grid 2) is 12'-0" [3600], yet the overall height of the Curtain System is 38'-0" [11400] after the last round of edits we made to the Mass element. In the last step, we shifted the grid so the difference of 2'-0" [600] is equally split at the top and bottom (see Figure 9.45).

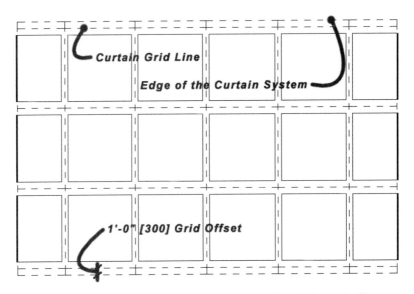

**Figure 9.45**  *With the spacing of 12'-0" [3600] between grids and the grid offset an equal space of 1'-0" [300] remains at top and bottom*

The preceding paragraph and figure adequately explain the presence of the extra Curtain Grid, but still do not explain why the two sets of Mullions are directly on top of one another. Revit treats interior Mullions differently than border types. Interior Mullions are centered on the Grid line. To change this default, you would need to edit the parameters of the Mullion type, not the Curtain Wall or Curtain System. Border Mullions, on the other hand, are always shifted to fall completely within the Curtain System (or Curtain Wall).

While we could potentially leave the Mullions as they are, it would cause us difficulties later when working with the corners. Solutions include removing the extra Curtain Grid at the top and bottom or removing the extra set of Mullions on either of the two Curtain Grids at top and bottom. In this case, the easiest thing to do is to remove the border Mullions in the Type Properties dialog.

40. On the Families branch of the Project Browser, right-click the **MRAC Front Façade** Curtain System type and choose **Properties**.

- Beneath Grid 2 Mullions, change Border 1 Type and Border 2 Type to None and then click OK.

A Revit warning will appear. The message indicates that some Mullions have become "non-type driven." This means that the Mullions in the model currently attached to the Curtain Grids at the top and bottom of the Curtain System are no longer being controlled by the Curtain System type. You have two choices. If you simply click OK, the Mullions will remain in the model but will no longer be pinned. The other option is to click the Delete Mullions button and remove them. This is what we will do here.

- In the Revit Warning dialog, click the Delete Mullions button (see Figure 9.46).

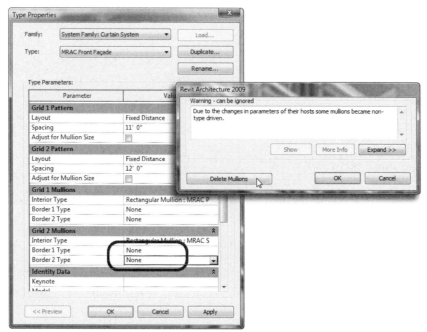

**Figure 9.46** *Remove the border Mullion assignments and then delete the Mullions in the warning dialog*

Click OK again if necessary to close any remaining dialogs.

## SUB-DIVIDE CURTAIN WALL BAYS

We can use virtually any of the techniques that we have covered so far to sub-divide the Curtain System that we have into smaller bays. Let's open the *South* elevation view and experiment with a few of these techniques.

41. On the Project Browser, double-click to open the *South* elevation view.

Our design currently has six bays—four equal ones in the center and two slightly smaller ones at the ends. Remember, the fact that the pattern is centered is a function of the element's instance properties and not something that we can set in the type. The spacing of the bays was determined by the type and is at this point rather large. We can sub-divide these panels into smaller ones by simply adding Grid lines as we did with the Curtain Wall at the start of the chapter.

42. On the Design Bar, click the Modeling tab and then click the **Curtain Grid** tool.

- Move the pointer near the left edge of the Curtain Wall.

A Grid line and temporary dimensions will appear.

- Using the temporary dimensions, click to create a horizontal grid line at the two-thirds mark (see Figure 9.47).

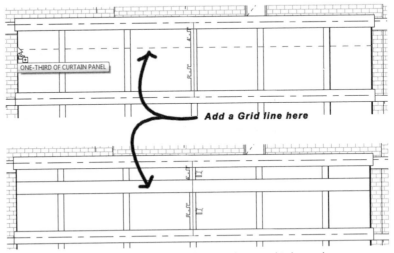

**Figure 9.47** *Create a Grid line that divides the bay at the two-thirds mark*

Notice that the new Grid line automatically creates Mullions in the *Spandrel* type. It would be better to use something a little less "heavy" for these intermediate Mullions. In fact, let's make a new Mullion type for these horizontal bands.

43. On the Project Browser, expand the *Families > Curtain Wall Mullions > Rectangular Mullion* node (as we did above).

44. Right-click on **2.5" × 5" rectangular [50 × 150mm]** and choose **Duplicate**.

- For the name, type **MRAC Horizontal Band Mullion** and then press ENTER.

- Right-click on **MRAC Horizontal Band Mullion** and then choose **Properties**.

- Set the "Thickness" to **1'-6" [450]**.

- Set both the "Width" parameters to **2" [50]** and then click OK.

This will make a 4" × 18" [100 × 450] band Mullion.

45. Pre-highlight one of the new Mullions (added automatically above to the new Grid line), right-click and choose **Select Mullions > On Gridline**.

This is a handy way to select an entire row of Mullions very quickly. Notice that when these Mullions highlight, that they are all pinned to the Curtain Wall. Before we can assign the new *MRAC Horizontal Band Mullion* to them, we must unpin them. You can click each of the pushpin icons individually, but this would be very tedious. Instead, use the command on the Edit menu.

46. From the Edit menu, choose **Unpin Position**.

**Tip:** The keyboard shortcut for Unpin Position is UP.

- From the Type Selector, choose **MRAC Horizontal Band Mullion**.

47. Repeat this process to add horizontal Grid lines and Mullions to the other two floors as well.

- On the Design Bar, click the Modeling tab and then click the **Curtain Grid** tool.

- Add a horizontal Grid line to each of the remaining floors at two-thirds the height.

- Select and unpin the Mullions and then assign the **MRAC Horizontal Band Mullion** type.

48. Open the {3D} view and repeat the process to add MRAC Horizontal Band Mullions on the other two faces of the Curtain System.

49. Save the project.

## SWAP PANEL TYPES

Using a process nearly identical to the above, we can select groups of Panels and swap them out with a different type. For this example, we'll use a custom Panel included with the dataset for this chapter. Later in the next chapter, we'll learn how this custom Panel Family was built.

50. On the View toolbar, click the Default 3D view icon (or open the {3D} view from the Project Browser).

51. From the File menu, choose **Load from Library > Load Family**.

- Browse to the *Chapter09* folder of the location where you installed the CD ROM files.

- Select the *MRAC Curtain Panel.rfa* file and then click Open.

52. Pre-highlight one of the larger panels along the bottom row of the Curtain System.

- Right-click and choose **Select Panels > Along Grid 2** (see Figure 9.48).

**Figure 9.48**    *Select one Panel then right-click to select the entire row*

- From the Edit menu, choose **Unpin Position** (or simply type UP).

- From the Type Selector, choose **MRAC Curtain Panel : 3 Bay**.

You should now have three equal divisions in each of the selected bays. The loaded Family subdivides the bay equally. This is a simple illustration of what is possible. As stated above, we'll learn how to build this Family in the next chapter. There are many other possibilities. Pay a visit to the AUGI forums (*http://forums.augi.com/index.php*), click the Revit forums link, and then use the search feature to search for "Curtain Panel." If you wish, you can also search for "Spider" for several examples of Curtain Panels that integrate spider clamps. Try downloading some of these and loading them into your projects to test them out. After you have completed the next chapter, try your hand at making some on your own.

Returning our attention to the façade at hand, the end bays might look better if they were returned to the simple panel. You can easily do this by selecting them and changing the type back again.

53. Select the two end bays, and on the Type Selector, choose **System Panel: Glazed**.

 **Tip:** TAB to pre-highlight one Panel, click to select it. TAB again to highlight the other, hold down the CTRL key, and then click to select it.

To repeat the process on the other two floors, you could repeat the selection process and then unpin the Panels again. However, try this tip: select the Curtain System and edit its Type Properties. Change the Curtain Panel to None and click OK. Notice that this does not remove any Panels that are already present in the model. It only affects how Panels are added automatically going forward (for example when adding another Curtain System of this type). Now when you select Panels, they will no longer appear pined, therefore, unpinning will not be necessary.

- Select the same four bays on each of the other floors and swap in the **MRAC Curtain Panel: 3 Bay** type (see Figure 9.49).

**Figure 9.49** *Apply the custom panels to each of the other floors*

All of the techniques covered here for Curtain Systems work in nearly identical fashion for Curtain Walls. You are therefore encouraged to experiment on your own with a Curtain Wall type and try to duplicate some of the techniques covered here. Unfortunately, you cannot use the same type for both a Curtain System and a Curtain Wall. If you needed the same type for both a Curtain Wall and a System, you would have to build the type twice—one for the Curtain Wall and another for the Curtain System.

## ADDING CORNER MULLIONS

If you open one of the floor plans, you will notice that we need to address the corner condition of this Curtain System. Currently, the panels have a gap at the corner. Revit includes several corner Mullion Families in common shapes. We can experiment with each one here to find the one that suits our taste.

    54. On the Project Browser, open the {*3D*} view.

- Zoom in on one of the corners of the Curtain System.

- On the Modeling tab, click the ***Mullion*** tool.

- From the Type Selector choose ***L Corner Mullion: 5" × 5" Corner*** [***L Mullion 1***].

    On the Options Bar, be sure that "Entire Grid Line" is selected.

- Click the Grid line at one of the corners.

A new corner Mullion will appear along the vertical grid. However, notice the way that the horizontal Mullions no longer join cleanly. Above in the "Edit the Vertical Mullions" topic and specifically at Figure 9.43, we adjusted the behavior of the join conditions. At the time, the "Border and Grid 2 Continuous" option gave us the best result. As we refine the design, it appears that changing back to the Grid 2 Continuous option will give us a nicer result. This sort of progressive refinement is part of the process when working with Revit. As your design ideas progress, you can easily adjust settings and options and see the results in real-time.

55. Edit the "Type Properties" for the **MRAC Front Façade** type again.

- For the "Join Condition" choose Grid 2 Continuous and then click OK (see Figure 9.50).

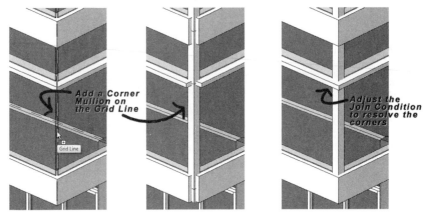

**Figure 9.50** *Add a Corner Mullion and adjust the Join Condition*

- Study the model in both the {*3D*} and *Level 2* plan views.

There are four Corner Mullion Families. To try an alternative, right-click one of the Corner Mullions and choose **Select Mullions > On Gridline**. Choose another type from the Type Selector. Try the others and choose the one you like best (see Figure 9.51).

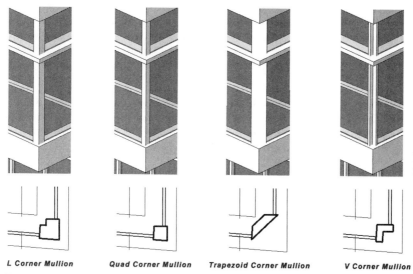

| L Corner Mullion | Quad Corner Mullion | Trapezoid Corner Mullion | V Corner Mullion |

**Figure 9.51**    *Four Corner Mullion conditions are included in Revit*

Further refinements are possible by creating new Corner Mullion types. Start by right-clicking the one you wish to start with on the Project Browser and then choose **Duplicate**. Give the new copy an appropriate name and then edit the properties. For example, if you want to vary the size of the L Corner Mullion, expand the Families > Curtain Wall Mullions item on the Project Browser. Right-click *5" × 5" Corner* [*L Mullion 1*] and choose Duplicate. Give the new type a good descriptive name. Right-click the new type and choose Properties. The size and shape of the Mullion is controlled by the Thickness, Leg 1 and Leg 2 parameters. You can also edit the material and identity data.

56. Add a Corner Mullion to the other side of the Curtain System.

   • Add *Rectangular Mullion: 2.5" × 5" rectangular* [*Rectangular Mullion: 50 x 150mm*] Mullions at the two Curtain Grid Lines where the Curtain System meets the building.

   If you get an error, click Delete Elements. This refers to the very top and bottom again.

   • Study the model in both the {3D} and *Level 2* plan views.

   • Save the model.

**BIM** *Manager Note:* Corner Mullions are System Families and cannot be edited. You can, however, create new types for each of the four built-in Families. (The concepts of Families and System Families will be discussed in detail in the next chapter.) When building custom types for any of the built-in Corner Mullion System Families, the parameters available for edit do not include any way to customize the shape or graphics used to portray the mullion. Therefore, technically, there is no way to create completely custom Corner Mullions. However, some very clever work-around solutions to the problem have been devised by members of the Revit user community. One of the most popular online communities is found at the Autodesk Users Group International (AUGI) web site. There are dozens of forums hosted at *www.augi.com* supporting nearly all Autodesk products. If you are not a member of AUGI, you can join for free to gain access to this very dynamic community and its information-packed forums.

To create a custom corner mullion, you must actually create a custom Curtain Panel Family that has a mullion "built-in." A very common mullion type overlooked by Revit is an angled mullion. An example of a Panel Family with integral custom angled corner mullion can be found on the AUGI forums. The thread can be found here: *http://forums.augi.com/showthread.php?t=42563&referrerid=90067*

This is simply one example. Spend some time searching the AUGI forums, the Autodesk discussion groups (*discussion.autodesk.com*) and the many other Revit sites on the web and you will find dozens of additional examples.

## EDIT WALL PROFILES

A few more refinements remain to finalize the front façade of our commercial building. The most obvious one is closing in the building behind the Curtain System.

57. On the Project Browser, double-click to open the *Level 2* floor plan.

58. On the Design Bar, click the **Wall** tool.

- In the "Element Properties" dialog, change the Family to: **Basic Wall**.

- Change the Type to: **Exterior – EIFS on Mtl. Stud**.

- Change the Base Constraint to: **Level 2**.

- Change the Top Constraint to: **Up to Level: Roof**.

- Change the Top Offset to: **4'-0" [1200]** and then click OK (see the left side of Figure 9.52).

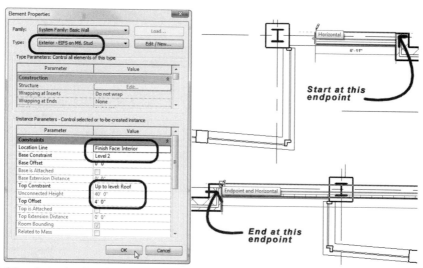

**Figure 9.52**  *Copy the first floor Curtain Wall to the second floor and configure it to be a Basic Wall*

Start on the right side of the plan.

59. Click the first point at the endpoint of the small vertical brick Wall.

   • Click the second point on the opposite side snapping to the endpoint of the other short brick Wall (see the right side of Figure 9.52).

60. Use the Align tool to make the new Wall's exterior face flush with the columns (see Figure 9.53).

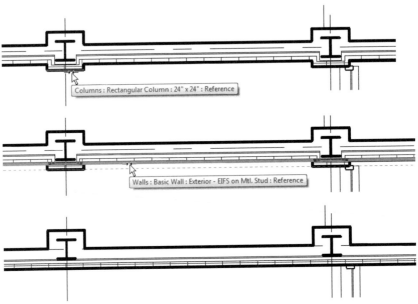

**Figure 9.53** *Align the Wall to the face of the columns*

You should now have a Wall filling in the space that was previously open on the upper floors aligned nicely to the columns; however, it fills in the space behind our Curtain System. To fix this, we'll use Edit Profile. This command was covered in Chapter 7—we'll review it here.

    61. On the Project Browser, double-click to open the *South* elevation view.

- Select the new Wall and on the Options Bar click the Edit Profile button.

- Change the model graphics style to wireframe.

    This will allow you to see the Columns and Roof to help create your sketch.

- Using Figure 9.54 as a guide, draw sketch lines on the inside of the Columns and lower edge of the Roof. Split out the unnecessary segment.

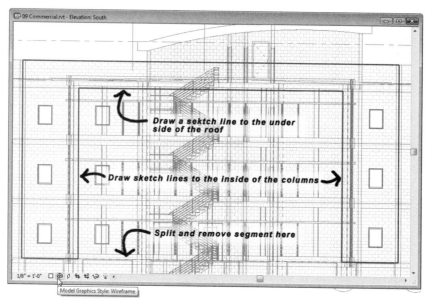

**Figure 9.54**   *Edit the sketch of the Wall profile*

- Click Finish Sketch and then change the model graphics style back to Hidden Line.

62. Add some punched Windows to the new Wall.

63. Edit the Floor sketches on each level to make them conform to the new front façade shape (see Figure 9.55).

   If Revit asks you if you want to attach Walls to Floor bottoms, answer No. This would work well for interior partitions but not exterior shell walls. However, if you are prompted to join with the exterior Walls and cut them, you can answer Yes to this.

To add a floor under the Curtain System on the front façade, you can sketch it manually, or you can add one to the bottom face of the Mass. To do this, show Masses, select the Front Façade Mass, and then on the Options Bar click the Mass Floors button. In the dialog that appears, choose Level 2 and then click OK. On the Massing tab of the Design bar, click the *Floor by Face* tool, select the floor plane created with the Mass Floors command, and then click the Create Floors button on the Options Bar.

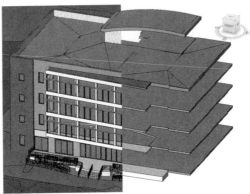

**Figure 9.55** *Edit the sketch of the Floors and Roof to conform to the shape of the exterior shell*

64. Save the model.

Whatever additional Curtain Wall explorations you wish to continue on to in this dataset, the specifics are left to you as an exercise. We will explore one more topic before ending this chapter—Stacked Walls.

## WORKING WITH STACKED WALLS

At the start of the chapter, we built a custom Wall type and applied it to the exterior Walls. We can go even further and create a special kind of Wall called a "Stacked Wall." A Stacked Wall simply uses two or more Basic Wall types stacked on top of one another to form a more complex design.

In the simple example that follows, we will create a Stacked Wall that combines the Wall type we created at the start of the chapter with two others provided. This will give more articulation to the exterior shell Wall forming a base condition at the first floor and a parapet condition at the roof.

### Create the Walls Types or Import them

**Import**—To save you time, both of the custom Wall types that we need in addition to the one that we already created are provided in a separate project file included with the Chapter 9 files. The name of the file is: *MRAC Wall Types.rvt* [*MRAC Wall Types-Metric.rvt*]. You can open this project file, select the Walls on screen and copy them to the clipboard. Return to the *09 Commercial* project, open the *Level 1* floor plan view, and paste the Walls. Be sure to actually place them in the model off to the side and then click the Finish button on the Options Bar to complete the paste. Once complete, you can safely delete the two Walls. The result will be that two new Wall types will be added to your project: **MRAC Exterior Base** and **MRAC Exterior Parapet**.

**Create**—If you prefer to create the Wall types yourself, they are very similar to the one that we created in the "Creating a Wall Type" and the "Edit Wall Type Structure" topics. Refer to those topics and create two new Wall types based on **MRAC Exterior Shell**. To create an **MRAC Exterior Base**, change the Finish 1 layer from brick to **Masonry – Stone** and make its thickness **7 5/8" [190]**. For **MRAC Exterior Parapet**, change the thickness of the Structure layer to **3 5/8" [90]**. With this layer selected, click the Insert button to add a second Structure layer within the core. Make the thickness of this one **4" [100]**. Change the material of both Structure layers to **Masonry – Brick** (see Figure 9.56).

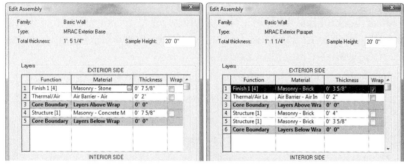

**Figure 9.56**   *Composition of the two custom Wall types*

Whether you choose to copy and paste the Wall types into the project from the provided file, or create them yourself, is not critical. In either case we will use these two Wall types next to create a custom Stacked Wall type.

## CREATE A STACKED WALL TYPE

1. On the Project Browser, expand the *Families > Walls > Stacked Wall* node.

2. Right-click on **Exterior - Brick Over CMU w Metal Stud** [**Exterior - Brick Over Block w Metal Stud**] and choose **Duplicate**.

- Right-click on **Exterior - Brick Over CMU w Metal Stud (2)** [**Exterior - Brick Over Block w Metal Stud (2)**] and choose **Rename**.

- For the name type **MRAC Brick Exterior with Stone Base** and then press ENTER.

- Right-click on **MRAC Brick Exterior with Stone Base** and then choose **Properties**.

- Beneath "Type Parameters" click the Edit button next to "Structure."

The Structure of a Stacked Wall is very simple. You designate one or more Wall types that you want to "stack" and how tall each should be. You can insert additional types by clicking the Insert button. Use the Up and Down buttons to move the various Wall types relative to one another. Each type will be assigned an explicit height except for one (use the "Variable" button to choose which is variable). Only one can be variable and it will adjust to the actual height of the individual Wall instance in the model.

3. Click in the Name field next to Type 1.

- From the pop-up menu, choose **MRAC Exterior Shell** (this is the type we created at the start of the chapter).

- For Type 2, choose **MRAC Exterior Base**.

- Change the height of Exterior Base to **6'-0"** [**1800**].

- At the bottom of the dialog, click the Preview button.

- In the Offset field next to each type, input **4"** [**100**].

Be sure to offset both. This shifts the Wall type overall to compensate for the additional thickness that the base material adds. In this way, the Wall will remain in the same relative position when we swap it into the model.

4. Select item number 1, and then click the Insert button.

This shifts the other two types down and inserts a copy of **MRAC Exterior Shell** above the original.

- Change the type to **MRAC Exterior Parapet**.

- Change the height to **4'-0"** [**1200**].

Notice how the parapet type does not line up with the others. This is because the core is not as thick and the default "Offset" setting at the tope of the dialog is to align the Wall types by their Core centerlines. We can change this.

5. From the "Offset" list at the top, choose **Finish Face: Interior** (see Figure 9.57).

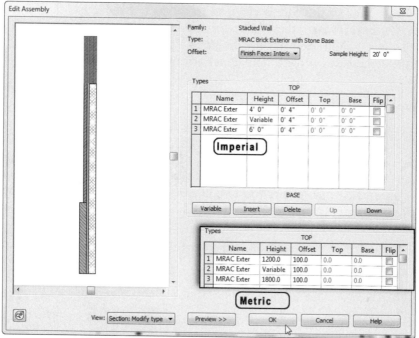

**Figure 9.57**  *Configure the Stacked Wall type*

- When you are done studying the settings, click OK twice to dismiss the dialogs.

6. Open the *Level 1* floor plan view and select one of the **MRAC Exterior Shell** Walls in the model.

- Right-click and choose **Select All Instances**.

- Use the CTRL key and add the masonry Wall at the stair core to the selection as well.

- From the Type Selector, choose **Stacked Wall: Brick Exterior with Stone Base** (see Figure 9.58).

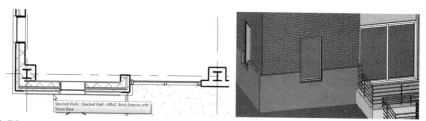

**Figure 9.58**  *Apply the Stacked Wall to the exterior shell walls*

You might get an Ignorable Warning. You may also need to re-cut the Wall Geometry with the Curtain Walls on the sides of the building. Make these and any other adjustments required.

When the back Wall of the core is changed to this new type, it has the parapet and the cap like the others. However, the Wall itself only goes up to roof level with no offset, and so the Wall above sits on the parapet. The best solution is to create another Stacked Wall type with the stone base but with no parapet or cap. The specifics are left to the reader as an exercise.

## ADD A WALL SWEEP

To complete the design of our shell Wall, we need to add a parapet cap at the top and it might be nice to add a soldier course. To do this, we'll use Wall Sweeps. You can add a Wall Sweep in two ways. They can be added to the Wall type parameters so that they automatically show on all instances of the Wall type throughout the model or you can use the Host Sweep tool which allows you to add Sweeps to the model manually at whatever location you designate. In this example, we'll add them to the *MRAC Exterior Parapet* Wall type.

7. On the Project Browser, expand the *Families > Walls > Basic Wall* node.

- Right-click the *MRAC Exterior Parapet* type and choose **Properties**.

  Make sure that the preview is showing and choose **Section: Modify type** from the Preview option.

- Click the Edit button next to Structure.

- At the bottom of the dialog, click the Sweeps button.

- In the "Wall Sweeps" dialog, click the Add button.

- From the Profile list, choose *Parapet Cap-Precast: 16" Wide [M_Parapet Cap-Precast: 450mm Wide]*.

- Choose *Concrete - Cast-in-Place Concrete* for the Material.

- Choose **Top** from the "From" list (see Figure 9.59).

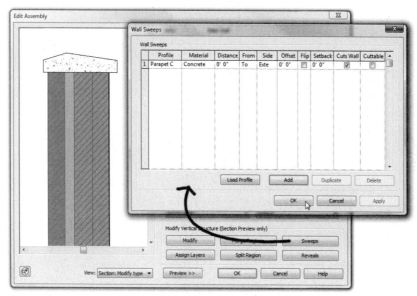

**Figure 9.59** *Add a Parapet Cap to the Wall type*

8. Click OK three times to complete the Sweep.

## ADD A HOST SWEEP

9. On the Project Browser, open the {3D} view.

10. On the Modeling tab of the Design Bar, click **Host Sweep > Wall Sweep**.

- On the Options Bar, click the Properties icon.

- Click Edit/New and then click Duplicate. Name the new type **MRAC Soldier Course**.

- Place a checkmark in the "Cut by Inserts" checkbox.

- Choose Wall **Sweep-Brick Soldier Course: 3 Bricks** [**M_Wall Sweep-Brick Soldier Course: 3 Bricks**] from the Profile list.

- Choose **Masonry – Brick Soldier Course** for the Material.

- Click OK twice.

11. Following the prompt on the Status Bar, mouse over the brick Wall and click when the sweep is where you want it.

- Click one or more Walls to add the soldier course (see Figure 9.60).

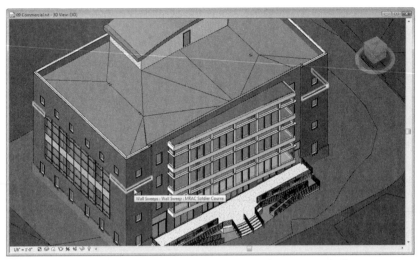

**Figure 9.60** *Add a Host Sweep soldier course to complete the exterior shell*

- On the Design Bar, click the **Modify** tool or press the ESC key twice.

Unlike the Sweep that we added to the Wall type, this one can be edited with the control handles that appear when you select it. You can add and remove Walls from the Host Sweep afterward using the command on the Options Bar. You can add Reveals the same way. Sweeps integral to Wall types cannot appear in schedules. Host Sweeps can appear in schedules. To learn more about Host Sweeps and Reveals, consult the online help.

12. In the *South* elevation view, click the Reveal Hidden element icon.

- Select one Column Grid line, right-click and choose **Unhide in view > Category**.

13. Save the model.

Continue to make any other refinements or experimentations that you wish. When you are finished, close and save your project file.

# SUMMARY

- Revit Architecture provides three types of Walls: Basic Wall, Curtain Wall, and Stacked Wall.

- You can create a custom Wall type and edit it to add and delete Layers (Components).

- Curtain Walls can be drawn like Walls or even embedded within other Walls.

- Use Cut Geometry to manually embed the Curtain Wall within the thickness of a Wall.

- Curtain Wall Grids can be sketched manually or defined within the Curtain Wall type.

- Mullions can be added to Curtain Grid lines.

- Mullion Joins can be edited to suit your preferences.

- Default joins can be assigned in the type or manually edited per Mullion.

- Walls can be used as panels for Curtain Walls.

- A Curtain Wall type can contain a pre-defined pattern of horizontal and vertical Grid lines.

- Curtain Systems have most of the same features as Curtain Walls but can conform to more free-form shapes such as the faces of Massing elements.

- Custom Curtain Panel Families can be used to create more complex Curtain Wall and Curtain System designs.

- Corner Mullions come in four pre-defined varieties.

- A Stacked Wall contains one or more Basic Walls stacked on top of each other.

- Wall Sweeps can be added to Wall types or added manually as Host Sweeps and used to represent bands, moldings, parapets, cornices, etc.

# Working with Families

## INTRODUCTION

All elements that you create and use in a Revit Architecture project belong to a Family. As noted in Chapter 1, a Family is an object that has a specific collection of parameters and behaviors. Within the limits established by the Family and its parameters, a potentially endless number of "Types" can be spawned. Understanding Families and how to manipulate them is an important part of learning Revit. Families are the cornerstone of the Revit parametric change engine and the topic of this chapter.

## OBJECTIVES

While there are several kinds of Families in Revit, the focus of this chapter will be "Component Families." Component Families are simply Families that describe some physical component in the model. Conceptually however, topics discussed in this chapter apply to any kind of Revit Family, including annotation Families such as tags and title blocks. After completing this chapter, you will know how to:

- Explore the Families contained in the current project
- Insert instances of Families in your model
- Manipulate Family Types
- Create custom Families
- Create custom Parametric Families

## KINDS OF FAMILIES

In the "Revit Architecture Elements" heading of Chapter 1, a detailed discussion of the various kinds of elements available in Revit is presented. To review, Revit elements include Hosts, System Families, Component Families, In-Place Families, Datums, Views and Annotations. Figure 1.1 presents a summary of these items graphically.

System Families include anything that is built into the software and cannot be manipulated by the user in the interface. This can include model components like Walls and Floors, but also includes less obvious items like Views, Project Data, and Levels. System Families cannot be created or deleted. Their properties are pre-defined in the software. However, many System Families like Walls, Floors and Roofs can have more than one Type. A Type is a named collection of variables associated with a particular Family and pre-assigned to certain values. A Family can contain one or more Types; each with its own unique user-editable settings. While we cannot, for example, create or delete Wall Families, we can add, delete, and edit the Types associated with each of the provided Wall Families. In the case of Walls, there are three Families: Basic Wall, Stacked Wall, and Curtain Wall. We worked with each of these in the previous chapter. Each of the Wall Families can (and often does) have more than one Type. For example, in the out-of-the-box template files, there are several predefined Basic Wall Types such as: Exterior – Brick on CMU, Generic 6 "and Interior – 5 ½" Partition (1hr). The Basic Wall definition simply means that the Wall has the same structure along its entire length and height. The actual make-up of this structure can vary widely as the sample names above imply. The Stacked Wall has the same structure along its length, but is allowed to vary in its height by "stacking" two or more Basic Walls on top of each other. Finally the Curtain Wall can vary in both length and height by defining detailed grid patterns directly into the structure of the Wall and/or including other Wall Types within regions of the Wall's mass.

Other System Families vary considerably in their specific composition and features, but at the conceptual level, they share the same basic characteristics—the overall behavior of the object is defined by the system and cannot be changed; however, the specific object-level parameters can be manipulated via the creation and application of Type and/or Instance variations.

As already noted, System Families include both things that are part of the physical model in your Revit Architecture projects (like Walls, Floors, and Roofs) and other items that are not (like Views, Project Data, and Levels). Another term for those System Families that are model elements is "Host." A Host is an element that can receive, support, or provide structure for other model elements. Model System Families or Host Families are the parts of the building that are typically assembled on site from a collection of raw materials. Examples include Walls, Floors, Roofs, Stairs, etc.

Component Families include everything that is not a System Family. Component families are typically model elements, but can also be annotation or other non-model elements. Component Families can be "Host Based" (require a Host), or they can be freestanding (not requiring a Host). Revit users can create, delete, and modify Component Families. This is accomplished in the Family editor interface and each Family thus created is saved to its own unique file. Like System Families, Component Families can contain one or more Types. Unlike System Families, Component Families are completely customizable. The user can create nearly unlimited parameters and behaviors for such a Family. Since they are created outside of the project editor context, they can be loaded into any project.

In addition to the System and Component Families, there is a third type of Family in Revit called the "In-Place Family." In-Place Families are very similar to the Component Families in terms of creation, editing, and strategy. However, an In-Place Family is created directly within a project (not in a separate Family file as Component Families are) and it *cannot* be directly exported to other projects. The only time you should consider creating an In-Place Family is for an element that is unique to a particular project with no possibility that you will ever want to reuse it in future projects. This could be effective for unique existing conditions in projects or very specialized design scenarios.

As has already been noted in the introduction, many of the concepts covered in this chapter might apply equally to System Families, Component Families, and even In-Place Families. However, for the purposes of the following discussions and tutorials, we will limit our discussion to the use and manipulation of Component Families. For techniques on manipulating editable variables of System Families, refer to the previous chapter where several examples of manipulating and/or creating Wall types were presented. The basic procedure for creating or editing a Type (whether belonging to a System or Component Family) is nearly identical in all cases. For an example of an In-Place Family, look back to the Fireplace creation example in Chapter 3.

 **Note:** For the remainder of this chapter the focus will be on Component Families. For simplicity's sake, we will refer to Component Families as simply "Families" for the remainder of this chapter.

## FAMILY LIBRARIES

A well-conceived template project is a critical component in successful Revit implementation (refer to Chapter 4 for more information). A well-stocked library of commonly used Families is equally important. A "Library" is nothing more than a series of folders on your hard drive that contain Revit content. Revit ships with a large library of ready-to-use items. These include Component Families, Annotation symbols, and Detail Component items. Some of these items are included in sample template files (see Chapter 4) and therefore automatically become part of each newly created project and others are provided in Family files (RFA) in the library folders.

The default United States installation creates the "Imperial Library" that is located by default in the *C:\ProgramData\Autodesk\RAC 2009\Imperial Library* folder (see Figure 10.1). Please note that the *"Program Data"* folder is hidden in Windows™ by default. To make it visible, click on the Start button and type: **Folder Options** in the search field. On the View tab, turn on Hidden Files

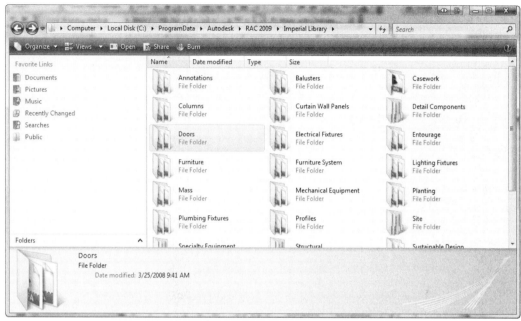

**Figure 10.1** *An example of one of the folders in the Revit Imperial Library*

If you are using Windows XP, the folder is: *C:\Documents and Settings\All Users\Application Data\Autodesk\RAC 2009\Imperial Library)*. Please note that the *"Application Data"* folder is hidden in Windows™ by default. To make it visible, choose **Folder Options** from the Tools menu in Windows Explorer and turn on Hidden Files).

 **Note:** The specific location of your library files will vary with your locality and particular version of Revit. Check the documentation that came with your product for the specific location.

There are many items in the library. For example, the Imperial library contains over 1,100 Family files. As we have seen in previous chapters and as we will discuss in more detail below, a Family can contain one or several "Types." Therefore, the 1,100 Imperial Family files represent potentially several thousand readily available component elements that can be added to our projects. It is a good idea to become as familiar as you can with the provided content. The reason is simple; it is always easier to use or modify something that exists than it is to create it from scratch. You may also find it useful to explore the libraries for other countries' components and unit systems since the items in Imperial and Metric libraries are not identical. You can browse through the library in Windows Explorer or directly from Revit. If you are in Revit, simply choose **Load from Library** > **Load Family** from the File menu. This should take you directly to your default library. In other open dialogs, you can click the shortcut icon on the left (i.e. *Imperial Library* or *Metric Library*) to open your library. Double-click a subfolder to view its contents. You can select a file (single-click) to see a preview on the right, or if you wish, you can use the Views icon at the top of the dialog to browse by Thumbnails (see Figure 10.2).

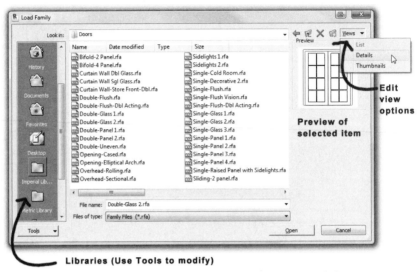

**Figure 10.2**   *Access a Library from within Revit from any load or open dialog*

No matter what method you use to browse your library, you can always open a Family file directly into the Revit interface. In Windows Explorer, simply double-click on the file you wish to edit. From the open command, double-click the file or select it and then click the Open button. The Family file will open into the Revit interface. The interface when a Family file is loaded is slightly different than when a project file is loaded. In such a state, the interface is often referred to as the "Family Editor." While in the Family editor, the Design Bar will show only one tab with Family-specific tools. Some of the other functions will appear differently as well. You will still have a Project Browser, but it typically includes only a few Views that have been developed for editing the current Family. These views are only seen while in the Family Editor and do not appear in projects to which the Family is loaded (see Figure 10.3).

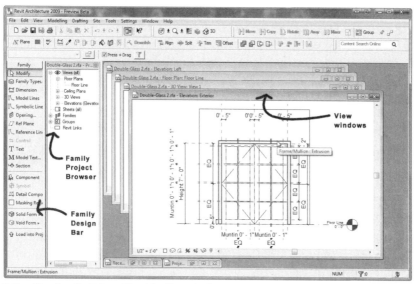

**Figure 10.3**    *The Family Editor (Double-Glass 2.rfa Door Family loaded)*

If you are following along in this passage within Revit, feel free to view and open as many Family files as you wish. However, please do not make any edits or save any changes to the out-of-the-box Family files. Treat this as an exploratory exercise. In the lessons that follow later in this chapter, we will have the opportunity to load, edit and build our own Families. The purpose of the current discussion is to give you familiarity with what has been provided with your software.

If you want to edit your library locations, you can do so by choosing Options from the Settings menu. In the "Options" dialog, click the File Locations tab. There you can change your default template file, the location of the Family template files, and access your "Places" which are the icons that appear on the left side of the open dialogs. In the "Places" dialog, you can add a path, edit a path, delete a path, and change the order in which they appear. You can also edit your places directly from any open dialog using the Tools pop-up in the lower left corner. You can see an example in the lower left corner of Figure 10.2 above. For more information on how to edit libraries and places, see the "Load Families from 'Preferred' Libraries" below.

In addition to any libraries that have been installed with your product on your local system, you may also have access to other libraries maintained by your firm's IT personnel. Libraries of this type are typically stored on the company network and accessible via your local or wide area network. Furthermore, Autodesk maintains a Web Library of Revit content. Several other online libraries can be found through a quick Web search. You can access the Autodesk Web Library directly from within Revit via the Recent Files page. If you do not have Recent Files open, choose it from the Window menu and then click the Web Library link on the right side. This opens a Web browser containing several content items organized in libraries by locality and/or manufacturer. We have already accessed this library a few times in the previous chapters. Another location hosted by Autodesk is the Autodesk Seek site. This can be accessed directly from Revit using the Content Search Online toolbar. Simply type a search in this toolbar and press ENTER. A Web browser will open and execute your search. Many other third-party web sites also maintain large libraries of Revit content that you can explore and download. Some are free and others charge a fee. Some of the more popular sites are listed in Appendix C of this book. You are encouraged to run a Google search for Revit, Revit Families, and Revit Content. The sites noted here and in the Appendix were accurate at the time of publication, but due to the ever-changing nature of the Web, it is always best to verify sites with a current search.

## FAMILY STRATEGIES

Do take the time to explore your Family library. As you analyze each of the library items, many of the items will seem useful. You will also find those that you will deem not useful. Still others will prove useful after they have been modified in some fashion. If you take the time to perform this analysis of the included library content before committing any time and resources to customizing existing items and/or building new items from scratch you can save yourself a great deal of effort and frustration. The other benefit to this process is that you will undoubtedly discover ways of performing certain tasks or representing certain items that you had not considered on your own. In other words, reverse engineering existing Families can prove to be a tremendous learning experience. Just remember not to save changes to existing files. Always make a copy or "Save As" first.

Nearly everything you do in Revit involves the use or manipulation of a Family. We have already seen several examples of this in the previous chapters. We "used" Families whenever we added something to our model. We manipulated Families whenever we edited the Type and or clicked the "Duplicate" button in the "Type Properties" dialog. Whenever you wish to add an element to your model, the element you add will be part of a Family. Always try to locate and use a Family that already exists and is available to you before you modify or create custom Families. The basic strategy of Family usage is to maintain the following priorities:

- **1st—Use Existing** See if what you want already exists somewhere (in the current project or in a library) and simply use it.

- **2nd—Modify Existing** If the precise Family (or Type) you want does not exist, find a Family (or Type) that is close to what you need, duplicate, and modify it.

- **3rd—Create New** If necessary; create a new Family to represent exactly what you need.

This basic approach is somewhat obvious and logical. However, it is surprising how often users will either resort immediately to building custom components without first checking the libraries, or do the opposite and settle for some less than ideal component in their models. Follow this simple three-step guideline, and you will always have the right Family for the job. If you follow the recommendations here and become familiar with the libraries available to you, it is likely that you will only need to "Use" or "Modify" a Family rather than "Create" a new one. The remainder of this chapter is devoted to tutorials that will illustrate this.

## ACCESSING FAMILIES IN A PROJECT

If you wish to use an existing Family in your project, the first step is to explore what is available. The first place you should look is the current project. You can quickly see all of the Families available in the current project in the Project Browser.

### INSTALL THE CD FILES AND OPEN A PROJECT

The lessons that follow require the dataset included on the Mastering Revit Architecture CD ROM. If you have already installed all of the files from the CD, simply skip down to step 3 below to open the project. If you need to install the CD files, start at step 1.

1. If you have not already done so, install the dataset files located on the Mastering Revit Architecture CD ROM.

   Refer to "Files Included on the CD ROM" in the Preface for instructions on installing the dataset files included on the CD.

2. Launch Revit Architecture from the icon on your desktop or from the *Autodesk > Revit Architecture* group in *All Programs* on the Windows Start menu.

   **Tip:** In Windows Vista, you can click the Start button, and then begin typing **Revit** in the "Start Search field. After a couple letters, Revit Architecture should appear near the top of the list. Click it to launch to program.

   • If the New Features Workshop dialog appears, choose "Maybe later" and then click OK.

3. On the Standard toolbar, click the Open icon.

   **Tip:** The keyboard shortcut for Open is CTRL + O. **Open** is also located on the File menu.

   • In the "Open" dialog box, click the *My Documents* icon on the left side.

   • Double-click on the *MRAC* folder, and then the *Chapter10* folder.

   If you installed the dataset files to a different location than the one listed here, use the "Look in" drop down list to browse to that location instead.

4. Double-click *10 Commercial.rvt* if you wish to work in Imperial units. Double-click *10 Commercial Metric.rvt* if you wish to work in Metric units

   You can also select it and then click the Open button.

The project will open with the last saved view visible on screen.

### ACCESSING FAMILIES FROM PROJECT BROWSER

There are a few ways to see which Families are already loaded into a project. Whenever you choose a tool from the Design Bar, the list of Family/Type combinations will appear in the Type Selector. The format is **Family Name : Type Name**. We have seen this already in several earlier chapters. While this method is effective for a particular category of element such as a Door, or a Window, the easiest way to see a list of loaded Families is via the Project Browser. The Project Browser has several major nodes or branches. The first branch is the Views branch. We have spent nearly all of our time in previous chapters on this node. Beneath this one are specialized view types such as Legends, Schedules and Sheets. The Families, Groups, and Revit Links nodes are at the bottom. (In the previous chapter, we used the Families branch to provide quick access to the Curtain Wall and Curtain System items we were editing). If you expand the Families node, you will see each element category currently loaded in the project. Expand any category, such as Doors, and you will see each Family of that category. Finally, expanding one step further reveals all of the Types for each Family.

5. On the Project Browser, double-click to open the *Level 3* floor plan view.

 **Note:** To make the contents of the Project Browser easier to read, drag the edge of the project Browser window to widen it.

6. On the Project Browser, collapse the *Views (all)* and *Sheets* nodes (click the minus (-) sign icon).

- Expand the *Families* Node, and then expand *Doors*.

- Expand **Single-Flush** [**M_Single-Flush**] (see Figure 10.4).

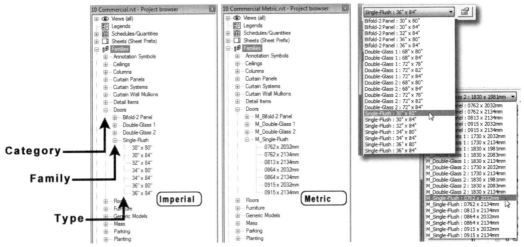

**Figure 10.4** *Expanding the Families branch of the Project Browser*

7. On the Design Bar, click the Basics tab and then click the **Door** tool.

- Open the Type Selector and compare the list of Family : Types to the list shown in Figure 10.4.

- On the Design Bar, click the **Modify** tool or press the ESC key twice.

As you can see, all of the Door Family/Type combinations will be listed in the Type Selector. In addition to simply taking inventory of Families and Types in the project, you can also use the list on Project Browser to manipulate and interact with Families. Use the right-click menu to do this. You can right-click on the *Family* and *Type* nodes of the Project Browser tree. (Right-clicking the *Families* or *Category* nodes will not produce a menu).

8.  Right-click on the Type **36" × 84" [0915 × 2134mm]** and take note of the menu that appears (see Figure 10.5).

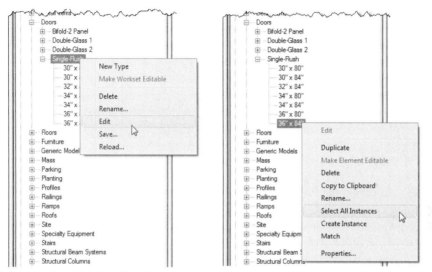

**Figure 10.5** *Right-click a Family or Type in Project Browser to access menu options*

**BIM** *Manager Note:* When you right-click on the Family name, you will only get a menu on Component Families. If you right-click a System Family, no menu will appear. A menu will always appear when you right-click types—either Component Family types or System Family types. This is one way you can tell if an item is a Component Family or a System Family. If a menu appears, it is a Component Family. If no menu appears, then it is a System Family.

-   From the right-click menu, choose **Select All Instances**.

All of the Doors in the model that use this Type will be selected. Be careful with this particular command as it selects *all* instances in the model, not just those visible in the current view. There are many other commands available on the right-click menu as well. You can Duplicate the current Type, Rename it, or edit its Properties. You can also create an Instance directly from this menu, rather than first clicking the tool on the Design Bar and choosing the Type from the Type Selector. In some cases this can be quicker; the end result is the same, it is simply a matter of personal preference.

-   On the Design Bar, click the **Modify** tool or press the ESC key twice to deselect the Doors.

Let's add some furniture to this plan using the right-click option on the Families branch.

9.  On the Project Browser, beneath Families, expand the *Furniture* node and then the *Desk [M_Desk]* node.

- Right-click on **72" × 36"** [*1830 × 915mm*] and choose **Create Instance**.
- Place the new Desk in one of the offices (see Figure 10.6).

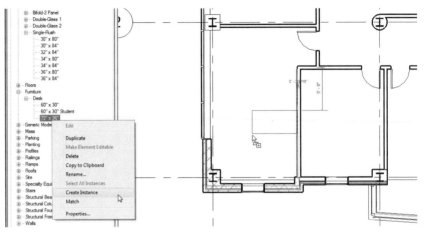

**Figure 10.6** *Use the Family Tree right-click menu to add a Desk to the plan*

- Repeat in other offices. Press the SPACEBAR to rotate the Desk before placement as required (see Figure 10.7).

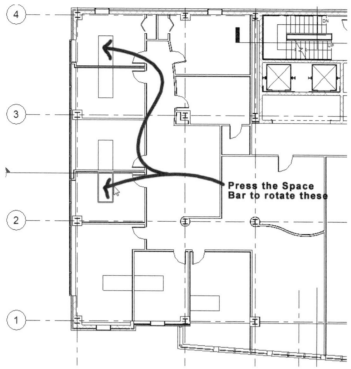

**Figure 10.7**   *Place Desk Families into the model from the Project Browser*

- On the Design Bar, click the **Modify** tool or press the ESC key twice.

## MATCH TYPE PROPERTIES

One of the useful functions on the Type's right-click menu is "Match." With this command, you can apply the Type parameters of the item highlighted in Project Browser to an element already in the model. This is a quick way to "paint" a Type's properties onto existing element. We can try this out on the entrance to the suite on Level 3. Currently there is a single Door there. Let's make it a double glass Door.

10. On the Project Browser beneath *Families*, expand *Doors* then *Double-Glass 1* [*M_Double-Glass 1*].

- Right-click on **72" × 84"** [*1830 × 2134mm*] and choose **Match**.

  When the cursor is moved into the plan view it will change to an eyedropper shape.

- Click on the Door at the entrance to the suite (see Figure 10.8).

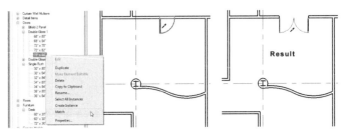

**Figure 10.8** *Match a Type from Project Browser to an element in the model*

The Door will change to a double glass Door.

- On the Design Bar, click the **Modify** tool or press the ESC key twice.

11. Save the model.

## ACCESSING LIBRARIES

So far we have limited our exploration to Families that are already part of our current project. As was mentioned however, you can find potentially limitless Families in external libraries. You have both the out-of-the-box library installed with your software and also remote libraries accessible via the World Wide Web. The process of using such resources is nearly identical to that already covered with the additional step of first locating and loading the Family in the appropriate library.

As we have already mentioned, a library is simply a collection of files and folders stored on your local system or a remote server. To access and place a Family in a project, you must first load it. There are several ways to do this. For Component Families, a Load button is presented on the Options Bar when you add the item. A Load button is also available within the "Element Properties" dialog for any Component Family. Click this button to open a dialog and browse to an appropriate library folder and Family file. You can also load a Family without first executing a particular tool. To do so, use the **Load from Library > Load Family** command on the File menu.

### LOAD FAMILIES FROM "PREFERRED" LIBRARIES

Continue in the same dataset as the previous topic.

1. From the File menu, choose **Load from Library > Load Family** (see Figure 10.9).

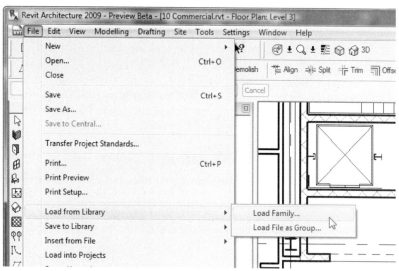

**Figure 10.9**   *Load a Family from a Library*

The "Load Family" dialog will appear starting in the default library folder as indicated in your "Options" dialog (Settings menu). Depending on the options you chose during installation, there may be additional libraries available to you. Each library that you have installed or added will appear as a shortcut on the left of the dialog. To add shortcuts, choose **Options** from the Settings menu, and then click the "File Locations" tab. Click the Places button to add or edit a Library shortcut, delete a shortcut, and move items up and down in the list (see Figure 10.10).

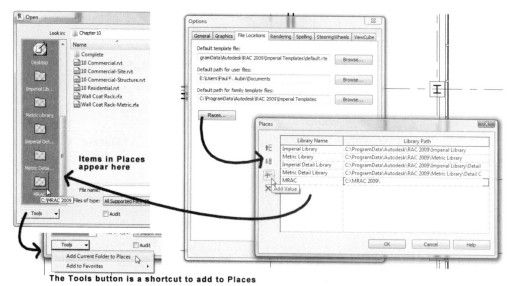

**Figure 10.10**   *Libraries show as shortcuts on the left of "Open" dialogs, add more in Options*

If your preferred library did not open automatically, use the shortcut icons on the left or the drop-down menu at the top to locate your desired library. The library folder will typically contain several sub-folders. Each folder can contain additional folders or Revit Family (RFA) files.

 **Tip:** If you access a folder that is not among your shortcuts, you can add it to Places quickly using the Tools button in the open and load dialogs (see the bottom left corner of Figure 10.10).

2. On the left side, click the *Imperial Library* [*Metric Library*] shortcut icon.

- At the top of the dialog, click the View Menu icon and choose **Thumbnails**.

This is not necessary, but it makes it easier to quickly locate the Family you need.

- Double-click the *Furniture* folder, select the *Credenza.rfa* [*M_Credenza.rfa*] file and then click Open.

 **Note:** If your version of Revit Architecture does not include either of the libraries mentioned herein, both Family files have been provided with the files from the Mastering Revit Architecture CD ROM. Look for them in the *Library\Imperial Library\Furniture* [*Library\Metric Library\Furniture*] folder wherever you installed the dataset files on your local system.

- On the Project Browser, beneath Families, expand *Furniture* to see the newly loaded Credenza Family in the list (see Figure 10.11).

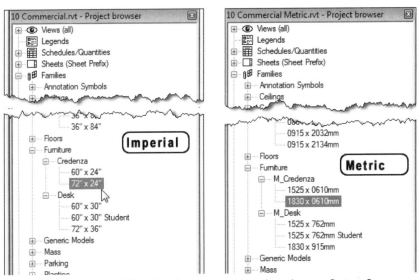

**Figure 10.11** *The newly loaded Family will appear among the others in Project Browser*

- Click on **72" × 24"** [**1830 × 0610mm**] and drag it to the model window.

This is an alternative to the right-click approach above that achieves the same end. Use whichever method you prefer.

- Place a Credenza in each office.

  Remember to press the SPACEBAR to rotate them as you place them.

 **Tip:** If you want to rotate to align with an angled edge such as the Curtain Wall at the front of the building, place the mouse over the Curtain Wall and then press the SPACEBAR. The Credenza will align to the angle of the Curtain Wall. Each time you press the SPACEBAR, it will rotate again. You may need to repeat a few times to get the correct angle.

## ACCESS WEB LIBRARIES

In addition to the libraries installed with the product, several Web sites are available containing Revit content and Families. Accessible from directly within the product is the Autodesk Web Library and Content Search. You can access the Web Library from the Recent Files page and Content Search from the toolbar.

3. From the Window menu, choose **Recent Files**.

Let's give our occupant in the lower left corner office a more fitting desk.

- On the right side of the screen, click the "Revit Web Content Library" link (see Figure 10.12).

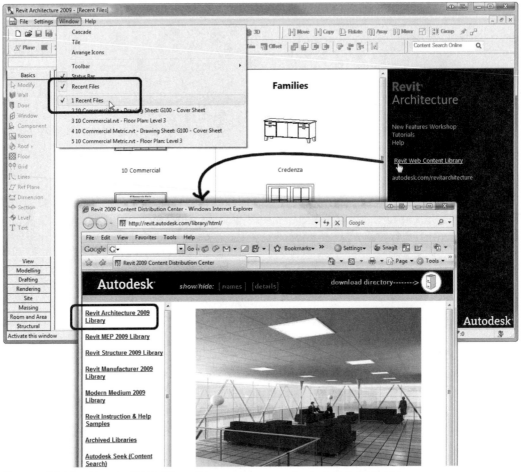

**Figure 10.12**  *Use the "Web Library" button to access the Autodesk Web Library*

This will open your Internet browser to the Revit Content Distribution Center web page where you can view and download available content. The web browser application may have opened behind Revit. Click on the browser's icon from the Windows Task Bar to make it the active window.

- On the left side of the page, click the "Revit Architecture Library" link.

  There are three libraries organized by region.

- Click the *US Library* link.

- Click the *Families* and then the *Furniture* link.

This will reveal a page full of furniture Families on the right. Each Family will have an image preview.

- Scroll as necessary and click the link beneath "Desk-Executive" to download this Family file.

If you prefer, you can locate the same file in metric units in the UK Library link instead.

- In the "File Download" dialog that appears, click the Open button (see Figure 10.13).

**Figure 10.13** *Download a Family from the Web Library and Open it in Revit*

This action will download the Family file and open it in your current Revit session. We briefly discussed the Family editor above. The Desk-Executive Family file opens in the Family Editor. If you wish, you can study each view, but please do not make any edits to this file.

 **Note:** In order to access this library using this method, you must have an Internet connection. If you do not have access to the Internet, the Family files have been provided with the files from the Mastering Revit Architecture CD ROM. Look for them in the *Library\Imperial Library\Furniture [Library\Metric Library\Furniture]* folder wherever you installed the dataset files on your local system.

4. On the Design Bar, click the Load into Projects button.

This will load the Family into the commercial project and make its view window active again. (Please note that the Family file is still open in the Family Editor in the background). When convenient you can switch from the project to Family Editor and close the Family file.

5. On the Project Browser, beneath Families, expand *Furniture* to see the new Desk-Executive Family in the list.

- Expand Desk-Executive to reveal that the only Type in this Family is also named **Desk-Executive**.

6. Right-click **Desk-Executive** and choose **Match**.

- Click the Desk in the corner office to change it to the new **Desk-Executive** Family.

- Move the Desk if required.

- On the Design Bar, click the **Modify** tool or press the ESC key twice.

 **Tip:** To make sure that the desks and credenzas are oriented correctly, you can cut some sections or create a third floor 3D view. To do this, copy the {3D} view, rename and orient it as instructed in the "Create a Working View" topic in Chapter 9. Mirror or rotate items as necessary.

7. Save the model.

There are many websites devoted to the distribution of Revit Families and general tech support and discussions. See Appendix C for a further list of websites offering such resources.

## EDIT AND CREATE FAMILY TYPES

Until now, we have only worked with existing Families and Types. In many cases, you will need to edit existing Types and create your own Types within existing Families. You should try both of these approaches before resorting to creating a completely new Family. There will certainly be situations where it is appropriate to create new Families. (We will look at examples in detail below.) For now, let's look at what we can do with Types first.

### VIEW OR EDIT TYPE PROPERTIES

In some cases, you will want to understand more about the particular Family or Type you are selecting before you place it in the model. You can view the Type Properties or edit the Family directly from Project Browser. This is also accomplished via the right-click menu.

1. On the Project Browser, beneath *Families*, expand *Furniture* then *Desk [M_Desk]*.

- Right-click on **72" × 36"** [*1830 × 915mm*] and choose **Properties**.

This method takes you directly to the "Type Properties" dialog. We have seen enough examples of this dialog in previous chapters to know that if we make any edits here, they would apply at the Type level and therefore affect *all* Desks currently in the model (since they all currently share this Type).

- In the "Type Properties" dialog box, click on the "<< Preview" button at the bottom left corner of the dialog.

This will expand the width of the dialog to include an interactive viewing pane on the left side.

- At the bottom of the viewing pane, click the "View" pop-up menu and choose **3D View:View 1** (see Figure 10.14).

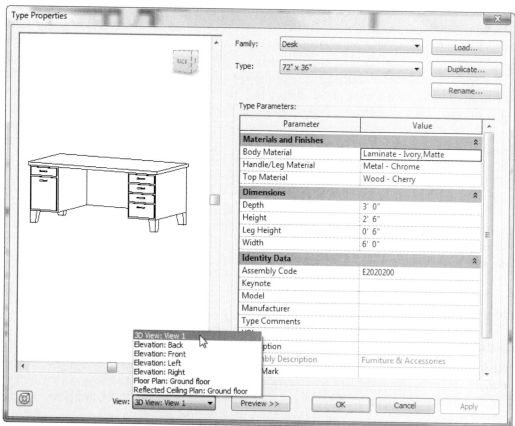

**Figure 10.14**   *Open the view pane and choose a 3D view point*

There are several other view options, such as plan, elevation and ceiling views. Try them all out if you wish and then return to the 3D view when finished. If you need to get a better look in any view, you can use standard navigation techniques. Right-click in the viewer to get a menu of standard zoom and scroll commands. There is also a "Steering Wheels" icon and ViewCube.

2. At the lower left corner of the dialog, click the "Steering Wheels" icon.

- Place your pointer over the Orbit area and then drag the mouse in the viewer to spin the Desk model.

- Try other modes if you wish and then close the Steering Wheels when finished.

Take a close look at the parameters available at the right of the dialog. First, you can switch the Family you are viewing by choosing another from the Family list. Likewise, you can choose any Type from the Type list at the top. Changing either the Family or the Type will be reflected immediately in the viewer. Leave the Family set to Desk, but try choosing each of the other two Types one at a time to see the viewer react. Return to the Type: *72" × 36"* [*1830 × 915mm*] when finished. Notice that all three Types look essentially the same. This is critically important to understand. Study the various parameters listed in the tables below the Family and Type lists. These are the "Type Parameters." The shape of this desk, with its two pedestals, four legs, closed back and top are all characteristics of the Family called: "Desk [M_Desk]." The overall width, height and even the height of the legs on the other hand are parameters of the each Type. As you choose a different Type from the list, one or more of these values will change accordingly. It is like choosing between several models of a certain make car. A Honda Civic comes in several trim models, some have higher quality wheels, or better sound systems, but they are all *Civics*. In this analogy, "Civic" is the Family and "LX" or "EX" is the Type. Returning to our Desk on screen, go slowly through each of the three types again and study the viewer and especially the change in Dimensions as you do. What this means is that if you simply want a different width or height Desk, or perhaps wanted brass hardware rather than the default chrome, you would simply choose or create a new Type. If however, you wanted a different "make" of desk, with only a right or left pedestal for instance, you would need to choose or create a new Family. For example, if you select the Executive Desk from the Family list on screen, you should see a much more dramatic change to both the viewer and the parameters at the right.

- Click Cancel in the "Type Properties" dialog to dismiss it without making any changes.

## CREATE A NEW TYPE FROM PROJECT BROWSER

We will be focusing on types for now, so let's create a new one. For the group of offices running vertically along the left, let's create a new Type in this Desk [M_Desk] Family that is a slightly smaller size, yet a bit larger than the other Types currently available.

3. Right-click on *72" × 36"* [*1830 × 915mm*] and choose **Duplicate**.

This will create a new Type named *72" × 36" (2)* [*1830 × 915mm (2)*] with the name highlighted and ready to be renamed. (Alternately, if you prefer, you can right-click the Family name and choose New Type, rename it, and the result will be the same).

- Name the new Type: **66" × 30"** [**1650 × 762mm**] and then press ENTER.

You can give the Type any name you want, but usually it is helpful to include in the name what makes this Type different from others, in this case the size. This is standard procedure in all of the provided Revit content and a good practice to follow.

4. Right-click on *66" × 30"* [*1650 × 762mm*] and choose **Properties**.

- In the "Type Properties" dialog, change the Depth to **2'-6"** [**762**].

Notice that when you click in this field, a temporary dimension appears in the viewer to indicate how this parameter will apply (see Figure 10.15).

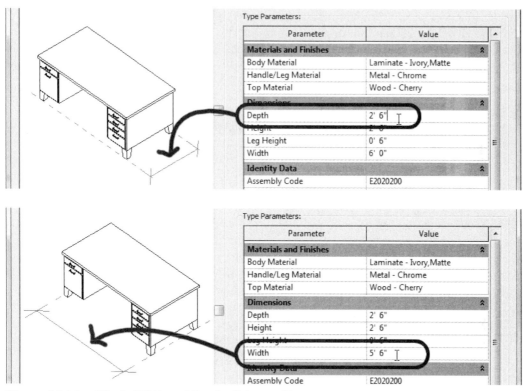

**Figure 10.15**    *Edit the Width and Depth of the new Type*

- In the "Type Properties" dialog, change the Width to **5'-6"** [**1650**] and then click OK.

5. Right-click on the new Type, and choose **Match**.

- Click on each of the Desks in the offices along the left side of the plan to change them to the new Type.

- Move the Desks if necessary to align with the Walls.

**Tip:** You can use the Align tool here and then apply a lock constraint to keep the Desk aligned to a particular Wall. Be careful not to "over" constrain the model however.

6. Save the model.

Because we can see that there is more than one Type in this family, with different names, we have a good indication that this family is parametric, and new types with different parameters, for example sizes or materials, can be created.

If there is only one Type, and it is the same name as the family, that typically means that particular family is not parametric. While you could make additional types from such a Family, you would be limited to editing identity data and materials. For an example, edit the properties of the Executive Desk. The only editable parameters are the Material of the desk and the standard identity data information. Changing the width, height, and depth is not possible at the type level for this Family. Another way to say this is that to edit the dimensions of the Executive Desk, you would be required to open and edit the Family itself.

## CUSTOMIZING FAMILIES

While a great deal of content is available to us in the installed and online libraries, we often need components in our projects for which a suitable Family is not readily available. In these situations, we can create our own custom Family. Custom Families can be simple or complex. In this topic, we will look at various examples of creating and using custom Families in our projects.

### DUPLICATE AN EXISTING FAMILY

When you decide to build your own Family, it can be useful to start with an existing Family that is close to the Family you wish to create. Doing so will only require you to save as and edit the existing Family which can be more expedient than starting from scratch. For example, flanking the corridor in the middle of our tenant suite is a secretarial space with room for two workstations. Perhaps the client would like desks in this area that have a CPU cubby rather than the two drawer pedestals occurring in the Family we currently have loaded. We can duplicate the existing Family, and then modify it to make this change. This is much more efficient than modeling the entire desk over again.

7. On the Project Browser, beneath *Families*, expand *Furniture* and then right-click on Desk [*M_Desk*].

- Choose **Edit** from the right-click menu (see Figure 10.16).

 **Tip:** As an alternative, select one of the desks in the model, right-click and choose **Edit Family** (or click the Edit Family button on the Options Bar).

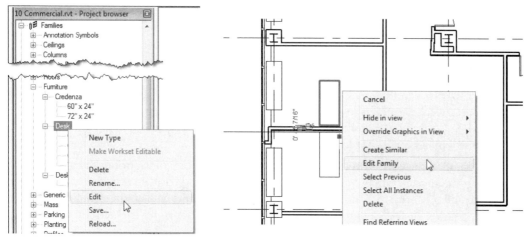

**Figure 10.16** *Edit a Family from Project Browser or the model*

- In the message dialog that appears, click Yes to confirm opening the Family file.

The *Desk* [*M_Desk*] Family will open into the Family editor with a 3D view active. (You are no longer in the commercial building project file. You are now in the Desk Family file where you can edit it directly).

8. Move your Modify tool over each part of the desk and pause for the tool tip to appear (see Figure 10.17).

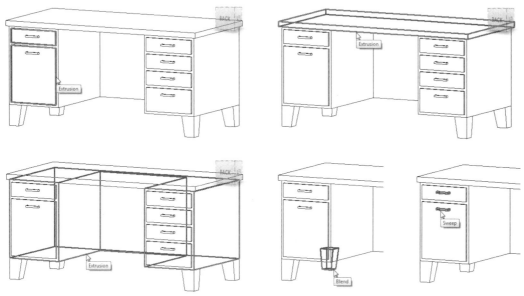

**Figure 10.17** *The Desk is made from various Solid Forms*

The desk is made from a collection of Solid Extrusions, Sweeps and Blends. An Extrusion is a 3D form created by a 2D closed shape that is "pushed" along a perpendicular path. A Sweep is an extrusion that follows a path, which can be any shape. A Blend is form that starts with one closed shape and then "morphs" into a second closed shape. The transformation from the bottom shape to the top shape occurs along a perpendicular path. In addition to the three solid forms used in this Family, we also have Revolve and Swept Blend. Revolve is a shape derived from rotating a 2D closed shape about an axis. Swept Blend is a new form in the latest release of Revit. It is basically a Blend that can have a non-perpendicular or curved path. The path is limited to one segment, but it can be any shape you wish.

Each of these five basic shapes can be made as solids or voids. Solids create physical material in the model, while voids carve away from the solid form in order to achieve more complex resultant forms. In addition to this "raw material," a Family can also contain other Families. These "nested" Families are added as Components to the Family within the Family Editor. (We will see an example of a nested Family below.)

## EDIT AN EXTRUSION

9. On the Project Browser, expand *Views (All)* and then expand *Elevations*.

- Double-click to open the *Front* elevation view.

As you can see, navigating the Views in the Family editor nearly the same as within a project, except the names of the views vary and the quantity of views is typically fewer.

10. Select the Extrusion on the left (that represents the drawer fronts) and then hold down the CTRL key and select the handles as well (see Figure 10.18).

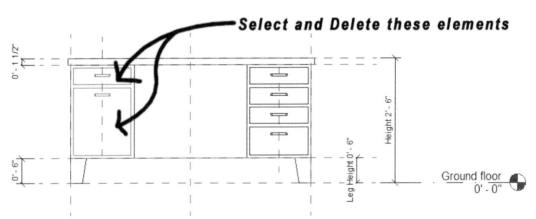

**Figure 10.18**   *Select and Delete the Drawer Fronts and Handles on the left*

- Press the DELETE key to erase both elements.

The author of this Family chose to create a single extrusion that contained two shapes—one for the top drawer and another for the bottom. While this is a perfectly valid approach, it forces us to edit both drawers together. In your own Families think about such issues carefully. This approach may ultimately be the best, but in some cases, you may decide to instead model each drawer separately.

The pedestal on the left will become a cubby for a CPU. This is why we have deleted the drawer fronts. We now need to make the cubby a bit narrower and then cut a void from it to complete the cubby.

11. Click on the Extrusion that comprises the major form of the desk.

Notice all of the shape handles that appear. We can simply drag one of these to make the cubby narrower (see Figure 10.19).

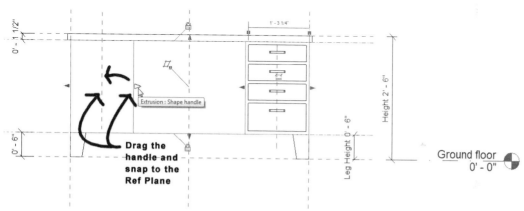

**Figure 10.19**  *Use the Shape Handles to resize the Cubby Pedestal*

12. Drag the shape handle (indicated in Figure 10.19) to the left and snap it to the Reference Plane (green dashed vertical line in the middle of the pedestal).

   • Click the small open padlock icon that appears to apply a constraint to this Reference Plane.

This action "locks" this side of the extrusion to this Reference Plane. If the Reference Plane later moves, the edge of the extrusion will move accordingly. Let's try it out.

13. Deselect the Extrusion and then select the Reference Plane.

   • Edit the temporary dimension that appears to **1'-8"** [**500**] (see Figure 10.20).

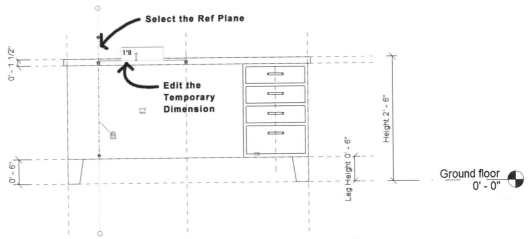

**Figure 10.20** *Move the Reference Plane to move the constrained Extrusion edge*

### ADDING A VOID FORM

Let's add the void next.

14. On the Design Bar, click the **Void Form** tool and then choose **Void Extrusion** from the flyout.

Since we are working with solids and voids that are three-dimensional forms, we need to indicate to Revit our preferred working plane. A Working Plane establishes a 2D surface in the model in which we can sketch our 2D shapes. In this case, since we are working on the Front elevation, a Work Plane parallel to Front makes the most sense (see Figure 10.21).

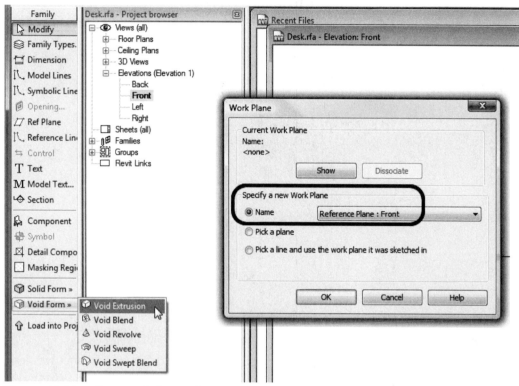

**Figure 10.21**    *Choose a Reference Plane in which to sketch the Void shape*

- In the "Work Plane" dialog, choose **Reference Plane : Front** from the "Name" list and then click OK.

- On the Options Bar, click the Rectangle icon and in the "Offset" field, type **-3/4" [-19]** (Note the value is negative).

- Using Figure 10.22 as a guide, snap to opposite corners of the CPU pedestal.

- On the Options Bar set the Depth to **1'-0" [300]**.

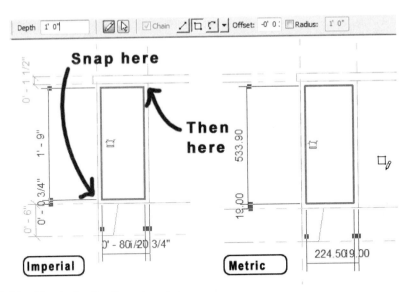

**Figure 10.22**  *Sketch the Cut Extrusion shape*

- On the Design Bar, click the Finish Sketch button.

15. On the Project Browser, double-click to open the *View 1* 3D view.

From this view you can see that the Void is cutting away from the overall form of the desk. However, you can also see that it does not project back very far into the desk's depth. We can easily adjust this. (To see the Cut Extrusion more clearly, pass the Modify tool over the cubby to pre-highlight it—it will appear as a buff colored outlined Cut element). The floor plan view might be a good location to work for this edit.

16. On the Project Browser, double-click to open the *Ground Floor* plan view.

The Cut Extrusion will appear among the other shapes.

- Click the Cut Extrusion shape in plan.

- Use the Shape Handle to stretch the Void back to the rear plane of the desk (it should snap automatically) (see Figure 10.23).

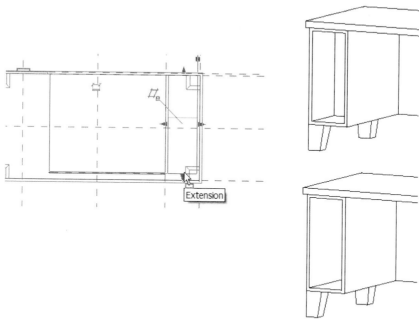

**Figure 10.23** *Stretch the depth of the Void using Shape Handles*

17. Return to the 3D view to see the results.

With the Cut Extrusion still selected note the Depth field on the Options bar shows the current dimension. The Depth can also be edited here as well.

## SAVE A NEW FAMILY

We are ready to save the results and load our new Family into the project.

18. From the File menu, choose **Save As**.

Revit Architecture imposes no limitations on where you can save Family files, however the location where you save Family files is a very important consideration particularly in team environments. Check with your IT support personnel regarding the preferred location for saving Family files. Sometime firms have a "check-in" process for newly created content. Follow whatever guidelines or practices are in place in your company. For this exercise, we will simply save the Family file (RFA) to our *Chapter10* folder. If you decide later that you wish to use this Family in real projects, you can copy it from this location to a suitable location on your company network.

- In the "Save As" dialog, browse to the *Chapter10* folder in the location where you installed the Mastering Revit Architecture CD ROM files.

- In the "File name" filed, type: **MRAC Desk-Secretary.rfa** for the name and then click Save.

**Caution:** *It is very important to use Save As rather than Save. If you simply save the Family, it will overwrite the existing Desk [M_Desk] Family from which we started.*

Before we load this Family back into the project, let's edit its Types. Remember, the Family controls the available parameters and physical form of the element. However, the Type(s) can have specific values for the established Family parameters. Following the lead of the Family from which we created this one, our Types will have predefined values for the sizes.

19. On the Design Bar, click the **Family Types** tool.

The "Family Types" dialog will appear. At the top is a drop-down list showing each of the existing Types inherited from the *Desk [M_Desk]* Family. Some of these are no longer necessary.

20. From the "Name" list, choose **60" × 30" Student [1525 × 762mm Student]** and then click the Delete button.

- Repeat for the **72" × 36" [1830 × 915mm]** Type (see Figure 10.24).

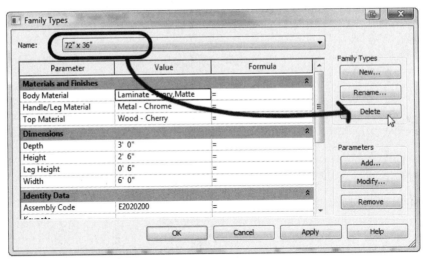

**Figure 10.24** *Delete unneeded Types*

The remaining Types can stay for this Family.

21. Click OK to dismiss the dialog.

- On the Toolbar click the Save icon (or choose **Save** from the File menu).

 **Note:** Since we have already saved this file as *MRAC Desk-Secretary.rfa*, we can simply choose Save this time to update the file. Save As would create a second copy of the file, which would be unnecessary.

## LOAD THE FAMILY INTO THE PROJECT

22. On the Design Bar, click the Load into Projects button.

As we saw above, this will return the commercial project to the front and add this Family to the Project Browser ready to be placed in the model.

> If a dialog appears listing more than one project, choose only the Commercial project and then click OK.

23. Add two of the new desks to the model as shown in Figure 10.25.

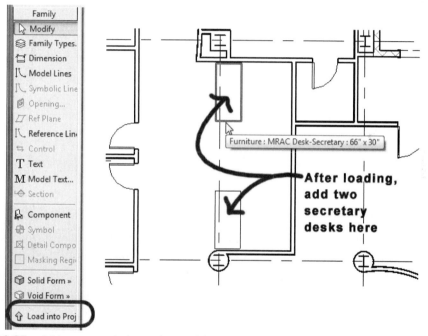

**Figure 10.25** *Add two new desks to the model*

From the floor plan view, we cannot really see any of the edits we made. Let's add a Camera view to the project so we can get a look at our new Component Family.

24. On the Design Bar, click the View tab and then click the **Camera** tool.

- Click a point behind the desks and then drag toward one of the desks.
- Click again in the center of one of the desks (see Figure 10.26).

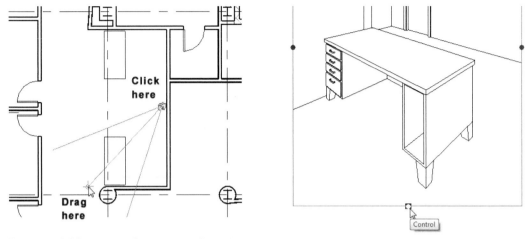

**Figure 10.26** *Add a Camera view from the plan*

- Use the Control handles to adjust the size of the 3D perspective view so you can see the whole desk.

**Tip:** If your desk is facing the wrong way, return to the *Level 3* view and mirror them. Be sure to remove the checkmark from the "Copy" option on the Options Bar.

## EDIT A FAMILY

Since this desk is now a computer-type desk, we should probably add a keyboard shelf in the middle. To do this, we simply repeat the process above and return to the Family Editor. We can model the shelf with a simple Solid Extrusion.

25. On the Project Browser, double-click to open the *Level 3* floor plan view.

- Select one of the Secretary Desks, right-click and choose **Edit Family**.

- In the dialog, click Yes.

26. On the Project Browser, double-click to open the *Ground floor* plan view.

27. On the Design Bar, click the **Solid Form > Solid Extrusion** tool.

- Using the rectangle option on the Options Bar, sketch a rectangular shape for the keyboard shelf in plan (see Figure 10.27).

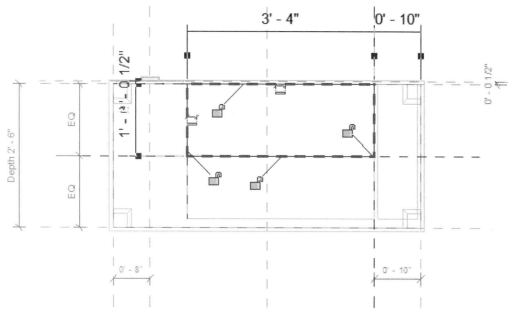

**Figure 10.27** *Sketch the keyboard shelf*

- On the Design Bar, click the Extrusion Properties button.
- In the "Element Properties" dialog, set the "Extrusion Start" to **2'-1"** **[625]** and the "Extrusion End" to **2'-1 3/4"** **[644]** and then click OK.
- On the Design Bar, click the Finish Sketch button.

28. Return to the 3D view (see Figure 10.28).

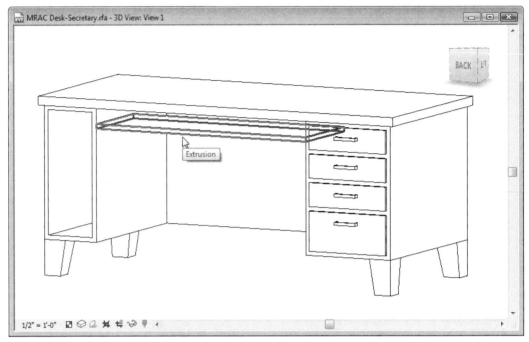

**Figure 10.28** *The secretary desk with the keyboard shelf added*

29. Save the Family, click Yes to overwrite the existing family if prompted, and then click the Load into Projects button.

- In the "Reload Family" dialog, click Yes to confirm reloading the Family and updating the project.

- On the Project Browser, right-click the *3D View 1* and choose Rename.

- Type: **Camera at Secretarial** for the new name.

- Double-click to open the *Camera at Secretarial* 3D view.

Notice how the update to the desk now appears in the project.

- Save the project.

Congratulations. You have completed your first custom Family. In the next topic, we'll create a new Family from scratch and get into more advanced Family creation techniques.

## BUILDING FAMILIES

In most cases, the preceding process will enable you to produce the Family you need by leveraging your existing library content. However, sometimes it will be necessary or easier to start from scratch. In this case, you will simply create a new Family file and begin modeling the element you require. All new Families are created from pre-defined Family templates. Revit Architecture ships with a large collection of pre-made Family templates from which to choose. It is important to select the Family template which best corresponds to the kind of Family you wish to create. This is because the template you choose determines the category of the Family and whether or not it requires a host. While possible to modify the category later, you cannot change the hosting behavior of a Family once it has been created, so choose your template carefully.

## CREATE A NEW FAMILY FILE

Keeping with the furniture layout on the third floor of the commercial project a little longer, let's create a custom reception desk for the lobby to the suite.

    1. From the file menu, choose **New > Family**.

This will open the "New Family" dialog box to the folder that contains all the available Family templates. Familiarize yourself with this list—there are many from which to choose. Remember the choice of template is important, so take the time to make a good choice (see Figure 10.29).

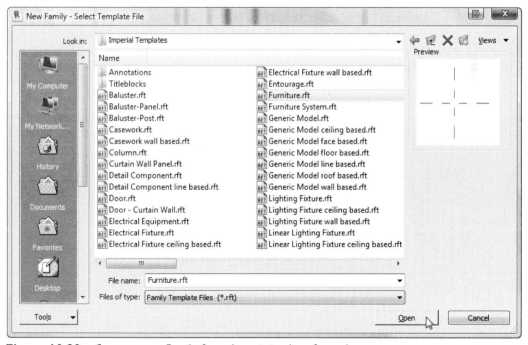

**Figure 10.29** *Create a new Family from the existing list of templates*

Notice that some categories have more than one template. For example, in the figure you can see that there is *Electrical Fixture.rft*, *Electrical Fixture ceiling based.rft*, and *Electrical Fixture all based.rft*. All three create Families of the electrical fixture category, but the ceiling based one requires a Ceiling host object and the wall based requires a Wall host. *Electrical Fixture.rft* requires no host and can be placed freestanding in a project.

As you can see, your choice of template determines two important characteristics of the new Family: its category and if it requires a host. Any of the templates whose name ends in "based" require a host. The most logical category for our reception desk is furniture. Therefore, we will select the *Furniture.rft* template. Unlike electrical fixtures, there is only one furniture template and it does not require a host.

> Browse to you *Templates* folder if necessary.

- Select *Furniture.rft* [*Metric Furniture.rft*] and then click Open.

 **Note:** If your version of Revit Architecture does not contain either of these templates, they have been provided in the *Templates* folder with the files installed from the Mastering Revit Architecture CD ROM.

As with the previous exercise, we are now in the Family Editor. The difference here is that the current view contains only two Reference Planes and no geometry. It is often easier when building Families to work with tiled windows. Most Family templates open several Views at once—a plan, a 3D and two elevations. However, if we were to tile the windows now, in addition to these four Views of the Family, our screen would also tile all of the Views we have open from the commercial project and the Desk-Secretary Family file. Therefore, we'll have to clean up our open windows a bit first.

2. From the Window menu, choose **Desk-Secretary.rfa - 3D View: View 1**.

- From the File menu, choose **Close**.

3. From the Window menu, choose **10 Commercial.rvt - Floor Plan: Level 3** [**10 Commercial-Metric.rvt - Floor Plan: Level 3**].

- From the File menu, choose **Close**.

> If prompted to save, choose Yes.

If you open your Window menu again, only the four Family views mentioned above and Recent Files will be listed (see Figure 10.30).

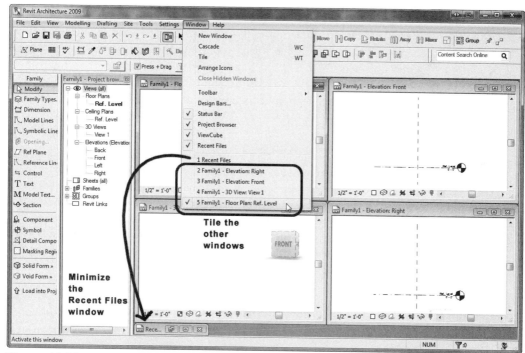

**Figure 10.30** *Close other projects and Families so that only the current Family file's views show in the Window menu*

4. From the Window menu, choose Recent Files.

- Minimize this window.

5. From the Window menu, choose **Tile**.

- From the View menu, choose **Zoom > Zoom All To Fit**.

**Tip:** The keyboard shortcut for Zoom All To Fit is **ZA**.

While it is not necessary to close the other projects and tile the Views, it will be easier to work on the Family in this environment. As we make changes in any view, we will see the results immediately in the others. Since we will be making many three-dimensional edits, this will be very helpful.

6. From the File menu, choose **Save**.

- In the "Save as" dialog, be sure you are in the *Chapter10* folder, type **MRAC Reception Desk** for the Name and then click Save.

## BUILD FAMILY GEOMETRY

As we have already seen, there are two methods to create objects in our Family: adding Solid forms and adding Void forms. Solid forms represent the actual physical materials from which our Family object is created. Void forms carve away from Solid forms to help us create more complex shapes. Both Solid and Void forms can be sculpted in five ways—Extrusion, Blend, Revolve, Sweep, and Swept Blend. We will see examples of each of these in this exercise.

**BIM** *Manager Note:* Typically, the first step in building a custom Family is to lay out a series of Reference Planes. The Reference Planes provide structure and form to the Family's geometry. All geometry added to the Family is typically constrained to one or more Reference Planes in some way. In this way, the Reference Planes can be moved (flexed) by a change in parameters and they will in turn affect the shape of the geometry. This is how parametric Families are created. You may recall seeing some examples of this in the desk Family that we modified earlier. While we will build a parametric Family below that fully utilizes Reference Planes in the way outlined here, the following Family example will not make use of Reference Planes to drive geometry. The focus of the next exercise is to acquaint you with the various solid geometry forms available.

## CREATE A SOLID EXTRUSION

The first form we will build is a simple extrusion for the work surface.

Make sure that the *Floor Plan : Ref. Level* view is active. (The title bar will appear bold—click the title bar to make it active).

7. On the Design Bar, click the **Solid Form** tool and then from the flyout that appears, choose **Solid Extrusion** (see Figure 10.31).

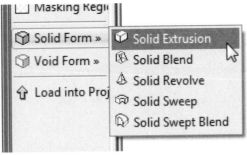

**Figure 10.31**   *Create a Solid Extrusion*

In the center of the plan view are two green dashed lines. These are Reference Planes that were part of the original Furniture template. We will use them to center our geometry.

8. On the Design Bar, be sure that the **Lines** tool is selected.

• On the Options Bar, type **-0' 3/4" [-19]** in the "Depth" field.

Be sure to use a negative value for the "Depth." We will discuss the reason below.

- Click on the Rectangle icon and draw a rectangle that is roughly centered on the two Reference Planes. (The exact size is not important yet)

- Edit the temporary dimensions to make the rectangle **6'-0"** [**1800**] × **2'-6"** [**750**] (see Figure 10.32).

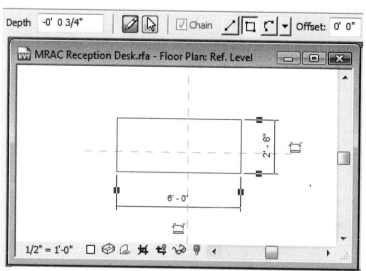

**Figure 10.32**   *Sketch a rectangle near the center of the plan and edit the temporary dimensions*

9. On the Design Bar, click the **Modify** tool.

- Select all four of the sketch lines.

- On the toolbar, click the Move icon.

- For the Move Start Point, click the Midpoint of one of the edges.

- For the Move End Point, click the intersection with Reference Plane (see Figure 10.33).

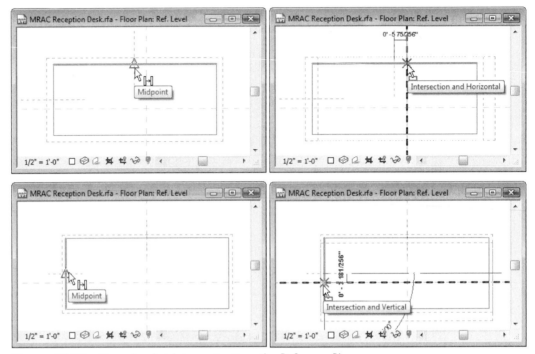

**Figure 10.33** *Move the sketch to center it on the Reference Planes*

10. Repeat the Move operation in the other direction.

The rectangular sketch should now be centered on the Reference Planes. You can use the Tape Measure icon to verify if necessary.

• On the Design Bar, click the "Finish Sketch" button (see Figure 10.34).

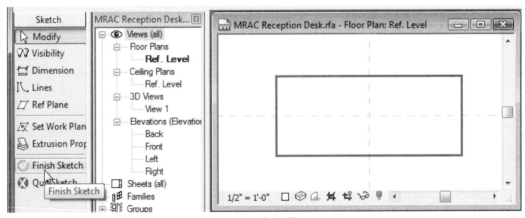

**Figure 10.34** *Finish the sketch to complete the Solid Extrusion*

## ADJUST THE VIEW WINDOWS

Now that we have some geometry built, we can adjust our view windows to show it better.

11. From the View menu, choose **Zoom > Zoom All To Fit** (or type **ZA**).

Notice that the 3D view shows the model from the side.

- Click in the *3D View : View 1* window and then hold down the SHIFT key and drag the middle (wheel) button to spin the model.

  If you prefer, you can use the Steering Wheels or the ViewCube to rotate the 3D view.

- Zoom in closer or otherwise fine-tune the view in each window to your liking.

- On the View Control Bar (for the 3D View), change the View Graphics to Shading with Edges (see Figure 10.35).

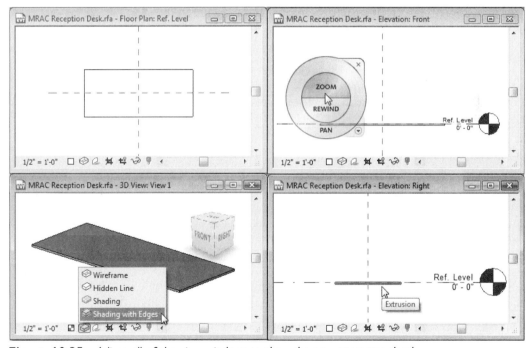

**Figure 10.35**  *Adjust all of the view windows to show the new geometry clearly*

- Save the Family.

## CREATE A REFERENCE PLANE

If you study the model, you will notice that the desktop surface is sitting on the floor. While it is possible to edit the parameters of the extrusion to set it at the correct height, we will get more control if we create a new Reference Plane and designate it as the Work Plane for the Solid Extrusion.

12. Click on the title bar of the *Elevation : Right* view window to make it active.

13. On the Design Bar, click the **Ref Plane** tool.

- Click two points horizontally above the Ref. Level (floor) plane.

- Edit the temporary dimension to **2'-6"** **[750]** (see Figure 10.36).

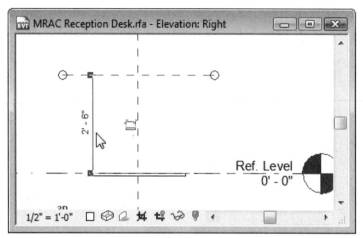

**Figure 10.36** *Add a Reference Plane in elevation and set it to a appropriate height*

- On the Design Bar, click the **Modify** tool or press the ESC key twice.

This will give you a Reference Plane that is 2'-6" [750] off of the floor. This is the height of the top of our work surface. This is the reason we used a negative Depth for the work surface extrusion above. It will make more sense to the users of this Family should height adjustments be required, that they are adjusting the top of the work surface. We now need to name this Reference Plane in order to make it eligible to be the Work Plane for the extrusion.

14. Select the new Ref Plane and on the Options Bar, click the Properties icon.

- For the "Name" parameter, type **Worksurface Height** and then click OK.

You will see the name near the end of the Reference Plane while it is selected.

15. Select the work surface Extrusion element.

- On the Options Bar, click the "Edit Work Plane" button.

- In the "Work Plane" dialog, from the "Name" list, choose **Reference Plane : Worksurface Height** and then click OK (see Figure 10.37).

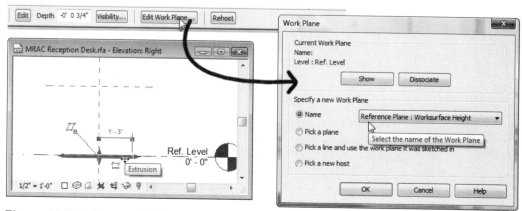

**Figure 10.37**  *Change the Work Plane of the work surface Extrusion to the new Reference Plane*

Notice the work surface extrusion will shift up to the new height in the elevation and 3D Views. If you move the Reference Plane, the work surface will move with it. Try it out, but be sure to return it to the 2'-6" [750] height when done experimenting.

16. Zoom all Viewports to Fit.

- Save the Family.

## CREATE A SOLID BLEND

Now that we have our desktop surface and it is sitting at the desired height, let's add some legs. For the main part of the leg, we will use a Solid Blend to give them a bit of a taper. We will square them off at the top later using a simple extrusion.

Make the in the *plan* view window active.

17. Using Zoom In Region (**View > Zoom** menu or right-click) zoom in on the top left corner of the work surface.

18. On the Design Bar, click the **Solid Form** tool and then from the flyout that appears, choose **Solid Blend**.

- On the Options Bar, type **1'-10" [550]** in the "Depth" field and then click the Rectangle icon.

- Sketch a simple square near the corner of the desk.

- Using the temporary dimensions and the left side of Figure 10.38 as a guide, make the shape 2" [50] square set 2" [50] away from the corner as shown.

**Tip:** If you have trouble with the temporary dimensions, add some permanent dimensions. They will become editable when the sketch lines are selected. You can delete them after you complete the shape.

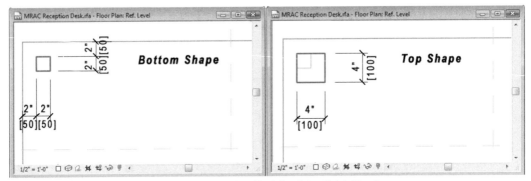

**Figure 10.38** *Sketch the Base (Left) and Top (Right) of the Solid Blend*

A Solid Blend is basically an extrusion that transforms from one shape to another along the length of the extrusion. In this case, we will use two squares of varying sizes that share a common offset from the desktop corner. This will give a tapered shape to the final leg form. When you build a Blend, you first sketch the shape of the bottom, then the shape of the top.

19. On the Design Bar, click the **Edit Top** tool.

- Using the Rectangle option again, sketch a **4"** [**100**] square with the top left corner aligned to the square at the base (see the right side of Figure 10.38).

 **Tip:** Select the sketch before editing the top, copy it to the clipboard. Edit the top, paste aligned to get a copy of the sketch, and then select an edge and use the temporary dimensions to edit the size of the square.

20. On the Design Bar, click the Finish Sketch button (see Figure 10.39).

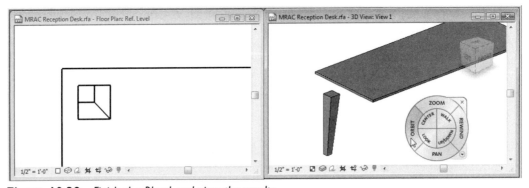

**Figure 10.39** *Finish the Blend and view the results*

There is a noticeable gap between the top of the leg and the work surface (orbit the 3D view as necessary to see this). We will fill this in with an extrusion.

21. Click on the title bar of the *3D View : View 1* window to make it active.

22. On the Design Bar, click the **Solid Form** tool and then from the flyout that appears, choose **Solid Extrusion**.

- On the Tools toolbar, click the Work Plane icon (see Figure 10.40).

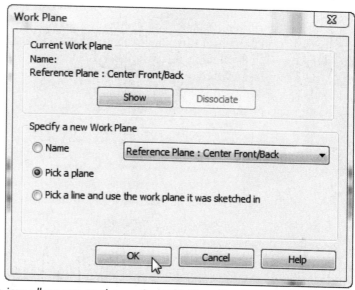

**Figure 10.40** *The Work Plane icon allows you to change the Work Plane for the current Solid form*

While it is most common to use Reference Planes and Levels as Work Planes, you can also use the surfaces in the geometry of your model as a Work Plane. In this case, we will use the top of the Blend.

- In the "Work Plane" dialog, choose the "Pick a Plane" radio button and then click OK.

- In the 3D view, click on the top of the Solid Blend leg (see the left side of Figure 10.41).

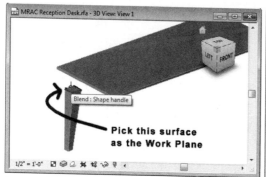
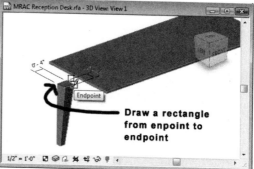

**Figure 10.41** *Pick a Work Plane from the model in the 3D view*

- On the Design Bar, click the **Lines** tool.

- On the Options Bar, click the Rectangle icon and then sketch a rectangle on the top of the leg (see the right side of Figure 10.41).

After sketching this shape, a lock icon will appear on each edge. You can close these locks to constrain the shape of the Solid Extrusion to the top shape of the Blend. This can be handy if you anticipate making edits to the Blend. This would keep the extrusion at the top of the leg coordinated with these changes. However, in some cases, such a constraint can have adverse affects. This could occur for example if you changed the shape of the Blend top to something other than rectangular. In this case, Revit might have trouble maintaining the constraints. If this were to occur, a warning dialog would appear at the time of edit. For the purposes of this tutorial, the decision is not critical. Regardless of your choice for this exercise, do keep this option in mind for future reference in your own Families.

- On the Design Bar, click the Extrusion Properties button.
- In the "Element Properties" dialog, set the "Extrusion End" to **4"** **[100]**.

 **Note:** There are two ways to set the height of the Extrusion, we can use the Depth field as we did earlier, or you can set the Extrusion Start and the Extrusion End in the Element Properties dialog. The end result is the same.

- On the Design Bar, click the Finish Sketch button.

The extrusion will appear at the top of the leg but will be too short to reach the work surface. In the next step, we will stretch the height of this element and constrain it to the work surface.

23. Click on the title bar of the *Elevation : Front* window to make it active.

- Select the Extrusion and using the Shape Handle at the top, stretch it up to the bottom edge of the work surface.
- Click the lock icon that appears to apply the constraint (see Figure 10.42).

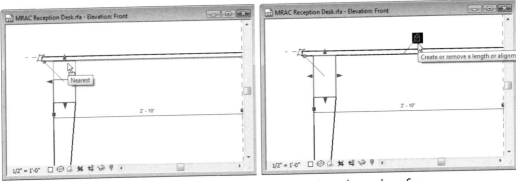

**Figure 10.42**   *Stretch the top edge of the leg and constrain it to the work surface*

## MIRROR THE LEG

Now that we have one completed leg, we can mirror it to create the one on the other side.

24. Click on the title bar of the *Elevation : Left* window to make it active.

25. Select both pieces of the leg (the Blend and the Extrusion).

   • Click the Mirror icon on the toolbar.

   • Click the vertical Reference Plane (Center Front/Back) as the Axis of Reflection (see Figure 10.43).

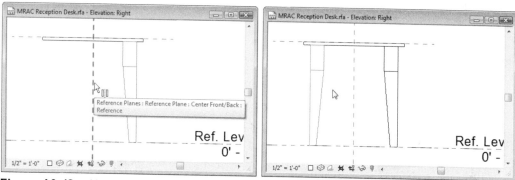

**Figure 10.43**   *Mirror the leg to create one on the other side*

## USING SWEPT BLEND TO MAKE A FANCIER LEG

We could simply mirror both of the legs again in the *Front* elevation. Before we mirror them, let's vary the Blend of one of the legs to make a slightly "fancier" design.

26. Click on the title bar of the *Plan* view window to make it active and zoom in on the top left corner of the work surface.

27. On the Design Bar, click the **Ref Plane** tool.

   • Draw a 45° Reference Plane through the middle of the leg as shown in Figure 10.44.

   • Select the new Reference Plane, edit its Properties and name it: **Leg Path**.

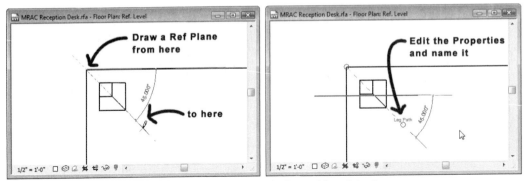

**Figure 10.44** *Draw a 45° Reference Plane through the middle of the leg*

28. Click on the title bar of the *3D View: View I* window to make it active.

- Move your mouse over the ViewCube and click the vertical corner between the left and front faces (see Figure 10.45).

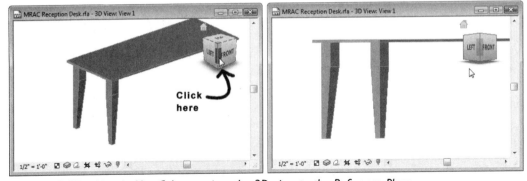

**Figure 10.45** *Use the ViewCube to orient the 3D view to the Reference Plane*

We now have the 3D view looking directly at the plane that the Reference Plane was drawn in above. Since we named the Reference Plane, we can make it a Work Plane to assist us in building the next object.

29. On the Tools toolbar, click the Work Plane icon.

- In the "Work Plane" dialog, choose the "Leg Path" from the Name list and then click OK.

30. On the Design Bar, click the **Solid Form** tool and then from the flyout that appears, choose **Solid Swept Blend**.

A Swept Blend is nearly identical to a Blend. It also has two sketched shapes, one at the top and the other at the bottom. However, instead of connecting these two shapes along a perpendicular path whose height we specify, we instead draw the path using either a straight line of any angle or a curve. In this case, we'll draw a gentle curve. The only limitation of the Swept Blend is that the Path can only be a single segment. To work around this, we'll use a spline curve in this example.

- On the Design Bar, click the **Sketch 2D** Path tool.

- On the Options Bar, click the drop down icon next to the arc tool and choose the Spline icon.

- Click the start point at the bottom of the existing leg.

- Click the next point straight up along the leg slightly above the first.

- Click the third point to left of the existing leg about half way up.

- Click the last two points on the existing leg near the top of the existing blend (see the left side of Figure 10.46).

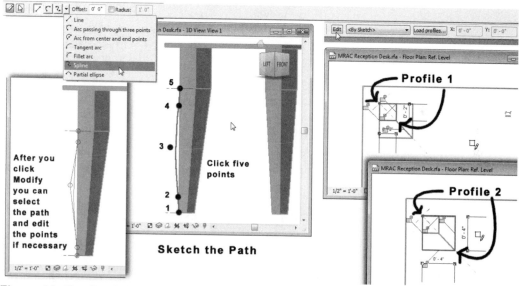

**Figure 10.46** *Sketch the path and profiles of the Swept Blend*

- On the Design Bar, click the Finish Path button.

31. Click on the title bar of the *Plan* view window to make it active.

- On the Design Bar, click the **Profile 1** tool.

- On the Options Bar, click the Edit button.

- With the rectangle tool, trace the small square and then click Finish Profile.

- Click the **Profile 2** tool, click the Edit button, trace the large square and then click Finish Profile (see the right side of Figure 10.46).

32. On the Design Bar, click the Finish Swept Blend button.

- Delete the original Blend (see Figure 10.47).

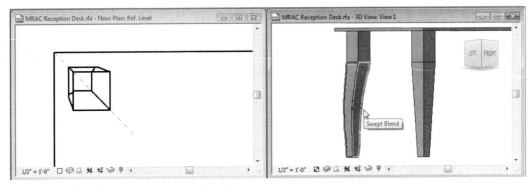

**Figure 10.47** *Complete the Swept Blend*

33. Click on the title bar of the *Elevation: Front* window

- Select both legs and mirror about the (Center Left/Right) Reference Plane.

- Save the Family file.

## CREATE A PENCIL DRAWER

Using another extrusion, we can add some geometry to represent the structural support of the work surface and a pencil drawer.

34. Click on the title bar of the *Floor Plan : Ref. Level* window to make it active.

Zoom as required to see the whole desktop.

35. On the Design Bar, choose **Solid Form > Solid Extrusion**.

- Using the Work Plane icon, set the Work Plane to **Reference Plane : Worksurface Height**.

- On the Options Bar, set the "Depth" to **-6 3/4" [-169]** and click the Rectangle icon.

- Type **-3" [-75]** in the "Offset" field.

- Trace the desktop from top left corner to lower right corner (see Figure 10.48).

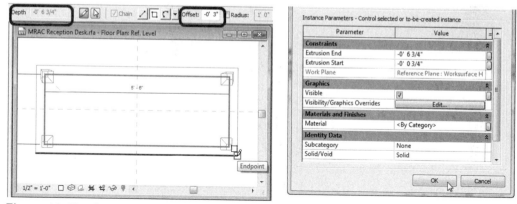

**Figure 10.48** *Sketch a rectangle offset from the desktop*

- On the Design Bar, click Element Properties, type **-3/4" [-19]** for the "Extrusion Start" and then click OK.

- On the Design Bar, click the Finish Sketch button.

36. Working in the *Elevation : Front* window create a 1/2" [12] deep extrusion for a pencil drawer using the face of the extrusion just drawn as a Work Plane (see Figure 10.49). The exact size and position on the drawer are not important.

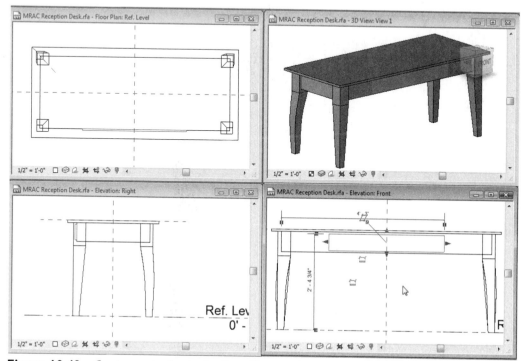

**Figure 10.49** *Create a extrusion in the Front Elevation for a Pencil Drawer*

## CREATE A SOLID REVOLVE

To add a drawer pull to the pencil drawer, we will use a Solid Revolve. A Solid Revolve is a form in which you sketch the profile of the form and then spin the profile around an axis. To create a simple drawer pull, you draw half of the cross section of the pull, and then place the revolution axis at the center of the pull.

37. Click on the title bar of the *Floor Plan : Ref. Level* window to make it active.

   - Zoom in on the Pencil Drawer in the plan view.

38. On the Design Bar, click the **Solid Form** tool and then from the flyout that appears, choose **Solid Revolve**.

   - On the Design Bar, verify that the Lines tool is active and on the Options Bar verify that "Chain" Is selected.

   - Starting at the midpoint of the pencil drawer, sketch the form shown in Figure 10.50. The depth should be about 3/4" [19] but the exact dimensions and shape are not critical.

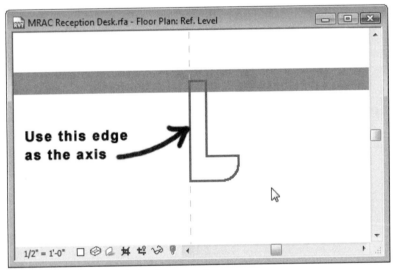

**Figure 10.50** *Sketch the shape of the drawer pull*

39. On the Design Bar, click the **Axis** tool.

   - Trace the vertical edge of the shape drawn above.

   - On the Design Bar, click the Finish Sketch button.

If you study the result in all Views, you will notice that the drawer pull is at the height of the work surface. This is because Revit used the "Worksurface Height" Reference Plane as the Work Plane for this element. As a result, you cannot move the drawer pull to the correct height—try it in the Front view to see for your self. If we had lots of hardware elements for this desk, we could create a new Reference Plane and associate the elements to it instead. In this case, we will simple disassociate the Work Plane from the Revolve which will allow us to move it freely.

40. Click on the title bar of the *Elevation : Front* window to make it active.

- Select the drawer pull.

   A small Work Plane icon appears attached to the element.

- Click the Work Plane icon to disassociate the Work Plane.

- Move the drawer pull to center it on the height of the drawer (see Figure 10.51).

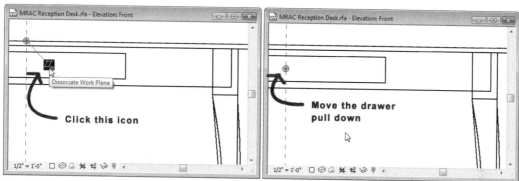

**Figure 10.51** *Disassociate the Work Plane and move the drawer pull*

## CREATE A SOLID SWEEP

Now let's create a privacy screen using the Solid Sweep command.

41. Make the *Floor Plan: Ref. Level* window active.

- Zoom out enough to see the whole desk.

42. On the Design Bar, click the **Solid Form** tool and then from the flyout that appears, choose **Solid Sweep**.

- On the Design Bar, click the **Sketch 2D Path** tool.

- On the Options Bar, click the "Pick Lines" icon.

- In the "Offset" field, type **2"** [**50**].

As with the "Offset" option in other commands, as you hover over a line in the model, a green dashed line will appear temporarily indicating the side to which the sketch line will offset. As you create sketch lines here, be sure to offset to the inside of the desk surface.

43. Offset the three outside edges of the desk (not the one with the pencil drawer) to create a 'U' shaped sketch (see Figure 10.52).

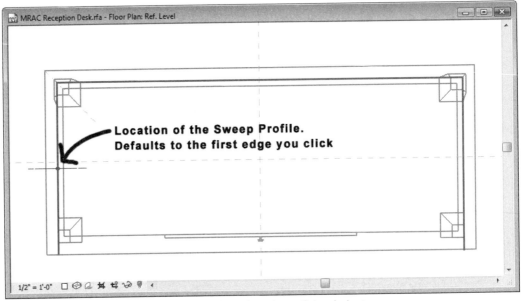

**Figure 10.52** *Create Sketch lines by picking the edges of the desktop*

- On the Design Bar, click the Finish Path button.

These three lines will be the "path" of the Sweep. We have seen Sweeps in other chapters. A Sweep "pushes" a profile along a path. It is similar to an extrusion except that the path does not have to be a simple straight line as it does in an extrusion.

44. On the Design Bar, click the **Profile** tool.

- On the Options Bar, click the Edit button.

The profile travels perpendicular to and along the path. This means that since we sketched the path in plan, we must sketch the profile in elevation. When you click the Edit button, Revit will prompt us to select an appropriate view in which to sketch the profile. Please note that it is also possible to load an existing Profile when creating a Sweep instead of sketching it. This allows you to create "Profile" Families and store them in your library so that you can quickly create Sweeps from your most common profiles. In this example, we will sketch the profile as a custom shape just for this desk.

- In the "Go To View" dialog, choose *Elevation: Front* and then click Open View.

- Zoom the *Front* elevation view as necessary to center the red dot on screen.

The red dot indicates the start point of the profile. It appears automatically on the first segment of the Path. If you are not satisfied with the default location of the profile, you can move it, but be sure to move it before you start sketching the profile.

45. Using the Lines tool and the "Chain" option, create a profile similar to the one shown in Figure 10.53.

**Note:** The exact shape of the profile is not critical. Just make sure to close the shape.

 **Tip:** Remember that you can zoom in to make sketch lines automatically snap to a smaller increment. The Profile in the figure has a thickness of 1/2" [12].

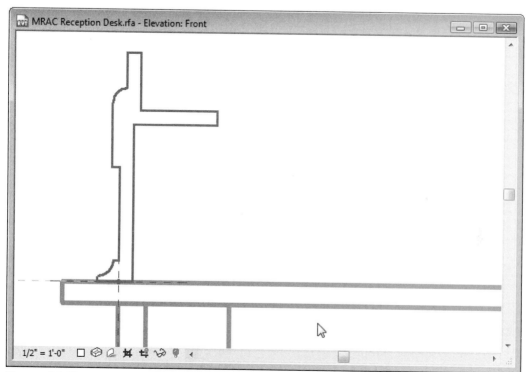

**Figure 10.53**   *Sketch the Profile of the Sweep*

- On the Design Bar, click the Finish Profile button.
- On the Design Bar, click the Finish Sweep button.
46. Zoom All Viewports to Fit (see Figure 10.54).

 **Tip:** Try setting the Viewports to hidden line display to see the final product more clearly in plan and elevation. You can also use the Join Geometry tool to merge solid forms together.

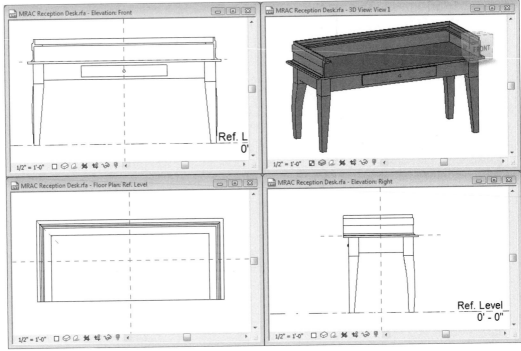

**Figure 10.54** *Completed reception desk with privacy screen*

> 47. Save the Family file.

## LOAD THE CUSTOM FAMILY INTO THE PROJECT

At this point, we have completed the geometry of our Custom Family and are ready to load it into our commercial project. The process has been covered before. Let's review it now.

### LOAD THE COMMERCIAL PROJECT

We closed the project above to make it easier to tile the windows in the Family Editor. We now need to reopen the project.

> 1. Open the Commercial project for your choice of units:
>
>    Open *10 Commercial.rvt* if you wish to work in Imperial units. Open *10 Commercial Metric.rvt* if you wish to work in Metric units
>
> 2. On the Project Browser, double-click to open the *Level 3* floor plan view.
>
> • Zoom in on the reception space (in the middle of the plan).

### LOAD A FAMILY

The Family file is still open. We can switch over to it and then load the Family back into the project. We could also use the "Load Family" command on the File menu, but since the Family file is still open, it is easier to load from there.

3. Hold down the CTRL key and then press TAB.

This will cycle through open windows.

Repeat the CTRL + TAB if necessary until the Reception Desk Family file comes into view.

4. On the Design Bar, click the Load into Projects button.

The screen will switch back to the commercial project. If you look at the Furniture node beneath Families on the Project Browser, you will see that the Reception Desk Family is now loaded.

5. Add an Instance of the Family to the reception space (see Figure 10.55).

 **Tip:** Remember to use the space bar as you place it to rotate it.

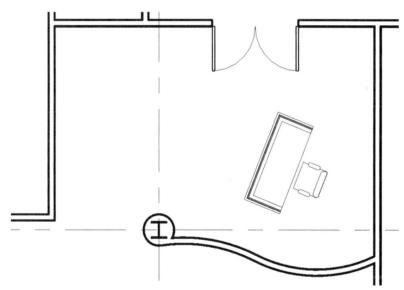

**Figure 10.55** *Load the Family and then add an Instance to the commercial project*

6. Use the **Component** tool on the Design Bar to load and then place a task chair with the reception desk.

The reception desk tutorial was designed to introduce you to the basics of the modeling forms available in the Family Editor. All five forms were explored, and even though we did not create any Void forms for the reception desk, we saw an example above in the "Adding a Void Form" topic. Voids can be made in any of the five forms just like Solids. The reception desk Family is a simple Family that cannot be edited in the project at all. To make even simple edits like changing the width, height or depth, we would need to return to the Family Editor. However, if we chose to, we could make the reception desk a "parametric" Family. Doing so would allow us to expose certain dimensions and other parameters to the users of Family directly in the project editor. For example, if our reception desk could be ordered in multiple sizes, we could make the dimensions of the desk parametric in the Family Editor and then adjust our geometry to be controlled by these parameters. In this way, a user of the Family could manipulate the exposed parameters without needing to open the Family Editor, nor understand the complexities of creating or editing Families. Our next topic is devoted to the subject of parametric Families.

## BUILDING PARAMETRIC FAMILIES

Throughout the course of this chapter you have been exposed to a variety of Family editing and creation scenarios. As you have seen, Families can be very simple or very complex. When you are first learning their scope and potential, it is helpful to start with simple examples and work your way up to more complex ones. This is exactly the approach this chapter has been following. As such, we have yet to cover one of the most powerful and useful aspects of Families. That is their ability to be "parametric." Like much terminology that is associated with Building Information Modeling, the term *parametric* is often misused or misunderstood. Since it is nearly impossible to avoid hearing reference to the term in nearly any document, seminar or training session on Revit Architecture, it will be helpful to take a moment to properly define the term.

Parametric is the adjective form of the noun "Parameter." Browsing "Merriam-Webster Online Dictionary" we find the following definition for the term:

> **Parameter**—any of a set of physical properties whose values determine the characteristics or behavior of something.

This particular definition of "parameter" was selected from a few available variations because it is the most appropriate in the context of its use in describing the behavior of elements in Revit Architecture. When we describe something in software such as Revit as being "parametric" we are therefore simply saying that the thing in question is characterized by its associated set of "physical properties" each of which hold "values determining characteristics of the element's behavior." In the specific case of Revit, each element has one to several available parameters. We input values into these parameters to determine the specific characteristics of the element (Wall, Door, Roof, Family, etc) which we are editing. Not specifically mentioned in the Webster's definition but implied by the use of the term *parametric* in software, is the ability to modify parameters at any time. Therefore, the ability for a particular parameter's value to determine the characteristics of behavior is not a limited to the point of creation, but rather is a "living" parameter with ongoing influence over the element. This behavior is what makes the notion of "parametric" so significant in software like Revit Architecture. This dynamic interaction across the whole system, (not just with Model and Annotation Elements, but Views and Schedules as well), is often referred to in Revit as its "Parametric Change Engine."

Having outlined our definition of the term "parametric;" a Parametric Family is simply a Family that has editable parameters. Typically, such a Family will also have Types—though it is not required. Therefore, as we will see, Family Parameters can be Type- or Instance-based. We could certainly continue in our Reception Desk Family file and add parameters to the desk, but for simplicity, we will explore these topics, in a new Family file.

## CREATE A NEW FAMILY

As we did above, we will create a new Family file based on one of the provided Revit Family templates. We are going to create a simple binder bin to hang on the wall behind the reception desk. This object will be geometrically simpler than the reception desk, so we can focus on adding and working with parameters.

1. On the Window menu, switch to the Reception Desk Family File and then from the File menu, choose **Close**.

- In the Commercial project, make sure that the *Level 3* floor plan view is open.

- From the Window menu, choose **Close Hidden Windows**.

    The *Level 3* floor plan view and the Recent Files windows should be the only open windows now.

- Minimize the *Level 3* floor plan view.

- From the Recent Files window, click the New link beneath Families (see Figure 10.56).

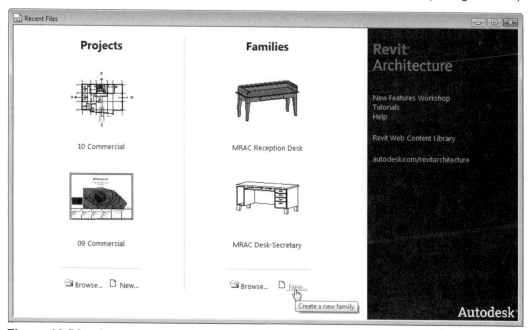

**Figure 10.56** *Create a new Family from the Recent Files window*

Browse to you Templates folder if necessary.

- Select *Specialty Equipment wall based.rft* [*Metric Specialty Equipment wall based.rft*] and then click Open.

**Note:** If your version of Revit Architecture does not contain either of these templates, they have been provided in the *Out of the Box Templates* folder with the files installed from the Mastering Revit Architecture CD ROM.

2. Minimize the Recent Files windows and then from the Window menu, choose **Tile**.

- From the View menu, choose **Zoom > Zoom All To Fit**.

**Tip:** The keyboard shortcut for Window Tile is **WT** and for Zoom All To Fit it is **ZA**.

- From the File menu, choose **Save**.

- In the "Save as" dialog, browse to the *Chapter10* folder, type **MRAC Binder Bin** for the Name and then click Save.

Like the Furniture Family template used above, this Family opens with four view windows and some existing Reference Planes on screen. In addition, a temporary Wall element is included in this template. This Wall is used for reference while working in the Family Editor and will not be included with the binder bin when it is loaded into a project. The Wall is provided since this is a "wall-based" Family. We have chosen the "Specialty Equipment" template in this exercise for its simplicity. Because there are not many Reference Planes, parameters or settings in this template, we can learn to build them ourselves. We could make the argument that the wall-based Casework template would be more appropriate for a binder bin. It would be difficult to refute this argument. In some cases it is simply a judgment call and a clear understanding of what the design/usage intentions for the Family are that will help you make the final decision. In this case, a wall-based Furniture category might have been ideal, but Revit does not provide such a template. Remember: when you build your own Families, study the list of provided templates carefully before making your choice.

**BIM** *Manager Note:* For the educational value of learning to create/use Reference Planes and parameters, we will stick with Specialty Equipment template here. If you open the Casework templates, you will note that these items are already provided. When you build your own Families, choose the template that gives you the most overall benefit. Having many of the required Reference Planes and parameters pre-defined can be a big time-saver. Also, it is important to note that you can start your Family using the Specialty Equipment template, and later choose **Family Category and Parameters** from the Settings menu and choose a different category. In some cases, it will make sense to do this. However, use caution with this approach since changing the category also resets the subcategories including any custom ones you may have added (see "Creating Subcategories" below). Therefore, where possible, it is best to make this decision early in the process of editing your Family.

## CREATE REFERENCE PLANES

We have used Reference Planes in several places throughout this book. Their inclusion in Families serves a vital purpose. If we wish to create a parametric Family, it is common and best practice to associate the parameters with various Reference Planes (or Reference Lines—which are similar but finite in length) and then lock the geometry to these

Reference Planes. In this way, we can vary the dimensions that control the locations of various Reference Planes and thereby manipulate the geometry constrained to them. This makes the Family's geometry parametric! Before we can demonstrate this here, we must add some Reference Planes to our Family.

3. Click on the title bar of the *Elevation : Placement Side* window to make it active.

> **Note:** View names in the various Family templates are deliberately generic. For example, the plan view is often called "Ref. Level" because the Level included in most Family templates is there for reference only. In the Furniture template, we had elevation names like "Left" and "Front." Here we have "Placement Side" and "Back Side" instead of "Front" and "Back." This is typical of a wall-based template and helps you orient yourself relative to how the Family will later insert within projects.

4. On the Design Bar, click the **Ref Plane** tool.

• Create four Reference Planes as shown in Figure 10.57.

Do not be concerned with the precise locations. Simply sketch the Reference Planes in the approximate locations indicated in the figure. We will adjust the dimensions later.

• With the Modify tool select the top Reference Plane, on the Options Bar click the Properties icon, and then in the "Name" field type: **Top**.

• Click OK to complete the change, and then repeat for each of the other three Reference Planes, using the names indicated in the figure.

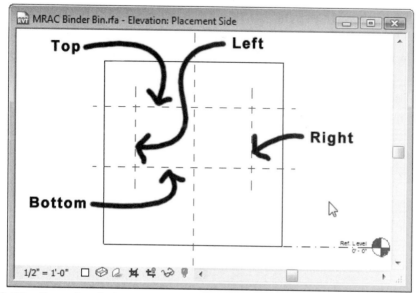

**Figure 10.57** *Create four Reference Planes in elevation and then name them*

5. On the Design Bar, click the **Solid Form > Solid Extrusion** tool.

- On the Options Bar, set the "Depth" to **-1'-2"** [**-350**] and then click the Rectangle icon.
- Snap to the intersection of the Top and Left Reference Planes for the first corner.
- Snap to the intersection of the Bottom and Right Reference Planes for the other corner (see Figure 10.58).

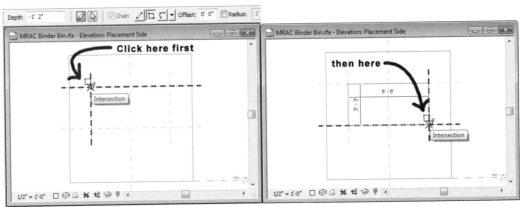

**Figure 10.58**  *Sketch a Rectangle snapping to the Reference Planes*

Four open padlock icons will appear—one on each edge of the rectangle.

- Close each of the lock icons to constrain the rectangle shape to the Reference Planes.
- On the Design Bar, click the Finish Sketch button.

Take note of the extrusion in each view window.

6. In the 3D view window, change the display to Shaded with Edges.

7. In the *Elevation: Placement Side* view window, move one of the Reference Planes (see Figure 10.59).

The exact amount of the move is not important. What is important is that the shape of the Solid Extrusion will adjust when you move the Reference Planes. This is because we constrained the edges of the sketch to the Reference Planes.

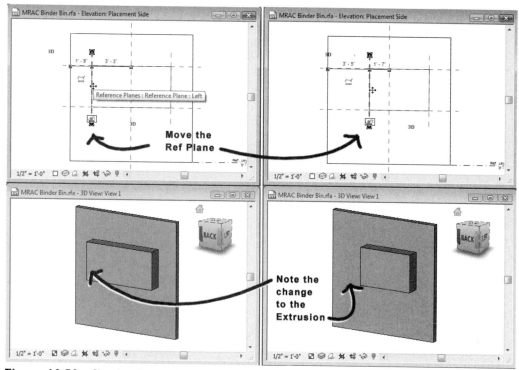

**Figure 10.59** *Sketch a Rectangle snapping to the Reference Planes*

- Undo the Move to return the Reference Plane to its previous position.
- Save the Family file.

## CREATE DIMENSION PARAMETERS

The first step in creating our parametric Family is complete—we created some geometry that is constrained to our Reference Planes. The next task is to create dimension label elements and associate them to parameters.

8. Make the *Elevation: Placement Side* window active.

- Maximize the view window (you can do this with the icon in the top right corner of the window or just double-click the title bar).

- Zoom the window to fit (type **ZF** on the keyboard or choose the command from the **View > Zoom** menu).

9. On the Design Bar, click the **Dimension** tool.

- Highlight and then click the Ref. Level line.

- Highlight and then click the Bottom Reference Plane (see Figure 10.60).

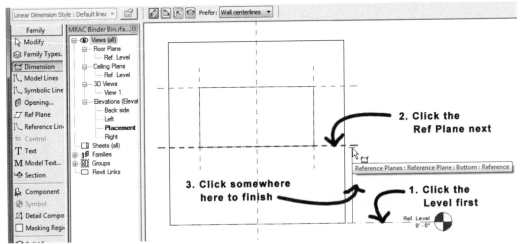

**Figure 10.60** *Place the first Dimension element*

- Click in a blank space between the two references to place the Dimension.

 **Note:** It is important to make sure you do not highlight any objects on your last click, or the dimension will assume you want to add that item to the dimension chain. Click in empty space to complete a dimension.

10. Repeat the process to add another Dimension between the Top and the Bottom Reference Planes (see Figure 10.61).

We want two separate dimension elements here, not one continuous one.

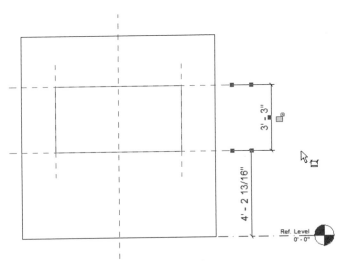

**Figure 10.61** *Place another vertical Dimension between the Top and Bottom Reference Planes*

Create a new horizontal Dimension.

11. Highlight and then click the Left Reference Plane.

- Highlight and then click the Center Reference Plane.

- Highlight and then click the Right Reference Plane.

    This will give you three witness lines and two dimensions in a continuous string.

- Click in a blank space above the binder bin to place the Dimension.

- Click the "Toggle Equality" icon (see Figure 10.62).

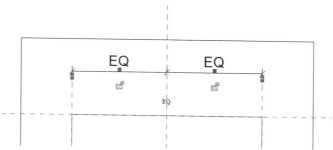

**Figure 10.62**  *Toggle on Dimension Equality for the horizontal string*

After you click this toggle, you should see the Center Reference Plane shift and the dimension values will be replaced with the letters "EQ." This applies a constraint that will keep the three Reference Planes equally spaced. If you prefer not to move the center Reference Plane, you can first manually move both the left and right Reference Planes so that they are the same distance to the center. Then apply the equality dimension.

12. Add one final Dimension horizontally between the Left and Right Reference Planes only—do *not* include the Center one in this string (see Figure 10.63).

**Caution:** Be very deliberate about highlighting the Reference Planes for the dimension witness lines NOT the Extrusion edges or the Wall. We want the parameters that we will apply to the dimensions to control the Reference Planes. The Reference Planes will in turn manipulate the geometry.

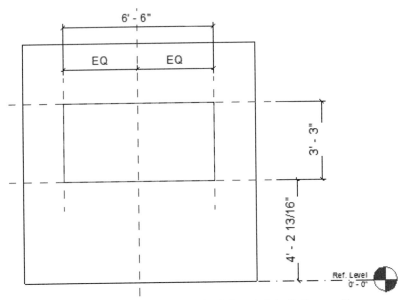

**Figure 10.63** *Add the final dimension between the left and right Reference Planes*

- On the Design Bar, click the **Modify** tool or press the ESC key twice.

Now that we have created several Dimensions, we will create parameters that will ultimately control them. Recall that you can manipulate model geometry in Revit by editing the values of permanent dimensions and temporary dimensions. When you apply parameters to Dimensions in a Family file, you are doing the same thing. Let's take a look.

13. Select the overall horizontal Dimension element (the one between left and right).

Take note of the "Label" item on the Options Bar. We use this drop-down list to assign parameters to the selected Dimension. You can also right-click and choose **Edit Label** if you prefer.

14. On the Options Bar, click the drop-down list next to "Label" (currently <none>) and choose **<Add parameter>**.

The "Parameter Properties" dialog will appear; verify that the "Family Parameter" radio button is selected at the top.

- In the "Parameter Data" area, type **Width** in the "Name" field.

- From the "Group parameter under" drop-down list, choose **Dimensions**.

- Choose the "Type" radio button (see the left side of Figure 10.64).

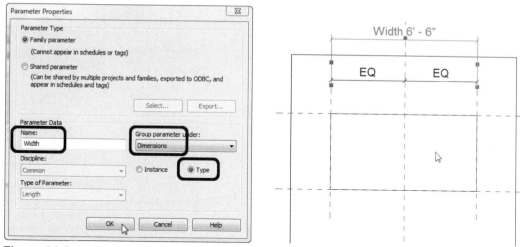

**Figure 10.64**    *Create a custom "Width" parameter for the Family's types*

These settings create a parameter named "Width" that will appear beneath the "Dimensions" grouping of the "Type Properties" dialog box. This means that if the value is edited, it will affect all elements in the model that use that type. If we chose "Instance" instead, the parameter would apply to the individual elements in the project and could be unique for each one—we will create an instance parameter below.

- Click OK to complete the parameter.

Notice that the Label "Width" now appears in front of the Dimension text. This tells us that this Dimension is controlled by this parameter (see the right side of Figure 10.64).

15. Repeat the process on the vertical Dimension between the Top and Bottom Reference Planes.

- Name the parameter **Height**, group it under Dimensions and make it a Type parameter as well.

16. Select the vertical Dimension between the Ref. Level and the Bottom Reference Plane and repeat the process again.

- Name the parameter **Mounting Height**, group it under Dimensions and make it an Instance parameter this time (see Figure 10.65).

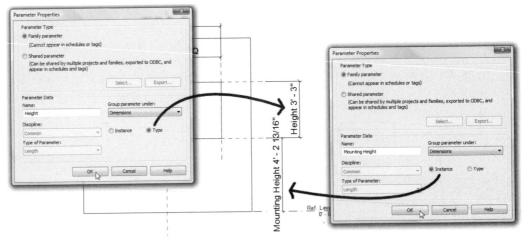

**Figure 10.65** *Create another Type parameter and one Instance parameter*

   17. Save the Family.

## ADD FAMILY TYPES

All that remains now is for us to test our Family parameters. In Revit, this is typically referred to as "flexing" the Family. When you flex the values of the parameters, you can see if the Reference Planes and geometry are moving as expected. We do this with the Family Types command. This command allows us to not only test to see if our parameters are behaving properly, but to also create some Types that will load into projects when we load the Family. Remember, just like the out-of-the-box Families that we have used so far, users will always be able to add new Types to our custom Family later. But it is still a good idea to stock it with a few preferred sizes or types ahead of time.

   18. On the Design Bar, click the Family Types button.

Notice that the three parameters are listed at the top of the dialog beneath a "Dimensions" grouping exactly as we specified. To flex the parameters, move the dialog to the side so you can see the view window in the background. Type numbers into the "Value" field next to each parameter and then click the Apply button. The Dimensions in the view window will adjust to the new sizes.

- Change the Width to **5'-0"** [**1500**].

- Change the Mounting Height to **4'-6"** [**1350**].

- Change the Height to **2'-0"** [**600**] (see Figure 10.66).

**Note:** In this exercise, we have waited until after adding several parameters to "flex" them. It is very good practice, however, to perform this procedure immediately following the creation of each parameter. In this way, you can catch mistakes or issues before they become potentially compounded by other parameters and/or geometry. Some Families can get very complex. Making sure that you test it thoroughly as you build will help you avoid hours of potential frustration.

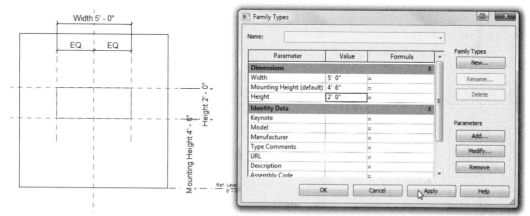

**Figure 10.66** *View the custom parameters in the Family Types dialog and edit the values to flex the model*

**Note:** If something did not work correctly, close the dialog and go back through the previous steps to find your error and then try again.

On the right side of the "Family Types" dialog, you can add Types and also add or edit Parameters. Let's add some Types.

19. On the right side of the dialog, click the "New" button beneath "Family Types."

   • In the "Name" dialog, type **60 wide 24 high [1500 wide 600 high]** and then click OK.

The new name will appear in a list at the top.

20. Click New again to create **60 wide 18 high [1500 wide 450 high]**.

   • Set the Height to **1'-6" [450]**, leave the other dimensions the same.

   • Create additional Types if you wish. Click Apply after each one to test them.

   • Click OK when finished.

## ADD HARDWARE

We are almost ready to load our Family into our project. Before we do, our bin could use some hardware.

21. Tile the view windows again and create the hardware.

22. In the *Elevation: Placement Side* view, create a Reference Plane 2" [50] above the Bottom Reference Plane.

- Edit the properties of the new Reference Plane and name it **Hardware Height**.

- Add a Dimension element between the Bottom Reference Plane and the new one.

- Click to close the lock icon on the dimension (see Figure 10.67).

When you toggle the equality or close the padlock icon, you are applying a constraint to your Family rather than a parameter. Remember it this way: a parameter is a value that a user can manipulate in the project without editing the Family. A constraint is "built-in" to the Family can only be changed by editing the Family.

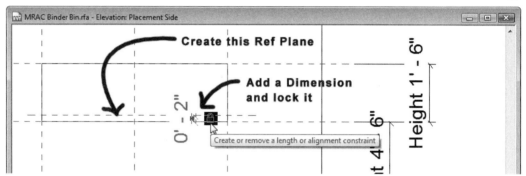

**Figure 10.67**   *Add a new Reference Plane, dimension it, and lock the dimension to constrain it*

The specifics of creating the hardware will be left to you as a practice exercise. You can make a knob using a Solid Revolve like the one we created for the reception desk above, or you can make a wire pull from a Solid Sweep. Just remember to constrain the position of the hardware to the appropriate Reference Plane(s). For example, if you create a Sweep, start in the plan view, edit the Work Plane, and choose the Hardware Height Reference Plane from the list. Sketch the path centered on the bin in plan, then switch to the elevation and create the profile. A simple square profile is sufficient. When you have finished the Solid, open the Family Types dialog and choose each of your predefined Types in succession from the list to test it out. The wire pull should remain 2" [50] from the bottom edge and centered on the bin as you flex the model (see Figure 10.68).

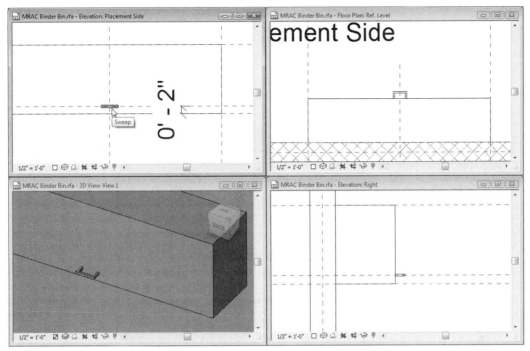

**Figure 10.68** *Add a Solid Sweep wire pull for a handle*

23. Zoom all Views to Fit when finished.

## CREATING SUBCATEGORIES

Revit classifies elements in a fixed hierarchy starting with Categories, then Families, then Types, and finally Instances. We have been working with Families, Types, and Instances throughout this chapter. Categories however are the broadest classification of elements in Revit. They are often sub-divided into Model, Annotation, and Imported object groupings for convenience. Each Family belongs to a Category. The list of Categories available is predetermined by Revit—we cannot rename, add, or delete Categories. However, we can add and delete "Subcategories." A subcategory gives us more detailed control over the visibility of Model and Annotation elements by allowing parts of an element to appear differently from the whole. For example, the "Doors" Category includes subcategories for "Elevation Swing," "Frame/Mullion," and "Panel" (among others). Each of these subcategories gives us visibility control over these common sub-components globally for all Door elements. When you build or modify a Family, you can assign the various components within the Family to any of the available subcategories. You can see a list of available categories and subcategories in the "Object Styles" dialog.

24. From the Settings menu, choose **Object Styles** (see Figure 10.69).

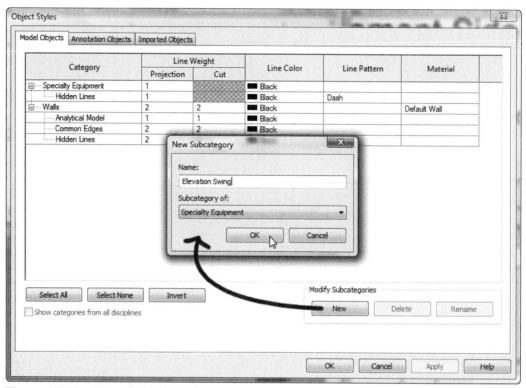

**Figure 10.69**  *Object Styles shows the available categories and subcategories*

Since we are currently in a Family file, the list of available categories includes only those relating to the Family we are editing—in this case specialty equipment. (If you wish to see a more complete list, you will need to repeat this command later in a project file.) Specialty equipment contains one subcategory called, "Hidden Lines." You should consider adding subcategories for large groupings of similar items. For example, if you were to edit a Door Family, you would notice that there is a subcategory for "Plan Swing" and another for "Elevation Swing." Since most drawings need to represent the swing of the doors regardless of the kind of door, the swing subcategories provide a useful way to control all door swings universally throughout the project without needing to edit several Families individually. If you study most of the subcategories provided with the software, you would find they share this kind "global edit" strategy. While most of the subcategories you will likely require are already included in most Family templates, there will be situations when you decide to create new ones. In the case of our binder bin Family, it might be useful to have a subcategory to indicate the door swing much like the Door Families do.

25. At the bottom of the dialog, in the "Modify Subcategories" grouping, click the "New" button.

- Name this new subcategory **Elevation Swing**.

- Change the Line Pattern to **Dash** and then click OK.

To use the new subcategory, you edit an element's properties and choose the subcategory you wish. If you do not assign a subcategory to an element, it simply associates with the parent category. In this case, everything we have added to the current Family belongs to the Specialty Equipment category and no subcategory.

## ADD SYMBOLIC LINES

Symbolic Lines are special Revit elements that can be added to Families to embellish the 2D views. In projects, we have model elements and annotation elements. Model elements appear in all views, and annotation (including detail) elements appear only in the view to which they were added. In Families, model elements behave exactly as they do in projects. However, the views that we see in the Family Editor are not transferred over to the project when we load the Family. Therefore, drafting lines would not be transferred either. Instead, we use Symbolic Lines in the Family Editor. Symbolic Lines will appear in all views parallel to the view in which they are created. So if you add a Symbolic Line to a front elevation, it will appear in all sections or elevations that face the same direction. The use of Symbolic Lines allows us to add detailing and embellishment that is view-specific like Detail Lines, but also somewhat parametric like model geometry. In this example, we will use Symbolic Lines to indicate the swing direction of the binder bin when it opens in the front facing elevation.

26. Click on the title bar of the *3D View : View 1* window to make it active.

   - On the Tools toolbar, click the Work Plane icon (shown in Figure 10.40 above).

   - In the "Work Plane" dialog, choose the "Pick a Plane" radio button and then click OK.

   - In the 3D view, click on the front face of the bin.

   - On the Tools toolbar, click the Work Plane Visibility icon (to the right of the Work Plane icon).

This allows you to see the plane that was selected.

27. Make the *Elevation: Placement Side* window active.

   - Zoom in on the bin.

28. On the Design Bar, click on the **Symbolic Lines** tool.

   - From the Type Selector, choose **Elevation Swing**.

Notice how the new subcategory added in the previous topic now appears here. If you forget to choose it while you are adding the element, you can edit the properties later as noted above.

   - On the Options Bar, select "Chain" and then sketch two lines as indicated in (see Figure 10.70).

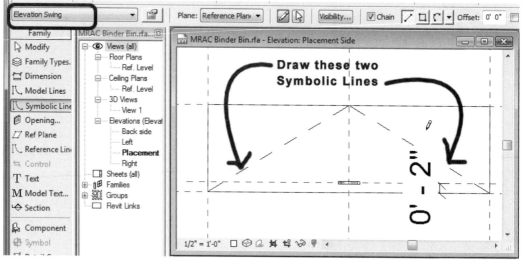

**Figure 10.70** *Add Symbolic Lines to the front elevation using the new Elevation Swing subcategory*

- On the Design Bar, click the **Modify** tool or press the ESC key twice.

If you recall the default Mounting Heights that we assigned above were 4'-6" [1350] for each Type. While we did make these Instance Parameters (which means that each bin we add to our model can have its own Mounting Height), initially when we place them, they will fail to show in most plan Views. This is because the default cut height in plan is 4'-0" [1200] which is below the lowest point of our bin's geometry. To deal with this situation, we can add additional Symbolic Lines in the plan view that show an outline of our bin in a dashed line style to indicate that it is mounted above the cut plane. We will also need to add an additional element to ensure that Revit recognizes our Family when it cuts the floor plan.

29. Make the *Floor Plan: Ref. Level* window active.

- Check the Work Plane, choose **Level: Ref. Level,** and then click OK.

30. Open Object Styles again and create a new subcategory of Specialty Equipment named **Overhead Items**. Make its Line Pattern **Overhead**.

31. On the Design Bar, click on the **Symbolic Lines** tool.

- From the Type Selector, choose **Overhead Items**.

- On the Options Bar, click the Rectangle icon and then trace the bin geometry.

- Close all four padlock icons (see the left side of Figure 10.71).

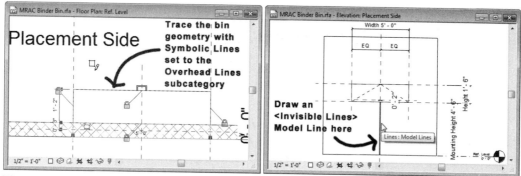

**Figure 10.71** *Trace the plan with Symbolic Lines using the new Overhead Lines subcategory and add an <Invisible Line> in elevation*

- On the Design Bar, click the **Modify** tool or press the ESC key twice.

## MAKING FAMILIES ABOVE THE CUT PLANE DISPLAY IN PLAN

We now have the Symbolic Lines that we need, but as noted above, we need to instruct Revit to show them. Since none of the geometry of the bins is being cut by the cut plane, Revit would simply not show any geometry in plan Views. To trigger it to display our Symbolic Lines, we can draw an invisible line element passing through the Cut Plane. In this way, Revit will recognize that we wish display something in plan. It is an effective work-around, conveying our intent to the system.

32. Click on the title bar of the *Elevation : Placement Side* window to make it active.

- On the Design Bar, click the **Model Lines** tool.

- From the Type Selector, choose **<Invisible lines>**.

33. Draw a Model Line from the Ref. Level to the bottom of the bin (see the right side of Figure 10.71).

- On the Design Bar, click the **Modify** tool or press the ESC key twice.

With the placement of this Model Line, Revit will "see" this object when the plan is cut at 4'-0" [1200]. Once the display of the object is triggered by the cut plane, the Family will display the bin's symbolic dashed lines.

The final step is to make sure that only the Symbolic Lines display in plan and not the bin geometry (which would appear like solid lines—instead of dashed overhead lines). We can use the visibility settings in the Family to instruct the bin and hardware geometry not to display in plan views.

34. Select the bin Extrusion.

- On the Options Bar, click the Visibility button.

- In the "Family Element Visibility Settings" dialog, clear the checkmark from the "Plan/RCP" box and then click OK (see Figure 10.72).

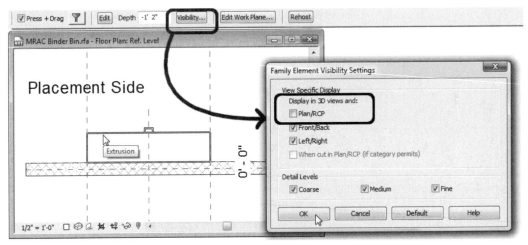

**Figure 10.72** *Turn off the visibility of the extrusion in plan views*

35. Repeat the process for any hardware geometry.

- Zoom all Views to fit.

36. Save the Family file.

Congratulations, you have just created a parametric Family from scratch! For the final test, let's load it into the project.

## LOAD THE NEW FAMILY INTO THE PROJECT

Make sure you are satisfied with the geometry and have tested the various Types. Be sure to save the Family before you continue.

37. On the Design Bar, click the Load into Projects button.

This will restore the *Level 3* view of the Commercial Project. Check the Families branch of the Project Browser under *Specialty Equipment* for the newly loaded Family.

38. Zoom in on the reception space.

39. From the Specialty Equipment node of the Families branch, drag one of the Family Types into the view. (You can also use the **Component** tool on the Design Bar if you prefer.)

- Since this is a wall-based component Family, you will need to click on a Wall to place it (see Figure 10.73).

At this point, we can test several of the behaviors of this new Family. The first test is verifying that it displays correctly in plan. It should appear as a dashed rectangle. Add a camera view to the reception area to look at the binder bin in 3D. Return to the plan and cut a section through the reception space looking at the binder bin. Open the new Section view. You should see the Symbolic Elevation Swing Lines in that view. The mounting height parameter is an instance parameter, so try adding more than one bin to the project, and then edit the properties of one of them to set the mounting height differently. Open

Object Styles and beneath Specialty Equipment, experiment with changing the settings—turn off components and/or change Line Styles. Remember you can also edit these settings per view by editing the View Properties and then clicking the Edit button next to "Visibility" (or just type **VG**). Finally, you can add, edit or delete Family Types on the Project Browser as we did above.

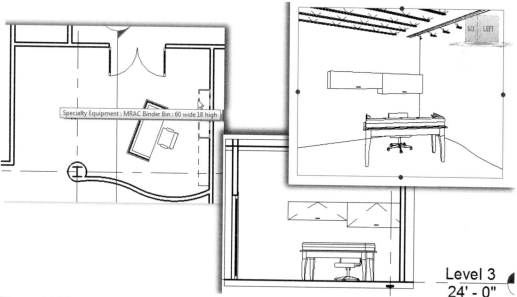

**Figure 10.73**  *Add a binder bin to the commercial project third floor reception space*

40. When you have finished experimenting with the binder bin, close the Family file.

41. Save the commercial project.

## CREATE A CURTAIN PANEL FAMILY

In the previous chapter, we loaded a custom Curtain Panel Family into the commercial project and used it in the Curtain System design at the front of the building. In the short exercise that follows, we will walk through the process used to create that Family.

### UNDERSTANDING CURTAIN PANEL FAMILY TEMPLATES

The process begins the same as any other new Family. The only difference is the template we'll choose.

1. On the Recent Files windows, click the New link beneath Families (or choose **New > Family** from the File menu).

There are three Curtain Panel templates: *Curtain Wall Panel.rft* [*Metric Curtain Wall Panel.rft*], *Door - Curtain Wall.rft* [*Metric Door - Curtain Wall.rft*] and *Window - Curtain Wall.rft* [*Metric Window - Curtain Wall.rft*]. All three will create a Curtain Panel that will appear on the Type Selector when working with Curtain Walls and Systems. The Door and Window templates are a little more specific and share characteristics of Curtain Panels and Doors or Windows respectively.

- Choose the *Window - Curtain Wall.rft* [*Metric Window - Curtain Wall.rft*] template and then click Open.

- Tile the windows and Zoom All to Fit.

Look at both the plan and exterior elevation views. Notice that several Reference Planes are already present as well as an EQ dimension. There is not, however, an overall width dimension. This is because the width of a Curtain Panel is determined automatically by the Curtain Grids. This is true for the height as well. If you look at the exterior elevation, the panel size is the space defined by the left and right Reference Planes horizontally, the top Reference Plane, and the Reference Level vertically. Therefore, we simply need to draw our panel geometry within this space.

2. From the Settings menu, choose Object Styles.

In the topic above, we learned about "Creating Subcategories". Notice that this Family actually belongs to the Windows category and as such has all of the Window subcategories. The Door Curtain Panel template would likewise be categorized as a Door.

- Click Cancel to exit the "Object Styles" dialog.

## ESTABLISH REFERENCE PLANES, CONSTRAINTS, AND PARAMETERS

3. In the *Floor Plan: Ref. Level* view, draw two vertical Reference Planes in the middle of the panel.

The spacing is not important yet. We'll correct this later.

4. Add a dimension including all Reference Planes *except* the center one.

- Toggle the equality on (see Figure 10.74).

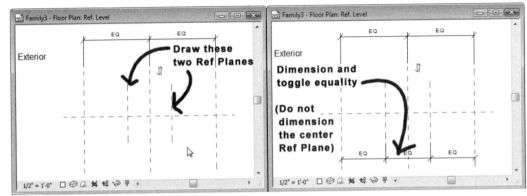

**Figure 10.74** *Add two Reference Planes and center them*

The next several steps are illustrated in Figure 10.75.

5. Click the Ref. Plane tool, choose the Pick icon on the Options Bar and set the Offset to **1"** [**25**].

- Offset a Reference Plane on each side of the two just created (four total).

- Change the Offset to **1 1/4"** [**31**] and Offset the Center (Front/Back) Reference Plane (running horizontally) up.

- Name this new horizontal Reference Plane **Panel Offset**.

6. Create a dimension between the Center (Front/Back) Reference Plane and the Panel Offset Reference Plane.

- Label this dimension with a new Parameter named **Panel Offset**.

- Group it under dimensions and leave it a type parameter.

- Finally, dimension the vertical reference planes and lock them as shown in Figure 10.75.

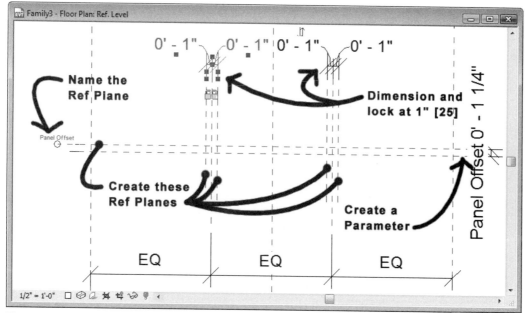

**Figure 10.75** *Add additional Reference Planes, constraints, and parameters*

7. Save the Family as **MRAC Window Panel**.

## BUILD PANEL GEOMETRY

8. In the Elevation: Exterior view, add a Solid Extrusion.

- Click the Work Plane icon, choose **Reference Plane: Panel Offset,** and then click OK.

- Sketch three rectangles that snap to the Reference Planes and lock the sketch lines (see Figure 10.76).

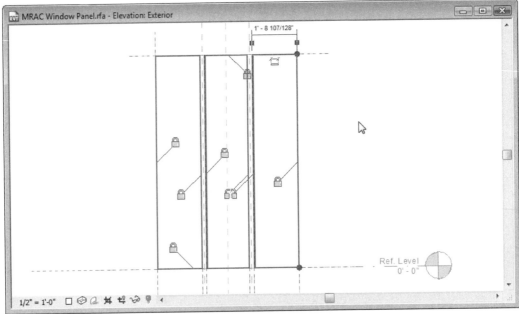

**Figure 10.76** *Snap the sketch lines to the Reference Planes and lock them*

9. On the Design Bar, click the Extrusion Properties button.

- Set the Extrusion End to **1/2"** [**12**].

- Set the Extrusion Start to **-1/2"** [**-12**].

- Click on <By Category> next to Material then click the small browse icon.

- In the "Materials" dialog, choose Glass and then click OK twice.

- On the Design Bar, click the Finish Sketch button.

You can best see the results back in the plan view. The extrusion appears like three panes of glass. If you open the 3D view and turn on shading, the glass will be transparent. To complete the Panel Family, we need to add some mullions between the panes of glass. You may also want to lighten the weight of the glass in plan. As you can see, it is being cut, so it displays bold. To make it lighter, use the Symbolic Lines again. Select the extrusion, click the Visibility button and turn it off in Plan and also deselect the "when cut in plan" box too. Next trace the three shapes in plan with Symbolic Lines set to the Glass (projection) subcategory. Be sure to lock the lines to the extrusion.

## BUILD OR LOAD A MULLION PROFILE FAMILY

Profile Families are simple 2D Families that contain a closed profile shape. The shape can be drawn using any of the standard line and arc tools. Profile Families give us a convenient way to save commonly used shapes. Profile Families can be loaded into other Families and used to create Sweeps.

10. From the File menu, choose **Load from Library > Load Family**.

- In the *Imperial Library\Profiles\Curtain Wall [Metric Library\Profiles\Curtain Wall]* folder, choose *Curtain Wall Mullion-Rectangular-Center.rfa [M_Curtain Wall Mullion-Rectangular-Center.rfa]*

If you prefer, you can build your own mullion profile. To do so, start a new Family from the *Profile-Mullion.rft [Metric Profile-Mullion.rft]* template. In the Family, sketch a closed 2D shape and save the Family. Then use Load into Projects to load it into the Panel Family. You can also find another example in the *Chapter09* folder.

Continue in the Elevation: Exterior view with the Work Plane set to: **Reference Plane: Panel Offset**.

11. Create a Solid Sweep.

- Sketch the 2D Profile in elevation along the Reference Plane centered in the gap between glass panels.

- Finish the Path.

- On the Options Bar, click on the drop-down menu labeled <By Sketch>.

- Choose one of the three sizes listed (see Figure 10.77).

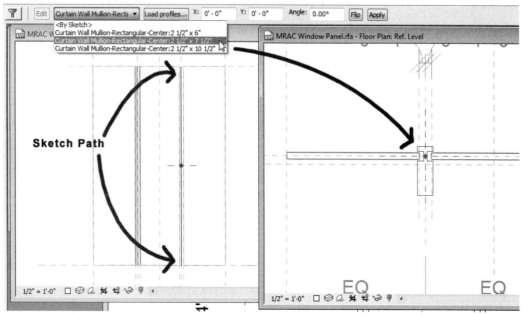

**Figure 10.77** *Create a Sweep using a Profile for the mullion*

- Click Finish Sweep.

- Repeat for the other mullion.

12. Save the Family (see Figure 10.78).

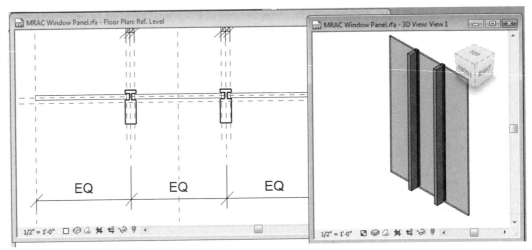

**Figure 10.78** *Complete the Panel Family*

- Load the Family into the commercial project and swap it in for some of the Panels in the Curtain Walls or Systems.

Perform any further experiments you wish in the commercial project before continuing. Try editing the Curtain Grid spacing and see how the mullions in the Panel Family stay equally spaced.

13. Close the Panel Family.

14. Save and close the commercial project.

## ENHANCE FAMILIES WITH ADVANCED PARAMETERS

In the binder bin example above, we made a basic parametric Family by adding labels to dimension strings that in turn controlled the size of the object. This is the most basic way to add parameters to a Family. Adding such parameters allows you to create multiple Types from a single Family. Each Type simply saves alternate values of each parameter.

Dimensional parameters are the most common parameters used in Families and are the most straightforward to define. However, the potential of parameters goes well beyond the controlling of dimensions. In this topic, we will explore several "advanced" parameters. They require a little more planning and often a few more steps to create, but once you get the hang of them, they can add much functionality and flexibility to the Families you create.

In this exercise, we will cover nested Families, visibility parameters, array parameters and the use of formulas in parameters.

### CREATE A COAT HOOK FAMILY

For use in this exercise, a Family file has been provided with the files on the CD. The file contains a wall-based coat rack. It was built from scratch from the same template used above for the binder bin. Similar geometry, Reference Planes and Dimension Parameters have already been set up in that file. In this exercise, we will create a new Family for an individual coat hook. We will then open the provided wall coat rack Family file and load our coat hook Family into it. From there we will explore the various parameters mentioned above.

1. From the Recent Files windows, create a New Family (or from the File menu, choose **New > Family**).

- Browse to you Templates folder if necessary.

- Select *Specialty Equipment.rft* [*Metric Specialty Equipment.rft*] and then click Open.

Be sure *not* to select the "wall based" template this time. The hook is not wall based, but by choosing "specialty equipment" again, we ensure that both Families share a common Category.

**Note:** If your version of Revit Architecture does not contain either of these templates, they have been provided in the *Out of the Box Templates* folder with the files installed from the Mastering Revit Architecture CD ROM.

2. Save the Family and name it **MRAC Coat Hook**.

3. Make the *Floor Plan: Ref. Level* window active.

- Zoom in a bit toward the center.

4. From the Design Bar, choose **Solid Form > Solid Revolve**.

- Draw a vertical line starting at the intersection of the two Reference Planes and moving up **4"** [**100**].

- Zoom in more if necessary and draw a horizontal line to the right **1/2"** [**12**].

- Draw a shallow arc vertically back down to the Reference Plane and then close the shape with a small **1/2"** [**12**] long horizontal line (see Figure 10.79).

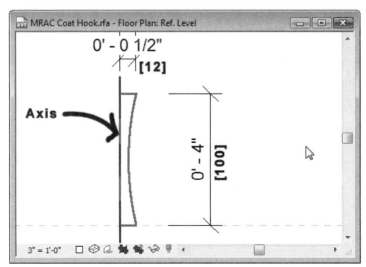

**Figure 10.79**　*Sketch the profile of the coat hook*

5. On the Design Bar, click the **Axis** tool.

- On the Options Bar click the Pick Lines icon and then click the vertical Reference Plane.

- On the Design Bar, click the Finish Sketch button (see Figure 10.80).

 **Note:** It will be necessary to zoom the various view windows to study the results.

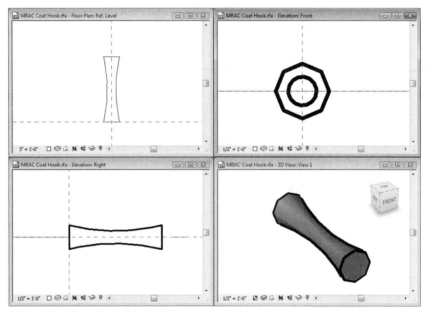

**Figure 10.80** *Complete the Solid Revolve*

 **Note:** If you wish to use Zoom to Fit, you will have to resize the Reference Planes to make them shorter. Zoom to Fit includes Reference Plane items in the extents of the model.

6. Save and Close the Family file.

## WORKING WITH NESTED FAMILIES

We now have the provided Wall Coat Rack Family and the Coat Hook Family that we just built. Let's add the hook to the rack Family. This will "nest" the hook Family into the rack Family.

7. In the *Chapter10* folder, open the *MRAC Wall Coat Rack.rfa* [*MRAC Wall Coat Rack-Metric.rfa*] Family file.

- Click on the title bar of the *Elevation : Placement Side* window to make it active.

At the top of the shelf are two Dimensions, one on each side reading 5" [125]. We need to create a new Dimension Parameter for these. We will create one parameter named "Hook Inset" and apply it to both dimensions.

    8. Select one of the 5" [125] dimensions at the top.

- Using the procedure in the previous lesson, create a new parameter named **Hook Inset**. Include it in the Dimensions grouping and make it a Type parameter.

- Select the other dimension and apply the same parameter.

- Use the ***Family Types*** tool on the Design Bar to flex the Hook Inset parameter—set the value to **6" [150]**.

    9. Make the *Floor Plan: Ref. Level* window active.

  10. On the Design Bar, click the ***Component*** tool.

An alert will appear indicating that no Families are loaded and requesting that you load one.

- In the dialog, click Yes to load a Family.

- Browse to the location where you saved *MRAC Coat Hook.rfa* and load it.

  11. Place the hook anywhere on the Placement side of the Wall in the *Floorplan: Ref. Level* view window.

Place it randomly; do not snap it to anything at this time.

    Using the left side of Figure 10.81 as a guide:

  12. Click the Align icon on the Tools toolbar and align the hook to the back plate of the Coat Rack in plan. Lock the padlock to constrain it.

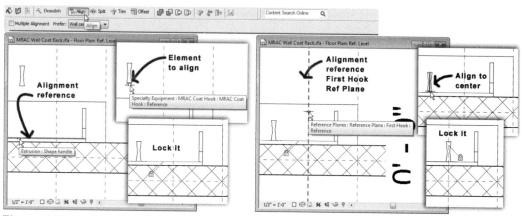

**Figure 10.81** *Align the Hook to the back board of the Coat Rack*

    Using the right side of Figure 10.81 as a guide:

- Align the center of the hook to the "First Hook" Reference Plane (on the right side of the rack) and lock it.

**Tip:** Identify the correct Reference Plane by highlighting it with the Modify tool (it will say "First Hook" on the tool tip and on the Status Bar).

When you align, be sure to click the Reference Plane first, then the middle of the hook. You may need to zoom in to the hook. When you zoom in on the hook, you will be able to highlight its centerline. Do *not* use align for the height of the hook. This we will achieve the correct vertical location with a new formula parameter.

- On the Design Bar, click the **Modify** tool or press the ESC key twice.

## CREATE A FORMULA PARAMETER

If you were to study the elevation and open the "Family Types" dialog, you would notice that the height of the back plate of the coat rack is controlled by a "Height" parameter. We can create a formula based upon this variable height that keeps our hooks centered vertically on the back plate.

13. On the Design bar, click on Family Types (see Figure 10.82).

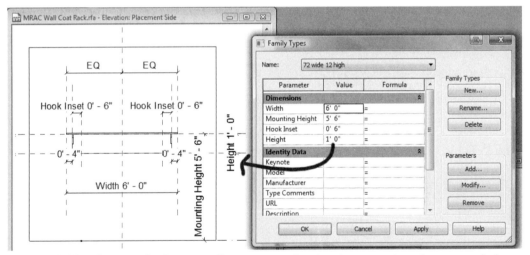

**Figure 10.82** *Compare the Dimension Parameters to the elevation to see how they are applied*

Move the "Family Types" dialog on screen so that you can see the *Placement Side* elevation in the background. In particular, you will see that the Height parameter is applied to the height of the coat rack unit. Also take note of the Mounting Height and Hook Inset parameters.

- In the "Parameters" area of the "Family Types" dialog, click the Add button.

Until now, we have added all of our parameters directly in screen in the Family Editor. Using the "Family Types" dialog is an alternative approach. Both methods yield the same result.

- Name the new parameter **Hook Height** and group it under **Constraints**.

- From the "Type" list, choose **Length** and leave it a Type parameter.

- Click OK to complete the parameter.

14. Back in the "Family Types" dialog, click in the formula field to the right of the Hook Height value field (see Figure 10.83).

- For the formula, type: **Mounting Height - (Height / 2)**.

 **Note:** These label names *are* case-sensitive.

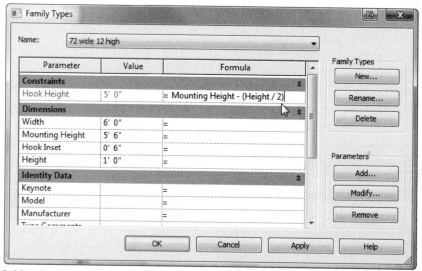

**Figure 10.83**   *Type a formula that subtracts half of the Height from the Mounting Height*

You will notice that once you click Apply, the result of the formula will be input read only in the Value field (it grays out). Study the formula we just typed. We are asking Revit to take the Mounting Height, which you can see in Figure 10.82 goes from the floor to the shelf of the coat rack and subtract from it half of the coat rack height. This means that no matter what values the Mounting Height and Height parameters assume, the value of this new Hook Height parameter when measured from the floor will place the hooks centered on the back board of the coat rack.

15. At the top of the "Family Types" dialog, choose **72 wide 8 high** [**1800 wide 200 high**] from the Name list.

Two types are pre-defined in this Family, each with different Heights. Note the change in the Hook Height's automatically calculated Value.

- Click OK to dismiss the dialog.

16. Select the hook and then on the Options Bar, click the Properties icon.

In elevation, you'll find the hook sitting on the floor. It might be easier to select it in plan.

Beneath the "Constraints" grouping, in the "Offset" row, in the far right column, is a very narrow column (it has an equal sign (=) in the header). In this column is a parameter button (see Figure 10.84).

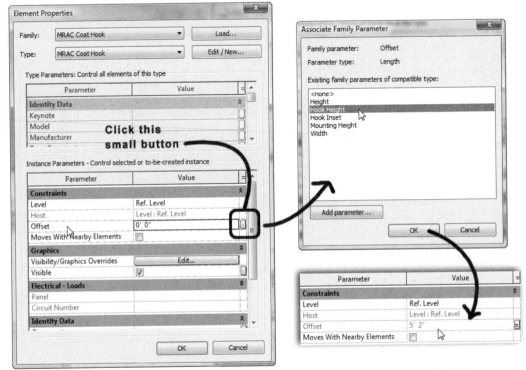

**Figure 10.84** *A parameter button appears in the formula column next to the Offset field*

- Click this parameter button and in the "Associate Family Parameter" dialog, select the "Hook Height" parameter and then click OK.

Again, notice the read only value of the resolved formula fills in automatically.

- Click OK again to dismiss the "Element Properties" dialog.

The hook will move to the center of the back plate height. You can re-open the "Family Types" dialog and test the formula by changing Types and then clicking apply. If you wish, you can create additional types with varying Mounting Heights and Heights to fully flex the new parameter.

17. Save the Family file.

## ADD A VISIBILITY PARAMETER

Sometimes you add details to a Family that you don't want to display in all views. We can create a Visibility parameter that will allow us to control the display of components in each

Type within the Family. In this case, we will make it possible to turn the hooks on and off in each Family Type.

18. Select the Hook component and on the Options Bar, click the Properties icon again.

- Beneath the "Graphics" grouping click the parameter button (in the same far right column) next to the "Visible" parameter (*not* the Edit button next to "Visibility").

- In the "Associate Family Parameter" dialog, click the "Add parameter" button.

Here is another alternative way to create parameters. Again, the method of creation is a matter of preference with the results being the same in each case.

- Name the parameter: **Show Hook**, group it under **Graphics**, leave it a "Type" parameter and then click OK twice.

19. Return to Family Types, and create a new Type. Name it: **Shelf only-no hooks**.

- Remove the checkmark from the Show Hook checkbox for this Type and then click Apply (see Figure 10.85).

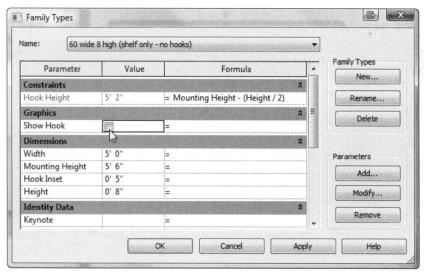

**Figure 10.85**   *Create a type that hides the hooks by unchecking the Show Hook parameter*

- Click OK to exit the dialog.

 **Note:** While you are working on the family in the Family Editor, even if the visible checkbox is unchecked, it will still be visible. This parameter only becomes active when the Family is used in a project.

## CREATE AN ARRAY OF HOOKS

By default, when you use the Array command in Revit it groups the arrayed elements and maintains some of the parameters used to create the array. We can utilize this feature to make a parametric array of hooks in our Family.

20. Make the *Elevation: Placement Side* window active.

21. Select the hook and then on the Tools toolbar, click the Array icon.

- On the Options Bar, be certain that a checkmark appears in the "Group and Associate" checkbox.

- For the "Move To" option, choose the "Last" radio button.

Following the prompts at the Status Bar (bottom left corner of the Revit screen), we must indicate the start point and then the end point of the Array. We will use the Reference Planes for these.

- For the First Point, click the "First Hook" (left side) Reference Plane.

- For the End Point, click the "Last Hook" (right side) Reference Plane (see Figure 10.86).

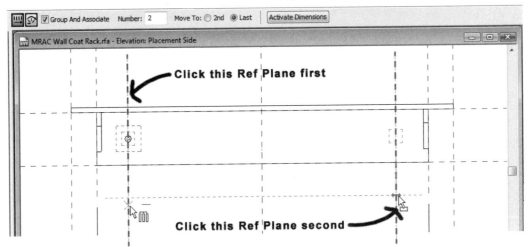

**Figure 10.86** *Click each of the Reference Planes as the start and end of the Array*

A temporary dimension will appear with the Array prompting you for the quantity of items.

- Type **3** for the "Array Count" and then press ENTER.

- On the Design Bar, click the **Modify** tool or press the ESC key twice.

There are now 3 hooks on the back plate of the coat rack. Click on any hook to edit the Array Count or to gain additional Group options on the Options Bar. To test this out, try changing the Array Count with the temporary dimension if you wish. Earlier, we constrained the hook on the left to the "First Hook" Reference Plane using the Align command. We need to also constrain the last hook (the one on the right) to the "Last

Hook" Reference Plane. Otherwise, if you choose a different Family Type changes the width of the coat rack, the last hook will no longer line up properly.

22. Select the last hook (the one on the right) and then use the Align command (like above) to Align it to the "Last Hook" Reference Plane and lock it.

    Zoom in on the hook as required.

23. On the Design Bar, click the Family Types button.

    - Create a new Type named **84 wide 12 high** [**2100 wide 300 high**].

    - Edit the Width and Height parameters accordingly, turn on Show Hook, move the dialog out of the way, and then click the Apply button to see the change.

If everything has flexed properly, the coat rack will have gotten larger but the hooks should still be positioned properly with the fixed offset at the ends and the height being in the middle of the back board.

24. Click OK to close the dialog and then Save the Family file.

## MAKING THE ARRAY COUNT PARAMETRIC

At this point, we have a coat rack that keeps our three hooks equally spaced even as we change its dimension parameters. As a finishing touch, let's change the Array Count into a parameter.

25. Select one of the hooks.

Like before, a temporary dimension indicating the current Array Count will appear. This dimension includes a horizontal line and a text field in which we can edit the count. If you select the horizontal dimension line, rather than the text field, you expose the dimension options on the Options Bar.

    - Click on the Array line to select it.

    - On the Options Bar, next to "Label" choose **<Add parameter>** from the drop-down list.

    - In the "Parameter Properties" dialog, name the parameter **Number of Hooks** and group under Graphics.

    - Leave "Type" selected and then click OK.

Notice the label that now appears on the Array dimension. You will only see this with a hook selected.

26. Re-open the "Family Types," dialog and change the "Number of Hooks" value a few times.

    - Click Apply after each change and watch the hooks appear and re-space automatically

This parameter can be different for each type.

27. For the **84 wide 8 high** [**2100 wide 200 high**] Type, set the Number of Hooks to **5** and then Apply.

- From the "Name" list at the top, choose **72 wide 12 high [1800 wide 300 high]** Type, set the Number of Hooks to **4** and then Apply.

- Repeat for the other Types.

Don't worry about changing the quantity on the "No Hooks" Type. Even though they show here in the Family Editor, they will not appear when used in a project.

### USE A FORMULA TO SET THE ARRAY COUNT

Finally we can make the Number of Hooks a formula based on the Width of the coat rack.

28. If you closed "Family Types," re-open it now.

- In the formula field next to the "Number of Hooks" parameter, type: **(Width - (Hook Inset * 2)) / 10" [(Width - (Hook Inset * 2)) / 250 mm]** (see Figure 10.87).

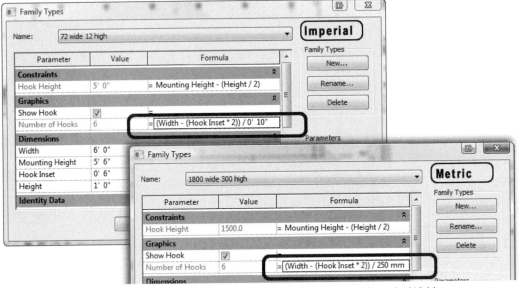

**Figure 10.87** *Use a formula to calculate the Number of Hooks based on the Width*

You can test the formula in the "Family Types" dialog by changing the Width parameter and clicking Apply. Test out several values until you are satisfied. Be sure that the values of Width and Height match the Type name before you close the dialog.

### MAKING THE FAMILY DISPLAY IN PLAN

We need to repeat the process used for the binder bins above in the "Making Families Above the Cut Plane Display in Plan" topic to make our coat rack Family display in plan views. The Symbolic Lines have already been included in this file. We only need to add the invisible line to the elevation view.

29. Add an <Invisible Lines> Model Line to the *Placement Side* elevation view.

- Turn off the "Plan/RCP" visibility of the items above the cut plane. This includes all of the extrusions and the hooks.

The hooks are inside an array Group, so you will need to edit the Group. Select one hook, click the Edit Group button and then select the hook again to see the Visibility button on the Options Bar. Since all of the array items are instances of the same Group, you only need to perform these steps on one of them. Refer to Chapter 6 for more information on Groups.

30. Save the Family file.

## LOAD THE RESIDENTIAL PROJECT

The Family file is complete and ready to load into a project. Let's load it into our residential project

31. Open the Residential project for your choice of units:

    Open *10 Residential.rvt* if you wish to work in Imperial units.

    Open *10 Residential Metric.rvt* if you wish to work in Metric units

The project will open in Revit with the last opened view visible on screen. By now we have learned a few ways to load Families into a project. You can start from the residential project, click the **Component** tool on the Design Bar and then click the Load button on the Options Bar to browse to and load the *Wall Coat Rack.rfa* Family file. You can choose Load from Library > Load Family from the File menu. You can also return to one of the Family file's view windows from the Window menu (or CTRL + TAB) and then click the "Load into Projects" button on the Design Bar. The exact method is up to you.

32. On the Project Browser, double-click to open the *First Floor* plan view.

- Load the *Wall Coat Rack* Family and place an Instance in the room on the left on the interior vertical Wall.

- On the Project Browser, double-click to open the *Entertainment Room Camera* 3D view (see Figure 10.88).

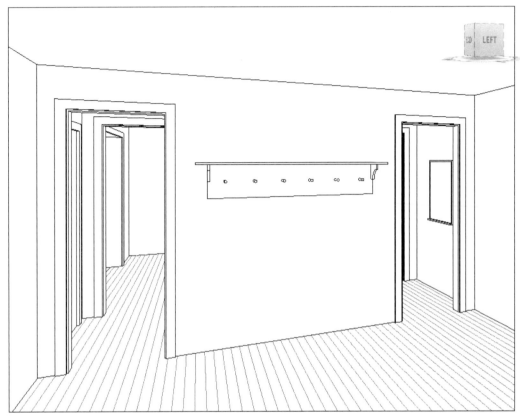

**Figure 10.88**  *View the Family in a 3D Perspective view*

From the Type Selector on the Options Bar, try the different Types to see them in the project. Note that the "No Hooks" Type now displays without hooks as we indicated. Remember to double-check each parameter and type. The final test of any Family is to flex it in a project. Some parameters, such as the hook Visibility parameter, can only be tested in a project. Thorough testing will prevent frustration later in your project cycle.

33. Save the project.

34. Switch to and close the *MRAC Wall Coat Rack.rfa* Family.

## FAMILIES FROM MANUFACTURER'S CONTENT

As the final exercise for this chapter, let's remain in the residential project and create a custom Family for a new whirlpool tub on the second floor. Rather than build this Family from scratch, we will use drawing files (DWG) downloaded from a manufacturer's website. (In this exercise, the hypothetical manufacturer's files have been provided with the other CD files, but the procedure would be the same if you visit and download from actual manufacturer's web sites).

**Note:** If any changes are required to the DWG files, you will need a copy of AutoCAD to open and edit them. Contact you Autodesk reseller for more information.

35. Create a new Family file.

- Choose the *Plumbing Fixture.rft* [*Metric Plumbing Fixture.rft*] Family template.

- Tile the windows and make the plan active.

36. From the File menu, choose **Import / Link > DWG, DXF, DGN, SAT**.

- Browse to the *Chapter10* folder and choose *Whirlpool_Tub_P.dwg* [*Whirlpool_Tub_P-Metric.dwg*].

- From the Colors list, choose **Black and White**.

- From the Positioning list, choose **Auto - Origin to Origin**.

- Click Open.

A two-dimensional drawing of the plan symbol will appear. You can import any AutoCAD or Microstation drawing this way. When you do, characteristics of the imported drawing, such as layers will be maintained. This particular file has only one layer, but if you link other drawing files, you can access their layers on the Options Bar and in the "Visibility/ Graphics" dialog. We only want this symbol to appear in plan Views. To do this, we will edit the visibility of the imported element.

37. Select the imported drawing.

- On the Options Bar, click the Visibility button.

- In the "Family Element Visibility Settings" dialog, clear the checkmarks from the "Front/Back" and "Left/Right" boxes and then click OK (see Figure 10.89).

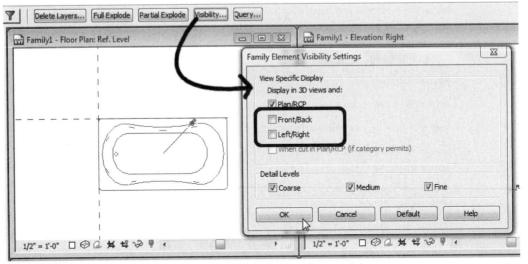

**Figure 10.89** *Turn off visibility in elevation Views for the plan symbol*

38. Switch to the 3D view.

Repeat the process to import another DWG file.

- In the *Chapter10* folder, choose *Whirlpool_Tub_M.dwg* [*Whirlpool_Tub_M-Metric.dwg*] this time.

Use the same color and positioning options.

This time the imported drawing contains a 3D model. We want this to show only in 3D and elevation Views.

39. Zoom all windows to fit. Select the imported 3D drawing.

- On the Options Bar, click the Visibility button.

- In the "Family Element Visibility Settings" dialog, clear the checkmark from the "Plan/RCP" box and then click OK (see Figure 10.90).

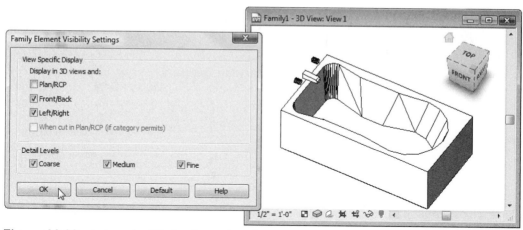

**Figure 10.90**  *Import the 3D drawing and configure its visibility*

40. Save the Family with the name **MRAC Whirlpool Tub**.

   • Click the Load into Projects button.

41. Back in the residential project, on the Project Browser, double-click to open the *Second Floor* plan view.

   • Demolish the bathtub at the top left of the plan.

   • Click the **Component** tool and place the **MRAC Whirlpool Tub** Family in the same room (Use the SPACEBAR if necessary to rotate it as you place it).

42. Save and Close the project and any open Family files.

## SUMMARY

We have covered quite a bit of ground in this chapter. By now, you should have a good grasp of the power and flexibility of Families in Revit Architecture. Despite the lengthiness of this chapter, we have only scratched the surface of the potential inherent in Families. Continue to explore and customize your own Families. If you have not already done so, read and work through the exercises in the tutorials provided with the software (access them from the Help menu). Each of the exercises provided explores different kinds of Families from the ones explored here. While the process is similar, the more examples you work through, the more comfort and confidence you will gain with this critical and powerful part of the Revit software package.

- An extensive Library of Family Content has been included with your Revit software.

- Familiarize yourself with the provided Library before embarking upon any customization.

- In addition to the included Libraries, extensive Libraries and resources are available on the Internet.

- Autodesk maintains a Web Library accessible directly from within Open dialogs in the software.

- The Autodesk Content Search Web site is accessible from the toolbar directly in Revit.

- The simplest way to customize Families is to add or edit Types.

- Before building a custom Family from scratch, determine if you can save a copy and modify an existing one first.

- Editing an existing Family is accomplished by opening it in the Family Editor and making modifications and then saving a copy of the Family file.

- You can build a completely custom Family from one of the many provided Family template files.

- Family templates establish the basic framework and behaviors of the Families you create—choose your template carefully.

- You can create a "singular" Family—a Family with only one Type, or a "extensible" Family—which has parameters allowing for multiple Types.

- Add dimension parameters to your Families to allow for various "sizes" of the same basic Family geometry.

- Advanced parameters such as visibility controls, formulas and parametric Arrays enable you to make very complex and robust Families.

- You can use manufacturer's drawing files directly in Families to create symbols and other items in your projects quickly.

# Construction Documents

While the benefits of Building Information Modeling descibe an architectural design and delivery process that may one day be "paperless" many firms and projects still rely heavily on traditional deliverables even as they look toward the future. Therefore, the creation of Construction Documents typically in the form of printed drawings will remain relevant and necessary for some time to come. In this section, we will explore tools that help us produce these deliverables in Revit Architecture.

Section III is organized as follows:

# CHAPTER 11

# Detailing and Annotation

## INTRODUCTION

In this chapter, we will look at detailing in Revit Architecture. We will explore the detailing process and tools in our residential project. As the design development phase gives way to construction documentation, details are created to clarify basic plan, section and elevation views of a project and assist in conveying overall design and construction intent. Before Revit, such details had been drafted independently of the overall drawings with perhaps some tracing to help minimize redundant effort. In Revit Architecture, most detailing begins within a fully coordinated model view like the other views (plans, sections, elevations) in a project.

The process is simple—create a callout or section view of the model at a large scale and then add additional drafted components, text, dimensions and other embellishments necessary to craft the detail and convey design intent. In most cases, such embellishments are at minimum drawn relative to an underlying building model view and in many cases remain automatically constrained or linked to model geometry in the view. Unlike the other Revit views, all drafting components appear *only* in the view to which they are added.

## OBJECTIVES

In this chapter, we will create detail drawings using several techniques. Working first from the Revit model, we add additional information to create a wall-floor-foundation section detail. A variety of tools will be explored to assist in this process. We will also create a detail in our project using a detail originally created in AutoCAD. This process allows you to utilize detail libraries that you may already have directly within Revit Architecture. Our exploration will include coverage of Detail Lines, Detail Components, Repeating Details, Filled and Masking Regions, and various annotations. After completing this chapter you will know how to:

- Modify Crop Regions and add View Breaks
- Add and modify Detail Lines
- Add and modify Detail Components and Repeating Details
- Add and modify Text and Leaders
- Add and modify Filled Regions, Masking Regions, and Break Lines

- Work with Drafting views
- Import legacy details into a Revit Architecture Drafting view

## MODIFY WALL TYPES

To prepare us for the detailing tutorial that follows, we will modify the Wall Types currently in use in the residential project for the exterior walls. We will unlock the outer two components (called Layers) of the brick Wall and add a brick ledge to the foundation Wall. Doing so will allow us to create a brick shelf on the concrete Wall and extend the brick down to sit on it. These steps will make the model more accurately represent the construction.

### INSTALL THE CD FILES AND OPEN A PROJECT

The lessons that follow require the dataset included on the Mastering Revit Architecture CD ROM. If you have already installed all of the files from the CD, simply skip down to step 3 below to open the project. If you need to install the CD files, start at step 1.

1. If you have not already done so, install the dataset files located on the Mastering Revit Architecture CD ROM.

    Refer to "Files Included on the CD ROM" in the Preface for instructions on installing the dataset files included on the CD.

2. Launch Revit Architecture from the icon on your desktop or from the *Autodesk > Revit Architecture* group in *All Programs* on the Windows Start menu.

**Tip:** In Windows Vista, you can click the Start button, and then begin typing **Revit** in the "Start Search field. After a couple letters, Revit Architecture should appear near the top of the list. Click it to launch the program.

- If the New Features Workshop dialog appears, choose "Maybe later" and then click OK.

3. On the Standard toolbar, click the Open icon.

**Tip:** The keyboard shortcut for Open is CTRL + O. **Open** is also located on the File menu.

- In the "Open" dialog box, click the *My Documents* icon on the left side.
- Double-click on the *MRAC* folder, and then the *Chapter11* folder.

    If you installed the dataset files to a different location than the one listed here, use the "Look in" drop-down list to browse to that location instead.

4. Double-click *11 Residential.rvt* if you wish to work in Imperial units. Double-click *11 Residential Metric.rvt* if you wish to work in Metric units

    You can also select it and then click the Open button.

The project will open with the last saved view visible on screen.

## ADD A BRICK SHELF

Let's start with the foundation Wall. By modifying the Wall Type, we can create a brick shelf to receive the bricks from the exterior Wall above. This will be achieved by adding a Reveal directly to the Wall Type.

5. On the Project Browser, double-click to open the *Basement* floor plan view.

There are four foundation Walls, three bounding the outside Walls and another framing the right side of the passageway to the existing basement. We do not want to apply a brick shelf to the Wall in the passageway. The easiest way to prevent this is to create a new Type for the Walls with the brick shelf.

6. Using the CTRL key, select the three exterior foundation Walls (two vertical and one horizontal) (see Figure 11.1).

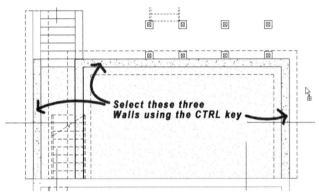

**Figure 11.1**   *Select the three exterior foundation Walls*

- On the Options Bar, click the Properties icon.
- Next to the Type list, click the Edit/New button.

 **Tip:** A shortcut to this is to press ALT + E.

The Type Properties dialog will appear.

- Next to the Type list, click the Duplicate button.

 **Tip:** A shortcut to this is to press ALT + D.

A new Name dialog will appear. By default "2" has been appended to the existing name.

- Change the name to: *MRAC - Foundation - 12" Concrete (w Brick Shelf)* [*MRAC - Foundation - 300mm Concrete (w Brick Shelf)*] and then click OK.
- At the bottom of the dialog, click the Preview button >>.

A viewer window will appear to the left attached to the "Type Properties" dialog.

- From the "View" list (bottom left), choose **Section: Modify type attributes**.

- Zoom in to the top of the wall.

- On the right side of the dialog, at the top, click the Edit button next to Structure (see Figure 11.2).

This will open the "Edit Assembly" dialog and show the Wall Layers included in this Type. We can edit these Layers here as well as add other parameters such as Sweeps and Reveals (below we will use the same process to "unlock" some of the Wall Layers of another Type).

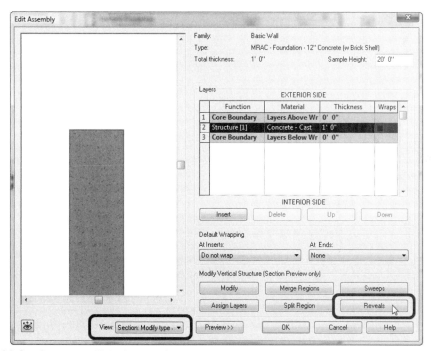

**Figure 11.2**   *Access the "Edit Assembly" dialog to edit the Wall Structure*

7. In the bottom right corner of the dialog, within the "Modify Vertical Structure" area, click the Reveals button.

 **Note:** The "Reveals" and other buttons in the "Modify Vertical Structure" area will not be available if you have not enabled the Section Preview as noted in the previous steps.

A Reveal is a profile-based extrusion that cuts away from the mass of the Wall. In this case, we will use a Profile that has been provided with the files from the Mastering Revit Architecture CD ROM.

- From the Reveals Dialog click the "Load Profile" button.

- Browse to the *Chapter11* folder where you installed the Mastering Revit Architecture CD ROM files.

- Select the file named: *MRAC Brick Shelf Reveal.rfa* [*MRAC Brick Shelf Reveal-Metric.rfa*] and then click the Open button.

8. In the "Reveals" dialog, click the Add button to add a Reveal.

- Click in the Profile cell and click again on the down arrow to display the Profile list, choose **MRAC Brick Shelf Reveal : 12" d × 6" w** [**MRAC Brick Shelf Reveal-Metric : 300 d × 140 w**].

- From the "From" cell, choose **Top** and then click the Apply button (see Figure 11.3).

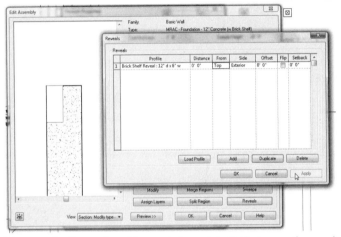

**Figure 11.3**  *Add a new Profile, set it to Top, and then click Apply to see the result*

In the "Edit Assembly" dialog in the background, you should see the Reveal Profile appear at the top left edge of the Wall in the viewer. Move the Reveals dialog out of the way if necessary.

- Click OK to return to the "Edit Assembly" dialog.

- Click OK three more times to return to the model view window.

9. On the Project Browser, double-click to open the *Longitudinal* section view.

- Zoom in on the left side to study the results (see Figure 11.4).

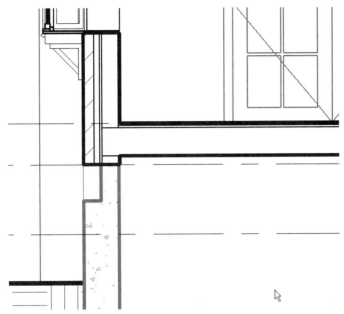

**Figure 11.4** *Open the Longitudinal Section view to see the results*

10. On the Design Bar, click the **Modify** tool or press the ESC key twice.

## UNLOCK WALL LAYERS

Now that we have a brick shelf in the foundation Wall Type, let's edit the brick Wall Type to unlock the outer material Layers (to allow their top and/or bottom offsets to move freely from the Wall's top and base offsets) and then project them down to sit on the foundation brick shelf.

11. Select the Left exterior brick Wall.

- On the Options Bar, click the Properties icon.

- Next to the Type list, click the Edit/New button.

- Open the Preview window again (if it is not already open), and then click the Edit button next to Structure.

The preview window should be showing the section view. If it is not, choose **Section: Modify type attributes** from the "View" list. You can use standard navigation techniques such as the wheel of your mouse or the right-click menu to zoom and scroll the model in the viewer.

12. Right-click in the viewer window and choose **Zoom in Region**. Zoom in on the lower portion of the Wall.

- In the "Modify Vertical Structure" area, click the Modify button (see Figure 11.5).

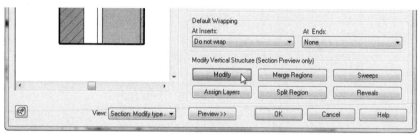

**Figure 11.5**   *Zoom in on the lower portion of the Wall and then click the Modify button*

13. Click the bottom edge of the Brick Layer.

The edge will highlight red to indicate that it is selected. A small padlock icon will appear on the edge. We can use this padlock icon to unlock the bottom edge of the Layer, which will allow it to be moved independently from the Wall itself in the model.

- Click the padlock (to open it) and unlock the bottom edge of the Layer.

- Repeat the process to unlock the bottom edge of the Thermal/Air Layer (next to brick) (see Figure 11.6).

**Figure 11.6**   *Unlock the Brick and Thermal/Air Layers*

- Click OK three times to return to the model.

Upon returning to the model view window, you will note that the Wall now has two Shape Handles at the bottom edge (see Figure 11.7). You can use the second handle to modify the bottom edge of just the unlocked Layers. The other handle will continue to modify the Base Constraint of the entire Wall.

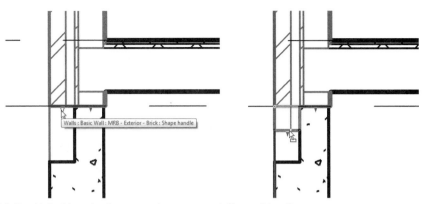

**Figure 11.7**    *Unlocking the Layers makes a second Shape Handle appear*

While the Shape Handle provides an easy way to edit the brick and air Layers, the Align tool gives a bit more control by allowing us to apply a constraint to the alignment after we make it.

14. On the toolbar, click the **Align** tool.

- For the Reference line, click the bottom edge of the brick shelf on the foundation Wall (use the TAB key as necessary to make the proper selection).

- For the Entity to Align, click the bottom edge of the brick Layer in the Wall above (see Figure 11.8).

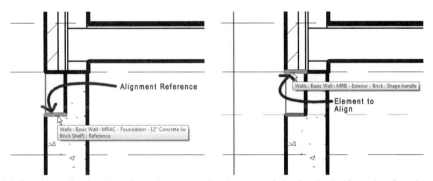

**Figure 11.8**    *Align the brick and air Layers to the bottom of the brick shelf on the foundation Wall*

- Click the small padlock icon that appears to lock the constraint and keep these elements aligned.

- On the Design Bar, click the **Modify** tool or press the ESC key twice.

15. On the toolbar, click the **Join Geometry** tool.

- Join the foundation Wall to the Brick exterior Wall.

- On the Design Bar, click the **Modify** tool or press the ESC key twice (see Figure 11.9).

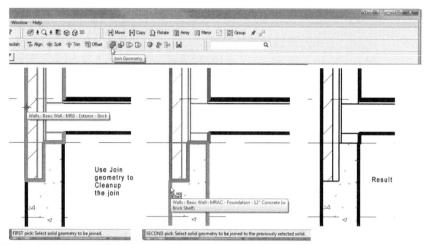

**Figure 11.9** *Join the two Walls to make them cleanup nicely*

 **Tip:** Be sure to click the Brick Wall first, and then join it to the Foundation Wall. Clicking in the opposite order will remove the customized bottom alignment.

• Save the project.

## DETAILING IN REVIT ARCHITECTURE

Detailing in Revit Architecture is in many ways similar to detailing in traditional drafting. This is true regardless of whether you compare it to drafting created by hand on a drafting board or created in Computer Aided Design (CAD) software on a computer. The major difference is that in Revit Architecture you rarely start from scratch because you can base your detail on views that are automatically generated from your Revit model.

The typical Revit model includes enough data to generate a majority of the drawings that will be required in an architectural document set at an appropriate level of detail and accuracy. This is true for most plans, sections, and elevations. In the case of details, however, while it is theoretically possible to model all of the bricks, fasteners, joints, hooks, and other items that will actually occur in the building, the amount of effort (in man-hours) and the sheer size of the resultant model (in computer memory and hard drive requirements) would typically not yield a sufficient return on investment.

To keep the size of our models reasonable and to avoid spending additional and often unnecessary time modeling every bolt, screw, and piece of flashing, the strategy to detailing in Revit Architecture is instead a hybrid approach. In nearly all details you may create in Revit, you will be able to start the process with a cut (callout) from the model. This live view of the model portrayed at the scale of the detail will give you a starting point upon which to add detail components and other view-specific 2D elements and annotations. By separating a detail into both live model elements and view-specific embellishments, we achieve the best of both worlds: we have an underlay that remains live and changes automatically with the overall building model and we have all of the additional data required to convey design intent occurring only on the specific detail view, thus saving on overhead and unnecessary modeling effort.

It is this process that will be discussed in detail in the following tutorial. In this exercise we will discuss the available tools and techniques using the wall and floor intersection from our residential project that we edited above.

## ADDING A CALLOUT VIEW

Continue from the previous exercise in the *Longitudinal* section view. If you closed the project or this view, reopen them now. Details are typically presented at larger scales than the drawings from which they are referenced. The Callout tools in Revit Architecture will allow us to create a detailed view at a larger scale of any portion of the building model.

> In the *Longitudinal* section view, be sure you can see the left exterior Wall from the first floor down to the footing. Zoom and Scroll as necessary.
>
> 1. On the Design Bar, click the view tab and then click the **Callout** tool.
>
> • From the Type Selector, choose **Section: Wall Section**.
>
> • Click a point outside the exterior Wall on the left above the first floor and then drag a callout around the Wall to beneath the foundation (see Figure 11.10).

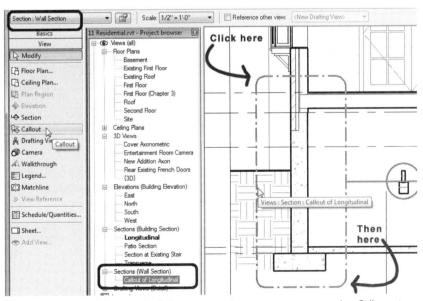

**Figure 11.10**  *Using the Callout tool, click two opposite corners to create the Callout view*

> Revit Architecture will create a new branch on the Project Browser and a new view called *Callout of Longitudinal*.

Like section and elevation markers, a callout marker will appear. When you deselect all elements, this callout will remain blue. As with the others, this indicates that you can double-click it to jump to the referenced view. You can also open the view from the Project Browser.

2. On the Project Browser, beneath *Sections (Wall Section)* right-click on *Callout of Longitudinal* and choose **Rename**.

   • Type: **Typical Wall Section** and then click OK.

3. Open up the callout view (you can double-click its name on Project Browser or its Callout symbol) (see Figure 11.11).

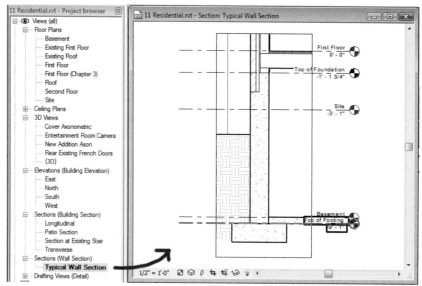

**Figure 11.11**   *Open the Callout view*

Notice that the boundaries of the crop region in the *Typical Wall Section* view match the extents of the callout boundary that we sketched in the *Longitudinal* section view. If this boundary is adjusted in either view, the boundaries in the other view automatically adjust. If you wish to see this, try tiling the *Longitudinal* section view and the *Typical Wall Section* view side by side and test it out. Remember to close hidden views or minimize other views first, so that when you tile, only the two sections will appear (see Figure 11.12).

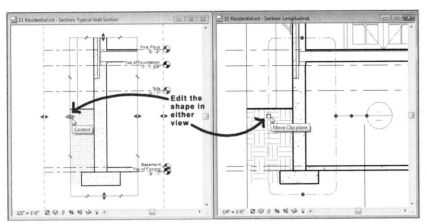

**Figure 11.12**    *Drag the Control Handles in either view to edit the extent of the Crop Region*

4. When you are finished experimenting, return the shape of the crop region to match approximately as shown above in Figure 11.11.

- Maximize the *Typical Wall Section* view.

## ADJUST SCALE AND ANNOTATION VISIBILITY

Annotation is separate from the model geometry shown in a view. While the level of detail and graphical display characteristics of the model may vary from view to view, the model will display in *all* views unless you specifically override the display settings to hide it. Annotation, on the other hand, is applied on top of the model and occurs only in the specific view in which it is created. Model and annotation elements also differ from one another with regard to scale. Annotation appears at a consistent height relative to its desired plot size, while the model geometry adjusts its size relative to the assigned scale. All of this behavior occurs automatically.

5. On the View Control Bar (bottom of the window) change the scale to ¾" = 1'-0" [1:20] (see Figure 11.13).

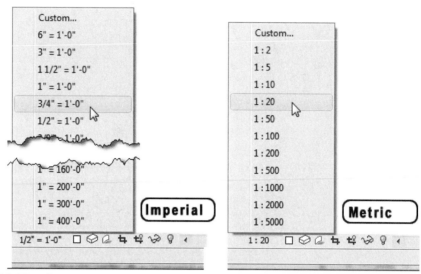

**Figure 11.13** *Choose a larger scale for the detail view*

6. Select the Site Level Line.

- Right-click and choose **Hide in View > Elements** from the menu that appears.

This operation does not have any effect on any other view. We have hidden this Level line in only the current view. The Site Level is not really relevant in the current Callout view, so by hiding the Level line, we eliminate potential clutter and confusion. In a similar fashion, we can adjust the location of the Level Heads and the length of the Level lines and again, the edits will be confined to *only* this view.

7. Click on any Level line.

- Using the Control Handle at the end, adjust the end points so the Level text is completely outside the Crop Region (see Figure 11.14).

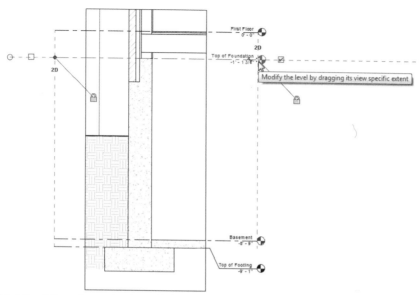

**Figure 11.14**  *Adjust the end points of the Level lines to move the text outside the Crop Region*

Notice that all Level lines move together when you drag one.

8. Save the project.

## DETAIL LINES

Now that we have cut our detail callout and configured the Level lines and scale to our liking, we are ready to begin adding embellishments. We can draw a variety of view-specific elements directly on top of the section view of our model. We will start with Detail Lines. These are simple drafted elements much like the sketch lines with which you are already familiar. When you add a Detail Line, it appears only in the view to which you add it. If you wish to draft a line that appears in multiple views, use a Model Line instead. We will use Detail Lines to sketch in some flashing at the bottom of the wall cavity.

9. On the Design Bar, click the Drafting tab and then click the **Detail Lines** tool.

Notice the choices on the Options Bar are very familiar and match those that we have seen in many sketch-based objects so far.

• From the Type Selector, choose **Wide Lines**.

• On the Options Bar, be sure that the "Draw" icon is chosen.

• Verify that there is a checkmark in the "Chain" checkbox and choose the "Line" icon.

10. Zoom into the bottom of the wall cavity (at the brick shelf).

11. Sketch the line segments shown in Figure 11.15.

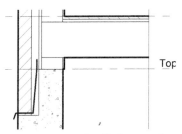

**Figure 11.15** *Sketch Detail Lines to represent the flashing in the wall cavity*

Unlike the sketch lines that we drew in previous chapters, these lines are complete "as is." They do not describe the shape of a more complex element like a Floor or a Stair. These are simply drafted lines placed on top of a model view, much like drafting directly on top of a Mylar background in traditional hand drafting.

12. On the Project Browser, double-click to open the *Longitudinal* section view.

- Zoom in to the same portion of the Wall and notice that this linework does not appear in this (or any other) view.

13. On the Project Browser, double-click to return to the *Typical Wall Section* view.

 **Tip:** You can hold down the CTRL key and then press TAB to cycle through the open views.

All of the parts of the detail that we are going to create next could be created with Detail Lines following the same process. However, several other detailing tools are available to us. Let's look at them now.

## DETAIL COMPONENTS

Detail Components are simply two-dimensional view-specific elements that (like Detail Lines) appear only within the view in which they are placed. They are more useful and more powerful than simple Detail Lines in that they are Families and can be parametric. Like other Families, a Detail Component Family can have many Types built into it. The parameters can be as simple as Length and/or Depth, or include dozens of parametric dimensions. For Example, a "Wide Flange" Family file included in the out of the box *Detail Component* folder contains hundreds of Types representing all of the commonly-available steel shape sizes. Another example that is a bit more pertinent to the detail that we are creating here is dimension lumber. While it would be possible to edit the Wall Type used in this project and begin adding the three-dimensional sole and sill plates, this would add a level of complexity to the model that is typically only needed in details. Instead, we can add predefined Detail Components to our Detail view to represent this information more efficiently only in the views that require it.

14. On the Design Bar, click the Drafting tab and then click the **Detail Component** tool.

Note the options that appear on the Type Selector and the Options Bar. Currently, there are no "Dimension Lumber" Families loaded in our project. Like other Components in Revit Architecture, we can simply load them from the library.

- On the Options Bar, click the Load button.

- In the "Open" dialog, browse to your default library folder (either the *Imperial Library* or the *Metric Library*) and then browse to the *Detail Components/Div 06-Wood and Plastic\06100-Rough Carpentry\06110-Wood Framing* folder.

 **Note:** If you do not have access to either of these libraries, the Family files mentioned in this tutorial have also been provided in the *Library\Detail Components* folder with the files installed from the Mastering Revit Architecture CD ROM.

- Select (do not double-click) the file named *Nominal Cut Lumber-Section.rfa* [*M_Nominal Cut Lumber-Section.rfa*].

 **Note:** Do not double-click this file, or simply open it. Unlike other Families, you must choose the specific Type from the matrix in the lower portion of the dialog to load the Type you need.

- From the matrix at the bottom of the dialog (called a "Type Catalog"), hold down the CTRL key; click **2×6 [50×150mm]** and then **2×10 [50×250mm]** to highlight them.

- Click Open to load the Types into the project (see Figure 11.16).

    This method allows you to load only the selected Types from this Family file into your project and not the other hundreds of Types associated with this Family file.

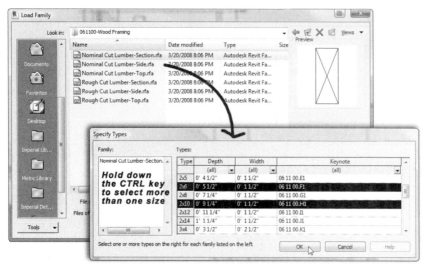

**Figure 11.16**  *Choose the Detail Component Family and Types that you wish to load*

15. From the Type Selector, choose **Nominal Cut Lumber-Section: 2 × 6** [M_Nominal Cut Lumber-Section: **50×150mm**].

- Press the SPACEBAR three times.

This will rotate it so the placement point is at the top right of the 2×6 [50×150mm].

16. Place two plates in the space between the floor joist and the foundation Wall (see Figure 11.17).

 **Tip:** Use the Move or Align tools to assist in accurate placement.

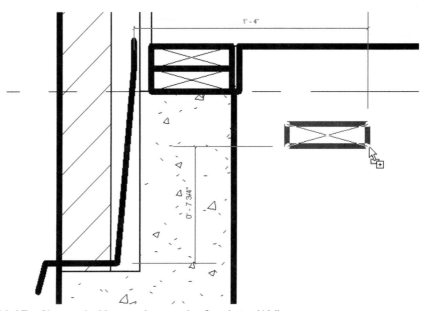

**Figure 11.17**  *Place a double top plate on the foundation Wall*

**Adjusting the Line Weight of Detail Components**

Perhaps you have noticed how heavy the outline is around this particular component. The line weight in this case is a result of the subcategory assigned to the outline geometry in the Detail Component Family. In this particular Family the subcomponent is "Heavy Lines," which is set to a line weight of 5. If you wish to make this line weight less bold in your detail, you have two options: you can edit Object Styles in the current project and reduce the line weight assignment of the Heavy Lines subcomponent, or you can edit the Family and modify the outline to use a different subcomponent. Editing the Object Styles in the current project is quicker and easier, but not considered "best practice." While the desired line weight will be achieved, the effect will apply to all Families that use the Heavy Lines subcategory and its name "Heavy Lines" will no longer be applicable.

The best practice approach is to edit the Family and reassign the outline to a lighter subcomponent. To do this, select one of the **Nominal Cut Lumber-Section: 2 × 6** **[M_Nominal Cut Lumber-Section: 50×150]** objects on screen and on the Options Bar, and click the Edit Family button (you can also right-click to find this command). In the dialog that appears, click yes. This will open the Family in the Family Editor. (The Family Editor was covered in detail in the previous chapter). Select the outline. The outline element is a Masking Region, which is a polygon object with an outline and solid opaque fill. On the Options Bar, click the Edit button to edit the sketch of the Masking Polygon. If you are working in Imperial units, you will need to create the Medium Lines subcategory. To do this, from the Settings menu, choose **Object Styles**. Click the New button in the lower-right corner to add a new subcategory. Name the new subcategory "Medium Lines" and set its line weight to 3. Click OK to finish. Select the entire outline (four lines) on screen and then choose Medium Lines from the Type Selector. On the Design Bar, click Finish Sketch and then save the Family. Finally, click the Load into Project button on the Design Bar. Answer yes when prompted to overwrite the Family.

A modified version of the Family named *MRAC Nominal Cut Lumber-Section.rfa* [*MRAC M_Nominal Cut Lumber-Section.rfa*] has been provided in the Chapter11 folder. You can make the edits listed here or load the provided Family using the **Load from Library > Load Family** command on the File menu.

17. Repeat the process (or copy) to add a sill plate above the joist at the first floor.

18. On the Design Bar, choose the **Detail Component** tool again.

- Change the Type to **Nominal Cut Lumber-Section: 2 × 10** [*M_Nominal Cut Lumber-Section: 50×250mm*].

- Use the spacebar to rotate if necessary and place a rim joist as shown in Figure 11.18.

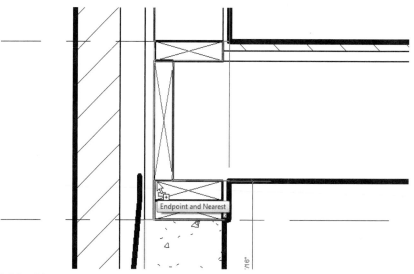

**Figure 11.18** *Place a rim joist using a 2×10 [50×250mm]*

You use the same process to load and place any Detail Component. Revit ships with a very large collection of pre-made Detail Component Families. As we discussed in the previous chapter, set aside some time to get acquainted with what is provided. You can use the components in the library, modify them, or build your own. It is usually best to start with those provided before endeavoring to create your own. Let's continue to add to our detail by repeating the load and add process to add an anchor bolt.

19. On the Design Bar, choose the **Detail Component** tool again.

- Click the Load button and browse to the *Detail Components\Div 05-Metals\05090-Metal Fastenings* folder.

- Open the *Anchor Bolt Hook-Side.rfa [M_Anchor Bolts Hook-Side.rfa]* file.

- Place the anchor at the midpoint of the lower plate.

- On the Design Bar, click the **Modify** tool or press the ESC key twice.

20. Select the bolt that you just placed; on the Options Bar, click the Properties icon.

- In the "Element Parameters" dialog, change the "Length" parameter to **1'-9" [525]**.

- Change the "Hook Length" to **3" [76]** and then click OK.

- Click the Mirror tool, clear the Copy checkbox, and mirror the bolt about its center.

- With the Align tool, use the top of the plate as reference and align the bottom of the bolt to it (see Figure 11.19).

Fine tune your placement to match the figure.

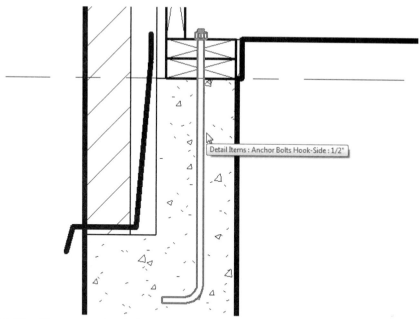

**Figure 11.19**  *Place an anchor bolt and adjust its location and parameters*

## REPEATING DETAIL ELEMENTS

Repeating Detail elements are Detail Components that automatically repeat about an invisible sketch line. This allows more rapid placement of Detail Components like studs, CMU, Brick, etc. In the detail that we are constructing, we can see the brick Layer of our Wall with the heavy cut line on the exterior and the diagonal fill pattern. This rendition is fine for general scales and overall plans and sections. However, at the scale of this construction detail, adding mortar joints will better delineate the brick veneer and suggest the individual bricks. While we could place one mortar joint and then array or copy it, a Repeating Detail Component is more expedient.

21.  From the File menu, choose **Load from Library > Load Family**.

  • Browse to the location where you installed the CD ROM files and open the *Chapter11* folder.

  • Select the *MRAC Mortar Joint with concave joint.rfa* file and then click Open.

   **Note:** Use the same Family in both Imperial and Metric.

Now that we have loaded a mortar joint Family, we will create a new Repeating Detail Type using this Detail Component.

22.  On the Design Bar, choose the **Repeating Detail** tool.

Only one Type is available on the Options Bar—*Repeating Detail: Brick*. We are going to use this as the basis for a new Type.

- On the Options Bar, click the Properties icon.
- In the "Element Properties" dialog, press ALT + E and then ALT + D to create a new Type.
- Name the new Type **MRAC Mortar** and then click OK.
- In the "Type Properties" dialog, choose **MRAC Mortar Joint with concave joint: Brick Joint** from the "Detail" list.
- Leave all of the other settings unchanged and then click OK twice to return to the view window.

23. Click at the bottom-left corner of the brick veneer and drag up past the top of the Crop Boundary and click again (see Figure 11.20).

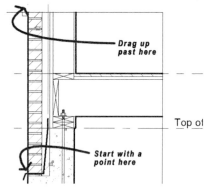

**Figure 11.20** *Place a Repeating Detail for the Mortar Joints*

- On the Design Bar, click the **Modify** tool or press the ESC key twice.
- Save the project.

## FILLED REGIONS

Filled Regions are two-dimensional shapes that contain boundary lines and fill patterns. You can draw them any shape you like and use them to create, hatch, or cover up parts of the detail drawing. We will use a Filled Region here to illustrate the filled trench on the exterior side of the foundation wall.

24. On the Design Bar, choose the **Filled Region** tool.

The Design Bar changes to sketch mode and shows only the Filled Region tools.

- From the Type Selector, choose **Wide Lines**.
- Be sure that the "Draw" tool is selected and that on the Options Bar the "Chain" checkbox and the "Line" icon are selected.

25. Zoom into the bottom of the foundation Wall near the footing.

26. Sketch the shape shown in Figure 11.21.

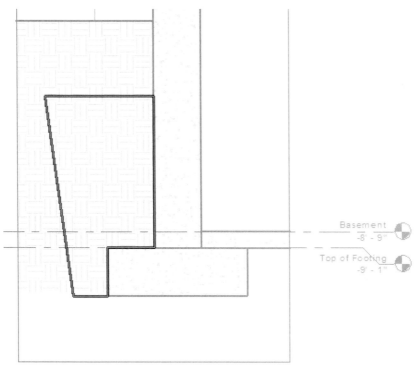

**Figure 11.21** *Sketch a Filled Region*

27. On the Options Bar, change the shape to Circle.

- Add a sketched circle with a **2" [50]** radius as shown in Figure 11.22.

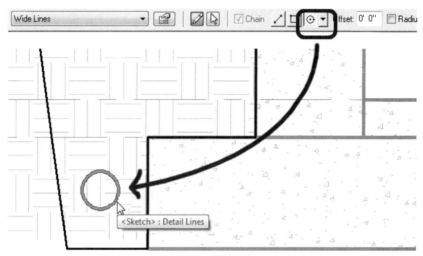

**Figure 11.22** *Sketch a circle in the Filled Region shape*

28. On the Design Bar, click the Region Properties button.

- From the Type list, choose **River Rock** and then click OK.

- On the Design Bar, click the Finish Sketch button (See Figure 11.23).

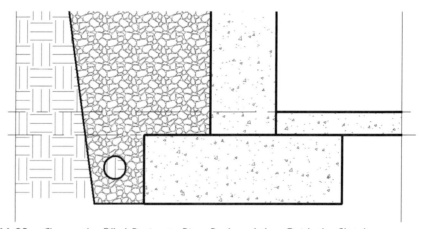

**Figure 11.23** *Change the Filled Region to River Rock and then Finish the Sketch*

 **Note:** If necessary, you can widen the crop region to allow more room to draw the Filled Region.

29. Repeat the same process for adding the finished grade with the **Earth Disturbed** Filled Region (See Figure 11.24).

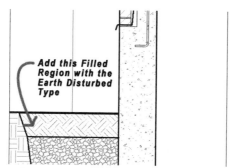

**Figure 11.24**  *Add another Filled Region*

## ADDING BREAK LINES

Next let's drop in some Break Line components to hide part of the model. Break Lines have Instance parameters so we can individually adjust their size to fit the Detail.

30. On the Design Bar, choose the ***Detail Component*** tool.

- On the Options Bar, click the Load button, browse to the *Detail Components\Div 01-General* folder, choose *Break Line.rfa* [*M_Break Line.rfa*], and then click Open.

31. On the Type Selector, verify that ***Break Line*** is chosen.

- Place a Break Line at the top of the detail to cover the top edge.

- Press the spacebar three times, and then place another Break Line covering part of the floor joist to the right (See Figure 11.25).

Break line components contain invisible Masking Regions that mask (cover up) the model objects beneath them. The concept of a mask is common in graphic design software and can be helpful in creating details.

- Use the Shape Handles to make adjustments as necessary.

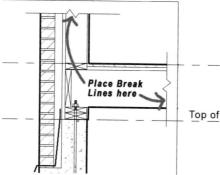

**Figure 11.25**  *Add Break lines with integral Masking Regions*

## BATT INSULATION

Next we'll place some batt insulation in the wall and floor.

32. On the Design Bar, choose the **Insulation** tool.

- On the Options Bar, set the Width to **5"** [**130**].

- Click the first point at the bottom midpoint of the stud space, move up vertically and then pick the second point above the Crop Region (See Figure 11.26).

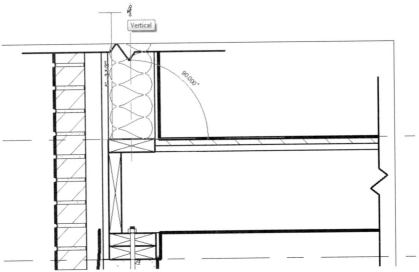

**Figure 11.26** *Draw Insulation in the Stud cavity*

33. On the Options Bar, choose "to far side" from the drop-down list.

- Click a point on the inside of the end joist and drag to the right past the Crop Region (See Figure 11.27).

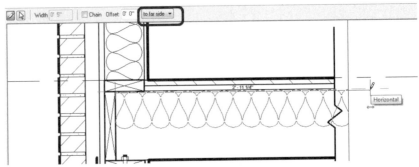

**Figure 11.27** *Draw Insulation in the Floor cavity*

- On the Design Bar, click the **Modify** tool or press the ESC key twice.

Notice that the insulation is not masked by the Breaklines. This is because there is a display order for the view-specific Detail Components and the insulation is currently on top because it was added to the view after the Breaklines. We can shuffle the display order now.

34. Select the Breakline component on the right with the Modify tool.

- On the Options Bar click the "Bring to Front" icon (See Figure 11.28).

**Figure 11.28**  *Use the Display Order icons to shuffle the order of Components in the view*

- Repeat this process on the other Breakline elements.

- Repeat once more on the flashing Detail Lines to bring them in front of the Mortar Joints.

## EDIT CUT PROFILE

Sometimes you encounter a situation where the automatically-created graphics do not suit your specific needs. One such example is the keyway locking the foundation Wall to the Footing. Revit Architecture provides us with the ***Edit Cut Profile*** tool. This tool gives us the ability to edit the path of the heavy cut line that Revit Architecture automatically generates. This type of edit is view-specific and two-dimensional. While it does not change the 3D shape of the model, it gives us a quick way to make the detail look the way we need without forcing us to model something that would have little or no benefit in other views. Since a key between the bottom of a foundation wall and the top of a footing would never be seen in any view other than a section or detail view, it would be difficult to justify the additional time or effort required to model it in 3D. Using the ***Edit Cut Profile*** tool we can make the section or detail appear as required more quickly and without the extra modeling overhead.

35. From the Tools menu, choose **Edit Cut Profile**.

- On the Options Bar, select the "Boundary between Faces" option.

This option allows us to edit two boundaries—in this case the Footing's boundary and the foundation Wall's boundary—with one sketch. If we used the other option, Face, we would have to first edit the bottom face of the foundation and then go back and edit the top face of the footings.

- Select the boundary line between the foundation Wall and the footing (See Figure 11.29).

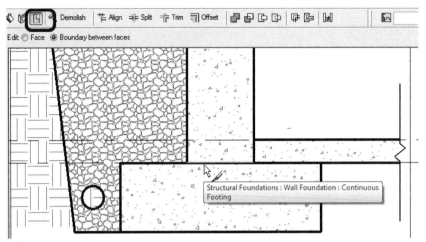

**Figure 11.29** *Using the "Boundary between faces" option, select the face to edit*

The Design Bar changes to sketch mode.

36. Using the **Lines** tool, sketch the new path as indicated in Figure 11.30.

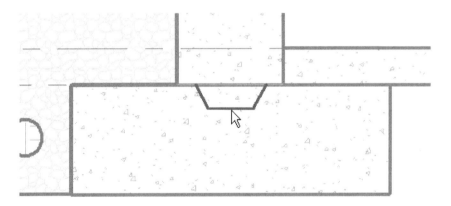

**Figure 11.30** *Sketch the new shape using a Chain of Lines*

- On the Design Bar, click the Finish Sketch button.

In this case the fill pattern is the same on both sides of the Cut line, but if they were different you would notice that the fill pattern for the footing receded and the fill pattern for the foundation wall extended to fill in the key shape.

## VIEW BREAKS

It is common that a detailed wall section will be too tall to fit on a Sheet. So it is typically broken into separate parts that crop out the repetitive areas. The crop boundary for any view includes "View Break" Controls and can be clipped to achieve this effect.

37. Select the Crop Boundary surrounding the section callout (it appears as a rectangle surrounding the drawing)

On each of the four edges of this Crop Region, a double triangle Control Handle appears at the midpoint and a "zig zag" break Control appears on either side of it. These allow us to truncate the view into smaller parts facilitating placement on a Sheet (see the left side of Figure 11.31).

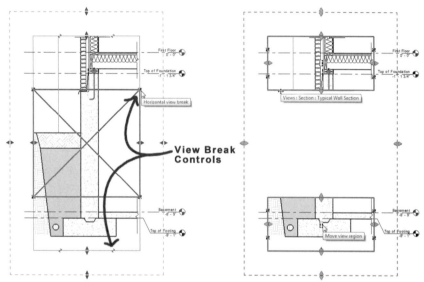

**Figure 11.31**　*View Break Controls allow you to crop out the middle portion of the section*

- Click one of the View Break Controls on a vertical edge (there are four total; you can pick any one) of the Crop Boundary (see the right side of Figure 11.31).

The view splits into two separate Crop Regions with a large gap in the middle. A blue arrow Control Handle appears in the middle of each View Break region. We can use these to move the two portions closer together. Notice that each of the two new Crop Boundaries includes the same types of Control Handles—but only on the vertical edges now. You can continue to break them into additional sub-views as necessary. But all breaks must be along the same direction as the first one—vertical in this case. In this example two is enough so we will not break it any further. However, we need to adjust the top View Break so we can see the entire Anchor bolt.

- Using the double triangle Control Handle at the bottom of the upper View Break, drag the edge down a bit to show all of the anchor bolt.

38. Click on the View Break Control arrow (in the middle) of the top View Break and drag it down so the Crop Region is a little above the upper crop boundary of the lower View Break (see Figure 11.32).

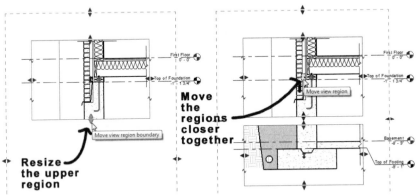

**Figure 11.32** *Drag the upper view closer to the lower one using the Move View Region Control*

If you continue to drag so that you over lap the two View Breaks they will join back into one. This is how you "remove" the break.

**Caution:** Be sure to move the portions of the View Breaks with the Control arrow in the middle. Do not drag the edge of the Crop Region. Doing so will actually move the area of the callout in both this view and the referring Longitudinal section view.

Although the sub-views are truncated and closer together, distances are dimensionally correct. Look at the Level lines to the right of the views. The First Floor is at elevation zero (0) and the Top of Footing is at elevation -9'-1" [-2700]. Therefore, if we were to add a dimension from the top of the finished floor on the first floor to the top of the footing, it should read a distance of 9'-1" [2700]. Let's try it out.

39. On the Design Bar, click the **Dimension** tool.

- Move the Dimension tool over the top edge of the Floor object at the First Floor.

Most likely the Level line will prehighlight. While we could dimension this point and still receive the correct value, we want to associate the dimension with the Floor element instead. We can use the TAB key here (like so many other places in Revit Architecture) to cycle to the element that we want.

- Press the TAB until the top edge of the Floor prehighlights and then click.

The selected edge will remain red while you complete the dimension operation.

- Move down with the Dimension tool and over the top cut line of the Footing.

If the top cut line of the Footing does not automatically prehighlight, use the TAB key again.

- With the top edge of the Footing prehighlighted, click to select it.

- Move to the left Crop Region Boundary and click next to it (in the white space) to place the Dimension string (see Figure 11.33).

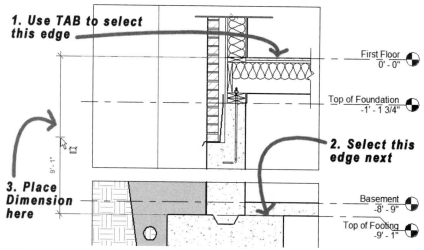

**Figure 11.33**   *Dimension the distance between the Floor and the Footing*

Notice that the Dimension displays the correct 9'-1" [2700] value from the top of the Footing to top edge of the finish Floor. As you can see, applying a View Break is a graphical convention only and has no impact on the dimensional accuracy of the model being displayed in each portion of the Crop Boundary.

- On the Design Bar, click the **Modify** tool or press the ESC key twice.

40. Save the project.

## ANNOTATION

Annotating a drawing with notes, dimensions, symbols, and tags is essential to communicating architectural design intent. Such annotations in Revit Architecture are view-specific elements. This means that these elements appear *only* in the view to which they are added. The exception to this is view indicators and Datums like section markers, elevation makers, Level Lines, Grids, and Callouts. These items are purpose-built to appear in all appropriate views and enhance the fully-coordinated nature of an Revit Architecture project.

Each view in Revit Architecture has a "View Scale" parameter and all annotations added to a particular view will scale and adjust accordingly. View indicators and Datums are included in this behavior. This means that no matter what the scale of the drawing, the annotation, view indicators and Datum symbols (level heads and grid bubbles) will be the correct size required for printed output. This behavior also applies to line weights and drafting patterns. Each graphical view you open can and often will have its own scale. In addition, if the scale parameter of a view is changed, the line weights and drafting patterns will automatically adjust.

The relative thickness of a particular line, or line weight, is controlled in a matrix of common plot scales. If desired, you can edit this matrix with the **Line Weights** command from the Settings menu. Drafting patterns will maintain their line spacing so the spacing always looks correct on printed output no matter what the scale is.

**BIM Manager Note:** Try using the out-of-the-box settings as-is for a while before making any changes. You will likely find the out-of-the-box settings to be adequate for most situations. If you do make changes, save these modified settings in a modified version of the standard Revit template file and make it your office standard. This is much more efficient than repeating your desired edits with each new project.

## CREATE A CUSTOM TEXT TYPE

A text element in Revit Architecture like other elements is associated with a Type. A text "Type" in this case is simply a grouping of parameters that control the look and formatting of the text. There are several parameters, many of which are similar to text in other computer software and are likely familiar to you.

Like other Families and Types, Text Types can be preconfigured and added to a Project Template. Additional Types can be added to the template or as a Project progresses. The process for creating a new text Type is nearly identical to the one used for duplicating other element Types; you simply duplicate an existing one and modify its parameters. Let's create a new text type for our project. In this example, we will create a "general note" text Type that is 1/8" [3] high and uses a different font. (Note: in Revit Architecture, the height is its final plotted height—you are not required to calculate text size relative to the model).

1. On the Design Bar, click the Basics tab and then click the **Text** tool.

- On the Options Bar, click the Properties icon.

2. Next to the Type list, click the Edit/New button.

**Tip:** A shortcut to this is to press ALT + E.

The Type Properties dialog will appear.

- Next to the Type list, click the Duplicate button.

**Tip:** A shortcut to this is to press ALT + D.

A new Name dialog will appear. By default "(2)" has been appended to the existing name.

- Change the name to **MRAC General Notes** and then click OK.

You can use any font that is installed on your system. Since the choice of fonts can vary widely from one computer to the next, your system may not have the same fonts as those indicated here. In an attempt to mitigate this, we will select a very commonly available Windows ™ standard font—Arial Narrow. Feel free to choose a different font if you prefer.

3. Beneath the "Text" grouping, from the "Text Font" list, choose **Arial Narrow**.

- In the "Text Size" field, type **1/8"** [**3**] (see Figure 11.34).

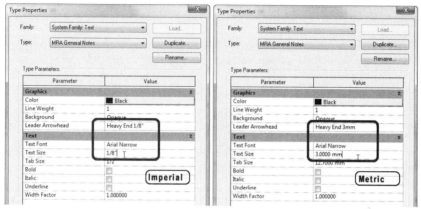

**Figure 11.34** *Create a new Text Type*

This is the size the text will be when printed out. Beneath this, you can choose to make the text bold, italic, or underline if desired. The "Width Factor" setting is used to compress or stretch the text. This is a multiplier. When set to 1, the text draws in the way it was designed in the font. A value less than 1 will compress the text and a value greater than 1 will stretch it out.

In the "Graphics" area, you can change the color of the text as well as assign an arrowhead to this Type. When you draw text, you can place just the text, or create it with a leader line attached. This is done on the Options Bar. The "Leader Arrowhead" parameter is used to assign an arrowhead Type to the Text Type.

- From the "Leader Arrowhead" list, choose ***Heavy End 1/8"*** [***Heavy End 3mm***].

**BIM** *Manager Note:* Arrowheads are System Families. You can add additional Types using the **Annotations > Arrowheads** command from the Settings menu.

- Click OK twice to complete the new Type.

The new Type will appear in the Type Selector list on the Options Bar.

## PLACING TEXT

To place text in a view, simply click down or drag a rectangle at the location where you want the text to appear. If you click the point, the text will flow in one continuous line without wrapping. If you click and drag two points, it will wrap to the width between the points. Regardless of your choice, you can always edit the wrapping of a text element later using the control handles on the text element. Pressing the ENTER key within a text element will insert a "hard" Return. This will move the cursor to the next line regardless of the automatic wrapping. It is similar to using a word processor. Typically, you simply begin typing and your word processor will automatically wrap the text to the next line when you reach the edge of the paper. If you want to start a new paragraph, you press ENTER. It is the same in Revit Architecture.

4. On the left side of the detail, click and drag a text region near the top, close to the crop region edge.

- Type **Standard Face Brick Veneer- see specification for color** (see Figure 11.35).

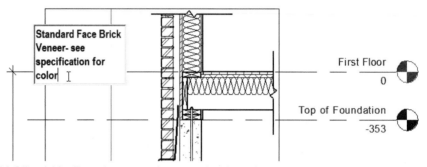

**Figure 11.35** *Add a Text element and type in the desired note*

- Click next to the note (in the white space) to finish typing.

**Note:** If a warning message appears and a text element disappears, you have created the text outside the annotation crop region (see Figure 11.36). The annotation crop appears as a dashed green boundary outside of the view crop region and hides any annotation that falls outside its boundaries. Review the next topic for a more thorough explanation of the annotation crop region.

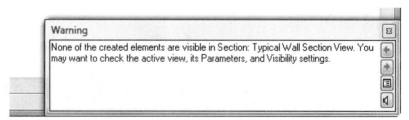

**Figure 11.36** *Revit warns you when a newly created element is invisible*

## ANNOTATION CROP REGION

In the exercises above, we made adjustments to the crop region of our detail callout views to fine-tune how much of the model's geometry was included in the view. In addition to the crop region, we also have the annotation crop region. This region falls outside the normal crop region and affects only the annotation elements added to the view. When any portion of an annotation element intersects the annotation crop, the entire element disappears. While the annotation crop can be enabled in any plan, section, or elevation view, the most effective place to utilize it is in drawings that contain matchlines. In this way, if you have text or other annotation that occurs near the matchline, Revit can show a limited amount of the duplicate annotations on each matchline sheet. To see this feature in action, explore the dependent views feature in Revit. An example is shown in Figure 14.5 in Chapter 14. You can also look up dependent views and annotation crop regions in the online help.

In our current detail, we have no need to crop the annotation. Therefore we will simply turn off the feature in the current view.

5. Right-click in the view window and choose **View Properties**.

- Beneath the "Extents" grouping, remove the checkmark from the Annotation Crop setting and then click OK (see Figure 11.37).

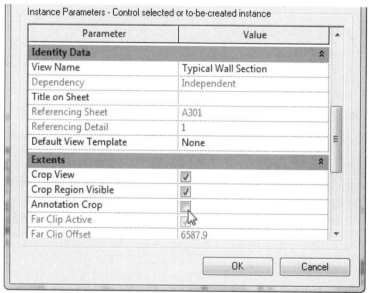

**Figure 11.37**  *Turn off the Annotation Crop*

If the notes you typed above were not showing, they should have now appeared.

**Note:** Another effective way to deal with the annotation crop in a detail view, such as the one we have here, would be to simply enlarge the annotation crop region using the control handles. However, since we used the View Breaks to split the view into two portions, we can no longer edit the width of the annotation crop region, only its height. Therefore, if you wish to enlarge the annotation crop instead of turning it off, be sure to do so before enabling the View Breaks.

Some blue control handles will appear attached to the text element while selected. You can use the one on the left to move the element (while leaving any arrow heads in place), the one on the right to rotate it, and the two small round ones on either side to resize and reshape the element and its word wrapping (see Figure 11.38).

6. Select the piece of text added above.

- Use any of the control handles to fine-tune its placement.

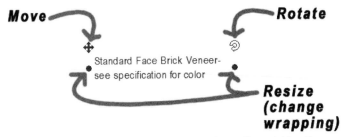

**Figure 11.38**  *Move, Rotate, or Resize a Text Element with its Control Handles*

## INCLUDING LEADERS WITH TEXT

To place a leader and arrowhead with a note, you can choose the appropriate option on the Options Bar.

7. Click the **Text** tool again.

- On the Options Bar click the "One Segment" icon next to "Leader."

- Click near the middle of the double top plate on the foundation Wall.

This is the location of the arrowhead for the Leader.

- Drag to the left and click beneath the first note (a temporary guideline will appear to assist you).

This is the end of the Leader. A text object will appear. (If you selected the "Two Segments" icon instead, you would place two segments of the Leader line before typing would begin.)

- Type the next note, **Double Top Plate**, and then click next to the note (in the white space) to finish typing (see Figure 11.39).

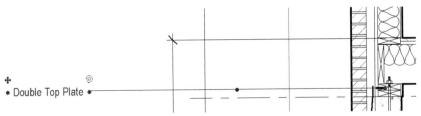

**Figure 11.39**  *Add a Text Element with a Leader*

- Using the Move handle, drag the text element to align it with the first one. A temporary guideline will appear to assist you.

When you drag a text element with a leader, be careful not to drag up or down as this will bend the leader line. This is because the leader's arrowhead stays attached to the element to which it points. If you want to move the entire thing (text and leader together), use the Move command (on the toolbar) or the arrow keys on the keyboard (to nudge) or the Modify tool and pick anywhere on the text Element's boundary except the double Move arrow icon. Make any fine-tuning adjustments that you wish to the position of either text element.

- On the Design Bar, click the **Modify** tool or press the ESC key twice.

The first text element we created does not have a leader attached to it. You can add leaders to existing text anytime.

8. Select the first text element (the brick veneer note).

- On the Option bar, click the "Add Right Leader" icon (see Figure 11.40).

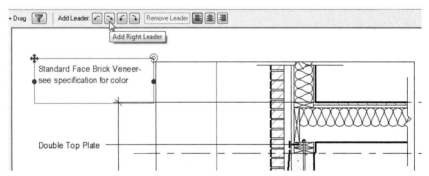

**Figure 11.40** *Add a Leader to an existing Text Element*

A leader will appear attached to the text. You can then use the drag handles to modify its shape and adjust the location of the arrowhead.

- Make adjustments with the drag controls as necessary to move the arrowhead to point at the brick.

9. Using the **Text** tool with a leader option, add a note pointing to the batt insulation that reads: **Batt Insulation**.

- Adjust the position of the note and leader as required.

Sometimes you want to have the same note point to more than one location in the detail. You can add additional leaders to an existing text element. To do this, you simply select the text element and then click the appropriate icon on the Options Bar. To remove a leader you no longer need, click the "Remove Leader" button.

- On the Options Bar, click the "Add Right Leader" icon.

- Position the leader and its arrowheads as necessary to point at the insulation in the floor (see Figure 11.41).

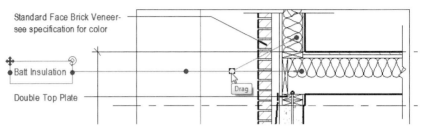

**Figure 11.41** *Add a second Leader line to the Text Element*

## ADDING KEYNOTES

Adding text is not the only way to add notes to a detail (and other views). We can also use keynotes. Keynoting allows you to annotate your details using a pre-defined list of notes. The notes are organized in a keyed list, which is why they are referred to as keynotes. If your firm uses a keynoting system such as the AIA ConDoc system, keynotes provide a means to simplify the application of keyed notes and the compilation of all keys into a keynote legend for inclusion on titleblocks.

However, it is not required that you actually utilize the keys in order to use the keynote functionality. Even if you do not currently use a keynoting system, you may still find the keynoting tools useful. This is because rather than being required to type out each note you add to a project, with keynoting you choose the note from a pre-defined list of standard notes. Furthermore, you can even pre-assign keynotes directly to your materials, templates, and Family Types in your office standard templates and library files.

Revit Architecture includes a sample keynote file organized in CSI format. You can use this list as is, edit it, or create your own. Creating or editing your own file is easy. The Keynote list is stored in a simple tab-delimited text file. If you wish to create your own file, search for the "Sample User Keynote Text File" topic in the online help for instructions and an example of the proper format. Of course using keynotes is optional, and to benefit fully from them, a certain amount of setup is required. You will have to decide if the benefits of doing so prove valuable enough to justify the initial configuration effort.

Before we begin adding keynotes to a project, you must choose a keynote file. You can use the same file for all projects in the office, or have different files for each project.

10. On the Settings menu, choose **Keynoting**.

Use the Browse button to load an existing file. You can use the provided file or create your own. The path to the file can be set to absolute, relative, or set by library location. An absolute path writes the complete path back to the drive letter. A relative path assumes that the keynote file is located in the same location as the project file and therefore only writes the path relative to the location in which the project is saved. Using the "At Library Locations" option writes the path relative to the locations defined on the "File Locations" tab of the Options dialog. The Options command is on the Settings menu.

Keynotes can be numbered using the keynote defined in the file or by sequential number relative to each Sheet in your document set. In other words, the "By keynote" method will use a fixed and predefined key. The "By sheet" method will compile the numbering uniquely for each Sheet of the set based on the notes actually used (see Figure 11.42).

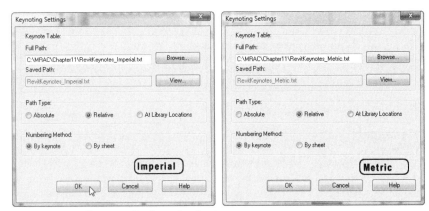

**Figure 11.42**  *Load a Keynote file from the Settings menu*

Copies of the out-of-the-box keynote files have been included with the book dataset files installed from the CD ROM. For this reason, the "Relative" path type is configured for our project (as shown in the figure).

- Verify the settings and then click OK without making any changes.

11. On the Design Bar, click the **Keynote > Element** tool.

- Following the prompt in the Status Bar, click the Anchor Bolt element.

- Click a point for the leader and then a point to place the keynote tag.

The keynotes dialog will appear. You will notice at the top of the dialog that the file and path indicated in the previous figure will appear in the title bar. A list of major categories will appear. Each contains additional sub-categories and notes. You can choose any appropriate note from the list for the item you are noting.

- From the 05090 section, choose an appropriate note such as 05090.E1 3/8" Hooked Anchor Bolt [M10 Hooked Anchor Bolt] and then click OK.

The note will appear within the keynote tag with a leader pointing to the anchor bolt detail item. Continue to keynote other items if you wish. You will only be prompted to select a note the first time you keynote an item. After assigning the note the first time, Revit will simply display that note on each subsequent instance you keynote.

## UNDERSTANDING KEYNOTE TAGS

The default keynote tag has three Types. Two Types display the key and the third displays the text of the note. Using the text display option, you can use keynote tags to speed up data entry without being required to actually use "keyed" notes (see Figure 11.43).

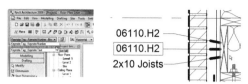

**Figure 11.43** *The out-of-the-box Keynote Tag includes three variations*

12. Select one of the keynotes you have added.

- From the Type Selector, choose **Keynote – Number** or **Keynote Text**.

The three variations are shown in the figure. If you prefer a variation not shown, you can edit the Keynote Tag Family.

**BIM** *Manager Note:* When you choose the Keynote Text option, you will notice that the text is center justified. To use right or left justified, edit the keynote Family, select the appropriate label element and change its properties to your preferred justification. A modified version is included with the files from the CD ROM in the *Chapter11\Complete* folder.

**Types of Keynotes**—Keynotes have three modes: Element, Material, and User. The Element option reads the keynote assigned to the element in the model such as the keynote assigned to a Wall or Door, not the individual layers or sub-components of the Wall or Door. To keynote the layers of a Wall or components of a Door, you would use the Material keynote option. This will read the keynote assigned to the Material of the selected component. When you wish to override the pre-defined keynote setting, choose the User option. This option will always display the "Keynotes" dialog and prompt you to choose a note. Since this option is an override, it will not update if you edit the Type or Material of the selected element.

Keynotes offer some compelling features, but they are not as mature as other features in the software. For example, certain items cannot be keynoted, like Repeating Details and Batt Insulation. Furthermore, keynotes have not been pre-assigned to the out-of-the-box content. This means that to fully benefit from the power of keynotes, a great deal of effort will be required to go through the library and assign keynotes to both Families and Materials. While you might be tempted to abandon the keynote functionality altogether based on these limitations, remember that the alternative to keynotes is to *manually* type every note. Once set up, having keynotes assigned to elements will save a great deal of time in production and will help to standardize the verbiage and phrasing used on notes throughout the office.

**Keynote Legend**—If you want to compile a list of all the keynotes used on a particular sheet or throughout the entire project, you can create a keynote legend. A Keynote Legend lists all the keys and their corresponding notes. This can be a real time saver versus manually compiling such a list. You create a keynote Legend with the command from the **View > New > Keynote Legend** menu.

## FINALIZING THE DETAIL

Our detail is nearly complete. With a few final edits, it will be ready to place on a sheet.

13. Using the process covered here, add additional notes or keynotes to the detail.

14. Using the process covered above, add in two more Break Lines between the two portions of the split detail.

15. Using the **Dimension** tool, add dimensions to the footing and foundation Walls.

 **Tip:** Remember to use your tab key as needed to select the required edges to dimension.

The Crop Regions around the detail are becoming a bit distracting. We can turn off their display.

16. On the View Control Bar (at the bottom of the view window) click the Hide Crop Region icon (see Figure 11.44).

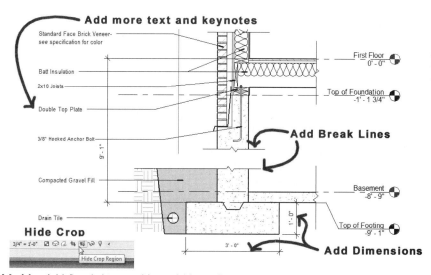

**Figure 11.44** *Add Break Lines, additional Notes, Dimensions, and hide the Crop*

## HIDING AN ELEMENT IN THE VIEW

On the left side of this detail we see a gray vertical line. This is the edge of the chimney beyond. In cases like this, where some piece of the model that we would rather not see displays, we can simply hide it in this view. This will be a view-specific override leaving the chimney unchanged in all other views.

17. Select the Fireplace element.

• Right-click and choose **Hide in view > Elements** (see Figure 11.45).

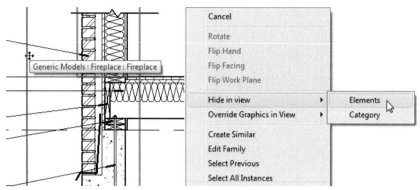

**Figure 11.45** *Hide the fireplace in the current view only*

The fireplace will disappear. Should you need to make it reappear, click the small light bulb icon on the view Control Bar. This will make all invisible elements reappear tinted red. You can then select the chimney (or any red element), right-click it, and choose **Unhide in view**.

## DETAIL THE REMAINDER OF THE WALL

To detail the rest of the Wall section, you can follow the same procedures as outlined here. Start by returning to the *Longitudinal* section view and create a new callout of the top portion. Use the View Break controls to crop the detail and remove the repetitive portions. Add masking break line Detail Components to each of the breaks. Hide the Crop Region of the view when finished. Begin adding Detail Components on top of the section cut as we did above, add drafting lines and edit the linework as required. Complete the detail with dimensions, notes and/or keynotes. Focus on the Wall connection at the second floor and the overall studs, rafters, joists, and insulation. When you are finished, the detail should look something like Figure 11.46.

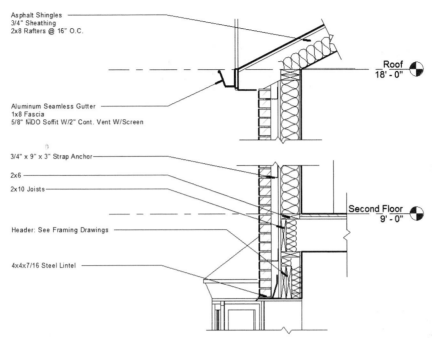

**Figure 11.46** *Create additional Details using the same process*

Most of the components that you will need are already loaded into this project; however, for items like the steel angle at the Window lintel, you can simply load them in from the appropriate library. At the roof eave, you will need to rely more on Filled Regions, Drafting Lines, and Edit Cut Profile. Using Figure 11.47 as a guide, add Filled Regions and Drafting Lines to create the blocking and roof vent. Use Edit Cut Profile to modify the Wall layers at the top of the detail (see below).

 **Tip:** Remember to use the "Send to Back" and "Bring to Front" icons on the Options Bar as required to achieve the proper look.

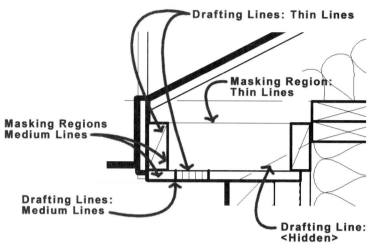

**Figure 11.47** *Using Detail Lines and Masking Regions, create a roof vent condition*

Take notice of the Detail Line:<Hidden> element near the bottom of the figure. This was drawn on top of the Filled Regions to represent the sloping rafter beyond. In many cases, you would use the Show Hidden Lines tool or the Linework tool to change the actual edge drawn automatically to the <Hidden> Type. However, in this instance, since the Wall is attached to the Roof, the Roof does not actually draw any graphics in this location automatically. This is one of many examples where judicious use of Detail Lines can make the detail read correctly in a specific view without requiring any edits to the model geometry.

## USING EDIT CUT PROFILE TO MODIFY WALL LAYERS

As you can see, you can use the technique of adding Filled Regions and Masking Regions to cover unwanted geometry and then sketching Detail Lines on top for almost any situation. There is nothing inherently wrong with the procedure, but if the underlying model should change, the Filled and Masking Regions may no longer cover the intended geometry leading to errors in coordination and intent. Another approach is to modify the underlying geometry as it is displayed in this view. To do this, we use the ***Edit Cut Profile*** tool as we did above for the footing.

18. On the Tools toolbar, click the Edit Cut Profile icon.

 **Tip: Edit Cut Profile** is also located on the Tools menu.

The cursor changes to a Cut Profile shape.

- Pass the cursor over the Wall and when the stud layer pre-highlights, click the mouse.

The Design Bar changes to sketch mode and shows some now familiar sketch tools. The existing boundary of the stud layer will show as an orange outline.

19. Using the Lines tool and the Pick Lines option (on the Options Bar), click the bottom edge of the top plate to create a sketch line (see Figure 11.48).

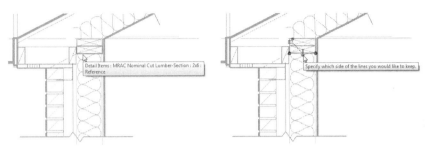

**Figure 11.48** *Sketch the new edge of the Cut Boundary*

A sketch line will appear on top of the selected edge. A small arrow handle will also appear. It should be pointing down to indicate that you wish to keep everything below the sketch line. If it points up, click it to point it down. Be sure that the line touches the edges of the stud component on both sides.

- On the Design Bar, click the Finish Sketch button.

The result should like something like Figure 11.49.

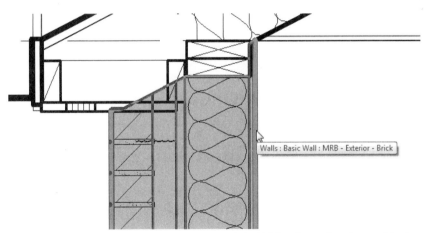

**Figure 11.49** *After the Cut Profile Edit, the shape of the Wall reflects the change (shading added for clarity)*

20. Repeat the procedure to modify the top of the brick, the air space, and the sheathing as well.

21. Make any additional edits and then save the project.

## ADD A DETAIL SHEET

Once we have created one or more Detail views, we can add them to Sheets in the same fashion as other views. We explored this process back in Chapter 4. Let's review the steps here to create a new Detail Sheet that contains our Typical Wall Section detail.

22. On the Project Browser, right-click the *Sheets (all)* node and choose **New Sheet**.

- In the "Select a Titleblock" dialog, choose **D 22 × 34 Horizontal [A1 metric]** and then click OK.

This will create "G101 - Unnamed." This is because the last Sheet we created was Sheet G100.

23. On the Project Browser, right-click on *G101 – Unnamed* and choose **Rename**.

- In the Name field, type **Details** in place of "unnamed."

- In the Number field, replace the existing value with **A601** and then click OK.

 **Tip:** You can also click directly on the (blue text) values in the titleblock and edit them directly on screen without opening the Properties dialog.

24. From the Project Browser, drag the *Typical Wall Section* detail view and drop it on the Sheet.

- Click a point to place the detail. Move it around as desired to fine-tune placement.

25. On the Project Browser, double-click to open the *Longitudinal* section view (if you prefer, you can also open *A301 – Sections* Sheet instead).

Notice that the callout annotation has automatically filled in to indicate that the detail is number one on Sheet A601. This will also remain coordinated automatically (see Figure 11.50).

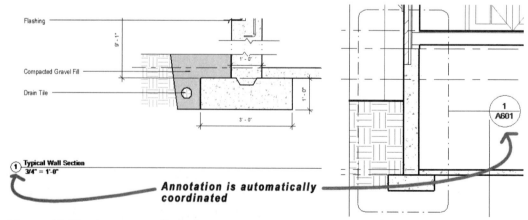

**Figure 11.50** *Annotation will coordinate automatically after adding the detail view to a Sheet*

26. Repeat the process to add the other detail to this Sheet.

If you want to align the views to one another on the sheet, it can be tricky sometimes to select and move the right thing. When you click the viewport, both the viewport and the title will highlight and move together. However, if you click just the title, you can move it independently. Revit will try to give you alignment guidelines as you drag items on screen.

In some cases, you will add details to the Sheet and then later wish to reorganize or renumber them. To do this, you edit the View Properties of the view in question. Edit the value of the "Detail Number" parameter in the "Element Properties" dialog for the view. Be sure to type a number not yet in use—Revit Architecture will not allow you to duplicate an existing number. To swap the numbers of two details, first edit one to a unique value, edit the other to the value originally used by the first, and then edit the first to the number originally used by the second.

If you make such a change, open the *Longitudinal* section view and note that the new numbers are reflected there as well. A change in one location is a change everywhere in Revit Architecture!

You can edit the view's Properties directly from the Sheet if you wish. Expand the Sheet entry on the Project Browser to see a listing of all views already placed on a particular Sheet. Right-click the name listed and choose **Properties** to jump directly to the "Element Properties" for that view. You can also double-click the view from there to open it.

## CREATE A CUSTOM VIEW TITLE

Some firms like to see the sheet where a detail is referenced. You can customize the View Title Family to include this information automatically.

27. Expand the *Families* branch on the Project Browser.

  • Expand *Annotation Symbols*, right-click the view *View Title*, and choose **Edit**.

  • In the confirmation dialog, click Yes to open the Family file.

Annotation Families are much simpler than the component Families we worked with in the previous chapter. Here you can add linework, text, and labels. Labels report the values of parameters. In the case of View Title Families, we can report the name and number of the detail, the scale and the referring detail, and the sheet.

28. On the Design Bar, click the **Label** tool (see Figure 11.51).

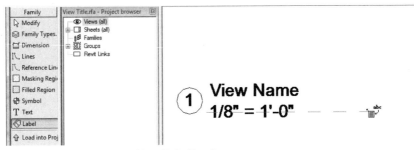

**Figure 11.51** *Add a Label to the View Title Family*

The line under the view title is actually part of the viewport object back in the project. Therefore, as you can see, there is no line here. You will have to approximate the correct location of the new label relative to the line. We'll place it next to the scale Label in this case.

- In the "Select Parameter" dialog that appears, choose Referencing Detail and then click OK.

- Use the control handles to reduce the width of the field.

- Repeat the process to add the Referencing Sheet Label next to it.

- Finally, add a piece of text in front to read "Referencing Detail:" and another one in between the two labels that contains a slash (/) (see Figure 11.52).

**① View Name**
**1/8" = 1'-0"**                    Referencing Detail:   **R / R101**

**Figure 11.52** *Add Labels for the Referencing Detail and Sheet*

If you wish, add additional graphics, text, or labels.

29. From the File menu, choose **Save As**.

- Browse to the folder with your project file and save the Family as: **MRAC View Title.rfa**.

- On the Design Bar, click Load into Projects.

30. On the *A601 – Details* sheet, select both viewports, right-click, and choose **Element Properties**.

- Click the Edit/New button and then the Duplicate button.

- Name the new Type: **MRAC Viewport with Referencing Title**.

- Change the Title entry to *MRAC View Title* and then click OK twice (see Figure 11.53).

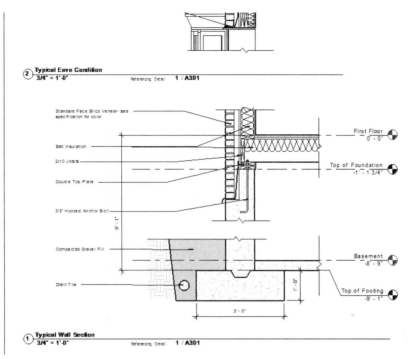

**Figure 11.53** *Load the Custom View Title into the project and apply to the viewports*

31. Save the project.

## DRAFTED DETAILS (NOT LINKED TO THE MODEL)

In some cases, you will want to add a detail to a project that does not require a callout underlay from the model. There might be several situations where this is appropriate. Examples include typical details that are generic in nature such as a typical head, jamb, or sill detail. Other examples might include a carpet transition, typical blocking condition, or just a simple diagram of something related to the project but not specific to a particular area in the model. To create these kinds of details in Revit, we use a drafting view. A drafting view is like a simple blank sheet of paper. You can draw your detail on this blank page using any of the tools covered so far like detail components, filled regions, masking regions, drafting lines, and text. You can even add view references from other views if appropriate.

### CREATING A DRAFTED DETAIL

In this example, we will create a drafting view and a simple carpet transition detail. This can be created either with or without a view reference callout in our floor plans. You can create it as a typical, unreferenced detail by creating a new **Drafting View** from the **View > New** menu. If you want to reference the drafting view from a particular area of the plan, you can create a drafting view from the section and callout tools. To do this, you choose the "Reference other View" setting on the Options Bar before drawing the section or callout. For this example, we will create an unreferenced detail. In the next sequence, we will create a referenced one using the section.

1. On the View tab of the Design Bar, click the **Drafting View** tool (or from the View menu, choose **New > Drafting View**).

- In the dialog that appears, type **Floor Transition Detail** for the name.

- Choose **3"=1'-0"** [**1:5**] and then click OK (see Figure 11.54).

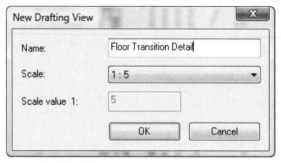

**Figure 11.54** *Create a new Drafting View*

A new drafting view will be created and opened. When Revit Architecture opens the new drafting view the most obvious characteristic is that the view is empty, showing no model geometry. A drafting view is like a blank sheet of paper. There are no automatically-generated graphics from the model.

2. On the Drafting tab of the Design Bar, click the **Detail Component** tool.

- On the Options Bar, click the Load button.

Your system should automatically take you to the Revit Library folder (usually the *Imperial* or *Metric Library* folder).

- Browse to the *Detail Components\Div 06-Wood and Plastic\06100-Rough Carpentry\ 06160-Sheathing* folder and load the *Plywood-Section.rfa* [*M_Plywood-Section.rfa*] file.

If you have trouble finding this file, or if you did not install the default library files, all of the Families noted in this section are provided with the files from the CD ROM. You will find them located in the same folder structure as noted here in the *MRAC\Imperial Library* [*MRAC\Metric Library*] folder. Feel free to load the required detail components from there instead.

- Following the prompts, create a horizontal length of plywood approximately 10" [250] long across the middle of the screen (see Figure 11.55).

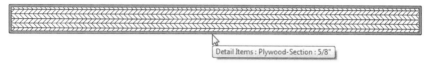

**Figure 11.55**  *Draw the plywood subfloor*

- Zoom in on the component after you draw it.

3. Click Load again, browse to the *Detail Components\Div 09-Finishes\09600-Flooring\09640-Wood Flooring* folder, and load the *Wood Strip Flooring-Section.rfa* [*M_Wood Strip Flooring-Section.rfa*] file.

- Place the item on the top edge of the plywood.

- Repeat the process to load four more Families from the *Detail Components\Div 09-Finishes\09600-Flooring\09680-Carpeting* folder:

  - *Carpeting-Section.rfa* [*M_Carpeting-Section.rfa*]

  - *Carpet Reducer at Flooring-Section.rfa* [*M_Carpet Reducer at Flooring-Section.rfa*]

  - *Carpeting Tack Strip-Section.rfa* [*M_Carpeting Tack Strip-Section.rfa*]

  - *Carpet Pad-Section.rfa* [*M_Carpet Pad-Section.rfa*]

With the first three, simply place them on screen in approximate locations for now. The *Carpet Pad-Section.rfa* [*M_Carpet Pad-Section.rfa*] Family behaves like the *Plywood-Section.rfa* [*M_Plywood-Section.rfa*] Family above did. You must click two points to place it. You can click two points along the top edge of the plywood for this component.

4. Move and copy the carpet and wood flooring components on screen to match Figure 11.56.

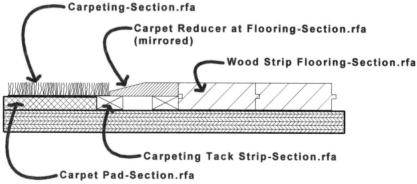

**Figure 11.56**  *Draw the plywood subfloor*

5. Add a Break Line Detail Component to the end of the detail.

- Edit the Element Properties of the Break Line and change the Dimensions parameters as follows:

| | |
|---|---|
| Jag Width | 1/2" [12] |
| Jag Depth | 3/4" [18] |
| Right | 1" [25] |
| Left | 1" [25] |
| Masking Depth | 2" [50] |

 **Tip:** Set the Jag Width and Jag Depth first and the Right and Left last. This will avoid Revit's displaying error messages.

- Mirror the Break Line to the other side.

6. Add notes or keynotes to complete the detail (see Figure 11.57).

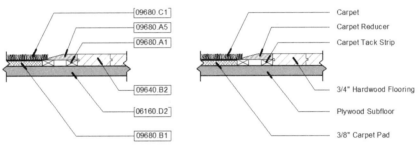

**Figure 11.57**   *The completed detail showing keynotes on the left and text notes on the right*

7. Add the detail to the *A601 – Details* sheet.

If you want this detail to be a typical detail, edit its View Properties and change the "Title on Sheet" parameter to "Typical Floor Transition Detail." Otherwise, if you prefer to call it out from the plan, you can open the *First Floor* plan view, zoom in on an appropriate area and then click the **Section** tool. Before you draw the section, check the "Reference other View" box on the Options Bar and choose Floor Transition Detail from the list of views. Draw the section line. The callout will read detail 3 on sheet A601 (see Figure 11.58).

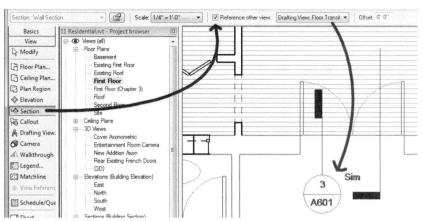

**Figure 11.58**  *You can optionally add a section callout that references the existing Drafting View*

8. Save the file.

## WORKING WITH LEGACY DETAILS

Often details are reused from one project to the next. These "standard" details are typically kept in libraries for easy reuse and retrieval. In the days before computer design and drafting software, such a library would be a three-ringed binder from which photocopies were made. With computers, these standard details are stored digitally. If your firm has been using CAD software for a while, you likely already have such a digital library of standard details. You can use these legacy files directly in your Revit Architecture projects. You simply import the DWG or DGN files and place them on Sheets like other details.

### CREATE A REFERENCED SECTION VIEW

In this tutorial we will assume that the handrail of the existing Stair will be replaced with a new one. To show this, we will create a "Referenced Section View" to create a Section marker callout of a handrail detail within a stair section view. However, instead of creating the actual section view in Revit Architecture or drawing an unreferenced drafting view as we did above, the Referenced Section will link to a drafting view containing an AutoCAD file.

1. On the Project Browser, double-click to open the *Section at Existing Stair* section view.

- Zoom in on the area of the Stair between the First Floor Level and the Second Floor Level.

2. On the Design Bar, click the Basics tab and then click the **Section** tool.

- From the Type Selector, choose **Detail View: Detail**.

- On the Options Bar, set the "Scale" to **6"=1'-0"** [**1:2**].

- Place a checkmark in the "Reference other View" checkbox, and verify that the menu is set to **<New Drafting View>** (see Figure 11.59).

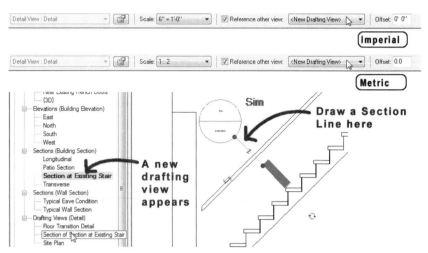

**Figure 11.59** *Create a new Section View set to reference a new Drafting view*

These settings instruct Revit Architecture to create a new Drafting view instead of the typical live section view of the model. The detail marker will point to this new drafting view.

- Drag the section line through the Railing as shown in Figure 11.59.

Notice that a new Drafting view was created on Project Browser beneath *Drafting* views.

3. On the Project Browser, right-click the new Drafting view and choose **Rename**.

- Name the view **New Railing Detail** and then click OK.

4. In the view window, double-click on the detail head to open this Drafting view (or double-click the name on the Project Browser instead).

We again have a blank page upon which to work. Drafting an image that makes sense relative to the detail cut location is up to you. The only reference back to the model is the callout. We have already seen how we can draft something from scratch. Now let's look at importing a legacy CAD file.

## IMPORT A DETAIL DRAWING

5. From the File menu, choose **Import/Link > DWG, DXF, DGN, SAT**.

- In the "Import/Link" dialog, browse to the *Chapter11* folder and choose *Typical Handrail Detail.dwg* [*Typical Handrail Detail-Metric.dwg*].

- In the "Layer/Level Colors" area, choose "Black and white."

- In the "Positioning" area, choose "Manually Place" and then the "Cursor at center" option.

- In the "Import of Link" area, be sure that the "Link" checkbox is cleared (see Figure 11.60).

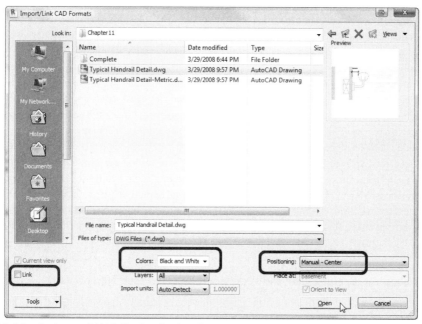

**Figure 11.60** *Link in a DWG file for the handrail detail*

If you check the "Link" box, it will link the drawing file into the project as opposed to importing it. Linking maintains a connection to the original file. If it should change, you can refresh the link in Revit Architecture to reload and display the latest version of the file. Import embeds the file into the Revit Architecture project and does not maintain a link. If the file were changed outside of Revit Architecture, you would not be able to display those changes and would need to re-import the modified file. To reload a linked DWG file, use the same "Manage Links" dialog (File menu) that we used in previous chapters to reload linked RVT files.

> 6. Click Open to import the detail, and then click a point on screen to place the detail in the view.
>
> • Verify that the scale of the current view is **6"=1'-0"** [**1:2**] as indicated above. If it is not, please change it.

Notice that if you change the scale, it has an impact on how the lineweights of the imported view display. If you wish to manipulate the way that the lineweights import, experiment with the **Import/Export Settings** command on the File menu.

This is a typical detail and there is no need for any changes. If we needed to make edits, we could select the detail, and then on the Options Bar, choose the Explode button to convert it to individual Revit Architecture Detail Lines and Text so we could edit it. However, it is best to make such edits in the original file using its native application instead of using Explode.

**BIM** *Manager Note:* If you choose to explode imported CAD files, you will discover that many element Types are added to your file beyond what you see on screen or what you would otherwise expect. For example, you will likely end up with many line styles, text styles, and other elements bearing names reminiscent of the original CAD file's layers. In addition, regardless of whether you choose to explode the file, you will get Materials bearing names like Render Material 63-0-255 in your Material list. In general, these items will not cause you difficulty but they can increase the size of your files and cause confusion among team members. If you have decided to explode a CAD file, consider the following procedure. First, if you have access to the CAD program that created the file, open the file there first and clean up the geometry as much as possible. This includes deleting unneeded geometry and layers, purging the file, and resaving it. Next, import the CAD file into a new Revit project. Explode the CAD file in this temporary project and perform additional cleanup. This will include reassigning linework to appropriate Revit Line Styles, changing text to Revit text Types, etc. Please note that CAD dimensions and text leaders will not become Revit dimensions or leaders, so if you want actual leaders and dimensions, you will need to recreate these items. Once you have cleaned up the file to your satisfaction, you can select all of the elements and copy and paste them back to a drafting view in your original project.

7. On the Project Browser, double-click to open the *A601 - Details* Sheet view.

- Drag the Drafting view and drop it on the Sheet (see Figure 11.61).

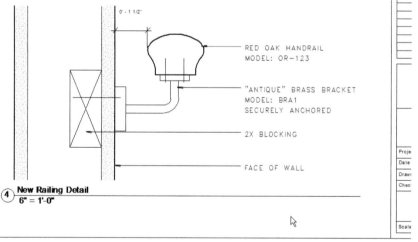

**Figure 11.61** *Add the Detail view to the Details Sheet*

This will become detail 4 on the sheet. If you return to the *Section at Existing Stair* view, you will see that this number and sheet reference have appeared automatically in the callout.

8. Save the project.

## ADDITIONAL DETAILING TECHNIQUES

Except for Drafting views, all views in the Revit Architecture project are generated directly from the building model. While Revit does a very good job of interpreting this model geometry into abstracted two-dimensional representations such as plans and elevations, there are often items that we wish to manipulate in order to create the Architectural drawings we are accustomed to producing. We have already seen all of the techniques that are used to perform such edits. Until now we have used these techniques and tools only on detail views. However, you can add drafting embellishment on any Revit view including plans, sections, and elevations. All such edits, including those made with the Linework tool, Filled Regions, Masking Regions, Detail Components, Edit Cut Profile, and Drafting Lines can be done on any view. More importantly, such edits apply *only* to the view in which they are applied.

### EMBELISHING MODEL VIEWS

Let's make a few enhancements to one of our elevation views.

9. On the Project Browser, double-click to open the *East* elevation view.

One common architectural drafting convention is to show the new foundation in an elevation as dashed below grade. We can achieve this using a combination of the Linework tool and adding Drafting Lines. Let's start with the footing.

10. Select the Terrain element, and then on the View Control Bar, choose **Hide Element** from the Temporary Hide/Isolate menu.

A cyan colored boundary will appear around the viewport. Remember that this is the temporary hide/isolate command. The cyan boundary appears as long as some elements are temporarily hidden. Temporary hide/isolate does not affect printing and is reset when the model is closed.

11. Select each of the Level Heads that do not have associated views (the ones that are black), right-click and choose **Hide in View > Elements**.

This will hide levels for "Top of Footing" or "Bottom of Stair" etc. This is the permanent hide command. These elements will stay hidden even after closing and re-opening the model. Permanently hidden elements also do not print. To reveal hidden elements and unhide them, click the light bulb icon on the View Control Bar. If you try this now, a maroon colored border will surround the screen, the three hidden Level Heads will appear maroon in color as well, and the temporarily hidden terrain, which is still hidden, will appear cyan in color. Click the light bulb again to exit the mode.

12. On the Tools toolbar, click the Linework icon.

- Choose **<*Hidden*>** from the Type Selector.

- Click on each of the edges of the footings to change them to <Hidden> lines.

- Do not change any of the vertical foundation lines yet (see Figure 11.62).

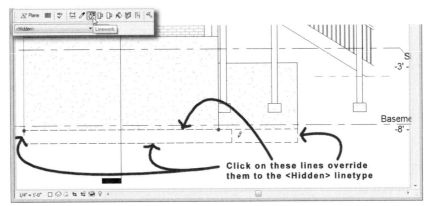

**Figure 11.62**   *Change the display of the footing lines to <Hidden> with the Linework tool*

You may need to pick more than once in the same general spot or a little to either side since there is more than one footing in the same spot in the elevation. If you are unhappy with the result, you can instead use the <Invisible lines> Type and then draw a continuous Drafting Line in on top. Be sure to lock the constraint padlock icon to keep the Drafting Line associated with the position of the footing

13. On the View Control Bar, choose **Reset Temporary Hide/Isolate**.

The terrain model will reappear. Notice that the footing still shows dashed through the terrain and is no longer hidden.

14. On the Tools toolbar, click the Linework icon.

- Choose **<Hidden>** from the Type Selector again.

- Click on one of the vertical lines of the foundation Walls.

   With it still highlighted, a drag handle will appear at either end.

- Drag the top handle down to the point where it intersects the terrain (see Figure 11.63).

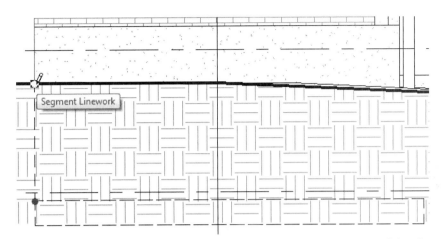

**Figure 11.63** *Change the Linework of the foundation walls and edit the extent of the change with the drag handles*

- Repeat for other vertical foundation Wall edges.

Later if you wish to change the linework to a different linetype or return it to its default setting, you can use the Linework tool again. To restore the default, use the <By Category> option from the Type Selector.

If you wish to modify the way that the terrain displays, you can use a Filled Region to trace over it. Draft additional linework as desired to complete the elevation. You can add notes, dimensions and tags as required.

15. On the Design Bar, click the Drafting tab, click the *Tag* tool, and then choose *By Category* from the flyout.

- On the Options Bar, clear the "Leader" checkbox.

- Click on each of the Windows in the new addition. (Do not tag the Windows of the existing house.)

16. Add some text or keynotes to the patio on the right or to indicate materials of the elevation such as brick veneer and roof shingles (see Figure 11.64).

**Tip:** Try the *Keynote > Material* option to keynote the materials in the Wall instead of the entire Wall.

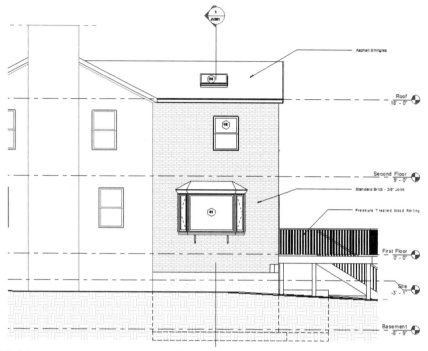

**Figure 11.64** *Add tags and notes to complete the elevation*

17. Perform similar edits in other elevations if you wish.

18. Save the project.

## CONTROLLING DISPLAY OF ITEMS BEYOND

To show depth in elevations, it is a common architectural convention to lighten the line weights of objects as they recede from view. Unfortunately, Revit does not offer an automated way to do this in elevations and sections. However, we can use the override graphics feature to manipulate the graphical display manually. Like all graphical overrides, edits you make are view-specific. So they will apply only to the view in which you make them. It may take you a little time and effort to fine-tune the elevations to display as desired, but you should be able to achieve acceptable results. We'll do a quick example here to illustrate the concept and process.

19. On the Project Browser, double-click to open the *North* elevation view.

• Feel free to repeat any of the previous edits (foundation display, notes, etc) on this elevation before you proceed.

• Select the gable Roof, its fascia boards and gutter, and the two upper Windows on the right.

• Right-click and choose **Override Graphics in View > By Element**.

- Expand the Projection Lines item and change the Weight to **1** and then click OK (see Figure 11.65).

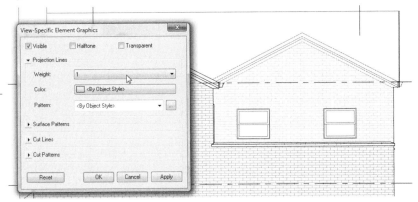

**Figure 11.65**   *Override the graphics of selected elements in the view*

This change may be hard to see without zooming in. We also want to lighten the brick hatch on the Wall beyond at the patio in the same area.

20. Select the Wall beyond at the patio in the same area.

- Right-click and choose **Override Graphics in View > By Element**.

- Check the Halftone checkbox and then click OK.

Notice that because of the way that the Wall joined, we are seeing the end of the perpendicular Wall on the right side. We can edit the join of this Wall on the Second Floor to fix this.

21. On the Project Browser, double-click to open the *Second Floor* plan view.

- On the Tools toolbar, click the Edit Wall Joins icon.

- Click on the intersection of Walls at the lower-left side of the patio.

- On the Options Bar, click the Miter radio button and then click the **Modify** tool to accept the change (see Figure 11.66).

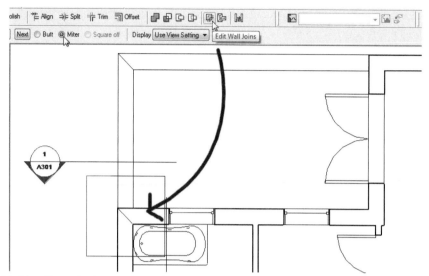

**Figure 11.66** *Override the graphics of selected elements in the view*

When you return to the *North* elevation, you will notice that selecting the Wall now gives better results.

> 22. Repeat any of these procedures on the remaining elevations and then save the project.

## LEGEND VIEWS

As our final exploration, we will look at another type of drafting view: the Legend View. This kind of view, as its name implies, is used to create symbol legends in your project. When working in a legend view, you can add Legend Components. Legend Components are symbolic versions of all Families and Types in your project.

> 23. On the Design Bar, click the View tab and then click the **Legend** tool.
>
> • In the "New Legend View" dialog, type Door Types for the name, choose 1/4"=1'-0" [1:50], and then click OK.

Like the other drafting views we have created, a blank page will appear. The unique feature of a legend view is the availability of the ***Legend Component*** tool on the Drafting Design Bar. We can use this tool to place a symbolic representation of any Family in the project. In this case, we are building a Door Types Legend, so we want to add elevation views of each kind of door, but do not want to add actual Doors, which would throw off the count in the Door Schedule later. This is where the Legend Component comes into play.

> 24. On the Drafting tab of the Design Bar, click the ***Legend Component*** tool.
>
> • From the Family list on the Options Bar choose **Doors: Single Flush: 36" × 80" [Doors: M_Single Flush: 0915 × 2032mm]**.
>
> • From the View list on the Options Bar choose **Elevation: Front**.

- Click a point on screen to place the symbol.

- Repeat the process to place each of the following:

  - Doors: Bifold-2 Panel: 36" × 80" [Doors: M_ Bifold-2 Panel: 0915 × 2032mm]

  - Doors: Bifold-4 Panel: 72" × 80" [Doors: M_ Bifold-4 Panel: 1830 × 2032mm]

  - Doors: Double-Glass 2: 68" × 80" [Doors: M_ Double-Glass 2: 1730 × 2032mm]

Line them up next to one another. You cannot tag the symbols because they are not real doors, but you can add text and dimensions where appropriate.

25. Add labels, notes, and dimensions as appropriate (see Figure 11.67).

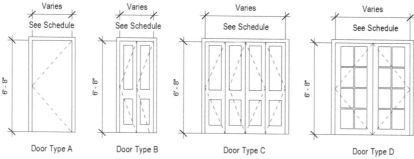

**Figure 11.67** *Add notes and labels to complete the legend*

To create the dimension shown in the figure, first add a dimension and then create a separate text object. Place the text on top of the dimension and it will mask out the actual value. You can drag this legend view onto any sheet like the other views. Perhaps the most appropriate sheet for this legend would be the door schedule sheet. Since we have not created a door schedule yet, we will wait until the scheduling chapter for that task.

26. Save the project and close the file.

## SUMMARY

Understanding the relationship between modeled elements and drafted elements is an important concept in Revit Architecture. Creating the basic model geometry can be accomplished in nearly any convenient view and as we have seen throughout this book, will remain coordinated as changes occur in all views. Drafting and annotation, on the other hand, occur in only the currently active view. This means that we can apply additional embellishments on top of an automatically generated model to explain and clarify design intent. We can modify the display of underlying model geometry using the Linework tool, element level graphic overrides, or view-specific display settings. Finally, we can create drafting views, which contain only drafting elements and no model geometry. Using a combination of these techniques, we can fine-tune any Revit view for inclusion in our complete set of architectural construction documents.

- Detailing occurs at many levels in Revit Architecture: as part of the model, as view-specific embellishment on top of the model, and as completely independent drafting views.
- You can use Wall Type edits such as unlocking layers, adding Sweeps, and adding Reveals to add details to the Walls that show throughout the model.
- Create callout views of any overall view to create the starting point for a construction detail.
- Each view has its own scale and visibility settings.
- Detail Components and Detail Lines are view-specific embellishments that are used to convey design intent.
- Repeating Detail Components save time by adding an array of Detail Components at a predefined spacing.
- Use Masking Regions in any view to mask unwanted portions of the model.
- Use Filled Regions to apply patterns to areas and draw view-specific embellishments.
- Add Break Lines and adjust Crop Regions to isolate "typical" portions of detail views.
- Keynotes can be used in addition to text to annotate drawings. Keynotes link to a standard list of notes and help to maintain consistency from one drawing to the next.
- Use Edit Cut Profile to modify the automatically created profile of building model elements within a particular view.
- Adding details to a sheet automatically numbers them and keeps the annotation coordinated.
- You can draw isolated two-dimensional details that do not link to the model. Use drafting views for this purpose.
- Import legacy CAD details and add them to drafting views.
- Edit any section or elevation view using similar techniques to those used to create and modify details.
- Use the Override Graphics in View command to indicate elements that appear "beyond" in an elevation or section.
- A legend view is a special kind of drafting view that allows symbolic representations of any project Family to be added and annotated.

# CHAPTER 12

# Working with Schedules and Tags

## INTRODUCTION

Schedules are an important part of any architectural document set. In traditional architectural practice, even when using computers, generating schedules is a laborious process of manually tabulating the sometimes hundreds of items in a project that require presentation in a schedule. For example, in firms that do not use BIM software, a door schedule involves the painstaking process of manually listing each door specified in a project and then typing in detailed information about each entry—even those bits of information (such as size) that should be easily queried from the plans. In Revit Architecture, Schedules are generated automatically from the building model data already in the project in real-time. A Schedule in Revit is simply another view of the project that differs from plans and elevations only in its presentation as tabular information rather than graphical information. You can create a Schedule from nearly any meaningful information in the model and like all Revit views; you can edit in one view and see the change instantly in all views, including the Schedule.

## OBJECTIVES

In Chapter 4, we added some typical Schedules to our commercial project. In this chapter, we will explore the workings of these Schedules as well as create additional Schedules not yet in our project. After completing this chapter, you will know how to:

- Add and modify a Schedule view
- Edit model data from a Schedule view
- Place a Schedule on a Sheet
- Work with Tags
- Work with Rooms and Color Fills

## CREATE AND MODIFY SCHEDULE VIEWS

In this chapter, we will return to our commercial project and look deeper into the Scheduling tools in Revit Architecture. As was mentioned above, we added Schedules to this project back in Chapter 4. However, we simply imported those

Schedules into the commercial project from an existing template project. We'll start with a quick review of those Schedules and then take a detailed look at how to create a new Schedule view from scratch.

## INSTALL THE CD FILES AND OPEN A PROJECT

The lessons that follow require the dataset included on the Mastering Revit CD ROM. If you have already installed all of the files from the CD, simply skip down to step 3 below to open the project. If you need to install the CD files, start at step 1.

1. If you have not already done so, install the dataset files located on the Mastering Revit Architecture CD ROM.

   Refer to "Files Included on the CD ROM" in the Preface for instructions on installing the dataset files included on the CD.

2. Launch Revit Architecture from the icon on your desktop or from the *Autodesk >* *Revit Architecture* group in *All Programs* on the Windows Start menu.

 **Tip:** In Windows Vista, you can click the Start button, and then begin typing **Revit** in the "Start Search" field. After a couple letters, Revit Architecture should appear near the top of the list. Click it to launch to program.

   • If the New Features Workshop dialog appears, choose "Maybe later" and then click OK.
3. On the Standard toolbar, click the Open icon.

 **Tip:** The keyboard shortcut for Open is CTRL + O. **Open** is also located on the File menu.

   • In the "Open" dialog box, click the *My Documents* icon on the left side.
   • Double-click on the *MRAC* folder, and then the *Chapter12* folder.

   If you installed the dataset files to a different location than the one listed here, use the "Look in" drop-down list to browse to that location instead.
4. Double-click *12 Commercial.rvt* if you wish to work in Imperial units. Double-click *12 Commercial Metric.rvt* if you wish to work in Metric units

   You can also select it and then click the Open button.

The project will open with the last saved view visible on screen.

## VIEW EXISTING SCHEDULES

You will recall that the Commercial project already has a few Schedules in it that were added back in Chapter 4. One of the most significant benefits of Revit Schedules is that once you have added them to the project, they maintain themselves. In other words, we have not looked at the Schedules since we added them back in Chapter 4. Let's have a look at how they have been shaping up.

5. On the Project Browser, beneath Schedules/Quantities, double-click to open the *Door Schedule* view.

Wow! Every Door added to the project since Chapter 4 has been automatically added to the Schedule. This is a live view of our project that is filtered to show only Doors and present them in a list rather than a drawing. That's all there is to a Schedule in Revit.

There is still plenty of work to be done to this Schedule, but a good deal of it is already done. Let's focus on the Door Number column for the time being. Notice that the numbers are not sequential and some are missing. Since a Schedule is a live view, you can simply click in the Door Number field and edit the value. The results are the same as if you were editing the item from a plan or other kind of view, but editing in a Schedule is typically faster and more efficient. You can work with a Schedule and plan view tiled next to each other to assist in editing. You can also select a line-item in the Schedule and then click the Show button on the Options Bar.

6. Open the *Level 3* floor plan view.

- Close or minimize all other windows and then tile the Schedule and the third floor windows.

- Select a row in the Schedule (see Figure 12.2).

    Notice how it highlights in the plan. You can also use the Show button on the Options bar. This zooms in on the selected item.

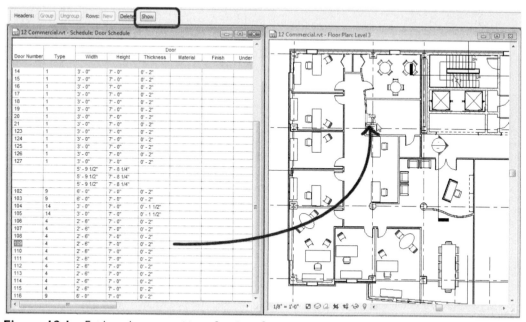

**Figure 12.1** *Explore the composition of various furniture items in the plan*

Spend some time going back and forth between the plan and Schedule views. Get comfortable with the interaction. If you wish, you can renumber the Doors so that they correspond better to the project. For example, you can renumber all the third floor Doors starting with the number 301 and incrementing from there. Open the other plans to edit their numbers as well.

7. After editing the numbers, right-click on the Schedule and choose View Properties.

- In the "Element Properties" dialog, click the Edit button next to Sorting/Grouping.

- On the Sorting/Grouping tab, choose **Mark** from the Sort by list and then click OK twice.

There are plenty of additional modifications that we could make to the Door Schedule. For now however, we will leave them for later and create a new Schedule. Before continuing, feel free to open any of the other existing Schedules and study how they have come along since we set up the project.

## ADD A SCHEDULE VIEW

8. On the Project Browser, double-click to open the *Level 3* floor plan view.

In Chapter 10, we added some furniture to this plan. As you can see, the layout has been refined a bit. In this topic, we will create a furniture Schedule. To add a new Schedule, look for the *Schedule/Quantities* tool on the View tab of the Design Bar or the **View > New** menu.

9. On the Design Bar, click the View tab and then click the *Schedule/Quantities* tool.

The "Category" list is a fixed list of categories built into Revit. In the "Name" field you can type any suitable name to describe the contents of the Schedule—usually the default is acceptable. When you choose the "Schedule building components" radio button, you get a list of building elements from the model with each one occupying a row of the Schedule. The "Schedule keys" option creates a user-defined list of common elements useful in filling in other Schedules. For example, you could define a certain kind of Door or chair, including its manufacturer, model number, finish, etc. When you begin the task of filling in your door or furniture schedule, you can choose a key and it will input all the other fields automatically saving you much of the redundant data entry. We will look at this option in the "Using Schedule Keys to Speed Input" topic below. In a multi-phase project, you can indicate which Phase the Schedule should report. It is most common to Schedule the New Construction Phase.

- In the "New Schedule" dialog, Select "Furniture" from the Category list.

- In the "Name" field, accept the default name of "Furniture Schedule."

- Accept the remaining defaults of "Schedule building components" and "New Construction" for the Phase and then click OK (see Figure 12.2).

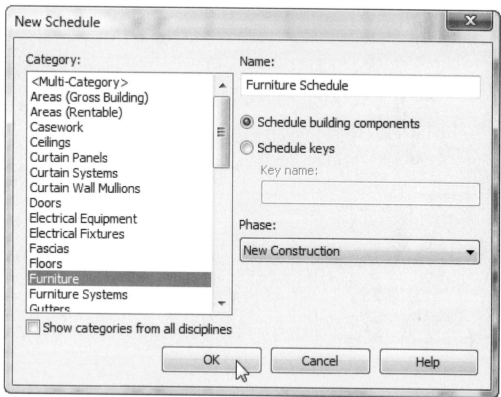

**Figure 12.2** *Create a Furniture Schedule*

The "Schedule Properties" dialog appears next. The first tab (Fields) presents a list of "Available Fields" on the left side and the fields actually appearing in the Schedule on the right. This list includes all of the parameters available for the items you are scheduling (furniture in this case). To add a field to a Schedule, simply select it from the list and then click the Add button to move it to the "Scheduled fields" list on the right. Once you add fields, you can adjust the display order on the right. Each field will become a column in the resultant Schedule view.

- In the "Available fields" list, choose Type Mark and then click the Add → button.

 **Note:** "Mark" is the term used for the "Number" or other unique designator of the items in the Schedule. Most categories of elements include both a Mark, which is unique per instance, and a "Type Mark" which like other type parameters is common to all elements of the type.

- Repeat for "Manufacturer," "Model," "Cost," "Count," "Family and Type," and "Comments."
- Be sure the order of fields is as listed in Figure 12.3.

  If you need to adjust the order, use the "Move Up" or "Move Down" buttons.

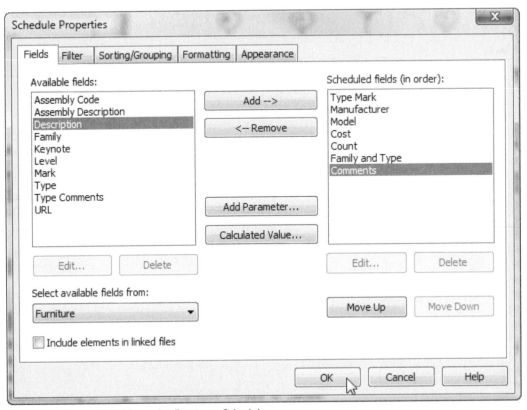

**Figure 12.3** *Add Fields to the Furniture Schedule*

10. Click OK in the "Schedule Properties" dialog to create the Schedule.

A new Schedule view will appear beneath the *Schedules/Quantities* node of the Project Browser and it will open on screen. The name of the Schedule will appear at the top and directly beneath it; each field appears as a column header. If you scroll the Schedule up or down, the title and header will remain at the top as hidden rows come in and out of view. (This is like "Freeze Panes" in Microsoft Excel.) In fact, the overall look of the Schedule view closely mimics Excel.

11. Close all project views except the *Furniture Schedule*.

- Open the *Level 3* floor plan view.

- From the Window menu, choose **Tile**.

Now that we can see the plan and the Schedule together on screen, we can explore some of the basic behaviors of Schedules in Revit.

## EDIT MODEL ELEMENTS FROM THE SCHEDULE

This has been stated before, but it is worth repeating: elements in a Schedule are the same elements that we see graphically in the model. A Schedule is just another view of the model—a tabular view rather than graphical view.

12. Click on any filed in the Schedule view.

Notice how the corresponding furniture element highlights in the plan view.

- Repeat on as many elements as you wish.

13. Click in the "Type Mark" column for any element.

Notice how this is a simple text field into which you could begin typing a value.

- Click in the Type Mark field and drag down through several fields.

Notice how the corresponding elements in the plan highlight simultaneously. The schedule can be an effective way to select elements.

- Click in the "Manufacturer" column.

- Click in each field in succession.

"Cost" is like "Mark" and is a simple text field. Notice that "Manufacturer" and "Model" however, appear with a drop-down list icon. If you try to open the list, it will appear empty. The way these fields work is that you can type in any value, as in the plain text fields, but these values you type will begin populating a list that you can choose from in subsequent edits. Let's try it.

14. Widen the Family and Type column.

- To do this, click and drag the edge between the Family and Type and Comments headers.

15. Scroll as necessary and locate the first instance of **MRAC Desk-Secretary** and then click in the "Comments" field next to it.

- Type: **Include Keyboard Tray and Footrest Option** and then press ENTER (see Figure 12.4).

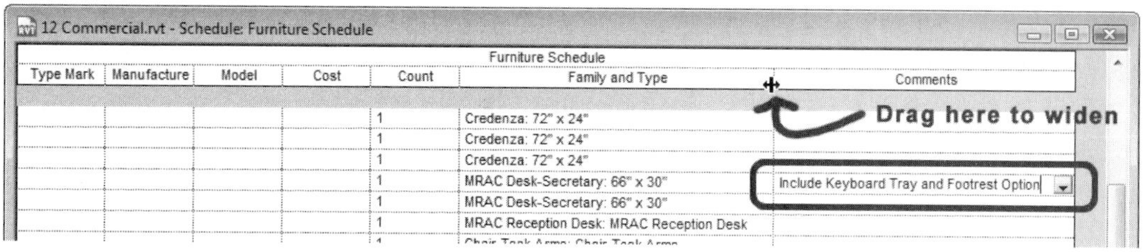

**Figure 12.4** *Input a value in a list field*

There is a second secretarial desk beneath the one we just edited.

- Click in the "Comments" field of the second secretarial desk and then click the drop-down list icon.

Notice that the note typed above now appears in the list. Each new item you type will be added automatically to this list making it easier to input existing values—simply choose them from the list. The parameter you are editing in the Schedule can be either an Instance parameter like the one edited here, or it can be a Type parameter. Instance parameters are applied independently to each instance of the item in the project. In this case, it would be possible to order a keyboard tray and footrest for one of the secretarial desks but not the other. With a Type parameter, the value applies to all instances of the Type in question. This is the case with "Manufacturer," "Model," and "Cost." The assumption here being that the Family and Type in question represents a particular item that can be ordered from a catalog, purchased, and installed in the project. Therefore, the Manufacturer, Model, and Cost would be the same for all instances of that item. For our purposes, these behaviors are perfectly logical. However, in your own projects, if you wish to modify these behaviors, you must edit the Families of the elements in question. Let's look at Type parameter next.

16. Click in the "Manufacturer" field for one of the desks.

- Type: **Acme Furniture** and then press ENTER.

    A dialog will appear indicating that this change will apply to all instances of this Type (see Figure 12.5).

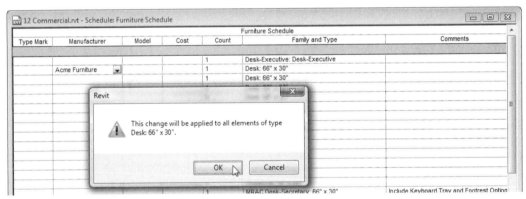

**Figure 12.5** *Applying a change to a Type parameter*

- Click OK to accept this change.

- Scroll down in the Schedule and notice that the change has applied to several desks.

- Repeat the process to add a Model designation and Cost to the same Desk.

17. Choose the "Acme Furniture" value from the "Manufacturer" column for the **Chair-Executive [M_Chair-Executive]** Family.

- Again, when you make this edit, it will apply at the Type level. Click OK to confirm the change.

- Add a Model and Cost to the Chair as well (see Figure 12.6).

| Type Mark | Manufacturer | Model | Cost | Count | Family and Type | Comments |
|---|---|---|---|---|---|---|
| | | | | | *Furniture Schedule* | |
| | | | | 1 | Desk-Executive: Desk-Executive | |
| | Acme Furniture | WorkDesk | 1200.00 | 1 | Desk: 66" x 30" | |
| | Acme Furniture | WorkDesk | 1200.00 | 1 | Desk: 66" x 30" | |
| | Acme Furniture | WorkDesk | 1200.00 | 1 | Desk: 66" x 30" | |
| | Acme Furniture | WorkDesk | 1200.00 | 1 | Desk: 66" x 30" | |
| | | | | 1 | Desk: 72" x 36" | |
| | | | | 1 | Desk: 72" x 36" | |
| | | | | 1 | Credenza: 72" x 24" | |
| | | | | 1 | Credenza: 72" x 24" | |
| | | | | 1 | Credenza: 72" x 24" | |
| | | | | 1 | Credenza: 72" x 24" | |
| | | | | 1 | Credenza: 72" x 24" | |
| | | | | 1 | Credenza: 72" x 24" | |
| | | | | 1 | Credenza: 72" x 24" | |
| | | | | 1 | MRAC Desk-Secretary: 66" x 30" | Include Keyboard Tray and Footrest Option |
| | | | | 1 | MRAC Desk-Secretary: 66" x 30" | Include Keyboard Tray and Footrest Option |
| | | | | 1 | MRAC Reception Desk: MRAC Reception Desk | |
| | Acme Furniture | WorkChair | 850.00 | 1 | Chair-Task Arms: Chair-Task Arms | |
| | Acme Furniture | WorkChair | 850.00 | 1 | Chair-Task Arms: Chair-Task Arms | |
| | Acme Furniture | WorkChair | 850.00 | 1 | Chair-Task Arms: Chair-Task Arms | |
| | Acme Furniture | WorkChair | 850.00 | 1 | Chair-Task Arms: Chair-Task Arms | |
| | Acme Furniture | WorkChair | 850.00 | 1 | Chair-Task Arms: Chair-Task Arms | |
| | Acme Furniture | WorkChair | 850.00 | 1 | Chair-Task Arms: Chair-Task Arms | |
| | | | | 1 | Chair-Corbu: Chair-Corbu | |
| | | | | 1 | Chair-Corbu: Chair-Corbu | |
| | | | | 1 | Chair-Corbu: Chair-Corbu | |
| | | | | 1 | Chair-Corbu: Chair-Corbu | |
| | | | | 1 | Chair-Executive: Chair-Executive | |
| | Acme Furniture | WorkChair | 850.00 | 1 | Chair-Task Arms: Chair-Task Arms | |
| | Acme Furniture | WorkChair | 850.00 | 1 | Chair-Task Arms: Chair-Task Arms | |
| | Acme Furniture | WorkChair | 850.00 | 1 | Chair-Task Arms: Chair-Task Arms | |
| | | | | 1 | Chair-Kinder: Chair-Kinder | |

**Figure 12.6**   *Changes to Type parameters apply to all instances*

18. Using the same process, assign a Manufacturer, Model and Cost to the Corbu and Breuer chairs.

19. Save the project.

## SORTING SCHEDULE ITEMS

Since many of the values in Schedule are identical for all instances of a particular item, it might be nice to sort and group some of these values in the Schedule instead of listing each element separately and showing so much repetition.

20. Right-click anywhere in the Schedule view window and choose **View Properties**.

 **Tip:** This can also be done from the Project Browser.

Beneath the "Other" grouping in the "Element Properties" dialog, are five items each with an "Edit" button next to them. These five items correspond to the five tabs of the "Schedule Properties" dialog (shown in Figure 12.3). We only looked at the "Fields" tab above. Each of these tabs can be accessed here anytime. Using the settings for Sorting/Grouping, we can modify the Schedule to group elements with common parameters into a single entry. We can also sort the Schedule in a variety of ways.

- Next to "Sorting/Grouping," click the Edit button.

**Note:** You can see an example below in Figure 12.8.

This returns us to the "Schedule Properties" dialog (that we saw previously) open to the Sorting/Grouping tab.

21. At the top of the dialog, from the "Sort by" list, choose **Manufacturer** and then click OK twice.

- Scroll to the bottom of the Schedule to see the results.

All of the blank fields (which are listed first in an alphabetic sort) are now at the top of the list, and the "Acme Furniture" items are next. The Desks and Chairs are still interspersed. If you wish, you can sort by more than one criterion.

22. Return to the Sorting/Grouping tab of "Schedule Properties" dialog (right-click and choose **View Properties**, then Edit Sorting/Grouping again).

- At the top of the dialog, beneath "Sort by" choose **Model** from the "Then by" list and then click OK twice (see Figure 12.7).

| | | | | | Furniture Schedule | |
|---|---|---|---|---|---|---|
| Type Mark | Manufacturer | Model | Cost | Count | Family and Type | Comments |
| | | | | 1 | Chair-Task: Chair-Task | |
| | | | | 1 | Chair-Task: Chair-Task | |
| | | | | 1 | Chair-Task: Chair-Task | |
| | | | | 1 | Chair-Task: Chair-Task | |
| | | | | 1 | Chair-Task: Chair-Task | |
| | | | | 1 | Chair-Task: Chair-Task | |
| | Acme Furniture | WorkChair | 850.00 | 1 | Chair-Task Arms: Chair-Task Arms | |
| | Acme Furniture | WorkChair | 850.00 | 1 | Chair-Task Arms: Chair-Task Arms | |
| | Acme Furniture | WorkChair | 850.00 | 1 | Chair-Task Arms: Chair-Task Arms | |
| | Acme Furniture | WorkChair | 850.00 | 1 | Chair-Task Arms: Chair-Task Arms | |
| | Acme Furniture | WorkChair | 850.00 | 1 | Chair-Task Arms: Chair-Task Arms | |
| | Acme Furniture | WorkChair | 850.00 | 1 | Chair-Task Arms: Chair-Task Arms | |
| | Acme Furniture | WorkChair | 850.00 | 1 | Chair-Task Arms: Chair-Task Arms | |
| | Acme Furniture | WorkChair | 850.00 | 1 | Chair-Task Arms: Chair-Task Arms | |
| | Acme Furniture | WorkChair | 850.00 | 1 | Chair-Task Arms: Chair-Task Arms | |
| | Acme Furniture | WorkChair | 850.00 | 1 | Chair-Task Arms: Chair-Task Arms | |
| | Acme Furniture | WorkChair | 850.00 | 1 | Chair-Task Arms: Chair-Task Arms | |
| | Acme Furniture | WorkDesk | 1200.00 | 1 | Desk: 66" x 30" | |
| | Acme Furniture | WorkDesk | 1200.00 | 1 | Desk: 66" x 30" | |
| | Acme Furniture | WorkDesk | 1200.00 | 1 | Desk: 66" x 30" | |
| | Acme Furniture | WorkDesk | 1200.00 | 1 | Desk: 66" x 30" | |
| | Exclusive Office Outfitters | Breuer | 135.00 | 1 | Chair-Breuer: Chair-Breuer | |
| | Exclusive Office Outfitters | Breuer | 135.00 | 1 | Chair-Breuer: Chair-Breuer | |
| | Exclusive Office Outfitters | Breuer | 135.00 | 1 | Chair-Breuer: Chair-Breuer | |
| | Exclusive Office Outfitters | Breuer | 135.00 | 1 | Chair-Breuer: Chair-Breuer | |
| | Exclusive Office Outfitters | Breuer | 135.00 | 1 | Chair-Breuer: Chair-Breuer | |
| | Exclusive Office Outfitters | Breuer | 135.00 | 1 | Chair-Breuer: Chair-Breuer | |
| | Exclusive Office Outfitters | Breuer | 135.00 | 1 | Chair-Breuer: Chair-Breuer | |
| | Exclusive Office Outfitters | Breuer | 135.00 | 1 | Chair-Breuer: Chair-Breuer | |
| | Exclusive Office Outfitters | Breuer | 135.00 | 1 | Chair-Breuer: Chair-Breuer | |
| | Exclusive Office Outfitters | Breuer | 135.00 | 1 | Chair-Breuer: Chair-Breuer | |
| | Exclusive Office Outfitters | Breuer | 135.00 | 1 | Chair-Breuer: Chair-Breuer | |
| | Exclusive Office Outfitters | Breuer | 135.00 | 1 | Chair-Breuer: Chair-Breuer | |
| | Exclusive Office Outfitters | Corbu | 649.00 | 1 | Chair-Corbu: Chair-Corbu | |
| | Exclusive Office Outfitters | Corbu | 649.00 | 1 | Chair-Corbu: Chair-Corbu | |
| | Exclusive Office Outfitters | Corbu | 649.00 | 1 | Chair-Corbu: Chair-Corbu | |
| | Exclusive Office Outfitters | Corbu | 649.00 | 1 | Chair-Corbu: Chair-Corbu | |

*This is our first sort* → (Manufacturer)  |  *Then here* ← (Model)

**Figure 12.7** *Sorting the Schedule first by Manufacturer, then by Model*

Several items in the Schedule still do not have parameters assigned. Let's edit another item and see how the Schedule responds with our current sort settings.

23. Locate an instance of the "Chair-Task" in the Schedule and edit the "Manufacturer" to **Furniture Concepts**.

• Click OK in the dialog that appears.

Notice that the Schedule immediately re-sorts to accommodate the new value.

• Input a Model and Cost as well.

## GROUPING SCHEDULE ITEMS

In addition to sorting the items in the Schedule, we can also group them when all of the values are the same. It would be easier to work with the Schedule if we simply had one listing for each type of chair that reported how many of them there were instead of showing each as a separate item. We need to make a simple change to display the Schedule this way.

24. Return to the Sorting/Grouping tab of "Schedule Properties" dialog.

• At the bottom of the dialog, clear the "Itemize every instance" checkbox.

• Click the Fields tab.

• Move the Count field to the bottom of the list and then click OK twice (see Figure 12.8).

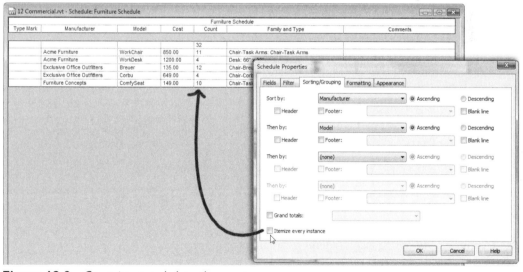

**Figure 12.8**   *Group items and show the count*

There are nearly limitless ways that we can format and display the same data. Notice the way the blank fields at the top (Count 32 in the figure) show none of the values even though we have data (particularly Family and Type) for some of the values. This is because we are sorting on the Manufacturer and Model fields and those fields are blank for the 32 items. If you edit the Schedule Properties and sort by Family and Type instead, the result will change dramatically (see Figure 12.9).

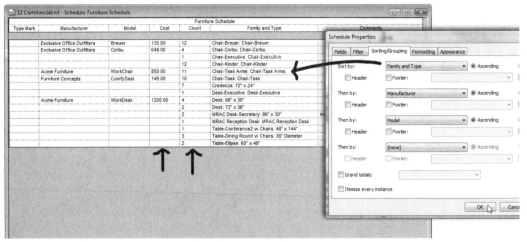

**Figure 12.9** *Sort by Family and Type instead of Manufacturer*

Add up the items with blank Manufacturer, Model, and Cost and you will see that the total is still 32. Only now, can we continue to edit them because each Family and type is listed separately.

## DUPLICATING A SCHEDULE VIEW

The current format of our Schedule view is useful for editing, but is perhaps not ideal for reporting furniture and making budgetary and purchasing decisions. Sometimes you will find it useful to have two versions of the same view—one for working purposes, the other for presentation and printing. This is true of any kind of view, Schedule, plan, or elevation, etc.

25. On the Project Browser, right-click *Furniture Schedule* and choose **Duplicate View > Duplicate**.

- Right-click *Copy of Furniture Schedule* and choose **Rename**.

- In the "Rename View" dialog, type **Furniture Schedule (Working)** and then click OK.

Now when we wish to modify an item, we can open the "Working" version of the Schedule to see each Family and type listed separately, which will make it easier to select and edit what we need. Regardless of where we make the edit, it will show in all views—both Schedules and the model.

26. Edit the remaining items to add Manufacturers, Models, and Costs.

- Add a code in the Type Mark column for each item. They can be any alpha-numeric codes you like (see Figure 12.10).

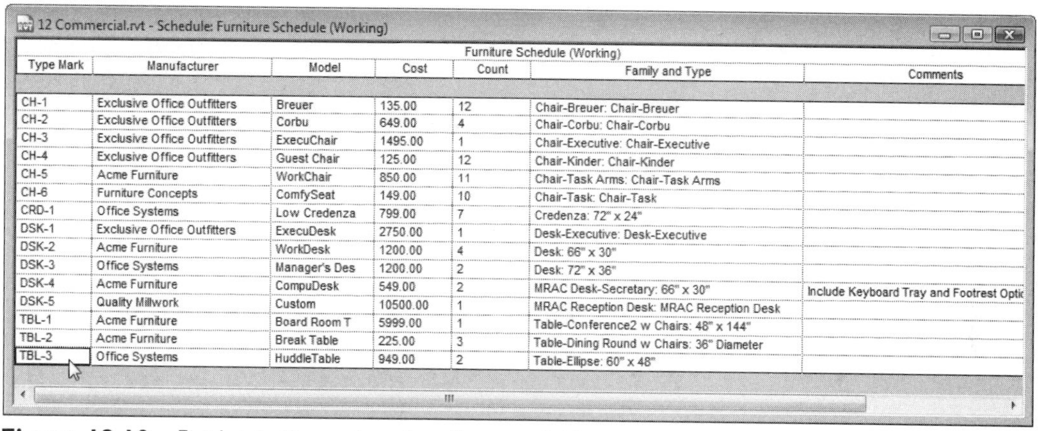

12 Commercial.rvt - Schedule: Furniture Schedule (Working)

Furniture Schedule (Working)

| Type Mark | Manufacturer | Model | Cost | Count | Family and Type | Comments |
|---|---|---|---|---|---|---|
| CH-1 | Exclusive Office Outfitters | Breuer | 135.00 | 12 | Chair-Breuer: Chair-Breuer | |
| CH-2 | Exclusive Office Outfitters | Corbu | 649.00 | 4 | Chair-Corbu: Chair-Corbu | |
| CH-3 | Exclusive Office Outfitters | ExecuChair | 1495.00 | 1 | Chair-Executive: Chair-Executive | |
| CH-4 | Exclusive Office Outfitters | Guest Chair | 125.00 | 12 | Chair-Kinder: Chair-Kinder | |
| CH-5 | Acme Furniture | WorkChair | 850.00 | 11 | Chair-Task Arms: Chair-Task Arms | |
| CH-6 | Furniture Concepts | ComfySeat | 149.00 | 10 | Chair-Task: Chair-Task | |
| CRD-1 | Office Systems | Low Credenza | 799.00 | 7 | Credenza: 72" x 24" | |
| DSK-1 | Exclusive Office Outfitters | ExecuDesk | 2750.00 | 1 | Desk-Executive: Desk-Executive | |
| DSK-2 | Acme Furniture | WorkDesk | 1200.00 | 4 | Desk: 66" x 30" | |
| DSK-3 | Office Systems | Manager's Des | 1200.00 | 2 | Desk: 72" x 36" | |
| DSK-4 | Acme Furniture | CompuDesk | 549.00 | 2 | MRAC Desk-Secretary: 66" x 30" | Include Keyboard Tray and Footrest Optic |
| DSK-5 | Quality Millwork | Custom | 10500.00 | 1 | MRAC Reception Desk: MRAC Reception Desk | |
| TBL-1 | Acme Furniture | Board Room T | 5999.00 | 1 | Table-Conference2 w Chairs: 48" x 144" | |
| TBL-2 | Acme Furniture | Break Table | 225.00 | 3 | Table-Dining Round w Chairs: 36" Diameter | |
| TBL-3 | Office Systems | HuddleTable | 949.00 | 2 | Table-Ellipse: 60" x 48" | |

**Figure 12.10** *Finish inputting values for all empty fields*

27. Save the project.

## HEADERS, FOOTERS, AND GRAND TOTALS

The Count Field is not the only way to have Revit tabulate the quantity of items. If you wish to show totals above or beneath the groupings, you can add Headers and Footers and Grand Totals to the Schedule instead.

28. Close the *Furniture Schedule (Working)* and return to the *Furniture Schedule*.
    - Edit the View Properties and click the Sorting/Grouping tab.
    - Change the first criterion back to **Manufacturer**.
    - Beneath Manufacturer, place a checkmark in the "Header" checkbox.
    - Change the next criterion to **Model** and then place a checkmark in the "Footer" checkbox.
    - From the drop-down list that appears choose **Title, count, and totals**.
    - Change the third criterion to **(none)**.
    - At the bottom of the dialog, place a checkmark in the "Grand totals" checkbox and accept the default of: **Title, count, and totals** (see Figure 12.11).

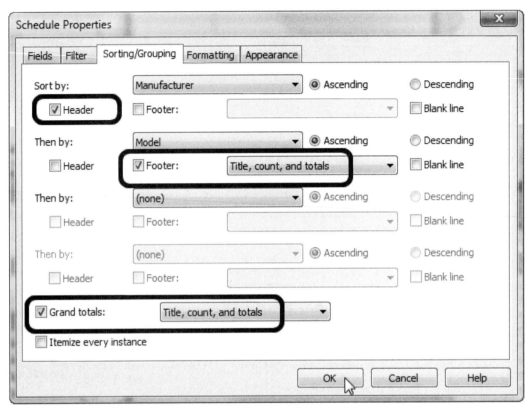

**Figure 12.11**  *Enable Headers, Footers and Grand totals*

29. Click OK twice to return to the Schedule view window and study the results (see Figure 12.12).

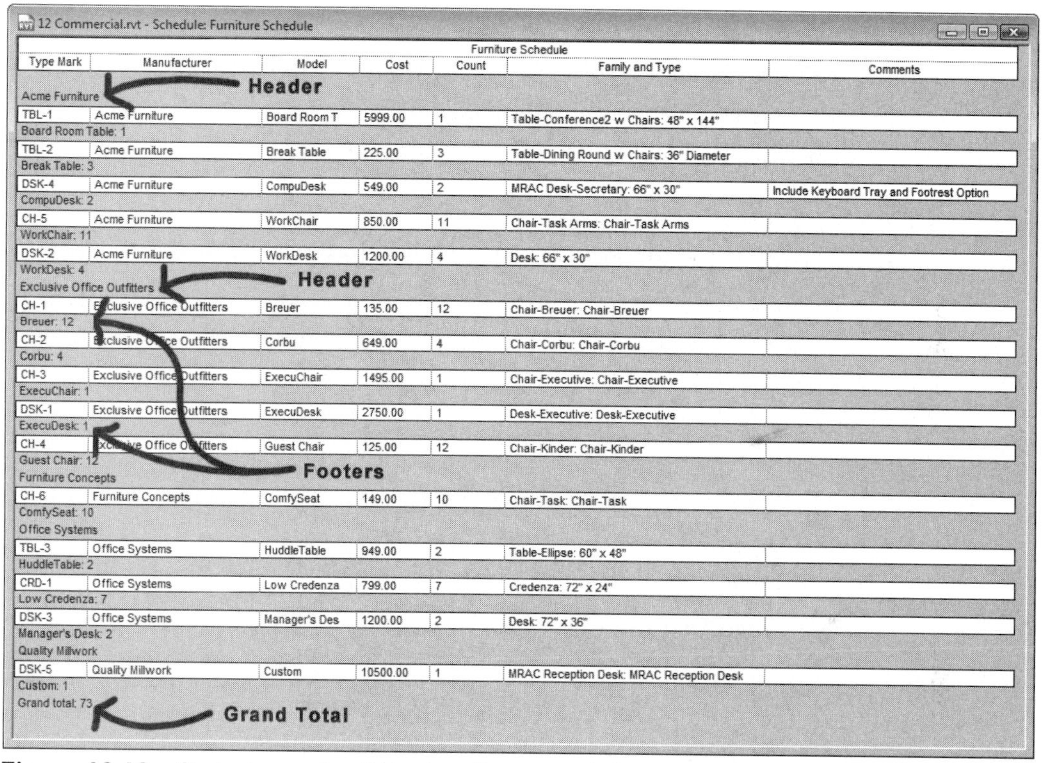


| Furniture Schedule | | | | | | |
|---|---|---|---|---|---|---|
| Type Mark | Manufacturer | Model | Cost | Count | Family and Type | Comments |
| Acme Furniture | | | | | | |
| TBL-1 | Acme Furniture | Board Room T | 5999.00 | 1 | Table-Conference2 w Chairs: 48" x 144" | |
| Board Room Table: 1 | | | | | | |
| TBL-2 | Acme Furniture | Break Table | 225.00 | 3 | Table-Dining Round w Chairs: 36" Diameter | |
| Break Table: 3 | | | | | | |
| DSK-4 | Acme Furniture | CompuDesk | 549.00 | 2 | MRAC Desk-Secretary: 66" x 30" | Include Keyboard Tray and Footrest Option |
| CompuDesk: 2 | | | | | | |
| CH-5 | Acme Furniture | WorkChair | 850.00 | 11 | Chair-Task Arms: Chair-Task Arms | |
| WorkChair: 11 | | | | | | |
| DSK-2 | Acme Furniture | WorkDesk | 1200.00 | 4 | Desk: 66" x 30" | |
| WorkDesk: 4 | | | | | | |
| Exclusive Office Outfitters | | | | | | |
| CH-1 | Exclusive Office Outfitters | Breuer | 135.00 | 12 | Chair-Breuer: Chair-Breuer | |
| Breuer: 12 | | | | | | |
| CH-2 | Exclusive Office Outfitters | Corbu | 649.00 | 4 | Chair-Corbu: Chair-Corbu | |
| Corbu: 4 | | | | | | |
| CH-3 | Exclusive Office Outfitters | ExecuChair | 1495.00 | 1 | Chair-Executive: Chair-Executive | |
| ExecuChair: 1 | | | | | | |
| DSK-1 | Exclusive Office Outfitters | ExecuDesk | 2750.00 | 1 | Desk-Executive: Desk-Executive | |
| ExecuDesk: 1 | | | | | | |
| CH-4 | Exclusive Office Outfitters | Guest Chair | 125.00 | 12 | Chair-Kinder: Chair-Kinder | |
| Guest Chair: 12 | | | | | | |
| Furniture Concepts | | | | | | |
| CH-6 | Furniture Concepts | ComfySeat | 149.00 | 10 | Chair-Task: Chair-Task | |
| ComfySeat: 10 | | | | | | |
| Office Systems | | | | | | |
| TBL-3 | Office Systems | HuddleTable | 949.00 | 2 | Table-Ellipse: 60" x 48" | |
| HuddleTable: 2 | | | | | | |
| CRD-1 | Office Systems | Low Credenza | 799.00 | 7 | Credenza: 72" x 24" | |
| Low Credenza: 7 | | | | | | |
| DSK-3 | Office Systems | Manager's Des | 1200.00 | 2 | Desk: 72" x 36" | |
| Manager's Desk: 2 | | | | | | |
| Quality Millwork | | | | | | |
| DSK-5 | Quality Millwork | Custom | 10500.00 | 1 | MRAC Reception Desk: MRAC Reception Desk | |
| Custom: 1 | | | | | | |
| Grand total: 73 | | | | | | |

**Figure 12.12** *Study the results of Headers, Footers, and Grand totals in the view*

Several other combinations are possible. You can change either Manufacturer or Model to display Headers or Footers or both. You also have control over whether it shows totals, counts, and/or titles. The Grand total can be turned on or off. Notice with these items enabled, that the information in the Count field is now somewhat redundant. You can return to the "Schedule Properties" dialog on the Fields tab and remove the Count field from the Schedule if you wish. Finally, if you wish to put a bit of space between each sort criterion, you can use the "Blank line" checkbox. In some cases, this will make your sub-totals easier to read, particularly when there are many items in the Schedule. If you would like to see for yourself, edit the properties once more and place a checkmark in the "Blank line" checkbox for the Manufacturer sort criterion only.

## THE FILTER TAB

So far we have explored the high level of flexibility in our two Schedules from just the Fields and Sorting/Grouping tabs. There are other settings that we can apply in the remaining three tabs of the "Schedule Properties" dialog. For example, sometimes you only need part of the information available. Suppose you were on the phone with your sales representative from Office Systems. You could create a version of your furniture Schedule that listed only the items that you are specifying from this manufacturer.

30. Edit the Schedule Properties.

- In the "Element Properties" dialog, click the Edit button next to "Filter."
- From the "Filter by" list, choose **Manufacturer**.

About a dozen operations are possible: Equal to, begins with, etc.

- Leave the next list set to "equals."
- From the list below Manufacturer, choose: **Office Systems** and then click OK twice (see Figure 12.13).

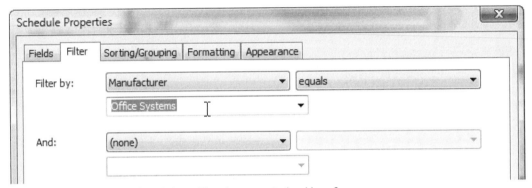

**Figure 12.13**  *Set the Schedule to Filter by a particular Manufacturer*

Notice that the Schedule now only shows the items from Office Systems. You can try other Filters if you wish. If you planned to leave the Schedule set this way, it would be a good idea to rename it more descriptively.

31. Return to the Filter tab and remove the Filters from the Schedule when you are done experimenting.

## FORMATTING AND APPEARANCE

In the interest of completeness, let's take a brief look at the remaining two tabs. The settings on these tabs are straightforward. On the Formatting tab you can configure the orientation and alignment of each individual field. On the Appearance tab, you configure the look of the overall Schedule.

32. Edit the properties of the *Furniture Schedule* again.

- Click the Edit button next to Formatting.

33. Click on the Cost field in the list at the left.

- Change the "Alignment" to **Right**.
- Place a checkmark in the "Calculate totals" checkbox.

34. Click on the Family and Type field in the list at the left.

- Place a checkmark in the "Hidden field" checkbox.

35. Click on the Count field in the list at the left.

• Change the "Header orientation" to **Vertical** and **Center** (see Figure 12.14).

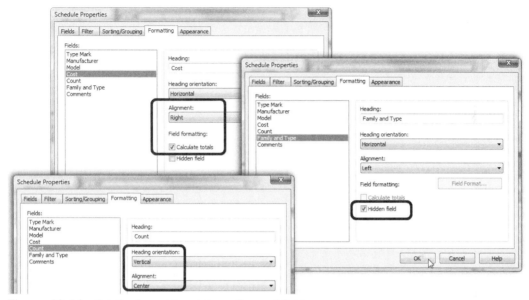

**Figure 12.14** *Set the Formatting options of several fields*

36. Click OK twice to return to the Schedule view window.

Notice that most of the formatting shows immediately in the view window. However, the changing of the "Header orientation" to Vertical for Count does not. It will only appear when we add this Schedule to a Sheet.

## ADD A SCHEDULE TO A SHEET

To see the vertical orientation of the Count field and to see the effects of most of the settings on the Appearance tab, we need to add the Schedule to a Sheet. You add a Schedule view to a Sheet the same way as any other view.

37. On the Project Browser, right-click the Sheets node and choose **New Sheet**.

The new Sheet will appear as G101 – Unnamed and should open automatically.

• Rename the Sheet: **A603 – Schedules**.

We have two Schedule Sheets (A601 and A602) but they have several Schedules on them already. Creating a new Sheet will give us plenty of room to work.

38. From Project Browser, drag *Furniture Schedule* and drop it on the Sheet.

• Click a point near the top left corner of the Sheet to place it.

Notice that with the Schedule still selected, there are control handles at the top of each column. You can use these to interactively change the width of each field (see Figure 12.15). Notice also that you can see that the Count field header is in fact oriented vertically.

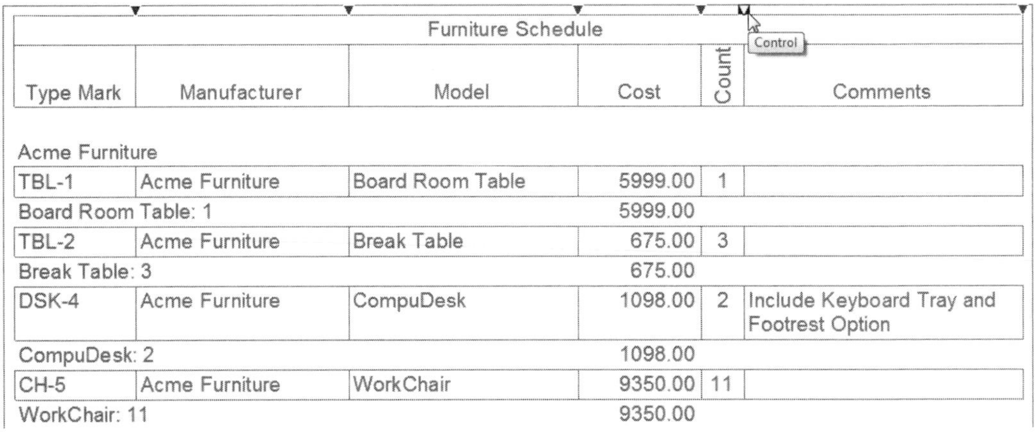

| Furniture Schedule | | | | | |
|---|---|---|---|---|---|
| Type Mark | Manufacturer | Model | Cost | Count | Comments |
| Acme Furniture | | | | | |
| TBL-1 | Acme Furniture | Board Room Table | 5999.00 | 1 | |
| Board Room Table: 1 | | | 5999.00 | | |
| TBL-2 | Acme Furniture | Break Table | 675.00 | 3 | |
| Break Table: 3 | | | 675.00 | | |
| DSK-4 | Acme Furniture | CompuDesk | 1098.00 | 2 | Include Keyboard Tray and Footrest Option |
| CompuDesk: 2 | | | 1098.00 | | |
| CH-5 | Acme Furniture | WorkChair | 9350.00 | 11 | |
| WorkChair: 11 | | | 9350.00 | | |

**Figure 12.15**  *Edit the width of the fields interactively on the Sheet*

Since we created a new Sheet and have plenty of room, let's resize each of the fields widths to prevent the data from wrapping to a second line as it does by default. To do this, you simply drag the control handles to the right. Work from left to right across the Schedule.

- Resize each column as required to prevent wrapping.

39. On the Project Browser, right-click the *Furniture Schedule* and choose **Properties**.

- Click the Edit button next to "Appearance."

If you wish to have the grid lines show continuously across the header and footer areas, check the "Grid in headers/footers/spacers" checkbox.

- Place a checkmark in the "Outline" checkbox and then next to it, choose **Wide Lines** from the list.

- Click OK twice to see the result (see Figure 12.16).

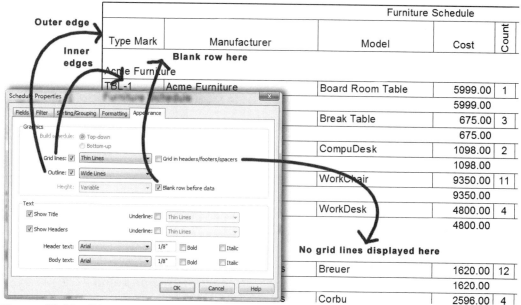

**Figure 12.16**   *Edits on the "Appearance" tab affect the look on the Sheet*

Return to the Appearance tab and make other edits if you wish to fine-tune the look of the Schedule on the Sheet. The figure shows that we can control the inner grid lines independently from the outline. Also, grid lines for headers and footers have a separate checkbox. The text settings should be self-explanatory. Notice that there are settings for the title, headers, and body text. You can change the fonts, heights, and linework drawn for each component.

## SPLIT A LONG SCHEDULE

In some cases, the data in the Schedule grows beyond the edge of the Sheet. You can break the Schedule into pieces to make it fit the Sheet better. (An example occurs on the existing *A601 – Schedules* Sheet).

40. On the Project Browser, double-click to open the *A601 - Schedules* Sheet view.

The Wall Schedule overruns the bottom of the Sheet. We can split it to make the Sheet legible.

41. Select the Wall Schedule on the Sheet.

- On the right side, is a small control handle (looks like an "Z") click this control to split the Schedule.

Using the other control handles that appear, you can make additional adjustments as necessary. The Schedule can be split multiple times if required. To remove a split and put the two pieces back together, simply drag and drop one piece on top of another (see Figure 12.17).

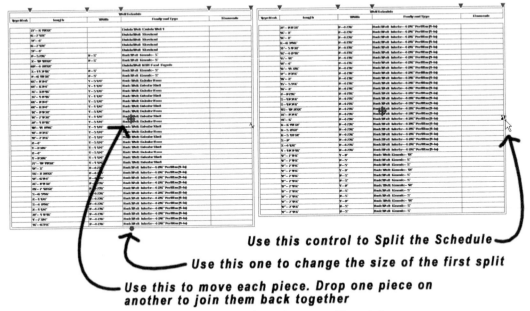

**Use this control to Split the Schedule**

**Use this one to change the size of the first split**

**Use this to move each piece. Drop one piece on another to join them back together**

**Figure 12.17** *Split the Schedule and make other adjustments if needed*

If after splitting a Schedule it is still too long for the Sheet, you can use the Filter tab to edit the Schedule view. Filter by a number range or by floor to shorten the overall Schedule. Then duplicate this Schedule view and Filter the new copy to include the rest. Drag the two (or more Schedules) to different Sheets.

42. Save the project.

## EDITING THE MODEL

Returning to our Furniture Schedule, let's see how it keeps up to date as we make modifications to the model.

### MODIFY ELEMENTS IN THE MODEL

Locate the conference room in the third floor plan (bottom middle). The current table has 12 chairs around it. However, the room looks like it could accommodate a larger table. Following up slightly on the lessons of the previous chapter, take a minute to understand how the conference table Family is constructed.

1. Repeat the process from the start of the chapter, close all views except *Level 3* and the *Furniture Schedule,* and then Tile them.

- In the *Level 3* view, zoom in on the conference table.

- Place your mouse over the conference table, note the way it pre-highlights, and note the name that appears in the Status Bar.

The conference table is a Family: Type named: ***Table-Conference2 w Chairs: 48" × 144"*** [***Table - Conference w Chairs: 1200 × 4500mm***].

- Place your mouse over one of the Chairs.

The table will continue to pre-highlight.

- Press TAB.

Notice how the chair now pre-highlights independently.

In the last chapter, we learned about nested Families when building the coat rack with nested coat hooks. This conference table Family is another similar example of a nested Family. The difference here is that the nested chairs are set to be "Shared" Families. When you make a nested Family shared, it is part of the host Family but can also be selected and scheduled independently. Another way to see this is to select the chairs for the Schedule.

2. Locate the conference chairs (the quantity is 10) in the Schedule and click that row.

The 10 conference chairs should highlight in the plan (see Figure 12.18).

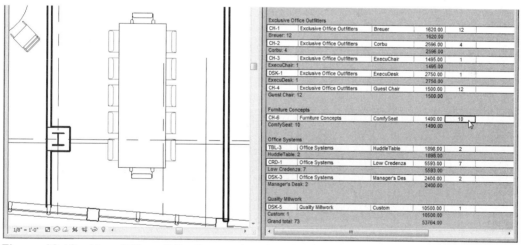

**Figure 12.18**   *Nested "Shared" Families can be selected and scheduled independently*

3. Select the conference table in the plan view.

- From the Type Selector, choose the next larger size - ***Table-Conference2 w Chairs: 54" × 168"*** [***Table - Conference w Chairs: 1200 × 5100mm***].

- Move it if necessary to fit the room better.

Notice the change in quantity of chairs both in plan and schedule (see Figure 12.19).

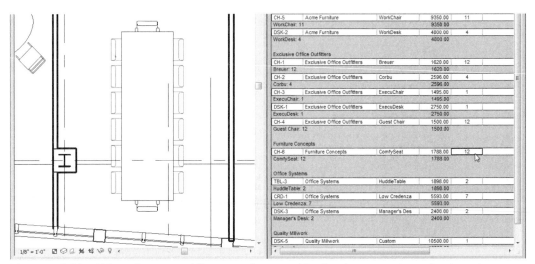

**Figure 12.19**  *When choosing a different conference table, the chair count immediately updates*

While the chairs took care of themselves, the conference table needs attention. The new type we selected does not have values for Manufacturer, Cost, and the other type-based fields. We can certainly edit the values in the Schedule as we did previously. However, we can also edit from the model.

### EDITING SCHEDULE PARAMETERS FROM THE MODEL

We edited all of the Manufacturer and other identity data from the Schedule. While this is typically the most efficient way to make such edits, it is possible to edit these data fields using the same methods that we have employed when editing other types. Namely we can select an object, edit its properties, and then click the Edit/New button, or we can right-click the item from the Families branch of the Project Browser.

4.  On the Project Browser, expand the *Families* branch and then the *Furniture* node.

Beneath the Furniture node, all of the Furniture Families used in the project are listed. *Chair-Task* [*M_Chair-Task*] is the chair that repeats along the sides of the conference table. *Chair-Task Arms* [*M_Chair-Task (Arms)*] is used at the ends. The conference table is *Table-Conference2 w Chairs* [*Table - Conference w Chairs*].

5.  Beneath the **Table-Conference2 w Chairs** [**Table - Conference w Chairs**] Family, right-click on the **54" × 168"** [**1200 × 3050mm**] type and choose **Properties**.

•  Input values for the Manufacturer, Model, and Cost (see Figure 12.20).

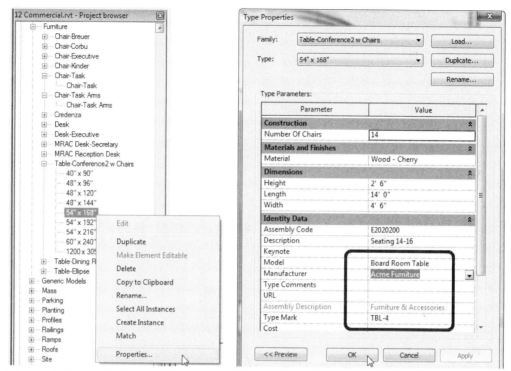

**Figure 12.20**    *Edit the type parameters of the new conference table*

6. Click OK to apply the change and update the model and Schedule.

## USING SCHEDULE KEYS TO SPEED INPUT

In the "Add a Schedule View" topic, Schedule Keys were briefly discussed. Schedule Keys allow you to define a named collection of parameter values that you can add to an item in a Schedule in one step. They are very useful in Door, Window, and Room Schedules where there are many instance-based parameters. Let's return to our pre-defined Door Schedule for a brief demonstration of this powerful feature.

7. On the Design Bar, click the View tab and then the **Schedule/Quantities** tool.

- In the "New Schedule" dialog, choose Doors for the category, change the Name to **Door Frame Style Schedule** and then select the "Schedule Keys" radio button.

- Change the name to **Frame Style** and then click OK (see Figure 12.21).

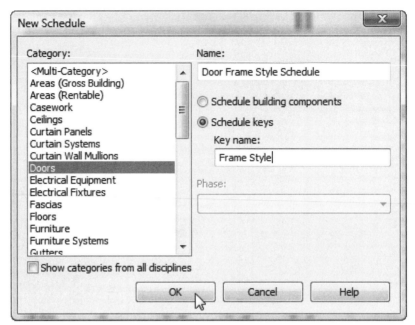

**Figure 12.21** *Create a new Door Frame Style Schedule Key*

Notice that the Available Fields list includes all of the instance parameters for Doors. Key Name is already added on the right. This field is required.

8. Add the five frame fields: "Frame Type," "Frame Material," "Frame Finish," "Jamb," and "Head."

- Click OK to finish.

- Adjust the widths of the columns in the Schedule window.

9. On the Options Bar, click the New Row button.

- Input appropriate values in the first three fields. Leave the Jamb and Head fields blank for now.

- Add at least one additional Row. Add more if you like (see Figure 12.22).

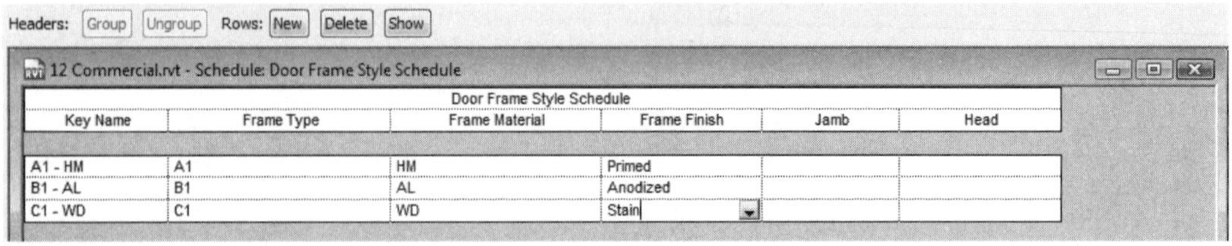

**Figure 12.22** *Add Rows and input standard values*

10. On the Project Browser, right-click the Door Schedule and choose **Duplicate View > Duplicate**.

- Rename the duplicate **Door Schedule (Working)**.

11. Edit the Properties of *Door Schedule (Working)* and then edit the Fields.

- Add the new Frame Style field and move it up to just before Frame Type.

- Click OK twice.

Your duplicated Door Schedule should now have a Frame Style column. All the setup is now complete. To use the Frame Styles, we simply choose them from the pop-up list in the Frame Style field and all the corresponding values will fill in automatically.

12. Select any Door and choose a Frame Style (see Figure 12.23).

| 12 Commercial.rvt - Schedule: Door Schedule (Working) | | | | | | | | | | | | | | | | |
|---|---|---|---|---|---|---|---|---|---|---|---|---|---|---|---|---|
| | | | | | Door Schedule (Working) | | | | | | | | | | | |
| | | | Door | | | | | | | | Frame | | | | | |
| Door Number | Type | Width | Height | Thickness | Material | Finish | Under Cut | Fire Rating | Hardware | Frame Style | Type | Material | Finish | Jamb | Head | Comments |
| 101 | 9 | 6' - 0" | 7' - 0" | 0' - 2" | | | | | | (none) | | | | | | |
| 102 | 9 | 6' - 0" | 7' - 0" | 0' - 2" | | | | | | (none) | | | | | | |
| 301 | 9 | 6' - 0" | 7' - 0" | 0' - 2" | | | | | | B1 - AL | B1 | AL | Anodized | | | |
| 302 | 4 | 2' - 6" | 7' - 0" | 0' - 2" | | | | | | A1 - HM | A1 | HM | Primed | | | |
| 303 | 4 | 2' - 6" | 7' - 0" | 0' - 2" | | | | | | A1 - HM | A1 | HM | Primed | | | |
| 304 | 4 | 2' - 6" | 7' - 0" | 0' - 2" | | | | | | A1 - HM | A1 | HM | Primed | | | |
| 305 | 4 | 2' - 6" | 7' - 0" | 0' - 2" | | | | | | (none) | | | | | | |
| 306 | 4 | 2' - 6" | 7' - 0" | 0' - 2" | | | | | | A1 - HM | | | | | | |
| 307 | 4 | 2' - 6" | 7' - 0" | 0' - 2" | | | | | | B1 - AL | | | | | | |
| 308 | 4 | 2' - 6" | 7' - 0" | 0' - 2" | | | | | | C1 - WD | | | | | | |
| 309 | 14 | 3' - 0" | 7' - 0" | 0' - 1 1/2" | | | | | | C1 - WD | C1 | WD | Stain | | | |
| 310 | 14 | 3' - 0" | 7' - 0" | 0' - 1 1/2" | | | | | | C1 - WD | C1 | WD | Stain | | | |
| 311 | 4 | 2' - 6" | 7' - 0" | 0' - 2" | | | | | | (none) | | | | | | |

**Figure 12.23**   *Choosing Frame Styles automatically fills in the pre-assigned values*

- Repeat for as many Doors as you wish.

Recall that our *Door Frame Style Schedule* is also controlling the Jamb and Head detail fields. However, at this time, we don't know which Sheet will have those details. A wonderful benefit of this process is that you can return to the *Door Frame Style Schedule* at any time and fill in the detail designation. The change will be reflected immediately by all Doors using that Frame Style. Give it a try now if you wish.

**Tip:** To apply your Frame Styles even faster, drag through several items in the Schedule to select the corresponding Doors. Switch to the floor plan view and then click the Properties icon. In the "Element Properties" dialog you can choose a Frame Style that will in turn apply to all Doors in the selection.

You can make another Schedule Key for the remaining Door instance properties. While it is possible to make a single Schedule Key control all the properties, having two Keys gives more flexibility to mix and match door and frame types.

**Tip:** Make your Door Schedule easier to read. Using the procedure covered in the "Headers, Footers, and Grand Totals" topic, edit Fields, add the Level field. In Sorting/Grouping, sort first by Level with a Header, and then by Mark (no header or footer for Mark). Finally, on the Formatting tab, make the Level field a Hidden Field.

## WORKING WITH TAGS

Tags serve an important role in Architectural Documentation. They provide a means of easily and uniquely locating and identifying elements in a project. All sorts of tags are needed in Construction Document sets: Room Tags, Door Tags, Window Tags, etc. Revit Architecture makes adding and managing Tags simple. Furthermore, the data displayed in the Tags come directly from the objects in the same way as the data that feeds Schedules. Therefore, one need only add Tags to a view and let Revit handle the rest.

### ADDING DOOR TAGS

We spent some time on our Door Schedule, but so far, we have not added any Door Tags. Adding Tags to existing elements in the model is simple.

1. Continue in the *Level 3* floor plan.

- On the Design Bar, click the Drafting tab and then click the **Tag > By Category** tool. (The **Tag** tool is a fly-out.)

- Slowly move the pointer around the model pausing over different kinds of elements. Do not click yet (see Figure 12.24).

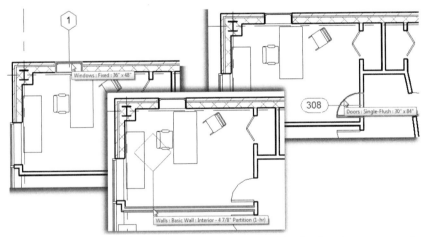

**Figure 12.24**  *Tag by Category automatically uses the correct Tag for the element you select*

Notice that the shape of the tag on your cursor changes with each element that you pre-highlight. Your project can have a different Tag loaded for nearly every type of element. Therefore, if you pre-highlight a Wall, it will automatically place a Wall Tag—pre-highlight a Door and you get a Door Tag and so on. You can place Tags with or without a leader attached, the default (on the Options Bar) is with a leader.

- On the Options Bar, clear the "Leader" checkbox.

- Pre-highlight and then click on one of the Doors to the offices at the left.

Notice that when the Tag appears, it already has a number in it. The Doors were numbered automatically when they were added to the model. If you followed the suggestion in the "View Existing Schedules" topic, then you have already renumbered the Doors more logically. The tag simply queries the element (the Door in this case) for the instance mark value and then displays the value in the tag. (You can create custom Tags that query and display nearly any parameter in an element.) You can continue placing Tags manually using the same process. However, there is a quicker method.

2. On the Design Bar, click the ***Tag All Not Tagged*** tool.

   • In the "Tag All Not Tagged" dialog, click on Door Tags and then click OK (see Figure 12.25).

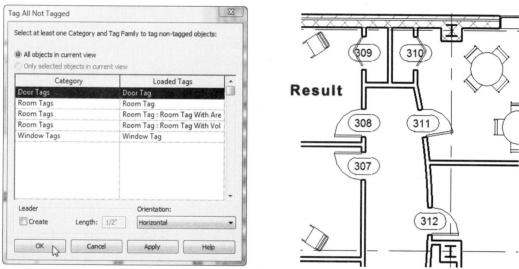

**Figure 12.25** *Tag all Doors that are not currently Tagged*

Door Tags will appear on all Doors in the currently active view.

## LOADING TAGS

Sometimes you want to Tag elements in your model, but an appropriate Tag is not loaded in your project. Like other Families, you can simply load a Tag from a library. Let's load a furniture Tag and then Tag some of our furniture.

3. From the Settings menu, choose **Annotations > Loaded Tags**.

Scroll through the list. All Tag types are listed. Tags that are already loaded are listed next to their respective category. Element categories with no Tag loaded will be indicated.

4. Click the Load button.

5. Browse to your library (Imperial or Metric) and open the *Annotations* folder.

6. Open the *Architectural* folder, select *Furniture Tag.rfa* [*M_Furniture Tag.rfa*] and then click Open.

**Note:** As with other library content in this book, you can find these items with the files installed from the CD in the *Library* folder.

7. Click OK to dismiss the "Tags" dialog.

8. On the Design Bar, click the *Tag > By Category* tool.

   • On the Options Bar, place a checkmark in the "Leader" checkbox.

   • Tag some of the chairs on the right in the reception area (see the left side of Figure 12.26).

If your Tag appears blank, you need to return to the "Duplicating a Schedule View" topic above and add values to the Type Mark fields for each piece of furniture. If you did complete this step, your Tag will appear with your Type Mark code already displayed.

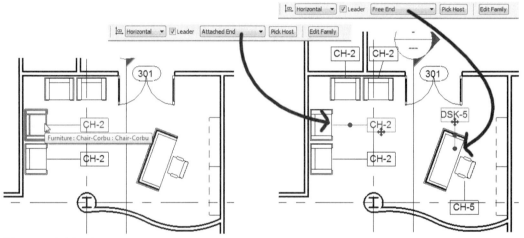

**Figure 12.26** *Add Tags to the furniture in the reception space*

After you place a Tag, you can click on it and use the control handles to move it or add elbows to the leader. If you want complete control over the leader, select the Tag and on the Options Bar, change the Leader type to Free End. This gives you an additional control handle at the end of the leader allowing you to drag it freely (see the right side of Figure 12.26).

## CREATE A FURNITURE PLAN VIEW

You do not have to Tag all pieces of furniture in this view. In fact, this might be a good time to create a separate Furniture Plan for the third floor and perhaps even create an enlarged "Typical Office Furniture Layout" view. Doing so at a larger scale would eliminate many redundant tags in this view and allow us more room to place the Tags. At this point in the text, you should have the skills required to complete this task on your own. Therefore, we will review the overall steps only.

1. On Project Browser, right-click the *Level 3* plan and choose **Duplicate View > Duplicate with Detailing**.

This command creates a copy of the view including all the Tags we have added so far. Duplicate would create a view of just the model geometry without any annotations.

2. Rename this plan **Level 3 Furniture Plan**.

3. In the original *Level 3* plan, delete the Furniture Tags added so far.

4. In the Level 3 Furniture Plan, delete all of the Door Tags.

The easiest way to do this is to maximize the entire plan and then click the Filter selection icon on the Options Bar. Click Check None, and then check only the Door Tags box. Click OK and then delete the Tags.

5. Tag all of the unique furniture in the view—reception, secretarial areas, conference room, etc.

6. Create a Callout plan at lager scale of one typical office—call it: **Typical Office Furniture Plan**.

7. Add Tags in the view to complete it (see Figure 12.27).

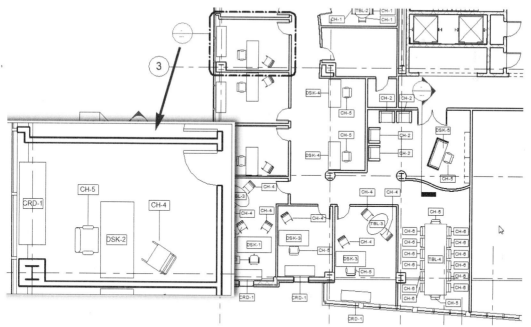

**Figure 12.27** *Create a separate furniture plan and enlarged plan Callout view and finish adding Tags*

## EDITING VIEW VISIBILITY GRAPHICS

Now that we have a separate furniture plan, we might not want to see the furniture in the *Level 3* plan view any more.

8. On the Project Browser, double-click to open the *Level 3* floor plan view.

- On the keyboard, type **VG** (You can also choose **Visibility/Graphics** from the View menu).

- Clear the checkbox next to "Furniture" (see Figure 12.28).

 **Tip:** If you only wish to hide an element category as we are doing here, a shortcut is to select one of the elements, right-click, and choose **Hide in view > Category**. This will achieve the same result as opening the "Visibility/Graphics Overrides" dialog and un-checking the element.

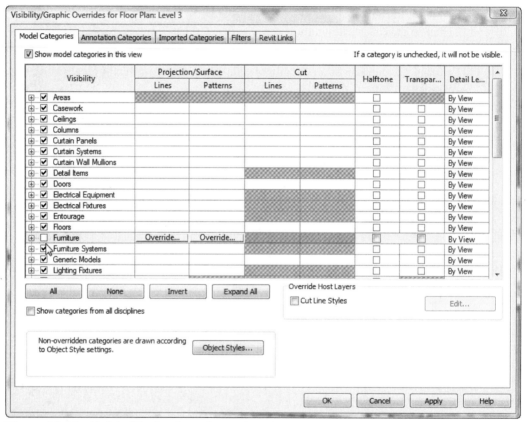

**Figure 12.28** *Visibility/Graphic Overrides apply to the current view only*

Visibility/Graphic overrides apply to the current view only. This extensive dialog lists all Revit elements grouped into major categories shown on each of the tabs at the top of the dialog. Take note of the message at the bottom left corner of the dialog. This entire dialog and all of its settings is devoted to overriding the default graphical settings. Anything that is not overridden is drawn according to the settings in the Object Styles dialog. This dialog can be accessed directly from Visibility/Graphic Overrides by clicking the Object Styles button. Object Styles can also be accessed from the Settings menu. Typically, the Object Style settings have been thoroughly configured in the project template ahead of time.

If you study just the Model Categories tab, you will see that in addition to being able to turn object categories on and off (using the checkbox); you can also override the line patterns and fills that each category uses. Furthermore, most categories have subcategories giving you even more fine control over visibility and graphic settings.

- Move the dialog out of the way so that you can see the plan in the background. Click the Apply button.

  Notice that the furniture is now hidden in this view.

Perhaps you would like to see the furniture after all, but make it less prominent.

- Check the box next to Furniture again to turn it back on and then place a checkmark in the Halftone column as well.

- Click Apply to see the result (see Figure 12.29).

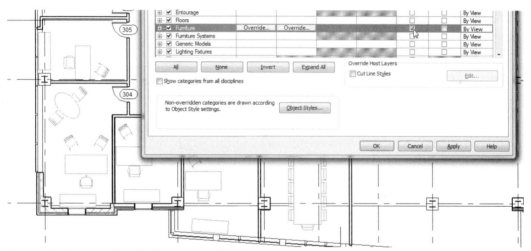

**Figure 12.29**　*Display the Furniture as Halftone*

While we will not go into detail on each item of the "Visibility/Graphic Overrides" dialog, a few additional tips are worth mention. If you wish to see all of the categories and their subcategories, click the Expand All button. This will expand the entire list in one step. Any item can be overridden. For example, if you wanted to change the lineweight, color, or pattern of the furniture, you select the Furniture item and then click one of the Override buttons that appear. A dialog will appear with additional override options (see Figure 12.30).

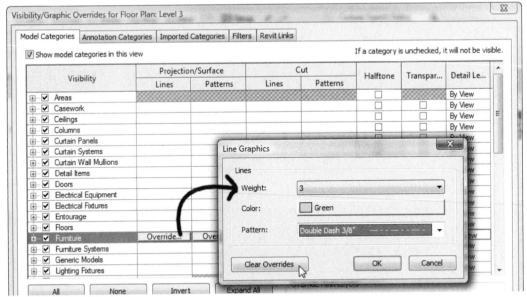

**Figure 12.30**  *Select an item and then click the Override button to change line pattern, color, weight, or fill patterns (not shown)*

You can make similar edits on the Annotation Categories tab. However, since annotation elements are simpler, there are fewer options to edit. If you have imported DWG files, you will be able to turn on and off the layers in those files on the Imported Categories tab. The Revit Links tab gives extensive control over the visibility of a linked Revit file. Unfortunately, we cannot explore this or the Filters tab in detail at this time. If you wish to learn more about any of the settings in this dialog and each of its tabs, please consult the online help.

9. Be sure that the settings for Furniture meet your preferences and then click OK.

The "Visibility/Graphic Overrides" dialog allows us to manipulate the display at the category level. In some cases, you will wish to modify the display of individual elements independently of their category. For example, in Chapter 10, we built a binder bin Family for the reception area and used the Specialty Equipment template to do so. As a result, the binder bins in the reception area did not respond to the display modifications that we just made because they are not furniture. We could edit the Family and change its category to Furniture, but this would be a bit of a drastic way to solve the problem. Instead, we can use element-level overrides.

10. In the Reception area, select both binder bins.

- Right-click and choose **Override Graphics in View > By Element**.

If you halftoned the furniture, do the following:

- Place a checkmark in the Halftone box and then click OK.

If you turned off the furniture, do the following:

• Remove the checkmark from the Visible box and then click OK (see Figure 12.31).

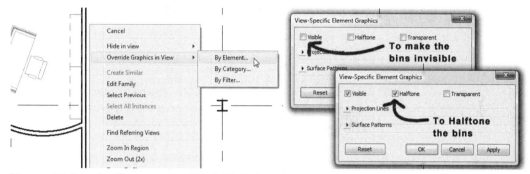

**Figure 12.31** *Halftone or turn off individual elements*

**Tip:** If you want to turn the elements off, a shortcut is to right-click and choose the **Hide in view > Elements** command. This makes them invisible without a dialog.

11. Save the project.

## ROOMS AND ROOM TAGS

A Room is a non-graphic element in your model representing a defined space in your project. Rooms are used for the obvious spaces that you would consider rooms like offices, kitchens, corridors, closets, and lobbies. However, you can define a Room for any program space you wish to label or quantify as such. Rooms are considered model elements even though they do not display graphically by default. Rooms are useful to generate room Schedules, reports, perform area calculations, volume calculations, finish Schedules, and of course for Room Tags. In most instances, Rooms can be created automatically from the bounding Walls and other elements already in your model. However, not all geometry can "bound" a Room, and in open plan situations Revit will be unable to correctly determine the proper shape and boundary of the Room. There are also situations where the shape of the Room you wish to define has no Walls or other "hard" boundaries. This may be the case in open plan layouts or outdoor spaces. In these cases, you can manually sketch Room Separations. We'll explore both techniques here.

1. On the Design Bar, click the Basics tab and then click the *Room* tool (see Figure 12.32).

Study the options on the Options Bar. Many options should be familiar from working with other tools. The first option is Tag on placement. While we previously turned this off when placing Doors and Windows, here we will leave it on.

• **Type Selector**—Use this drop-down to choose from a list of loaded Room Tags.

• **Tag on Placement**—Check this box to add a Room Tag as Rooms are added to the project. Even if you deselect this option, you can still add Tags to Rooms later.

- **Upper Limit** and **Offset**—These settings are used to determine the height of a Room. This will come into play if you enable volume calculations.

- **Tag Orientation**—When "Tag on Placement" is selected, this option sets the orientation of the Tag as either horizontal or vertical.

- **Leader**—Check this to add a Leader to your Tags as they are placed.

- **Room**—Rooms behave differently than other model elements in a few ways. You can add Rooms to the model graphically in a plan view or in a Schedule view. Adding Rooms to the model's database is not the same as placing them physically in the model. In some cases, it may be useful to create several Rooms based on a building program before the plans have been laid out or the Rooms placed. This is done in a Schedule. If you have previously added Rooms to a Schedule but have not yet placed them in a plan, they will appear in this list. If there are no unplaced Rooms in your model, the only choice here will be "New," which will create and place a new room in a single step.

- **Show Bounding Elements**—This button highlights all objects on screen that can bound Rooms. This gives you a quick way to determine if adding bounded Rooms will be successful. It is possible to create a Room without boundaries, but you will not be able to query area or volume calculations from it.

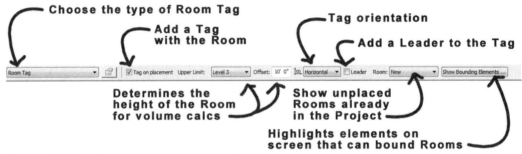

**Figure 12.32**  *The Options Bar for adding Rooms*

- Zoom In Region around the two spaces to the left of the Stair tower.

- Move the cursor into the room at the corner of the plan (see Figure 12.33).

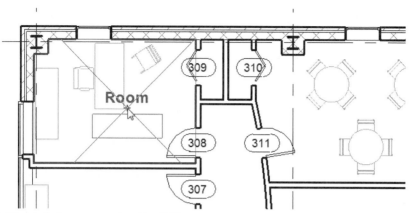

**Figure 12.33** *A Reference graphic (the "X") appears as you are placing the Room*

 **Note:** In the illustration above, if you wish to include the closet as part of the room area, you can select the Wall between the room and the closet, edit its Properties, and deselect the "Room Boundary" checkbox to have Revit ignore that Wall when finding the Room Boundary. To do this for the Wall at Door 309, you would first have to split the Wall. Otherwise, the entire length of the Wall would become non-bounding, affecting all of the offices below it.

- Click in the space to Create the Room and its Tag.

- Move into the space across the hall and then click again.

Notice that Revit attempts to align the Tags as you place Rooms. While the Room tool remains active, the existing Rooms will remain highlighted in blue on screen. This will make it easier to place the remaining ones. Also, note that the first Room was automatically numbered as "1" and the next one sequentially numbered "2." The problem with this is that you would likely want the reception space to be the first number and then number sequentially from there. Not surprisingly, you can renumber the Rooms at any time and their Tags will automatically update. However, knowing that Revit will number sequentially, and using a little strategy as we place Rooms can save us the effort of renumbering in many situations.

2. Zoom out and move the mouse into the reception space. Don't click yet.

Notice how the corridor flows into the reception, secretarial spaces, and conference room. In this open plan configuration, we need to create some manual boundaries before we can add proper Rooms.

- On the Design Bar, click the **Modify** tool or press the ESC key twice.

3. Right-click on the Design Bar and choose **Room and Area**.

If you already have this tab displayed, skip this step.

4. On the Design Bar, click the **Room Separation** tool. (You can also find this command on the Drafting menu.)

• Sketch a horizontal line segment above column line 2 to enclose the reception space.

• Sketch a vertical line segment next to column line C to enclose the conference room.

Snap both of these to the endpoints to the existing Walls and then to the intersection with the Column.

• To enclose the secretarial space, use a three point arc and sketch it as indicated in Figure 12.34.

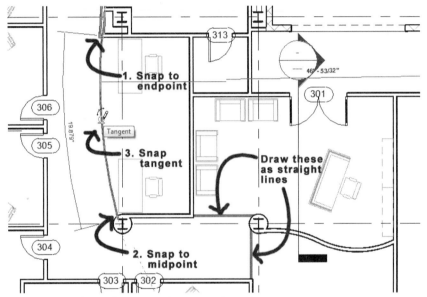

**Figure 12.34**  *Sketch Room Separation lines as indicated*

 **Note:** Room Separation lines in the figure have been enhanced for clarity. You can make this edit if you wish. Choose **Line Styles** from the Settings menu and edit the lineweight and color of the <Room Separation> line style.

5. Test your Room Separation by returning to the **Room** tool and placing a Room in the reception area.

It should now give the proper shape Room. Revit automatically uses the next number in the sequence as you can see (3 in this case). If you want a different pattern to your numbering such as 301, 302, etc. edit the first number to 301 and then return to the **Room** tool. It will pick up at 302 and continue sequencing from there.

• On the Design Bar, click the **Modify** tool or press the ESC key twice.

6. Click on the Room tag in the reception area.

The text label will turn blue to indicate that you can edit it.

- Click on the number 3 to edit it.

- Change the value to 301 and then press ENTER.

7. Click the **Room** tool again.

- Add the remaining Rooms starting with the conference room and continuing clockwise around the tenant suite (see Figure 12.35).

**Figure 12.35**   *Add Room Separations and then Tag the perimeter Rooms*

## EDIT ROOM NAMES

To edit the Room names, simply click on the Tag, click on the blue name value, and then type in the new value. The process is the same as editing the numbers.

8. Click on the room label of the Tag in the reception area.

An editable text field should appear on screen.

- Type **Reception** and then press ENTER.

- Rename the Conference room next (it is directly below Reception).

You can continue to rename Rooms this way, but in the case where several Rooms have the same name, like the offices, there is a quicker way.

Move your mouse around inside one of the offices and near the Tag.

As you move your mouse around inside a Room, you can make the reference indicator (the "X") pre-highlight. When you see this reference, you can click to select the Room. Once a Room is selected, you can edit it the same as any other model element. Simply right-click and choose **Element Properties** or click the Properties icon.

9. Pre-highlight and then select one of the offices.

10. Using the CTRL key, select all of the remaining offices (2 at the bottom, 1 in the bottom left corner and 4 at the left).

- On the Options Bar, click the Properties icon (or right-click and choose **Element Properties**).

- Beneath the "Identity Data" grouping type **Office** in the "Name" field and then click OK (see Figure 12.36).

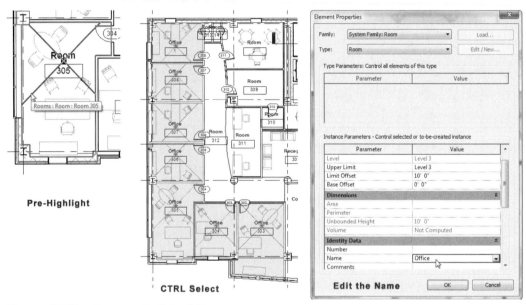

**Figure 12.36** *Rename all of the Offices*

### VIEWING A ROOM SCHEDULE

Back in Chapter 4, we added several Schedules to this project—among them a Room Schedule. At the start of this chapter, that Schedule was still empty. Let's take a look at that Schedule now to see how it reflects the data that we just added to the model.

11. On the Project Browser, double-click to open the *A602 - Schedules* Sheet view.

12. Zoom In Region on the Room Schedule (top right corner of the Sheet).

Earlier when we made adjustments to the Wall Schedule, and also on this Sheet, the Room Schedule was empty. Now that we have added Rooms to our model, this Schedule has been filled in to list these Rooms. Notice that this is really a room finishes Schedule and that currently there is no finish information listed. We can add finish data directly in the Schedule as we did in the furniture Schedule, or we can select a Room and edit the

Properties in the model. You cannot edit a view directly on a Sheet however. If it is a graphical view, right-click and choose **Activate View**. For a Schedule, right-click and choose **Edit Schedule**.

13. Select the Schedule on the Sheet, right-click and choose **Edit Schedule**.

    This will open the Schedule view on screen.

 **Note:** You can, of course, go directly to the Schedule on the Project Browser. It is not necessary to first open the Sheet, but is another alternative method.

14. Close or minimize all views except the *Room Schedule* view and the *Level 3* floor plan.

    • Tile the windows.

 **Note:** For the next several steps, refer to Figure 12.37.

15. Select one of the Rooms whose name still reads "Room" in the Schedule.

    The Room will highlight in the plan.

    • Type an appropriate name in the Schedule.

    • Repeat until all Rooms are named.

For the two closets at the top of the plan, we will need to move the Tags outside the spaces to make them legible. However, if you try to move a Room Tag outside of its Room boundaries, a warning will appear. To do this, you must first enable the leader option. Then you can safely move the Tag outside the Room.

16. Select one of the closet Tags.

    • On the Options Bar, enable the Leader option and then drag the Tag outside the space.

    • Repeat for the other closet.

The final adjustment to the Room Tags is the numbering of the first two Rooms we added. We need to renumber these to match the pattern established in the rest of the plan. This is best accomplished in the Schedule since you can easily see when you have duplicate numbers. If you type a duplicate number, Revit will warn you but will not prevent you from using the number. Therefore, you must be sure the Schedule is sorted by Room Number so you can catch and correct such errors.

17. Edit the Room Schedule to sort by Number.

18. Renumber Rooms 1 and 2 to fit within the pattern of numbering you established for the rest of the Rooms (see Figure 12.37).

 **Note:** Ignore any warnings that appear about duplicate numbers, but do correct them using the Schedule to assist you in renumbering other Rooms to correct redundancies.

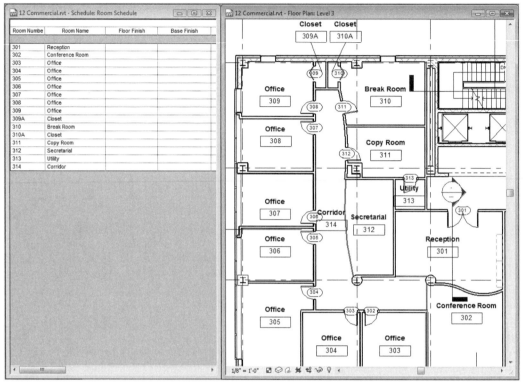

**Figure 12.37**  *Finish renaming, renumbering, and adjusting Tag locations*

## ADDING ROOM FINISH INFORMATION

As noted previously, the Room Schedule is really a room finish schedule. Using the procedure covered in the "Using Schedule Keys to Speed Input" topic, create a Room Style Schedule Key that includes all of the finish fields: "Base Finish," "Ceiling Finish," "Ceiling Height," "Floor Finish," and "Wall Finish." Create a few key styles such as those shown in Figure 12.38.

| Key Name | Base Finish | Ceiling Finish | Ceiling Height | Floor Finish | Wall Finish |
|---|---|---|---|---|---|
| Executive Office | WB-1 | ACT-2 | 9' - 0" | CPT-2 | P-2 |
| Public Area | VB-2 | ACT-1 | 9' - 0" | CPT-3 | P-3 |
| Standard Office | VB-1 | ACT-1 | 9' - 0" | CPT-1 | P-1 |

**Figure 12.38**  *Create a Room Style Schedule Key and use it to define various room finish schemes*

Once you have defined a few styles, you can add the Room Style key field to your Room Schedule and use it to input the finishes quickly on the basis of room type. For example, the two large offices at the bottom of the plan could be assigned the Executive Office key, while all the others get the Standard Office key. Define and assign as many keys as you wish. Remember, if you change a value in the Schedule Key later, the Room Schedule will automatically reflect the change. When you are finished, your Room Schedule should look something like Figure 12.39. Please notice in the figure that you can manually edit custom values in the fields as long as you leave the Room Style set to "none." If you prefer, you can define another Room Style, but in spaces that are unique like the reception or the conference room, it is just as easy to not assign a Room Style and instead assign the values of each finish directly.

12 Commercial.rvt - Schedule: Room Schedule

Room Schedule

| Room Numbe | Room Name | Room Style | Floor Finish | Base Finish | Wall Finish | Ceiling Finish | Ceiling Height | Comments |
|---|---|---|---|---|---|---|---|---|
| 301 | Reception | (none) | TER-1 | WB-2 | WC-2 | GWB-1 | 9' - 6" | |
| 302 | Conference Room | (none) | CPT-4 | WB-2 | WC-3 | ACT-1 | 9' - 6" | |
| 303 | Office | Executive | CPT-2 | WB-1 | P-2 | ACT-2 | 9' - 0" | |
| 304 | Office | Standard | CPT-1 | VB-1 | P-1 | ACT-1 | 9' - 0" | |
| 305 | Office | Executive | CPT-2 | WB-1 | P-2 | ACT-2 | 9' - 0" | |
| 306 | Office | Standard | CPT-1 | VB-1 | P-1 | ACT-1 | 9' - 0" | |
| 307 | Office | Standard | CPT-1 | VB-1 | P-1 | ACT-1 | 9' - 0" | |
| 308 | Office | Standard | CPT-1 | VB-1 | P-1 | ACT-1 | 9' - 0" | |
| 309 | Office | Standard | CPT-1 | VB-1 | P-1 | ACT-1 | 9' - 0" | |
| 309A | Closet | Public Are | CPT-3 | VB-2 | P-3 | ACT-1 | 9' - 0" | |
| 310 | Break Room | Public Are | CPT-3 | VB-2 | P-3 | ACT-1 | 9' - 0" | |
| 310A | Closet | Public Are | CPT-3 | VB-2 | P-3 | ACT-1 | 9' - 0" | |
| 311 | Copy Room | Public Are | CPT-3 | VB-2 | P-3 | ACT-1 | 9' - 0" | |
| 312 | Secretarial | Public Are | CPT-3 | VB-2 | P-3 | ACT-1 | 9' - 0" | |
| 313 | Utility | Public Are | CPT-3 | VB-2 | P-3 | ACT-1 | 9' - 0" | |
| 314 | Corridor | (none) | TER-1 | WB-2 | WC-1 | GWB-2 | 9' - 6" | |

**Figure 12.39** *Input values for the finishes of all of the Office spaces in the Properties dialog*

## TAG ALL NOT TAGGED

If you return to the furniture plan, you will note that it does not have Room Tags. At first this may seem strange, until we recall that annotation is always view-specific. This means that Tags and other annotation will appear *only* in the view to which it is added. However, it is very easy to add the Tags to any plan that needs them.

19. On the Drafting tab of the Design Bar, click the ***Tag All Not Tagged*** tool.

- Select the Room Tag item.

Since there are already many furniture tags in this view, it might be useful to add leaders to the Room Tags so we can move them around.

- Place a checkmark in the Create checkbox beneath "Leader" and then click OK.

You should get a Tag in every Room. You will need to move them around to make the plan legible, but as you can see, Tag All Not Tagged saves considerable time over manually tagging.

20. Save the project.

## QUERYING DATA

You should now be getting fairly comfortable with using Schedules. We have explored many examples and you are no doubt beginning to see just how powerful Schedules are. One of the biggest benefits to scheduling in Revit is all the robust data we can readily extract from our model. This data extraction is the "I" in BIM. While the potential data we can extract is virtually limitless, in this topic, we will focus on two very simple queries.

### REPORTING THE ROOM IN THE FURNITURE SCHEDULE

We have a detailed furniture Schedule and a detailed Room Schedule. While we have seen how easy it is to tile a Schedule and plan side by side and quickly find elements by selecting them in the Schedule, it might be useful to have Revit report directly in the Schedule in which Room each piece of furniture is located.

1. On the Project Browser, right-click the *Furniture Schedule (Working)* view and choose **Duplicate View > Duplicate**.

- Rename the view: **Furniture Location Schedule**.

- Right-click anywhere on screen and choose **View Properties**.

2. Edit the Fields.

- At the bottom left corner of the Fields tab, click the pop-up menu beneath "Select available fields from" and choose **Rooms**.

Notice that all of the Room fields are now available to add to the Schedule. All we need is Room Name and Number, so we'll add just these.

- Add Room: Name and Room: Number to the right side.

- Use the Move Up and Move Down buttons to locate the fields where you want them and then click OK twice.

Room names and numbers now appear in the Schedule, but because of the way that we are currently sorting and grouping, the data appears incomplete. A quick modification on the Sorting/Grouping tab will take care of this.

3. Return to the Schedule properties and edit the Sorting/Grouping.

- Change the first sort criterion to **Room: Name**. Enable the Header option and also the Blank Line option.

- Change the second criterion to **Room: Number**. Enable the Header option.

- Change the third criterion to **Model**.

- On the Formatting tab, make both the Room: Name and Room: Number fields hidden fields.

This removes the redundancy. When you use a field as a header, you can hide it as a column.

- Click OK twice to see the result (see Figure 12.40).

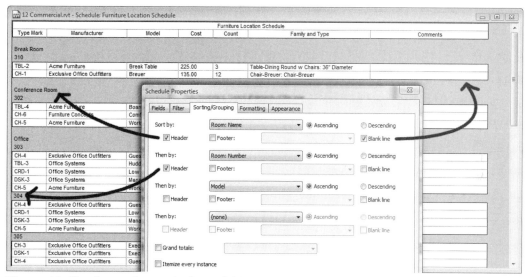

**Figure 12.40** *Sort and group by Room fields*

There are many other ways that this Schedule can be sorted, grouped, or formatted. Please feel free to experiment further. Remember, if you are uncertain how a particular combination of settings will look, make a duplicate of the Schedule first.

## MATERIAL TAKEOFFS

Most model elements have materials assigned to their components. These materials control how the components draw in various view types. For example a brick material controls the cross hatching in plan and the brick pattern in elevation as well as the brick photo texture in rendering. In addition to the display characteristics, materials can also have identity data assigned to them. A Material Takeoff is a Schedule that can query the material properties assigned to model elements. For example, using a Material Takeoff, we can ask Revit for the volume of concrete used in our project. Like other schedules, we can limit the query to a single category like Walls, or we can do a multi-category takeoff from many categories at once.

4. From the View menu, choose **New > Material Takeoff**.

- In the "New Material Takeoff" dialog, accept the <Multi-Category> choice on the left, name the takeoff **Concrete Takeoff** and then click OK.

- Add the following fields: "Family and Type," "Material: Name," "Material: Description," and "Material: Volume."

- Click OK to create the takeoff.

The resultant takeoff is very long and includes far more than concrete. Scroll through the entire list. Notice that our best chance for isolating just the concrete is the Material: Description column. Any of the Concrete materials use the word "Concrete" in the description. We could be tempted to say the same of the Material: Name column, but notice that doing so would also give us "Concrete Masonry Units," which we don't really want to include in the volume of concrete. To limit this Schedule to just the Material: Descriptions that contain the word Concrete, we can use the Filter tab.

5. Edit the properties of the *Concrete Takeoff*.

- Click the Edit button next to Filter.

- For the first Filter by criterion, choose **Material: Description**.

Several operators are available. The default is "equals."

- Change the operator to **Contains**.

- In the text field, type **Concrete** (see Figure 12.41).

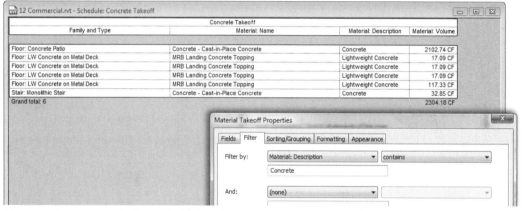

**Figure 12.41** *Add filtering to limit the takeoff to just concrete materials*

To fine-tune the takeoff further, edit the Formatting options. Set the alignment of the volume column to right and select the Calculate totals option. On the Sorting/Grouping tab, you have to turn on Grand totals for this to display (each of these enhancements are shown in the figure).

Filtering and totals work in any Schedule. The only difference between a Schedule and a Material Takeoff is the inclusion of the material fields in the Material Takeoff.

## ADD A COLOR SCHEME

When working with Rooms we noted how they only display on screen during creation and selection. In many cases, this is appropriate particularly in printing construction documents. There are times however when graphical indication of rooms or room types can be valuable, for example, when creating a programmatic color-coded diagram. Revit gives us this ability with Color Schemes.

### CREATE A NEW VIEW WITH ALTERNATE ROOM TAGS

Let's make a color-coded third floor plan.

1. On the Project Browser, right-click on *Level 3* and choose **Duplicate View >
   Duplicate with Detailing**.

   - Rename the new view **Level 3 Color Plan**.

2. Make a window selection around the entire plan.

   - On the Options Bar, click the Filter Selection icon.

   - In the "Filter" dialog, click the Check None button.

   - Place a checkmark in only the Door Tags box and then click OK.

   - Delete the selected Door Tags.

Remember, each view's annotation is unique to that view. We have only deleted the Door
Tags in the *Level 3 Color Plan*, not the *Level 3* plan.

3. Make the same selection again and return to the filter dialog.

   - Click the Check None button again.

   - Place a checkmark in only the Room Tags box this time and then click OK.

4. From the Type Selector, choose *Room Tag: Room Tag With Area [M_Room Tag:
   Room Tag With Area]*.

All of the Room Tags in this view will now display the area beneath the Tag (see Figure 12.42).

**Figure 12.42**  *Swap the Room Tags for a Type that displays the Area*

The Room Tag used here was part of the template from which the project was created. You
can create a Room Tag (or any Tag) that displays any of the parameters you wish. While we
will not create one here, building a Tag Family is not unlike creating any of the Families we
built in Chapter 10. In fact, creating a custom Tag Family is more like creating a custom title
block Family like we did in the "Create a Custom Cover SheetTitleblock Family" topic in

Chapter 4. If you wish to experiment on your own, select one of the Tags in this project that is similar to the one you wish to create. On the Options Bar, click the Edit button to launch the Family Editor and load the Tag for editing. Save the file as a new name and manipulate it to suit your needs. The text values are special elements called "Labels." You can add or edit Labels in the Family editor and have them report any of the parameters available. Look up "Labels" in the online Help for more information. If you wish to add a parameter that is not included on the list available, you must create a custom Parameter. This must be done with a "Shared Parameter" in a Shared Parameter File (**Shared Parameters** on the File menu). A Shared Parameter File is simply a text file saved on a hard drive or network server. The advantage of this file is that you can store your custom parameters in this file, save it to a common network server location and then all users in the firm can access and use these parameters in their projects, Schedules, and Tags. For more information on creating and working with Shared Parameters, see the online Help.

**BIM** *Manager Note:* There should only be one Shared Parameter file for the entire firm. The file should be stored on a network drive accessible to all users. These points are very important to guarantee that custom parameters work properly. Having more than one Shared Parameter file will almost certainly cause problems, annoyances, or even outright failures in the implementation of custom parameters in projects. The Shared Parameter file is merely the holding place for the parameter definition. It is not the parameter itself. A single Shared Parameter file can store hundreds of custom parameters for an unlimited number of projects and users. Therefore, please heed this recommendation and implement a single network-based Shared Parameter for your office.

## ADDING A COLOR SCHEME

Using the Room Tag with Area is a useful way to display the area of each Room directly in the plan, but in many cases it is useful to show this data more graphically using color coded shading. Adding a Color Scheme to a view will achieve this goal.

5. Right-click in the view window and choose **View Properties**.

- Locate the Color Scheme item and click the button (labeled <none>) next to it.

You can make a color scheme based on any of the Room's data fields. There are two predefined schemes: Name and Department.

- Click on the Name scheme on the left and then click OK twice (see Figure 12.43).

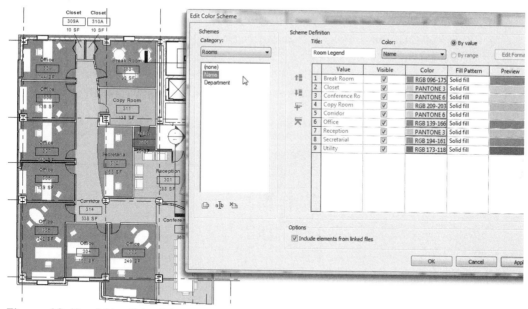

**Figure 12.43** *Add a Color Scheme to the Level 3 Color Plan view*

As you can see, it is easy to add a color scheme. However, a scheme based on Room Names does not convey much useful information. Let's learn to create our own scheme.

## CREATE A CUSTOM COLOR SCHEME

- Edit the View Properties again and then click the Color Scheme button again (it is now labeled "Name" instead of <None>).

6. Select the Name item in the list.

- At the bottom of the list area, click the Duplicate icon.

- In the "New Color Scheme" dialog, type **MRAC Area by Range** and then click OK.

- At the top of the dialog, beneath "Color," choose Area from the drop-down list.

- In the warning that appears, click OK.

There are two ways to create an Area scheme: by the actual values in the model (By value) and by ranges that you define (By range). We'll build a By range scheme.

7. Click the By range radio button.

- Click in the "At Least" column for the last item (currently 20.00 ft² [20m²])

- Change the value to **50.00 ft² [5m²]**.

- On the left side of the table, click the Add Value icon.

A new 100.00 ft² value will appear.

8. Select the 100.00 ft² [10m²] value, and then click the Add Value icon again.

A new 150.00 ft$^2$ [15m$^2$]value will appear.

9. Keep selecting the last item and clicking the Add Value icon until you reach 300.00 ft$^2$ [30m$^2$] (see Figure 12.44).

Make sure you select the last item in the list each time before you click Add Value or the new value will be added in between the one you have selected and the one after it.

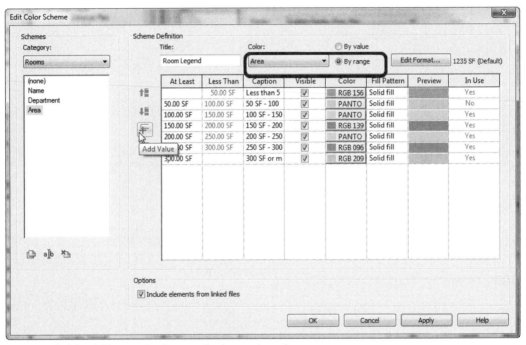

**Figure 12.44** *Add several values to the Area color scheme*

By default, the colors are assigned automatically. This is the easiest way to see immediate results. If you prefer to designate your own colors, you can click the color buttons next to each item. We will not do that for this exercise.

10. Click OK twice to accept the default color scheme and display it in the view.

## ADD A COLOR SCHEME LEGEND

It may not always be obvious what the colors in a plan represent. So to make it easier to understand the color scheme, we can add a legend.

11. On the Design Bar, click the Drafting tab and then click the ***Color Scheme Legend*** tool.

• Zoom out a bit allowing room to the left side of the plan.

• Click a point to the left of the plan to place the Color Fill Legend (see Figure 12.45).

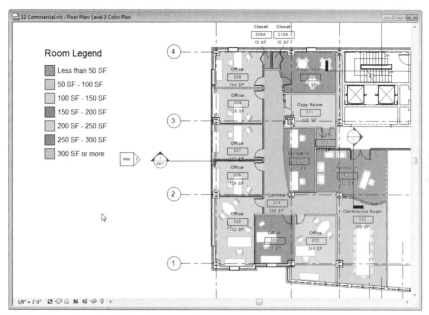

**Figure 12.45** *Complete the Color Scheme and add a legend*

## CREATE ANOTHER COLOR SCHEME

You can continue to experiment by creating new color schemes. Simply return to the View Properties dialog and click the Color Scheme button. There you can make new duplicates and configure them. Let's do one more.

12. Return to the View Properties dialog and click the Color Scheme button.

- Click the Duplicate icon and name the new Scheme **MRAC Floor Finish** and then click OK.

- From the "Color" list choose: **Floor Finish**.

- Click OK in the warning dialog.

- Click OK twice more to return to the plan view and see the results.

In this case, the colors show where the floor finish varies based on the values we assigned earlier in the chapter. Feel free to create other schemes and vary the colors or other settings.

## CREATE A ROOM/AREA REPORT

In some jurisdictions detailed area analysis with triangulated area proofs are required. You can generate such a report from Revit in HTML format.

13. From the File menu, choose **Export > Room/Area Report**.

- In the dialog that appears browse to a location where you wish to save the report, give it a name and then click Save.

14. Launch your web browser and open the HTML file created to view the report.

## SUMMARY

- A Schedule is simply a tabular view of your building model data.

- You can edit elements referenced in a Schedule directly from the Schedule or in the model

- Changes to model elements are reflected immediately in both the graphical views and the Schedules.

- Schedules can contain any combination of fields and be grouped, sorted, and filtered.

- Add headers, footers, and totals to your Schedules to break them up and make the data more legible.

- Adding Schedules to Sheets is a simple drag and drop process.

- You can adjust the size and formatting of Schedule columns on the Sheet.

- Add breaks to the Schedule on the Sheet to wrap a long Schedule into two or more columns.

- New elements added to the model will immediately appear in Schedules as appropriate.

- Schedule Keys allow you to efficiently manage repetitive groups of related property values shown in Schedules.

- Tags report data in similar fashion to Schedules.

- Add Tags manually or all at once using "Tag All Not Tagged."

- Create custom views to show different Tags and graphics.

- Create Room elements to generate Room Tags, Area takeoffs, and finish Schedules.

- Use Room Separations to manually define the shape of Rooms.

- Color Schemes can be used to color-code nearly any data that appears in the Rooms.

CHAPTER 13

# Ceiling Plans and Interior Elevations

## INTRODUCTION

The goal of this chapter is to round out our construction document set. Reflected ceiling plans will be the primary focus of the chapter with a brief exploration of interior elevations at the end. Reflected ceiling plan views are included in the default project template file used to start both projects in this book. Therefore, we simply need to open these views and add appropriate model data and annotation. The default templates include only exterior elevations. Therefore, we will need to indicate which rooms we wish to elevate and create the required interior elevation views and any embellishments they require.

## OBJECTIVES

Ceiling plans are very similar to other plans. The View Properties are the only major difference between the two. We will add Ceiling elements to model and ceiling specific annotation. After completing this chapter, you will know how to:

- Add and modify Ceiling elements
- Understand Ceiling Types
- Add and modify Ceiling Component Families
- Manipulate the View Properties of the Ceiling Plan
- Create Interior Elevation views

## CREATING CEILING ELEMENTS

Ceiling elements in Revit are used in Ceiling p lans to convey the material of the ceiling plane—such as acoustical tile ceiling, gypsum board or other ceiling treatment. Ceiling elements are created like Rooms and are also very similar to Floors and Roofs. The default tool when creating Ceilings allows you to pick a point within a closed space in the model (much like creating Rooms, demonstrated in the previous chapter). However, they are sketch-based elements, and can utilize all the typical functions like the "Pick Walls" and other sketch methods on the Design Bar during creation and editing.

## INSTALL THE CD FILES AND OPEN A PROJECT

The lessons that follow require the dataset included on the Mastering Revit CD ROM. If you have already installed all of the files from the CD, simply skip down to step 3 below to open the project. If you need to install the CD files, start at step 1.

1. If you have not already done so, install the dataset files located on the Mastering Revit Architecture CD ROM.

   Refer to "Files Included on the CD ROM" in the Preface for instructions on installing the dataset files included on the CD.

2. Launch Revit Architecture from the icon on your desktop or from the *Autodesk > Revit Architecture* group in *All Programs* on the Windows Start menu.

   **Tip:** In Windows Vista, you can click the Start button, and then begin typing **Revit** in the "Start Search field. After a couple letters, Revit Architecture should appear near the top of the list. Click it to launch to program.

   • If the New Features Workshop dialog appears, choose "Maybe later" and then click OK.

3. On the Standard toolbar, click the Open icon.

   **Tip:** The keyboard shortcut for Open is CTRL + O. **Open** is also located on the File menu.

   • In the "Open" dialog box, click the *My Documents* icon on the left side.

   • Double-click on the *MRAC* folder, and then the *Chapter13* folder.

   If you installed the dataset files to a different location than the one listed here, use the "Look in" drop down list to browse to that location instead.

4. Double-click *13 Commercial.rvt* if you wish to work in Imperial units. Double-click *13 Commercial Metric.rvt* if you wish to work in Metric units

   You can also select it and then click the Open button.

The project will open with the last saved view visible on screen.

## CREATING AN ACOUSTICAL TILE CEILING

Suspended acoustical tile ceilings in commercial office buildings are constructed in one of two ways. In the first method, walls are built past the height of the ceiling tiles (to a fixed height or all the way to the deck) and each room contains its own ceiling. In the second method, the walls are built only up to the height of the underside of the ceiling (underpinned) with the ceiling plane being continuous across the tops of the Walls.

In Revit, when each room contains its own ceiling and the walls continue past the ceiling plane height, Ceiling elements can usually be created quickly with the "Auto Ceiling" function of the *Ceiling* tool. This is the default function of the Ceiling tool. When you have an underpinned Ceiling, you can instead use the sketch option to draw the shape of the overall Ceiling plane using any of the available sketching tools. We will explore both ceiling types as we continue to refine the third floor of our commercial project.

Revit usually creates both a floor plan and a ceiling plan view whenever you create a new Level in a project. The template from which we originally created the commercial project includes such views. In this sequence, we will work in the ceiling plan views for the first time. Be sure that you work in a Ceiling plan view when adding Ceiling elements. If you do not, you will receive a warning like the one shown in Figure 13.1.

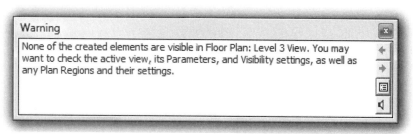

**Figure 13.1**    *Attempting to add Ceilings in floor plan views yields a warning*

5. On the Project Browser, expand the *Ceiling Plans* node.

• Double-click to open the *Level 3* ceiling plan view.

 **Note:** Again, please be sure that you are opening the *Level 3* ceiling plan view and *not* the *Level 3* floor plan view.

Don't let the vertical lines that appear confuse you. These are the joists in the linked Structural model. You can hide or unload the file if you find them distracting, but there is no need since the Ceiling elements will cover the joists when we add them.

6. On the Design Bar, click the Modeling tab and then click the **Ceiling** tool.

On the Status Bar, a message will appear "Click inside a room to create ceiling." On the Options Bar, a single button labeled "Sketch" will appear (see Figure 13.2).

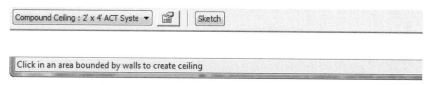

**Figure 13.2**    *Using the Ceiling Tool, "Sketch" is an option, but not the default*

• From the Type Selector, choose **Compound Ceiling : 2' × 4' ACT System [*Compound Ceiling : 600 × 1200mm grid*]** if it is not already selected.

• Move the cursor around the screen pausing within various Rooms—do not click yet.

Notice that the behavior is similar to the **Room** tool from the previous chapter. However, unlike the **Room** tool, this tool does not have the "Room Separation" option, nor can it recognize those Room Separation lines already in the model. Therefore, if the auto-detecting routine does not automatically recognize the Room you need, you can use the "Sketch" option noted above. We will try the sketch option a little later. For now, we will add a Ceiling in a Room that is recognized by the auto-detection routine.

- Move the cursor into the upper left corner office.

  The Room boundary should highlight automatically.

- Click in the Room to add the Ceiling (see Figure 13.3).

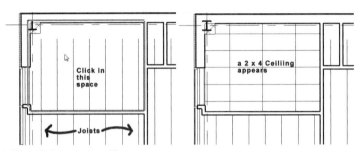

**Figure 13.3** *Add a Ceiling to an office*

- On the Design Bar, click the **Modify** tool or press the ESC key twice.

## EXPLORE CEILING TYPE PROPERTIES

7. Click on one of the Ceiling grid lines.

Notice that each line is individually selectable. However, they behave as a unit. The auto-creation routine will attempt to center the grid in the room left to right and top to bottom. You can move any grid line if you like, and the entire grid pattern will move with it. Use this technique to apply custom centering (see below).

- With a grid line selected, on the Options Bar, click the Properties icon.

Notice that the "Height Offset From Level" is to 8'-0" [2600] above the associated Level

- Change the value to **9'-0" [2900]** (see Figure 13.4).

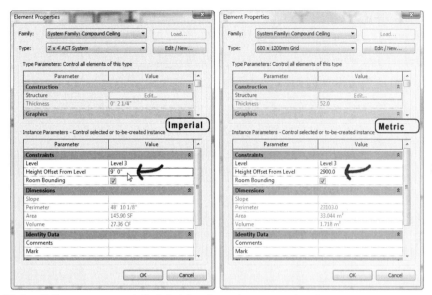

**Figure 13.4**  *Ceilings are set a default height above the current Level*

8. Next to the Type list, click the Edit/New button.

**Tip:** A shortcut to this is to press ALT + E.

The Type Properties dialog will appear.

Notice that a Ceiling element's Type has an Edit "Structure" button like many other Types. If you were to edit this Structure, you would notice that it is comprised of a simple Core element with a Finish Layer on the bottom. The grid pattern comes from the Material assigned to this Finish Layer (see Figure 13.5).

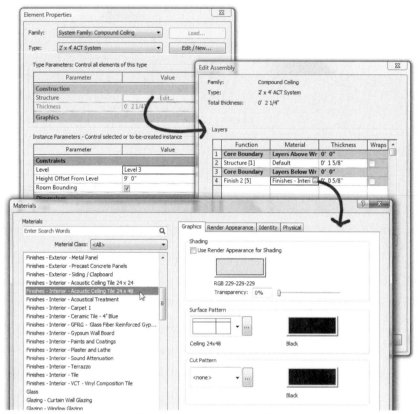

**Figure 13.5** *The Type Properties of the ACT Ceiling uses a "Ceiling Tile" Material*

We do not need to make any changes to the material, structure, or type properties. However, such exploration is educational, particularly when you realize that editing a Ceiling type is nearly identical to editing a Wall or Floor type. Keep this in mind if you need to create custom Ceiling types in your own projects.

9. Cancel the Materials, Structure, and Type Properties dialogs when finished.

- Make sure that the Ceiling Height Offset is still set to 9'-0" [2900] and then click OK.

- Select the linked structural model (click on the joists) and on the View Control Bar, choose Hide Element from the Temporary Hide/Isolate icon.

While noted as optional above, temporarily hiding this linked file will make it easier to work.

10. Add another Ceiling using the same technique in the Break Room (across the corridor on the right side).

- On the Design Bar, click the **Modify** tool or press the ESC key twice.

- Select the new Ceiling, edit its properties, change its Height Offset to **9'-0"** [**2900**] and then click OK.

## MOVE, ROTATE AND ALIGN CEILING GRIDS

We could continue and add additional Ceilings in the other Rooms, but for now we will work with just these two.

11. Select one of the gird lines of the Ceiling you just added.

• On the toolbar, click the Move icon and then move the grid line (see Figure 13.6).

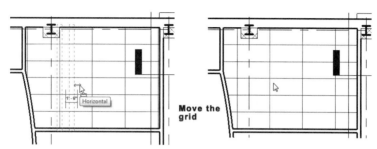

**Figure 13.6** *Move a Ceiling grid line*

As noted above, notice that the entire grid will reposition with this move, not just the selected grid line. Also, only the grid pattern is affected by this move, not the boundary of the Ceiling element, which still conforms to the shape of the Room. You can use other typical modification techniques as well, such as Rotate. If you need your grid pattern at a different angle, simply click the Rotate icon and rotate the grid line. The rest of the grid will follow (see Figure 13.7).

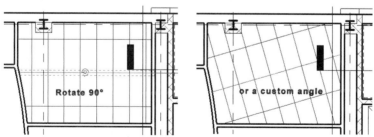

**Figure 13.7** *Rotate a Grid using the Rotate tool*

If you rotated the grid, please undo the change now to return it to a horizontal orientation.

You can use the Align command to create alignment between the grids in two Rooms.

12. Select one of the horizontal grid lines in either Room.

• On the toolbar, click the Align icon.

• For the "Point of Reference" click on one of the horizontal grid lines.

- For the "Entity to Align" click on a horizontal grid line across the hall in the other Room (see Figure 13.8).

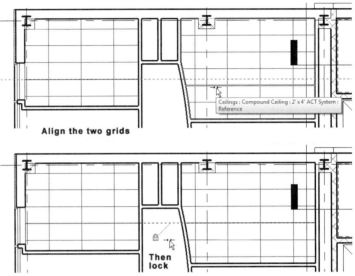

**Figure 13.8**   *Align two Ceiling grids*

13. Click the padlock icon that appears to constrain the alignment.

14. Select one of the horizontal grid lines.

- Move it up or down.

Notice that the Ceiling grids in both Rooms move together. When you move this way, you are simply moving the Model Pattern that is applied via a Material to the Ceiling, *not* the Ceiling itself. This is why the pattern remains clipped to the shape of the Room as you move it.

## CREATE A CEILING VIA SKETCH

Some of the spaces that we have will prove problematic for the automatic Ceiling generate option. The routine does not recognize Columns very well. To create the next several Ceilings, we'll use the Sketch option. In addition, we will sketch a single continuous underpinned Ceiling for the remaining non-office spaces in the suite.

15. Zoom in on the office on the left directly beneath the corner one to which we already added a Ceiling.

16. On the Design Bar, click the Modeling tab and then click the **Ceiling** tool.

- From the Type Selector, choose **Compound Ceiling : 2' × 4' ACT System [Compound Ceiling : 600 × 1200mm grid]**.

- On the Options Bar, click the Sketch button.

The Design Bar changes to sketch mode and shows only the Ceiling tools.

By now, the sketch tools in Revit should be very familiar to you. The basic steps are summarized here, but please feel free to use whatever techniques you prefer to sketch the inside shape of the Room.

- On the Design Bar, click the **Pick Walls** tool.
- Pick the Walls surrounding the Room.
- Use the Flip controls to change the side of the Sketch lines as required.
- Use Lines to sketch around the Column.
- Use Trim/Extend to cleanup the sketch (see Figure 13.9).

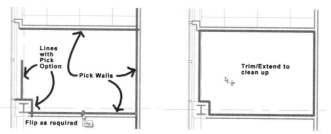

**Figure 13.9**   *Sketch the inside shape of the office space Ceiling*

17. On the Design Bar, click the Finish Sketch button.

A new Ceiling will appear in the office.

18. Repeat this process to add Ceilings to each of the remaining offices.

Notice that Revit will typically choose what it considers to be the best orientation for the ceiling grid. If you wish, you can rotate the grid 90° to orient the offices all the same. To do this, remember, simply select one of the gird lines, and then click the Rotate icon and rotate the line 90°. The remainder of the grid will match the new orientation. If you need to select the actual Ceiling element to edit the sketch, place the cursor near the edge of the Room and use the TAB key to pre-highlight and then select the Ceiling element.

19. Using the process covered above, Rotate and Align the Ceiling grids to one another in a logical pattern (see Figure 13.10).

**Tip:** To align several Ceilings to one another, choose the "Multiple Alignment" option on the Options Bar when using the Align tool.

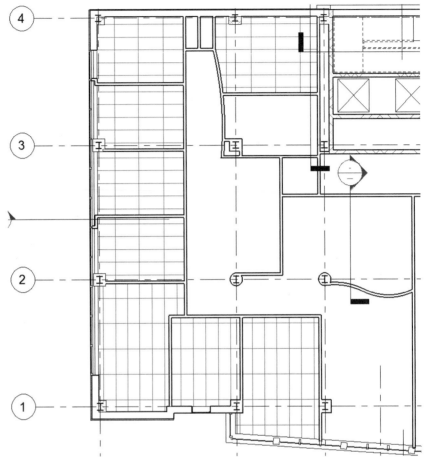

**Figure 13.10**  *Sketch the remaining office space Ceilings, align and rotate them as desired*

20. Make a window selection of all the Ceilings, use the Filter icon to filter everything except Ceilings.

   • Edit the Properties of the selected Ceilings, change the Height Offset to: **9'-0" [2900]** and then click OK.

21. Save the project.

## CREATE AN UNDERPINNED CEILING

The Ceilings that we just sketched were all contained wholly within single Rooms. You can use the same procedure, using sketch lines to encompass the overall perimeter of several spaces at once. When you do this, you will have a single continuous Ceiling grid that can represent an underpinned ceiling.

22. On the Design Bar, click the ***Ceiling*** Tool.

   - On the Options Bar, click the Sketch button.

   - Sketch the outline shown in Figure 13.11.

      Don't include the Conference Room. We will do this one later.

You can create most of the required edges with the Pick Walls tool. The rest you can build with the Lines tool and both the pick and draw options. Trim/Extend to clean up the shape.

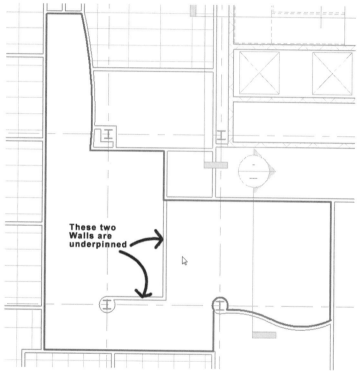

These two
Walls are
underpinned

**Figure 13.11**   *Sketch an underpinned Ceiling across several Rooms*

23. On the Design Bar, click the Ceiling Properties button.

   - Change the "Height Offset From Level" to **9'-6"** [**2950**] and then click OK.

   - On the Design Bar, click the Finish Sketch button.

## STUDY A CEILING IN SECTION

At this point we have several Ceiling elements in our project. We have studied their Type parameters, moved, rotated and aligned them with one another and discussed the difference in selecting and manipulating the Model Pattern vs. the Ceiling itself.

24. On the Project Browser, double-click to open the *Longitudinal* section view.

**Tip:** If you prefer, double-click the Section Head in the plan instead.

25. Zoom in on the third floor at the left side of the section.

Notice that the Ceiling in the offices at the left is a bit lower than the one in the public spaces. Also, notice that the Walls currently go all the way to the deck and join with the Floor. Since we have made a single continuous underpinned Ceiling in the public spaces, we might want to adjust these Walls to stop at the Ceiling height. The section view helps us spot the problem, but the edit is best accomplished in the ceiling plan view.

26. On the Project Browser, double-click to open the *Level 3* ceiling plan view.

- Use Split Wall at the back of the secretarial area as indicated in Figure 13.12.

- Select all of the Walls in the underpinned area (see the right side of Figure 13.12).

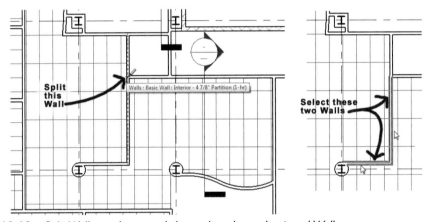

**Figure 13.12** *Split Walls as shown and then select the underpinned Walls*

27. On the Options Bar, click the Attach button.

- Move your mouse to the edge of the Ceiling.

  The edge of the Ceiling will pre-highlight.

- Click when the Ceiling pre-highlights to attach the selected Walls to the Ceiling.

**Note:** You could also edit the properties of the Walls and change their Top Constraint to Unconnected at a height of 9'-6" [2950]. However, if you anticipate that the Ceiling height will change at some point, the Attach option is better. Further, the graphics in the section will be nicer with Attach.

Now that we have adjusted the height of the Walls, the model more accurately reflects our design intent. You can see this best by returning to the *Longitudinal* section view (see Figure 13.13).

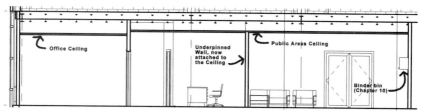

**Figure 13.13** *Study the results it the* Longitudinal *Section view*

If you return to the ceiling plan view, you will notice that even though the Wall now sits beneath the height of the Ceiling plane, it is still hiding the Ceiling as if it were going all the way to the deck. There is an easy fix for this.

28. Select the same two Walls again.

**Tip:** To quickly select the same elements again, right-click and choose **Select Previous**.

29. Right-click and choose **Override Graphics in View > By Element**.

• Place a checkmark in the "Transparent" checkbox and then click OK (see Figure 13.14).

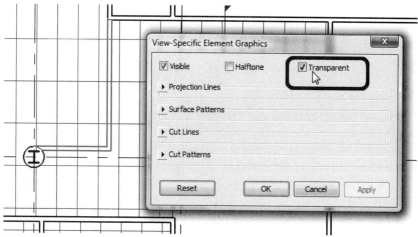

**Figure 13.14** *We can make the Walls display correctly in ceiling plan by making them transparent*

The transparent feature can also be useful in 3D views when you wish to display elements that would otherwise be obscured by elements in front of them.

## ADDING A DRYWALL CEILING

For the Conference Room, we will use the same basic process as the other Ceilings, but simply choose a different Type.

30. On the Design Bar, click the Modeling tab and then click the **Ceiling** tool.

   - From the Type Selector choose **Compound Ceiling : GWB on Mtl. Stud [Compound Ceiling : Plain]**.

   - On the Options Bar, click the Sketch button.

   - Follow the process outlined above to create a Ceiling in the Conference Room.

As you can see, except for choosing a different Type, the process is the same as with Ceiling grids.

### SWITCH TO A DIFFERENT GRID SIZE

Like other elements in Revit, you can change the Type used on an existing element at any time. Suppose you preferred a 2' × 2' [600 × 600] ceiling grid layout instead of the 2' × 4' [600 × 1200] one we used. With an existing Ceiling selected, you can simply choose a different type from the Type Selector. While we will not cover the steps here, you can also duplicate and modify an existing Type, assign a different pattern and achieve other Ceiling designs not included in this project file. Feel free to experiment with this on your own later if you wish. The process is nearly the same as creating a new Wall or Floor type.

31. Select one of the grid lines on the large underpinned Ceiling in the middle of the plan.

   - From the Type Selector, choose **Compound Ceiling : 2' × 2' ACT System [Compound Ceiling : 600 × 600mm grid]**.

32. Save the project.

### ADDING CEILING FIXTURES

Now that we have completed adding ceiling elements to our project, we can add some lights and other fixtures to the ceiling plan. The process for adding such elements is very similar to the detailing that we did back in Chapter 10. To achieve the optimal lighting layout, you may need to adjust the location of your grid lines. You can use your Move and Rotate commands as noted above to this. If you rotate a grid and an error appears indicating that constraints are no longer satisfied, simply click the "remove Constraints" button and then if desired, use the Align tool again to reapply constraints.

### ADDING LIGHT FIXTURES

Let's start with a simple fluorescent lighting fixture in the offices.

1. Make adjustments to the position of gird lines in the offices at the left to accommodate a suitable lighting layout (see Figure 13.15).

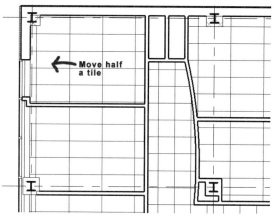

**Figure 13.15**   *Move grid half a tile to better accommodate the lighting*

2. On the Design Bar, click the Modeling tab and then click the **Component** tool.

- On the Options Bar, click the Load button.

- Browse to your library folder and open the *Lighting Fixtures* folder.

- Select the *Troffer - 2×4 Parabolic.rfa* [*M_Troffer - Parabolic Rectangular.rfa*] Family file and then click Open.

**Note:** As mentioned in previous chapters, if you do not have these Families in your installed libraries, you can find them in the *Library* folder with the files installed from the Mastering Revit Architecture CD ROM.

3. Place a light fixture in the office.

- Use the Align command to align it to the grid lines.

- Select and then Copy the light as appropriate for the space (see Figure 13.16).

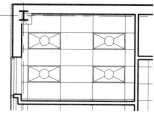

**Figure 13.16**   *Align the light to the grid and then copy it to create a lighting pattern*

If you are not satisfied with the position of your lights relative to the grid lines, you can move the grid line one-half a tile or one-quarter of a tile in either direction. The lights will move with them.

4. Select all of the lights in the office and copy them to the other offices (see Figure 13.17).

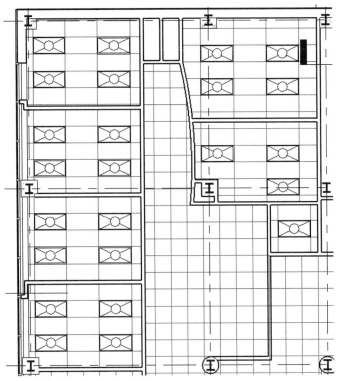

**Figure 13.17** *Copy the lights to other Offices*

### ADDING LIGHT SWITCHES

Some firms like to add light switches to ceiling plans instead of floor plans. You add switches the same as any Component Family.

5. On the Design Bar, click the **Component** tool.

- Click the Load button again, browse to the *Chapter13* folder and then open the *Switch-Single (rcp).rfa* Family file.

This Family file is a modification of the standard Revit Architecture library symbol with modifications to allow it to display in reflected ceiling plan. In Chapter 10 we explored the creation and editing of Families. We discussed Symbolic Lines in that chapter. However, in order for any elements to display in a plan view (ceiling plan in this case), there must be some element that passes through the Cut Plane height that triggers Revit to display its 2D plan representation. The two ways to deal with the situation are to lower the Cut Plane of the ceiling plan view (currently set at 7'-6" [2300]) or to add an element to the Family file that passes through this default Cut Plane. The Family file provided in the *Chapter13* folder has a simple Model Line drawn vertically and set to the <Invisible Lines> Line Style passing through the reflected ceiling plan Cut Plane. This triggers Revit to display the graphics for this symbol in the RCP.

If you prefer, you can edit the View Properties of the Level 3 ceiling plan view and then click the View Range Edit button. In the "View Range" dialog, edit the Offset to a lower height such a 4'-0" [1200] (see Figure 13.18).

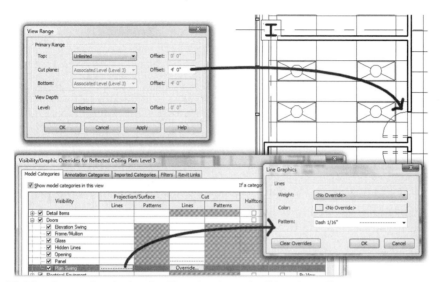

**Figure 13.18**     *Optionally edit the Cut Plane height in the "View Range" and the graphic display of Doors*

This would have the added benefit of showing the Doors in the reflected ceiling plan view. Some firms prefer for Doors to show in RCP to make it easier to locate light switches, exit signs and other RCP equipment relative to door swing. If you make this edit, you might also consider editing the Door graphic defaults in the Visibility/Graphics dialog.

6. Place a light switch adjacent to the latch side of the door.

- Place or copy additional switches in the other offices.

## ADDING WIRING

Revit Architecture does not include a wiring element. (Revit MEP does include wiring.) To show the wiring connection to the switches, simply draft Detail Lines on the view. Remember, Detail Lines are view-specific and will show only in this reflected ceiling plan.

7. On the Design Bar, click the Drafting tab and then click the **Detail Lines** tool.

- From the Type Selector, choose <Hidden>.

- On the Options Bar, click the "Arc passing through three points" icon.

8. For the first point, click near the switch.

- For the end point, click near the middle of a light fixture, and then click a third point somewhere in between (see Figure 13.19).

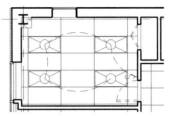

**Figure 13.19**  *Sketch a three-point arc Drafting Line for the wiring*

9. Continue drawing arcs to connect all the lights in the room to the switch.

10. Repeat in the other Rooms.

- On the Design Bar, click the **Modify** tool or press the ESC key twice.

## COMPLETE THE RCP

At this point, you can continue adding elements to the reflected ceiling plan as needed. Following procedures covered in previous chapters, you can add text, dimensions, and tags as necessary. If you wish to add Room Tags, they will "see" the Rooms that we have already added to this model in the previous chapter including the Room Separation lines. If the Room Tag does not fit comfortably in the Room, add it with the "Leader" option. Then you can move it outside the boundaries of the Room. The fastest way to add the Tags is with Tag All Not Tagged (see the "Tag All Not Tagged" topic in Chapter 12). If you add dimensions, you can click on the value and add manual edits, prefixes, and suffixes. For example, you may want to add a note indicating the point of beginning (P.O.B.) for the grid. Add a dimension, click its text, and then add the text above, below, or as a prefix or suffix (see Figure 13.20).

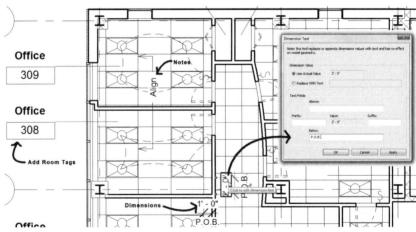

**Figure 13.20**  *Adding notes and tags*

11. When you are finished add elements to your reflected ceiling plan, save the project.

## CREATING INTERIOR ELEVATIONS

No construction documents set would be complete without interior elevations. Adding such views in Revit is easy to do. Our reception area, the corridor and the toilet rooms in the core are all good candidates.

### ADD AN INTERIOR ELEVATION

1. On the Project Browser, double-click to open the *Level 3* floor plan view.

2. On the Design Bar, click the View tab and then click the **Elevation** tool.

- From the Type Selector, choose **Elevation : Interior Elevation**.

- On the Options Bar, from the "Scale" list, choose **1/4"=1'-0"** [**1:50**].

3. Move the cursor into the reception area—do not click yet.

- Move the cursor around the room and watch the orientation of the elevation head change dynamically (see Figure 13.21).

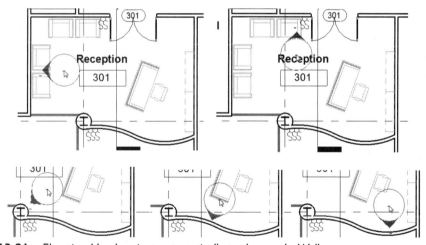

**Figure 13.21**   *Elevation Heads orient automatically to the nearby Walls*

The orientation of the elevation symbol automatically orients to a nearby Wall. This makes creating single elevations easy, especially to angled or curving walls. Move slowly along the curved Wall to see this clearly. In this case, we want four elevations—one in each direction.

4. When the symbol is pointing up, click to place the elevation marking in the room.

- On the Design Bar, click the **Modify** tool or press the ESC key twice.

You will notice that a new node has appeared in the Project Browser for the *Interior Elevations*. In addition, if you expand it, a new view associated with the elevation marker has also appeared named *Elevation 1 - a.*

5. Double-click on triangle part of *Elevation 1 – a* to open it.

Notice that the view is cropped nicely to the size of the Room. You can adjust the cropping if you need to fine-tune it. If you don't want to see the section markers, right-click them and choose **Hide in View > Elements**.

## ADD ADDITIONAL INTERIOR ELEVATIONS

We can add additional elevations of the reception space very easily.

6. On the Project Browser, double-click to open the *Level 3* floor plan view again.

7. Click on the Elevation Marker (the circle part of the marker) in the middle of the reception space.

A series of checkboxes will appear surrounding the symbol and ghosted arrows pointing to up to four possible elevation directions. To add another elevation view, simply check one of the boxes.

- Place a checkmark in each of the three other checkboxes surrounding the elevation symbol (see Figure 13.22).

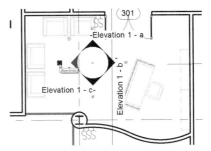

**Figure 13.22** *Add new Elevation views by checking the direction checkboxes*

The new views will appear on the Project Browser.

8. On the Project Browser right-click on the *Elevation 1 – a* view and choose **Rename**.

- Name it **Reception – North Elevation**.

- Repeat for each of the other three elevations.

You can complete these views in any way you wish. Add notes, dimensions or tags. Drop them onto a new Sheet to round out the set (see Figure 13.23).

To turn off the view name on the Elevation marker edit the Type properties of the marker. To do this go to the **Settings menu > View Tag > Elevation tags**. From the Type list, choose 1/2" Circle. Remove the checkmark from the "Show View Name" box and then click OK.

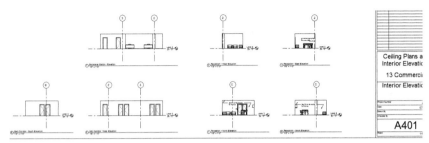

**Figure 13.23** *Annotate Interior Elevations and add them to a new Sheet*

## SUMMARY

Reflected ceiling plans and interior elevations are two important parts of a complete construction documentation set. In this chapter, we have taken a brief look at the steps to create both of these important document types.

- Ceiling elements are easily created from existing Walls by picking points within a Room.

- If bounding Walls are not available, or if you wish to create a custom Ceiling shape, you can Sketch the Ceiling using standard sketch tools.

- Ceiling Types control the structure of a Ceiling element.

- Acoustical Tile Ceilings are simply Ceiling Types that have a finish layer assigned to an appropriate Material and surface pattern.

- The grid lines in the Material use a Model Pattern.

- To create an underpinned Ceiling, use the Sketch option.

- Adjust the height of Walls to coordinate with the underpinned Ceiling after creation.

- Drywall Ceilings are created with the same process and simply use a different Type.

- You can change grid size by swapping the Type.

- Light fixtures and switches are component Families that you load and place in the Ceiling plan.

- Add wiring using Detail Lines.

- Add Interior Elevations using the Elevation View tool.

- Check more than one box on the interior elevation symbol to add additional elevations.

# CHAPTER 14

# Printing and Publishing

## INTRODUCTION

Printing (or plotting) from Revit Architecture is similar to printing from any Windows software application. You choose **Print** from the File menu; configure your options, and then print. In this chapter, we will look at the basic process of printing from Revit.

## OBJECTIVES

Creating printed output from Revit is simple. After reading this chapter, you will know how to:

- Configure Print Setup options
- Print to a hard copy printer
- Create a multi-sheet DWF file

## DATASET

For the purposes of the topics covered in this chapter, you can open any Revit project. Versions of the commercial and residential projects in Imperial units have been provided in a folder called *Chapter14*. You can practice and print from these projects if you wish, or open the "Complete" version of either project in metric or Imperial units from any chapter folder instead. Feel free to open and print your own Revit projects as well.

## PRINT SETUP

The "Print Setup" dialog box has many settings that can be configured to enhance the quality of printed output. Choose **Print Setup** from the File menu to access this dialog (see Figure 14.1).

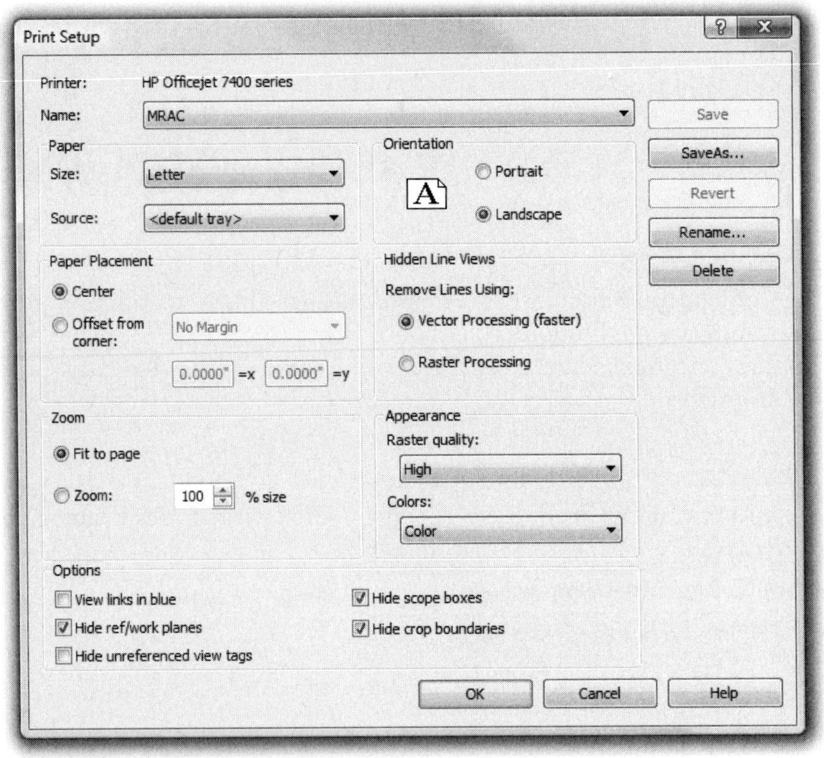

**Figure 14.1** *The Print Setup dialog*

If you make changes to any of the default settings in this dialog, you can click the SaveAs button on the right and give the configuration a new name. The Print Setups are stored in the project. This custom configuration will then be available to you when you print the project in the future (MRAC is shown in the drop-down in Figure 14.1 as an example). If you want to read a description on any element in the dialog, click the question mark (?) icon at the top right and then click it on the item in question. A help window will appear with a detailed description of the item.

Be certain that the printer you wish to use is listed at the top of the dialog before you configure the options in the dialog. If it is not listed, click Cancel, choose Print from the File menu, choose the desired Printer from the list and then click the Close button. Reopen the Print Setup dialog and continue.

**Paper:** Set the correct paper size. Only sizes that the printer driver supports will be available.

**Orientation:** Set to Portrait or Landscape.

**Paper Placement:** Center will work for most printers. Otherwise, use Offset from corner and its associated options. You may need to experiment with your printer and driver to find the right combination for correct placement. If you make such adjustments, be sure to save the configuration with a descriptive name to preserve your efforts.

**Zoom:** To print to the view's current scale it must be set to 100% of size. To print a half sized set change the value to 50%. Keep in mind that everything will be smaller including all annotations.

Fit to page will also potentially affect the actual size and scale of the printed output. If you want your print to be "to scale," choose 100%.

**Hidden Line Views:** Most construction document views in Revit are set to Hidden line. For example, plan views, elevation views, sections views, and many 3D views default to Hidden Line. Printing is frequently faster if you use "Vector processing" as indicated. The time it takes Revit to process the Vector print job depends on the size of the project, the type of model objects in the view, and the amount of hidden line removal that must be processed.

Run some test prints on actual projects in your environment to find the best settings.

**Appearance:** If you are using Raster Processing, choose an appropriate Raster Quality. The higher the quality you choose, the longer it will take to print. If you are printing to a color printer and want black lines, remember to set the colors to black lines. This will result in all lines printing black—even ones that are colored on screen. Setting the colors to grayscale will print any lines in the view that are black in solid black ink, and every other color will be a value of gray. If the gray printed output does not look exactly the way you want, try setting the colors to Color and then set the printer driver to black and white. This often gives slightly different grayscale results that might be useful.

**Options:** This tells Revit to print or hide certain elements in the view. Hide reference/work planes is very useful. You would not typically want these items to print; likewise for scope boxes and crop boundaries. The "Hide unreferenced view tags" option will hide any elevation, section, or callout tag view that has not yet been placed on a Sheet. This is a very useful feature (see Figure 14.2).

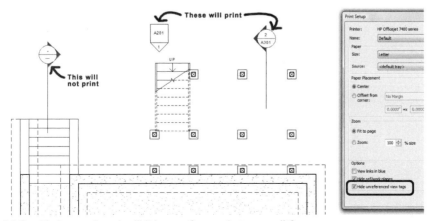

**Figure 14.2** *Understanding the "Hide unreferenced view tags" feature*

**Name:** (See the top of the "Print Setup" dialog.) This is the name of the Print Setup settings. Click Save or Save As to create a new Print Setup. This is retained in the project for future use.

## PRINT

When you are ready to print, choose **Print** from the File menu. You can print from any Revit view. Most often, you will want to print Sheets because they are designed for this purpose and include title blocks and borders. However, if you wish to print from another view you can do that as well.

Always choose your printer from the list first. Click the Properties button if necessary and make any edits to the printer's unique properties (these properties vary by printer device). For the "Print Range" you have the option of the current view, the "Visible Portion" of the current view, or a range of Views or Sheets. Click the Select button to choose a range of Sheets or views to print (this is sometimes referred to as batch printing). Under Options choose "Reverse print order" if you want to reverse the print order. If a physical printer is used, you may also set the number of copies to be printed and whether they are collated. Click the Setup button to open the "Print Setup" dialog and make additional edits. Click OK to print (see Figure 14.3).

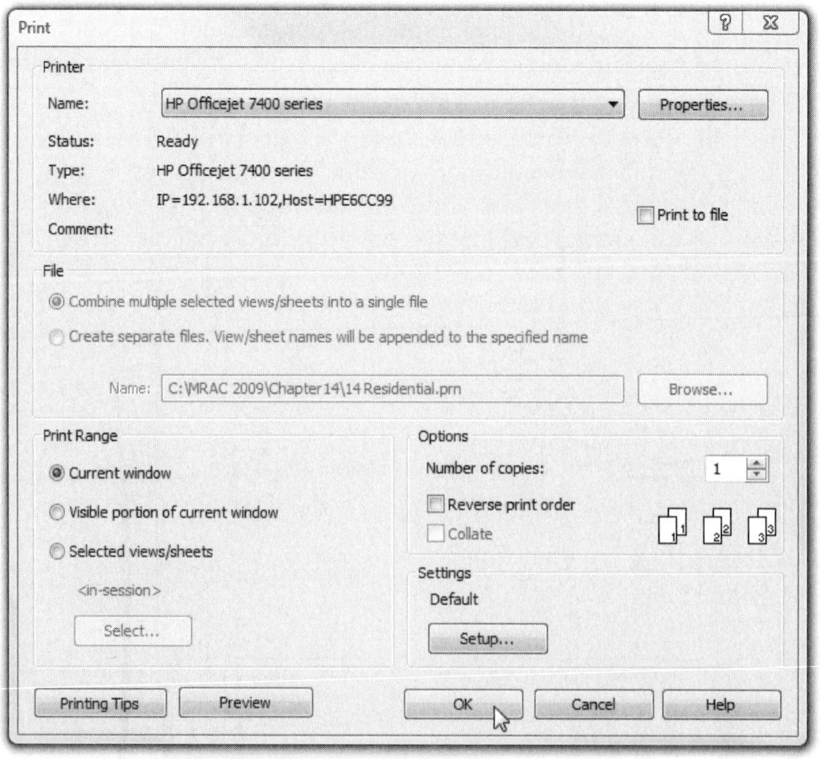

**Figure 14.3** *The Print dialog*

**Print to File:** If your output device requires it, you can select this option and then use the fields in the "File" area to create a plot file.

**Current window:** prints the entire extent of the currently active view or sheet on screen.

**Visible portion of current window:** prints only the part of the view that is displayed within the current window on screen. To set what you want to print, close the Print dialog, re-size, and zoom the view's window accordingly. The proportion of the window is also important and should be similar to the paper size and layout (portrait vs. landscape). If the window proportions do not match the paper proportions, the printed output may exclude part of the view or might include extra white space.

**Selected views/sheets:** is a mechanism for batch printing. It allows you to pick multiple views and or Sheet views and print them at the same time. You can save the selection of views and or Sheet views in named sets that can be re-used later.

## PUBLISH A DWF FILE

While many projects require output to actual paper sheets, digital submissions such as DWF files are becoming more popular. A DWF/DWFx (Design Web Format) file is a highly compressed, vector-based file format designed for viewing and distributing design files over the Internet and by email. What makes the DWF file so powerful is that it is a vector-based, high-quality graphics file that is read only. This means it can be distributed to consultants and clients without fear of unauthorized editing. DWF preserves access to critical design data and graphics from the original model. DWF files can be referenced as backgrounds in AutoCAD. DWF files can also be embedded in Web pages for viewing in a browser with the plug-in provided free from Autodesk. Anyone with a copy of the Autodesk DWF Viewer software, available as a free download on the Autodesk Web site (*http://www.autodesk.com/*), can view, zoom, pan, query, and print the DWF file. In Windows Vista, DWFx files can be viewed native without any additional software required. If the recipient has a copy of Autodesk Design Review, they can add redline comments to the DWF file, which can then be loaded back into Revit. Creating a DWF is simple, because it is the same as printing to a hard copy device.

In order to create a DWF file from Revit Architecture, choose **Publish DWF > 2D DWF** from the File menu. When using the 2D DWF option, it is possible to create several separate DWF files—one for each printed view—in the same print operation. However, it is usually more desirable to choose the "Combine multiple selected views/sheets into a single file" option and thereby create a single DWF file that contains several Sheets within it. Verify the location for the saved file and choose your naming option. Revit can name the file(s) automatically using a combination of the view and file name, or you can name the file manually.

From the "Range" list, choose the "Selected views/sheets" option and then click the browse button. In the "View/Sheet Set" dialog that appears, clear the "Views" checkbox (at the bottom) and then click the Check All button to select all of the Sheets in the list. If you wish to reuse this list again later, click the SaveAs button and give it a name (see Figure 14.4).

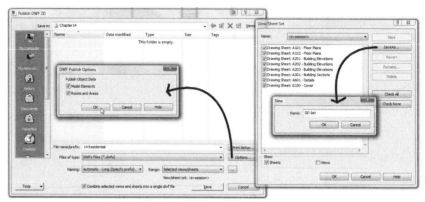

**Figure 14.4** *Create a Multi-sheet DWF file*

In the "Print" dialog, there is also a button to return to Page Setup and another that calls an Options dialog. In this dialog, you can choose to include model, Room, and Area data in your DWF file. This creates a DWF file complete with all of the rich data that you have built into your BIM model. Recipients will be able to query objects in your DWF and measure areas.

When printing is complete, open the DWF file with Autodesk DWF Viewer or Design Review software and view the results. Again, if you are using Vista, and you create a DWFx, you can just double-click it in Windows Explorer.

## PUBLISH A 3D DWF

The process to publish a 3D DWF is simpler. Open any 3D view and then choose **Publish DWF > 3D DWF** from the File menu. If you run this command while in a 2D view, Revit will prompt you to open a 3D view. Give the file a name and save it. Open the 3D DWF and study the results.

## PRINTER DRIVER CONFIGURATION

Since Revit prints using the Windows printing mechanism like most Windows software it is important that the correct printer driver be used. The correct driver is the driver specifically made for the physical printer and model used. To obtain the correct drivers for your printers contact the printer manufacturer and ask for the Windows printer driver for the model printer you have and follow the manufacturer's instructions for installing the driver. Most drivers are available for download from the printer manufacturer's web site.

Frequently Printing/Plotting Services or Bureaus will ask for a plot file in the HPGL format. This will not work with Revit. Inform the plotting service that you are printing from Windows-based Revit and must use the Windows printer driver for the printer model that they intend to use. Alternatively, you can ask if they can print a DWF file. Most service bureaus can print DWFs. In this case, simply print the DWF file and send it to them.

## PUBLISHING TOOLS

Revit includes several tools to make Sheet composition and printing easier. While space here does not permit detailed tutorial coverage of the following tools, a quick mention is appropriate.

**Dependent Views and Matchlines**: When a drawing is too large to fit on a Sheet, it is customary for the drawing to be split into sections and a matchline to be employed. To accommodate this need, Revit provides the Dependent view feature. With this feature, you can duplicate a large view into several dependent views. Unlike the standard duplicate options, dependent views share their annotation. Annotation added or editing in either the parent or child view is reflected in both. To set up dependent views, right-click the overall view and choose **Duplicate View > Duplicate as a Dependent**. Repeat for as many sections as you need. For example, to matchline a plan into east and west sections, you will have one overall view and two dependent views. The dependent views are shown indented beneath their parent on the Project Browser. Adjust the crop regions as appropriate in each of the dependents (see Figure 14.5).

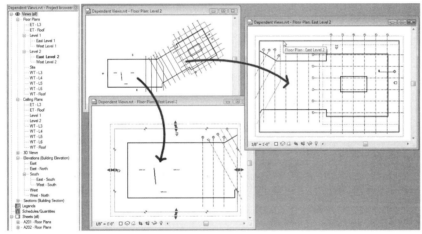

**Figure 14.5** *Simple example of a large plan split into two dependent views*

The figure also shows the dependent view rotated to match the Sheet orientation. To do this, select the crop region boundary and use the rotate tool. The crop region will remain horizontal when you are finished rotating and the model within it will rotate.

**Scope Boxes**: In complex buildings, you may have multiple Levels or Grids in different sections of the building. By default, all datums show across the full extent of the building. While it is possible to stretch the handles and adjust them manually, this can become tedious on large projects. A Scope box is a three-dimensional box that you sketch in plan. Use the handles to adjust its size. Give the scope box a unique name. All datums have a Scope Box property. If you have no Scope boxes, the value of this property remains None.

If you add Scope Boxes, you can edit the properties of the datum and assign it. The Grid or Level will now only display in views that intersect the Scope Box (see Figure 14.6).

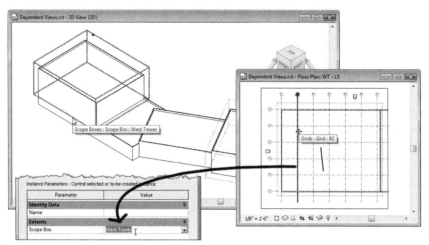

**Figure 14.6** *Simple example of a Scope Box*

**Revision Tracking**: Despite every Architect's best efforts, revisions are a part of any document set. Once you have issued your drawings for bid and the inevitable revision packages must go out, use the revision tool on the Drafting tab and the coordinated revision table available from the Settings menu to manage them (see Figure 14.7).

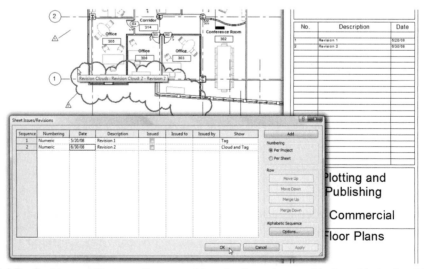

**Figure 14.7** *Revision tracking coordinates, revision clouds, and tags with the table in the title block*

You add each new revision in the "Sheet Issues/Revisions" dialog (choose Revisions from the Settings menu). When you add revision clouds (Drafting tab of the Design Bar), they will appear automatically in the table on the Sheet.

## TROUBLE SHOOTING

If your printed output does not look correct or complete, the first thing to check is whether the Print Preview looks correct. If the Preview looks correct that tells us that Revit, Windows, and the Printer Driver are all working correctly and the problem is at the physical printer. Note that the resolution of the Preview may not be correct because the high resolution is being displayed on your screen at a lower resolution and that is to be expected.

A common problem includes printed output that does not look complete. Either all the text will be missing or whole portions will not be printed. This is indicative of the printer's memory being overloaded. Revit often sends a combination of vector, raster, and other data formats. Raster data in particular is more memory intensive than vector data. As a result, it is common to have a 100 MB or larger print job. Some printers/plotters do not have enough onboard RAM or a built in hard disk to handle this.

Some printer drivers have and option that says something like "Process print job in the computer's memory". To resolve the issue, try using this setting. If your printer driver does not have this setting you can either install more memory in the physical printer, use a printer that has a built-in hard disk and more memory or add a separate RIP (Raster Image Processor) device to aid the printer. Many currently available new large format printers have built in RIPs.

## SUMMARY

Printing from Revit is simply and straight forward.

- Like other Windows software, choose Print Setup to configure the Print options.

- Save your choices for future use in the project.

- Choose Print from the File menu to print to paper.

- Choose Publish DWF from the File menu to create a 2D or 3D DWF file.

- You can create a multi-sheet DWF file of your entire document set in one step. This can be an excellent way to distribute document sets.

# SECTION IV

# Appendices

The Appendices include additional important exercises and resources. Worksharing is an important part of Revit Architecture workflow, enabling teams to work on projects efficiently and effectively. Appendix A offers some additional practice exercises that space did not permit in the main text of the book. Key concepts and issues about Worksharing are introduced in Appendix B. Appendix C provides additional resources that you might find useful as you begin using Revit Architecture in your daily work. Also be sure to explore the contents of the files installed from the CD for other useful files and resources.

Section IV is organized as follows:

| | |
|---|---|
| **Appendix A** | Additional Exercises |
| **Appendix B** | Worksharing |
| **Appendix C** | Online Resources |

# APPENDIX

# a

# Additional Exercises

## INTRODUCTION

Contained in this appendix are some additional exercises that you can perform to gain more practice with some of the basic Revit modeling techniques. With the exception of the existing conditions of the residential project, most of the attention in chapters 3 through 9 was paid to the exterior of the buildings. The dataset provided with Chapter 10 includes interior partitions and other refinements not detailed in the tutorials. This appendix addresses that gap by providing the minimal guidance required to assist you in creating the elements provided at the start of the Chapter 10 dataset.

## OBJECTIVES

The exercises presented in this appendix are intended for additional practice and are meant to be self-directed. As such you given the overall goal of the exercise, the critical dimensions and some guidelines and tips to assist you. The exercises are optional and if you prefer you can skip them. The dataset at the start of Chapter 10 matches the state of the model at the completion of these exercises. After completing this appendix you will:

- Gained additional experience with Wall layout

- Added Windows, Doors, and other embellishments to the residential dataset

- Completed the third floor tenant build-out on the commercial project

- Add plumbing fixtures to both projects

## RESIDENTIAL PROJECT

The residential project requires refinements on all three levels (Basement, First, and Second). The exercises that follow will progress floor by floor starting with the Basement.

1. Open *A Residential.rvt* if you wish to work in Imperial units. Open *A Residential Metric.rvt* if you wish to work in Metric units.

## BASEMENT PLAN

2. Start in the *First Floor* plan view.

- Select the three Walls of the addition, edit the properties and change the Base Constraint to **Site** with a **6" [150]** Base Offset.

- Copy and paste aligned the same three Walls to the *Basement* plan.

  Ignore any warnings.

- Select the Wall on the west and the north one at a time and click the Remove Sketch button on the Options Bar.

- Select all three Walls, edit their properties and change the type to **Foundation - 12" Concrete [Retaining - 300mm Concrete]**. Set the Base Offset to **0**, the Top Constraint to **Up to level: Site** with a **6" [150]** Top Offset.

3. Offset a copy of the west Wall inside the plan. Trim and or Split the Walls as required and use the handles to stretch the two west Walls to match the dimensions in the figure (see Figure A.1).

- Add a monolithic Stair from the Basement level to the height of the terrain at the rear of the house. Cut a section to assist in placement and sizing.

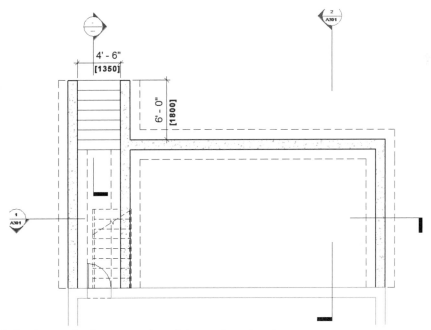

**Figure A.1** *Layout the Basement plan of the residential project*

- On the Structural tab of the Design Bar, click the Foundation > Wall tool and then click on each Wall to add strip footings.

- To display the footings, edit the View Properties of the Basement floor plan, click the View Range edit button, and the change the View Depth Offset. Try a value like **-4'-0" [-1200]**.

Then select each footing, right-click and choose **Override Graphics in View > By Element** and change the Projection lines Pattern to Hidden (see Figure A.2).

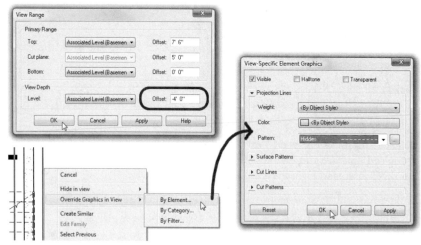

**Figure A.2** *Adjust View Range and then override footings to display properly*

If you want to hide the demolished Stair, you can open the View Properties and choose **Show Complete** for the Phase Filter.

## FIRST FLOOR PLAN

4. Open the *First Floor* plan view.

5. Create a new Wall type.

- Make the Core use a Structure layer of Wood Stud **3 1/2"** [**90**].

- Add a Finish layer of Finishes - Interior - Gypsum Wall Board on each side **1/2"** [**12**] thick.

- Draw the Walls in the center of the plan using the new type.

- Add Windows and Doors as shown. Assume that all inserts are centered unless dimensioned otherwise.

- Load the Bay Window from the web library or from the CD.

- Use the Demolish tool on the two Windows between the addition and the existing house (see Figure A.3).

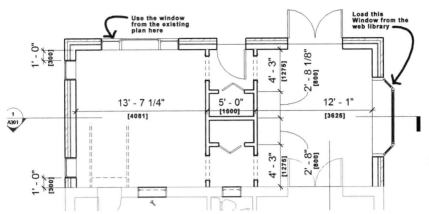

**Figure A.3**  *Add Walls, Windows and Doors to the First Floor*

## SECOND FLOOR PLAN

Demolition needs to take place in the bathrooms between the existing house and the new addition. However, the challenge is that currently, this is one continuous Wall from ground to roof. To demolish just the portion on the second floor, we need to copy and paste the Wall like we did for the basement, adjust the height parameters, and then demolish the portion we don't need.

6. Stay on the *First Floor* plan view.

- Copy and paste aligned (to the same place) the Wall between new and existing.

    Ignore the warning.

- With the Wall still selected, edit the properties and change the Base Constraint to **Second Floor**.

- When the Warning appears, click the Delete Instances button.

The warning is about the Doors and Windows on the lower portion of the Wall that no longer have a host. It is OK to delete them because we actually have two copies, but ultimately need only one.

- Remaining in the *First Floor* plan, select the Wall again.

- Edit its properties and change the Top Constraint to **Up to level: Second Floor**.

- You will also need to open the *Transverse* section and Detach the bottom Wall from the Roof.

When the process is done, you should have two Walls stacked on top of each other and the Doors and Windows on both first and second floors should still be intact.

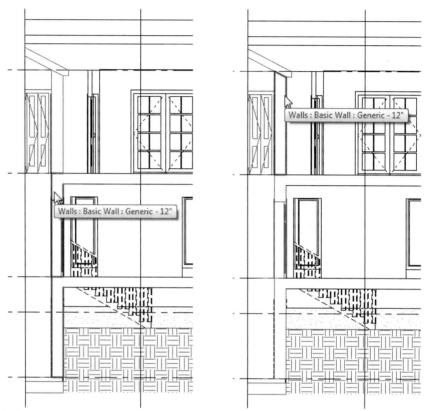

**Figure A.4** *Add Walls, Windows, and Doors to the First Floor*

7. Open the *Second Floor* plan view.

- Split the Walls between the new addition and existing house.

- Demolish the left side of the Wall and demolish the remaining Window in the Wall as well.

- Draw a new Wall using the custom type build for the first floor between the two bathrooms.

- Add a Wall between the bathroom and new bedroom.

- Add a closet to the new bedroom.

- Add Doors and Windows as shown. Choose shorter height Windows for the bathrooms and set their sill heights at **3'-8" [1100]**.

Assume that all Windows and Doors are centered unless noted otherwise (see Figure A.5).

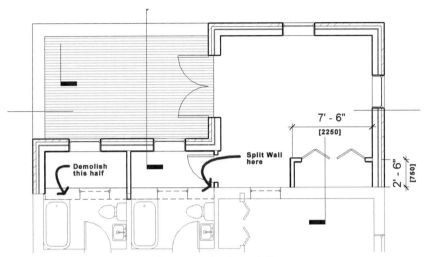

**Figure A.5** *Add Walls, Windows, and Doors to the Second Floor*

Feel free to add additional embellishments, such as upgraded fixtures in the bathrooms and counters, cabinets, and appliances in the kitchen. You can find such items in the out-of-the-box library and the web library.

8. Save and close the project.

## COMMERCIAL PROJECT

The commercial project requires a lobby plan, restroom layouts, and some tenant build-out on the third floor. The exercises that follow will progress through these tasks.

1. Open *A Commercial.rvt* if you wish to work in Imperial units. Open *A Commercial Metric.rvt* if you wish to work in Metric units.

### LEVEL 1 PLAN

2. Start in the *Level 1* plan view.

The lobby will follow the shape of the floor slab already in place on Level 2. The easiest way to do this is to add Level 2 as an underlay to Level 1.

- Edit the View Properties of Level 1 and add Level 2 as an underlay.

- Draw Walls using the pick option with an offset of approximately **1'-6"** [**450**] and click each of the gray lines from the Floor object on Level 2.

- Draw the additional Walls, Doors, and Openings indicated (see Figure A.6).

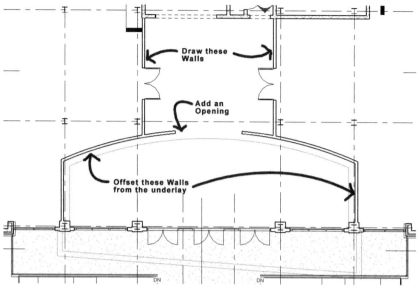

**Figure A.6** *Add Walls and Doors to create a lobby on Level 1*

## TOILET ROOMS

3. Add Walls and fixtures to the toilet rooms in the core area.

 **Note:** To find rest room fixtures, look in the Autodesk Web Library or use the Content Seek toolbar to search for them. Some samples have also been included with the files form the Mastering Revit Architecture CD ROM.

- Select all the components added, Group them, and then copy and paste aligned to the other floors (see Figure A.7).

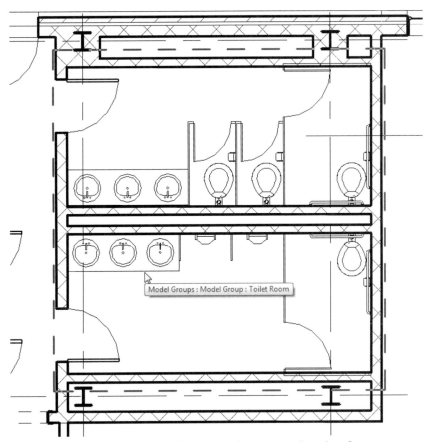

**Figure A.7**   *Create a toilet room layout, Group it, and copy it to the other floors*

## LEVEL 2 PLAN

4. Open the *Level 2* plan view.

- Add a guardrail to the open balcony.

## LEVEL 3 PLAN

5. Open the *Level 3* plan view.

- Using the ***Interior - 4 7/8" Partition (1-hr)*** **[*Interior - 123mm Partition (1-hr)*]** Wall type, add the Walls for the interior build-out as shown in the figure.

- Add Doors. Assume **6"** **[150]** offset from corners or centered in the room as appropriate. Load any Door types not currently resident in the project (see Figure A.8).

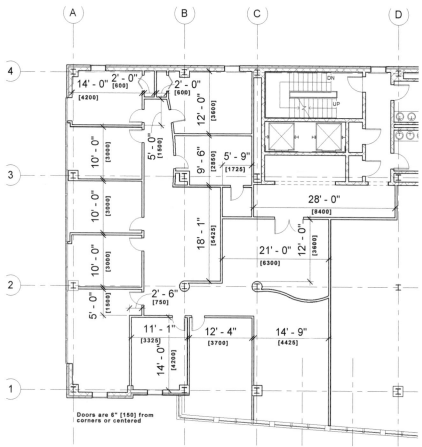

**Figure A.8** *Add Walls and Doors for interior build-out on the third floor*

 **Note:** dimensions shown in the figure are to the face of the Wall. Wall thickness dimensions are not shown to maintain clarity.

# Worksharing

## INTRODUCTION

The term "Worksharing" applies collectively to the various techniques used in Revit Architecture to work in teams of multiple individuals and firms. Worksharing utilizes processes we have already seen in this book such as linked Revit (RVT) files and linked AutoCAD (DWG) files, as well as the internal segmentation of a project using Worksets. Worksets is the Revit toolset that allows a single project file to be divided into smaller pieces with which team members can work independently without impeding the work of others. Care must be taken when enabling Worksets to ensure that a strategy for their use appropriate to the team dynamics is established. In this appendix, we will introduce the concepts and key terminology used in Worksets as well as briefly discuss general Worksharing issues.

## OBJECTIVES

Many resources are available to the reader for learning and understanding the concept of Worksets. This appendix will provide an overview of the salient concepts and suggestions for further reading in the Revit Architecture Help system and online resources. After completing this appendix you will understand:

- Key Worksharing tools available
- Workset Terminology
- Where to find additional Worksharing Resources

## TYPES OF WORKSHARING

Several forms of Worksharing are possible in Revit Architecture. Architectural projects usually involve teams of professionals either within the same firm (under the same roof) or dispersed among several companies and physical locations. Whether you simply need to load a CAD file as a background for your own design work or you need to manage a fully-coordinated Revit model among several members of your firm, Revit has tools and capabilities suited to the task.

## LINKING AND IMPORTING

In earlier chapters, we explored two forms of linking: linked RVT files and linked CAD files. If you need to simply keep track of work being done in another application such as AutoCAD or Microstation, then file linking is the appropriate solution. DWG (AutoCAD) and DGN (Microstation) files can be either linked or imported into your Revit model. If you wish to maintain the ability to update the file periodically as the original author of the file makes changes, choose to link the file. In this scenario, you simply reload the linked file when you receive an updated version from your consultant or teammate. If you instead need to use the geometry in the CAD file to assist you in creating your Revit model and have no need to reload changes in the future, you can Import the file instead. This places a static copy of the file within your Revit model. You can leave this imported file intact as a single element in the Revit model or even choose to explode it. If you explode it, Revit will convert the imported geometry into simple Revit Drafting and Model Lines. You should not explode these files unless necessary, as it will increase overhead and memory demands on your system depending on how large and extensive the imported files are. Examples of linking and importing DWG files can be found in Chapters 4, 8, and 11.

While linking to CAD files does help bridge the gap between your firm and those not using Revit Architecture, it is always better if you can get the entire project team working in the same file format and software. Therefore, wherever possible, having all team members using Revit Architecture, Revit Structure, or Revit MEP is preferable. You can still work in separate models and utilize linking in this scenario as we did in Chapter 6. In the case of RVT files, you will typically link the file. Should you decide that it is desirable to "merge" two RVT files into a single file, it is possible to bind a link. This converts it to a Group but breaks the link to the original file. You can also use Copy and Paste or Group Save and Group Load instead. Open one of the files, copy all of the elements that you wish to merge, and then paste them into the other project. Group all of the model, annotation, and datum elements you wish to merge, save the group file out to a RVT file and then load that RVT file into the other project. Remember, model and annotation elements will actually be stored in separate model and detail Groups. Chapter 6 covers the process in more detail.

A recent addition to Revit is the ability to import and export IFC files. The Industry Foundation Class (IFC) standard attempts to define a universal standard file format for the storage of building model data. IFC is currently the only bridge format that allows Walls, Doors, Windows, Roofs, Floors, and other building components to be preserved when imported and exported to and from Revit and other BIM software packages. While the technology is promising, it still cannot preserve all of the nuances that each software package introduces into its building models. If you absolutely must work in a team that uses Revit and other BIM packages together, then spend some time testing out the IFC import and export commands found on the File menu.

## COORDINATION MONITOR

When you link two Revit files together, (such as a Revit and Revit Structure file) you can use the Coordination Monitor tool to keep track of duplicate elements or elements that rely upon one another in each file. For example, if the Structural Columns are in the Revit Structure file and the partitions and the Architectural Columns are in the Revit file, these elements can get out of synch as users in each model make changes. Using the **Copy/Monitor**, **Coordination Review** and **Interference Check** items on the Tools menu, you can copy elements between files, watch them for changes and make updates to keep both linked projects synchronized. Look for more information about these tools in Chapter 6 and the online help.

## WORKSETS

Even if all work on a project is carried out within the physical walls of your own office, it is typical that more than one member of your firm will need to access and work in the project at the same time. In this situation, use "Worksets" to organize a project into smaller pieces, each enabling a different member of the project team to work simultaneously on the same project file. Each Workset remains linked back to a "Central file" that keeps all changes and team interactions coordinated. While Workset use is most common on a team project, it can be used in any Revit project and some Sole Practitioner firms find it useful as a tool for managing project data and visibility of elements. Whatever your specific situation, achieving success with Worksets requires a clear understanding of each tool, careful pre-planning and ongoing management.

 **Note:** The following is a brief introduction to the concept of Worksets. It is intended to introduce the concepts, define the terminology, and suggest some common scenarios for their usage. A comprehensive tutorial-based exploration of the topic falls out of the scope of this text. Explore the online help and various online resources on the topic and consider attending a formal training session at a local training provider or Autodesk reseller. To learn more about hands-on training opportunities, visit the author's web site at: www.paulaubin.com.

### GETTING STARTED WITH WORKSETS

Before using Worksets they must be enabled in a project. After enabling them, some configuration will be required. Typically this task should be performed by a single member of the project team knowledgeable in Worksets and their nuances. This person is often referred to as the "Project Coordinator" or some similar title. They might be a CAD Manager, BIM Manager, or simply a Project Architect on the project who has good knowledge of Revit Architecture.

To enable Worksets the first time, choose the command from the File menu, or display the appropriate toolbar (see Figure B.1).

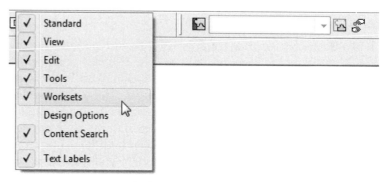

**Figure B.1**   *Display the Worksets toolbar*

Every member working on the team should have an understanding of the basic concepts of a Workset-enabled project. In addition, it is highly recommended that each member of the team completes the tutorial on the subject provided with the software and multiple team members practice together on a sample Workset project. To access the tutorial, choose **Tutorials** from the Help menu. You may need to follow instructions to download the files. In the Help file, expand the "Using Advanced Features" topic, then select the "Sharing Projects" topic. Read and complete all exercises beneath this topic (see Figure B.2).

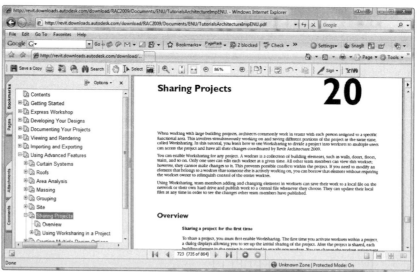

**Figure B.2**   *Worksharing Tutorials accessed from the Revit Help menu*

You may need to return to your installation CD to install the required tutorial files. Depending on the options that you chose when you installed your Revit software, these files may or may not already be on your system. You can also download them from Autodesk. You will find a link when you choose Tutorials from the Help menu. Several other topics and tutorials are also included in the Help system. It is worthwhile to spend some time exploring this valuable resource.

## UNDERSTANDING WORKSETS

The concept behind Worksets is simple. Normally, in a computer environment, only one user at a time can access a particular file. Since by default, Revit places all project elements within a single file, this would naturally impede teamwork. Using Worksets, there is still a single "Central" file that houses all project data, but from this Central file, each team member can save a local copy that maintains its association with the Central file. In this way, when any team member saves work from their local copy back to the Central, those changes are merged into the Central file and become available to all members of the team. Simply having the Central and Local files is not enough however, the vast collection of potentially editable elements within the project files must be managed. Worksets are used for this purpose. A Workset is basically a named group of elements that can be "checked out" by an individual user. Once checked out, a Workset becomes "read-only" to other members of the team. This allows those members to view any part of the project that they wish, but only edit those parts that they have checked out. When saves back to Central are made, users can choose to "relinquish" their Worksets and/or check out other ones. The concept is similar to a public library or video rental store. If you wish to check out a particular book or video, it must be in the store at the time of your visit. If someone else has already checked out the item, you must wait for her to return it before you can check it out.

The biggest challenge involved in a Revit Workset-enabled project is deciding how to organize the elements into Worksets. You want to have enough Worksets for the quantity of team members, but don't want to have so many that management of them becomes problematic. Certain Worksets are created and maintained automatically by Revit. These include "Views," "Families" and "Project Standards" Worksets.

- **View Worksets:** For each view, a dedicated View Workset is created. It automatically contains the view's parameters (scale, visibility, graphics style, etc.) and any view-specific elements such as text notes, dimensions, detail elements, etc. view-specific elements *cannot* be moved to another Workset.

- **Family Worksets:** one Workset is created for each loaded Family in the project.

- **Project Standards Worksets:** one Workset for each type of project setting, such as Materials, Line Styles, etc.

These Worksets are created and maintained automatically by Revit. In addition, Revit also allows "User Defined" Worksets. These contain all of the building model elements in a project. By default, Revit will create two such Worksets when Worksets is first enabled: "Shared Levels and Grids" and "Workset1." All of the Levels and Grids are moved automatically to the "Shared Levels and Grids" Workset and all of the rest of the geometry is moved to "Workset1."

An element can only belong to one Workset at a time. You can move a model element from one Workset to another, but you cannot assign it to more than one at the same time. This is why careful planning is important. The Project Coordinator must therefore attempt to anticipate the needs of the project team and create Worksets to house model elements in a way that supports these needs. While there are no "rules" regarding this, there are common best practices.

The factors to consider include:

- Project size

- Team size

- Team member roles

- Default Workset visibility

Each of these factors may have an impact on the use and composition of each Workset. Larger projects and larger project teams will typically have more Worksets. This stands to reason. However, even small projects can benefit from Worksets, so this is not the only factor. When you open a project that has Worksets enabled, you can choose at that time which Worksets to load. Choosing to open only those Worksets needed for a particular task can help files load more quickly and preserve valuable computer resources. You can also take advantage of Workset visibility to hide entire Worksets in one or more views as appropriate. Therefore, any or all of these factors can play an important role in determining the ultimate composition of Worksets in a particular project. Also, remember that each project is unique and while you may follow many common strategies from one project to the next, you must always be flexible enough to allow for specific circumstances that may arise in a particular project.

## ENABLING WORKSETS

As was mentioned above, it is typically advisable for a single experienced member of the project team to take responsibility for Workset management and setup. Worksets must be enabled before they can be used. Then a Central file must be saved in a location that all team members can readily access. This location is typically on a network server on the company LAN (Local Area Network).

 **Note:** While it is possible to host a shared project over a WAN (Wide Area Network) it is only advisable if considerable bandwidth is available to all team members.

To Enable Worksets, choose the Worksets command from the File menu, or click the icon on the Worksets toolbar. A dialog like the one in Figure B.3 will appear.

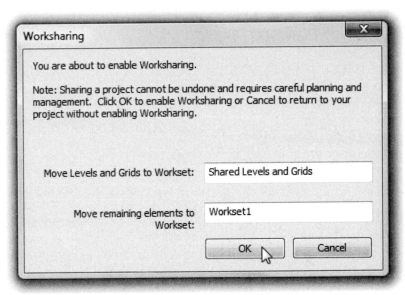

**Figure B.3**  *Worksharing must be enabled. Revit will suggest two User Defined Worksets to start*

By default, two User-Created Worksets will be suggested by Revit: One for Shared Levels and Grids containing those elements, and another called "Workset1" that will contain all of the other elements. You can accept these names or choose alternate names. You can also rename a Workset later (provided that no users are accessing the project at the time you choose to rename). It will take a moment for Worksharing to be enabled; once complete, the "Worksets" dialog will appear and list each Workset in the project. Additional User-created Worksets can be added at this time, or later. You can also use this dialog to list all of the "Views," "Families", and "Project Standards" Worksets (see Figure B.4).

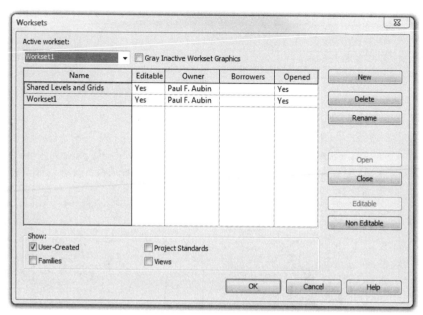

**Figure B.4**  *Use the "Worksets' dialog to view and edit Worksets*

## CREATE A CENTRAL FILE

Before users can begin working in a Workset-enabled project, there must be a Central file. This file is the main "hub" of the project. It should be located on a network server accessible to all team members. After Worksharing has been enabled, you use the **Save as** command on the File menu to save a Central file. The first time that Save as is chosen after enabling Worksets, the file saved automatically becomes a Central file. After that, if you wish to create a new Central file for any reason, you must choose Save as from the File menu, click the Options button and then place a checkmark in the "Make this the Central location after save" checkbox (see Figure B.5).

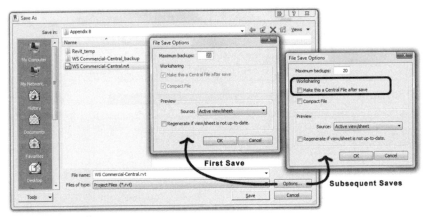

**Figure B.5**  *Making a file the Central file*

**Note:** While it is possible to move or make a copy of the Central file it is highly recommended that you maintain only one Central File to avoid confusion to project team members and potential loss of work. You can and should regularly back up the Central file using whatever method is currently in place in your firm, but avoid simply making a separate copy to another location on the network as users may mistakenly open this file and create Local files from it. Also, avoid "Detach from Central" except in situations where you are positive you will not need to save changes back to the Central File.

## CREATING ADDITIONAL WORKSETS

After the Central file has been saved, you can return to the "Worksets" dialog and create additional Worksets required by the project team. In the "Worksets" dialog, simply click the New button to add a Workset. You can name each descriptively. An important consideration when creating new Worksets is their default visibility. If you wish to have the Workset automatically visible in all project views by default, then check the "Visible by default in all views" box when creating it. However, to help increase performance on large projects, it can be better to leave this setting disabled and allow users to control the visibility of each Workset manually. This can (see Figure B.6).

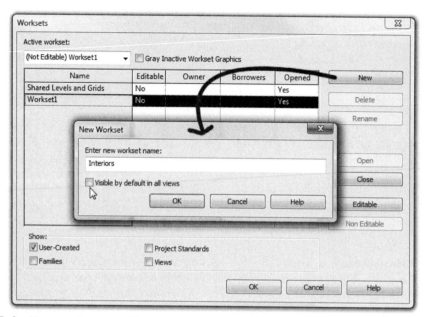

**Figure B.6** *You can disable visibility by default of a new Workset if desired*

When setting up the initial Worksets, try to divide building elements into logical working groupings. These will often be "task-based" to support the work of the team member who will author its contents. For example, in a typical commercial office building like the project constructed in this book, you might create a Workset for the exterior shell of the building, the core elements, the lobby, and one for each floor's interior elements. Furniture might also be separated as might other specialty equipment (see Figure B.7).

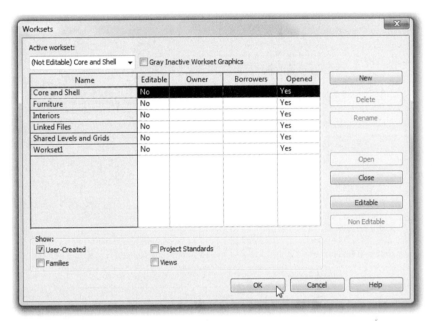

**Figure B.7** *Some Worksets added to the commercial project*

You can add new Worksets at any time in a project. It is however a good idea to try to establish the basic Workset organization as early as possible. This will make it easier for project team members to become comfortable with the project and its organization. As new Worksets are added later in the project, be sure that these and their functions are clearly communicated to the project team.

## MOVING ITEMS TO WORKSETS

Once Worksets have been created in the Central file, existing elements in the model must be moved to the appropriate Workset. For instance, if a Workset named "Interiors" has been created, then any interior elements such as partitions, doors, etc. should be moved to this Workset. This way, when a team member opens their local file and checks out this Workset, they will gain access to all of these elements.

To move existing elements to a different Workset, select one or more elements, click the Properties icon on the Options Bar, and then choose the desired Workset from the "Workset" list (see Figure B.8). When Worksets are enabled in a Revit project a Workset parameter will be automatically added to every model element and appear on the Element Properties dialog.

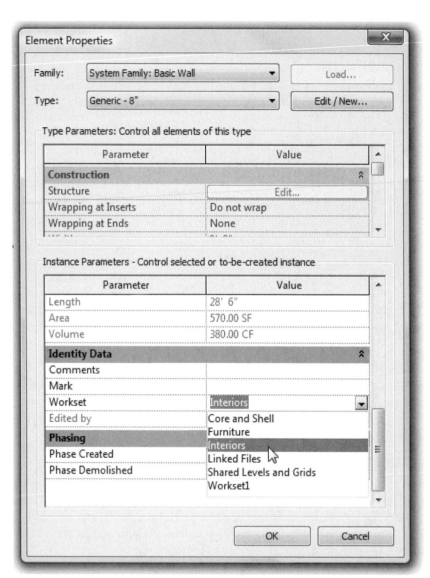

**Figure B.8**   *Move a selection of elements to a different Workset*

When you work in an Workset-enabled project, the tool tips that appear when elements are pre-highlighted will now report the Workset in front of the usual Element Class : Family : Type designation (see Figure B.9).

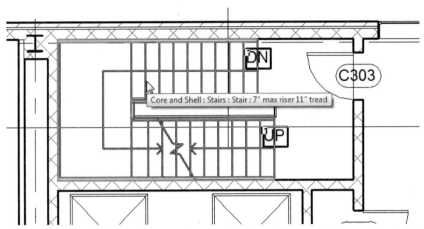

**Figure B.9**    *Tool tips in Workset projects will report Workset : Element Category : Family : Type*

## WORKSET VISIBILITY

Worksets can be set to display or be hidden in each view of the project. When Worksets are created, the default visibility setting is assigned to the Workset. In the topic above, this issue was noted as one of the considerations to factor into Workset creation. To see which Worksets are visible in a particular view, open the Visibility/Graphics dialog. In a Workset-enabled project, a Worksets tab will appear in the dialog. On this tab, you can choose which Worksets you wish to see in the current view (see Figure B.10).

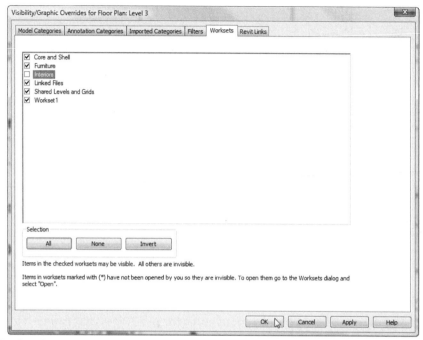

**Figure B.10** *Choose which Worksets you wish to display in a particular view in the Visibility/Graphics dialog*

This functionality offers a powerful way to manage the specific visibility of elements in a project. To fully realize the potential of this functionality, take care in the planning stages to determine the default visibility of each Workset in each view of the project. Users can change the settings later, but it is a good idea to establish effective defaults ahead of time.

## CREATING A LOCAL FILE

Each member of the project team must create and work in a Local copy of the Central file. From this Local copy, users can check out the Worksets in which they need to work and can work disassociated from other team members. When a user checks out a Workset, it is locked in the Central file and becomes read-only to other team members until that user relinquishes it.

Creating a Local file is easy. In Windows Explorer, copy the Central file to a local drive. The Local file can be saved to the hard disk of your local computer. This will increase performance when saving and loading the Local file. Rename the file before opening it. It is best practice to add your name or initials and the date to the file name. Open this file in Revit and a message like the one shown in Figure B.11 will appear. While a bit ominous looking, this message simply informs you that the copy you made will be treated like a local file linked back to the original Central file. This is exactly what you want, so simply click OK.

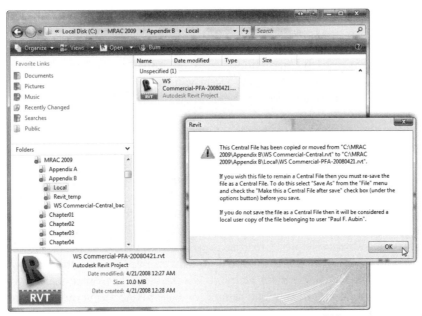

**Figure B.11**  *Copy the Central file in Windows Explorer and rename it. When you open it, Revit will treat it like a local file*

When you open your Local file, you can choose which Worksets you wish to open. In larger projects, this can make load times quicker. To do this, choose the Specify option from the drop-down next to the Open button. After choosing "Specify" an "Open Worksets" dialog will appear. Choose the Worksets that you wish to open by using the SHIFT and CTRL keys to select multiple items. With various Worksets highlighted, click the Open or Close buttons as desired (see Figure B.12).

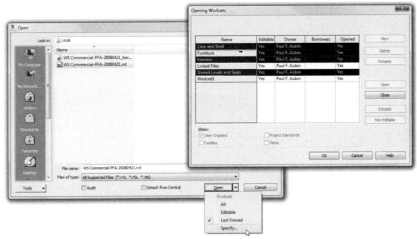

**Figure B.12**  *Open and Close Worksets in the "Opening Worksets" dialog*

When you click OK on this dialog, the model will load with only those Worksets that you specified. You can see this by opening views that require a Workset that you did not load. The required elements will not appear in that view. Should you realize that you need to open a Workset that you did not choose to open initially; you can simply launch the "Worksets" dialog, and then select the Workset you need and click the Open button. You can Close Worksets that you no longer need in the same fashion.

## EDITING WORKSET ELEMENTS

When you first open your Local file and choose the Worksets you wish to Open, none will be editable. Opening a Workset does not automatically make it editable. This action is achieved in the "Worksets" dialog. There are a few approaches you can take to editing the model. In general, if an element that you wish to edit is not being edited by another user, Revit will allow you to edit it, even if you do not have the associated Workset checked out. If another user is actively editing this element, or has the Workset checked out, you will need to issue an "Editing Request" to that user. The other user can then choose to allow your edit or refuse it. In any case, a good line of communication between you and your team is crucial. When you need an element checked out by someone else, you should contact the user via phone, instant messenger, or face-to-face to coordinate the edits.

You can "check out" a Workset by making it "Editable" in the "Worksets" dialog. Launch the "Worksets" dialog, select one or more Worksets and click the Make Editable button (see Figure B.13).

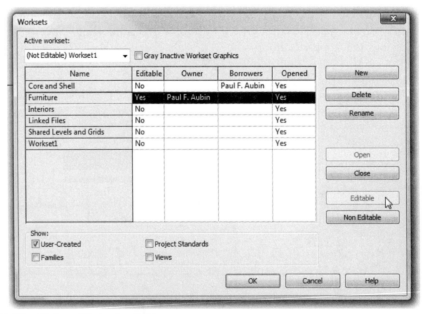

**Figure B.13** *Make a Workset editable*

Notice in the figure that the user "Paul F. Aubin" is listed as a "Borrower" to the Core and Shell Workset. Borrowing occurs in real-time as you edit elements in the model. If no one else is editing the element already or if the Workset to which it belongs is not locked for editing, Revit will "borrow" the element from the Workset and allow you to make changes. From that point on, no one else on the team will be able to edit that element until you relinquish it. When you Save to Central, the element will be relinquished automatically (unless you specify otherwise) and the changes updated to the Central file.

Any new elements that you add to a model will be added to the active Workset. You can choose the active Workset from the drop-down list on the "Worksets" toolbar. In Figure B.13, Workset 1 is the active Workset (see the top left corner). You can make a Workset active even if it is not editable (as you can see in the figure). This means that you will be adding elements to the Workset, but once you save to Central, you may not be able to edit them anymore if someone else has the Workset checked out. Always pay attention to the active Workset! This is very important since you don't want to add several objects to the wrong Workset inadvertently.

## SAVE TO CENTRAL

You should save your work at regular intervals regardless of whether you are working in a Workset-enabled project or not. When you work in a Local file, there are two types of save: Save and Save to Central. When you choose **File > Save**, you are simply saving your Local copy of the project file. As mentioned, it is a good idea to do this regularly, for example, every 15 minutes. Also at regular intervals, (but perhaps not quite as frequently) you should also **Save to Central**. Every 30 minutes to up to 2 hours would be a practical choice. This command updates the Central file with all of the changes that you have made in your Local file. It also retrieves changes made to the Central file by other team members since you opened your Local copy (basically synchronization), or last Saved to Central, or performed a Reload Latest. When you Save to Central, you also have the option to relinquish your checked out Worksets and borrowed elements. You can add comments to your Save to Central as well. This will be recorded in the log file maintained by the Central file and might be useful if there is ever a reason to roll back changes to a previous version (see Figure B.14).

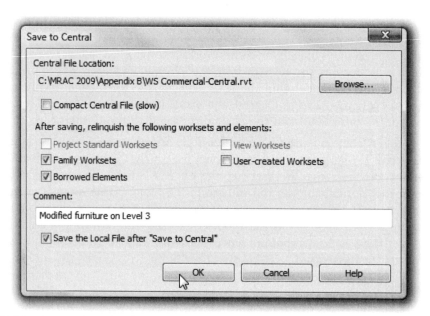

**Figure B.14** *Save to Central with options to relinquish borrowed elements, Worksets and to save your Local copy after save*

As a general rule of thumb, you should always check the 'Save the local file after "Save to Central" completes successfully' box in the "Save to Central" dialog to have Revit also save your Local copy when finished. This helps keep both files synchronized. You should also relinquish your borrowed elements. You may wish to keep your Worksets checked out if you are still actively editing them. Otherwise, you should relinquish those as well. Comments are also a good idea as noted above.

## WORKSET TIPS

Keep the following tips in mind when working in Workset projects.

- Never work in the Central file. If you ever have to open the Central file for any reason, perform your required tasks quickly and get out quickly. Inform your team before opening the Central file and have them close their files first.

- Save often! Save both your local copy and Save to Central often.

- Revit projects can get large. Central files can get even larger. While it is important to save often, Save to Central times can be very long on big projects. Plan your Saves to Central to take advantage of down times.

- Always relinquish all borrowed elements and Worksets before leaving work for any extended period of time: going to lunch, leaving at the end of the day, going on vacation, etc.

- If you have been away from your computer for a while, and left a project file open, re-read the previous tip. After that, perform a Reload Latest (File menu). This command synchronizes your local copy with the latest saved changes of the Central file.

- Check the Review Warnings dialog (Settings menu) on a regular basis. Perhaps once a week. Sometimes excessive warnings can have a negative impact on performance and even in extreme cases corrupt your Central file.

- Recreate your Local file on a regular schedule. You can recreate it as often as you like, (even every day) but in general you ought to create a new Local file once a week.

- You can never have too many backups of your Central file. Hard drives are much cheaper than hours of recreating lost work.

- Placing linked files on a dedicated Workset can be an effective way to manage performance. While you can certainly unload a linked file, this change would affect all users the next time you save to Central. Alternatively, each user can control which Worksets are active and displayed in their own work session. Therefore you can turn off the "Links" Workset on your local file without affecting your colleagues.

- Be careful with using the Activate View feature on Sheets. While this is a handy command, sometimes annotation added in this mode can loose its association to the proper View Workset and not display properly or disappear altogether.

- In most cases, trying to load and save Central files over a WAN is prohibitively slow. Consider setting up a remote desktop connection instead.

- When archiving a Central file, do not just copy it to an archive folder. Remember, the way we make a Local file is to copy it and rename it in a new location. To make an archive copy, open the copy and choose Save As. Click Options and choose the "Make this a Central file after save" option as shown in Figure B.5 above.

- If you or a team member wishes to open a project "read only" to take measurements, print, or perform other work that they do not wish to save, open the file with the "Detach from Central" box selected in the "Open" dialog. Use this option with caution however. There is no way to change your mind later and "re-connect" a detached file with the Central. Once you have detached it, it cannot be re-attached!

## GOING FURTHER

This appendix explains the basic concepts of Worksharing and Worksets. The best way to learn and implement this multi-user mechanism is to practice with at least one other individual. Remember, you are encouraged to work through the tutorials provided with the software. If you can work with another individual on these tutorials, you will get a better sense of the nuances of working in a Workset-enabled project. A copy of the commercial project has been provided with the files from the Mastering Revit Architecture CD ROM in the *Appendix B* folder. If you wish to experiment with this file, make a copy of it in the Local folder. Rename the copy and open it. This will be your local file. Try the features discussed in this Appendix.

Feel free to experiment in these files. If you have a colleague who can help you, each of you can create a Local copy and work simultaneously. Each of you should make changes, Save to Central, then reload from the Central file. Try borrowing elements, try to edit the same element as your colleague, and see what happens.

## SUMMARY

Working in teams in an important part of Architectural production. Using linked files and Worksets, Revit provides the means to accomplish sharing of data and managing coordinated team projects. When implemented with care, Worksets provide an invaluable toolset to the extended Revit project team.

- AutoCAD and Microstation files can be linked to Revit models.

- Import CAD files when you do not need them to update.

- Worksets provide the means to sub-divide a project into parts so that multiple team members can work simultaneously.

- Enable Worksets in the project and create User-created Worksets.

- Move existing elements to appropriate Worksets.

- Save the project file as a Central File on a network server accessible to all users.

- Each user creates a Local file from the Central file in which to work.

- Decide which Worksets to Open when opening the Local file.

- Decide which Worksets to make Editable.

- You can edit elements that are not in an editable Workset if they are not being edited by other users.

- Editing an element in a non-editable Workset is called "borrowing."

- Borrowed elements and Editable Worksets can be relinquished during Save to Central operations.

- Save your Local file and Central files often.

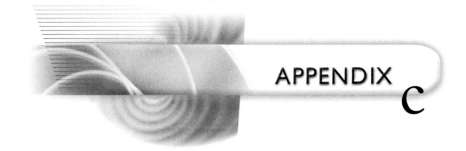

# APPENDIX C

# Online Resources

In this appendix are listed several online web sites and other resources that you can visit for information on Revit and related topics.

## WEB SITES RELATED TO THE CONTENT OF THIS BOOK

### http://www.paulaubin.com

The web site of the author. Includes information on ordering this book and Aubin's other books like *Mastering AutoCAD Architecture*. Check there for ordering information and addenda.

Paul F. Aubin is available for training and consulting services at your office. Please visit the web site and use the contact form to inquire about services offered.

### http://www.autodeskpress.com

Web site for Autodesk Press. Visit for information on other CAD titles, online resources, student software, and more.

### http://www.rgmaarchitecture.com

Web site of technical editor, Robert Guarcello Mencarini, Architect, AIA. Includes information on Architectural Design services and Revit consulting and implementation services.

### http://www.studiovim.com/

Web site of Velina Mirincheva, editorial and dataset contributor to this edition of *Mastering Revit Architecture*. Studio VIM provides architects and engineers with design and CAD consulting, online and live support for Revit and AutoCAD Architecture.

## AUTODESK SITES

### http://www.autodesk.com

Autodesk main web site. Visit often for the latest information on Autodesk products.

### http://discussion.autodesk.com

Autodesk Discussion Groups main page: Online community of Autodesk users sharing comments, questions, and solutions about all Autodesk products.

## USER COMMUNITY

### http://www.augi.com/revit/default.asp

Autodesk Users Group International Revit focused forums and information.

### http://www.revitcity.com

Web site devoted to all things Revit.

### http://forums.cadalyst.com/index.php

Online user forum hosted by CADalyst Magazine and moderated by Paul F. Aubin.

### http://www.cadalyst.com/cadalyst/

Main home page for CADalyst Magazine. View magazines online or subscribe to print edition.

### http://modocrmadt.blogspot.com/2005/01/bim-what-is-it-why-do-i-care-and-how.html

Excerpts from Chapter 1 come from an essay on this Blog (Web Log) maintained online by Matt Dillon. Matt is a registered Architect and expert in AutoCAD Architectural, Revit Architecture, and other BIM technologies. Matt was gracious enough to provide permission to quote his essay in this book.

## ONLINE RESOURCES FOR PLUG-INS

### www.GreenBuildingStudio.com

Green Building Studio is a free web-service that provides architects and engineers using Autodesk Revit, Autodesk Architectural Desktop, or Autodesk Building Systems with early design stage whole building energy analysis and product information appropriate for their building design.

### www.e-specs.com

Built around our e-SPECS technology, which links the project drawings to the specification documents, InterSpec has a variety of products and services to help you manage your construction specifications in the most accurate, efficient, and cost-effective manner possible.

# INDEX

**IMPORTANT!  READ CAREFULLY**: This End User License Agreement ("Agreement") sets forth the conditions by which Cengage Learning will make electronic access to the Cengage Learning-owned licensed content and associated media, software, documentation, printed materials, and electronic documentation contained in this package and/or made available to you via this product (the "Licensed Content"), available to you (the "End User"). BY CLICKING THE "I ACCEPT" BUTTON AND/OR OPENING THIS PACKAGE, YOU ACKNOWLEDGE THAT YOU HAVE READ ALL OF THE TERMS AND CONDITIONS, AND THAT YOU AGREE TO BE BOUND BY ITS TERMS, CONDITIONS, AND ALL APPLICABLE LAWS AND REGULATIONS GOVERNING THE USE OF THE LICENSED CONTENT.

**1.0    SCOPE OF LICENSE**

1.1    Licensed Content. The Licensed Content may contain portions of modifiable content ("Modifiable Content") and content which may not be modified or otherwise altered by the End User ("Non-Modifiable Content"). For purposes of this Agreement, Modifiable Content and Non-Modifiable Content may be collectively referred to herein as the "Licensed Content." All Licensed Content shall be considered Non-Modifiable Content, unless such Licensed Content is presented to the End User in a modifiable format and it is clearly indicated that modification of the Licensed Content is permitted.

1.2    Subject to the End User's compliance with the terms and conditions of this Agreement, Cengage Learning hereby grants the End User, a non-transferable, nonexclusive, limited right to access and view a single copy of the Licensed Content on a single personal computer system for noncommercial, internal, personal use only. The End User shall not (i) reproduce, copy, modify (except in the case of Modifiable Content), distribute, display, transfer, sublicense, prepare derivative work(s) based on, sell, exchange, barter or transfer, rent, lease, loan, resell, or in any other manner exploit the Licensed Content; (ii) remove, obscure, or alter any notice of Cengage Learning's intellectual property rights present on or in the Licensed Content, including, but not limited to, copyright, trademark, and/or patent notices; or (iii) disassemble, decompile, translate, reverse engineer, or otherwise reduce the Licensed Content.

**2.0    TERMINATION**

2.1    Cengage Learning may at any time (without prejudice to its other rights or remedies) immediately terminate this Agreement and/or suspend access to some or all of the Licensed Content, in the event that the End User does not comply with any of the terms and conditions of this Agreement. In the event of such termination by Cengage Learning, the End User shall immediately return any and all copies of the Licensed Content to Cengage Learning.

**3.0    PROPRIETARY RIGHTS**

3.1    The End User acknowledges that Cengage Learning owns all rights, title and interest, including, but not limited to all copyright rights therein, in and to the Licensed Content, and that the End User shall not take any action inconsistent with such ownership. The Licensed Content is protected by U.S., Canadian and other applicable copyright laws and by international treaties, including the Berne Convention and the Universal Copyright Convention. Nothing contained in this Agreement shall be construed as granting the End User any ownership rights in or to the Licensed Content.

3.2    Cengage Learning reserves the right at any time to withdraw from the Licensed Content any item or part of an item for which it no longer retains the right to publish, or which it has reasonable grounds to believe infringes copyright or is defamatory, unlawful, or otherwise objectionable.

**4.0    PROTECTION AND SECURITY**

4.1    The End User shall use its best efforts and take all reasonable steps to safeguard its copy of the Licensed Content to ensure that no unauthorized reproduction, publication, disclosure, modification, or distribution of the Licensed Content, in whole or in part, is made. To the extent that the End User becomes aware of any such unauthorized use of the Licensed Content, the End User shall immediately notify Cengage Learning. Notification of such violations may be made by sending an e-mail to delmarhelp@Cengage.com.

**5.0    MISUSE OF THE LICENSED PRODUCT**

5.1    In the event that the End User uses the Licensed Content in violation of this Agreement, Cengage Learning shall have the option of electing liquidated damages, which shall include all profits generated by the End User's use of the Licensed Content plus interest computed at the maximum rate permitted by law and all legal fees and other expenses incurred by Cengage Learning in enforcing its rights, plus penalties.

**6.0    FEDERAL GOVERNMENT CLIENTS**

6.1    Except as expressly authorized by Cengage Learning, Federal Government clients obtain only the rights specified in this Agreement and no other rights. The Government acknowledges that (i) all software and related documentation incorporated in the Licensed Content is existing commercial computer software within the meaning of FAR 27.405(b)(2); and (2) all other data delivered in whatever form, is limited rights data within the meaning of FAR 27.401. The restrictions in this section are acceptable as consistent with the Government's need for software and other data under this Agreement.

**7.0    DISCLAIMER OF WARRANTIES AND LIABILITIES**

7.1    Although Cengage Learning believes the Licensed Content to be reliable, Cengage Learning does not guarantee or warrant (i) any information or materials contained in or produced by the Licensed Content, (ii) the accuracy, completeness or reliability of the Licensed Content, or (iii) that the Licensed Content is free from errors or other material defects. THE LICENSED PRODUCT IS PROVIDED "AS IS," WITHOUT ANY WARRANTY OF ANY KIND AND CENGAGE LEARNING DISCLAIMS ANY AND ALL WARRANTIES, EXPRESSED OR IMPLIED, INCLUDING, WITHOUT LIMITATION, WARRANTIES OF MERCHANTABILITY OR FITNESS FOR A PARTICULAR PURPOSE. IN NO EVENT SHALL CENGAGE LEARNING BE LIABLE FOR: INDIRECT, SPECIAL, PUNITIVE OR CONSEQUENTIAL DAMAGES INCLUDING FOR LOST PROFITS, LOST DATA, OR OTHERWISE. IN NO EVENT SHALL CENGAGE LEARNING'S AGGREGATE LIABILITY HEREUNDER, WHETHER ARISING IN CONTRACT, TORT, STRICT LIABILITY OR OTHERWISE, EXCEED THE AMOUNT OF FEES PAID BY THE END USER HEREUNDER FOR THE LICENSE OF THE LICENSED CONTENT.

**8.0    GENERAL**

8.1    Entire Agreement. This Agreement shall constitute the entire Agreement between the Parties and supercedes all prior Agreements and understandings oral or written relating to the subject matter hereof.

8.2    Enhancements/Modifications of Licensed Content. From time to time, and in Cengage Learning's sole discretion, Cengage Learning may advise the End User of updates, upgrades, enhancements and/or improvements to the Licensed Content, and may permit the End User to access and use, subject to the terms and conditions of this Agreement, such modifications, upon payment of prices as may be established by Cengage Learning.

8.3    No Export. The End User shall use the Licensed Content solely in the United States and shall not transfer or export, directly or indirectly, the Licensed Content outside the United States.

8.4    Severability. If any provision of this Agreement is invalid, illegal, or unenforceable under any applicable statute or rule of law, the provision shall be deemed omitted to the extent that it is invalid, illegal, or unenforceable. In such a case, the remainder of the Agreement shall be construed in a manner as to give greatest effect to the original intention of the parties hereto.

8.5    Waiver. The waiver of any right or failure of either party to exercise in any respect any right provided in this Agreement in any instance shall not be deemed to be a waiver of such right in the future or a waiver of any other right under this Agreement.

8.6    Choice of Law/Venue. This Agreement shall be interpreted, construed, and governed by and in accordance with the laws of the State of New York, applicable to contracts executed and to be wholly preformed therein, without regard to its principles governing conflicts of law. Each party agrees that any proceeding arising out of or relating to this Agreement or the breach or threatened breach of this Agreement may be commenced and prosecuted in a court in the State and County of New York. Each party consents and submits to the nonexclusive personal jurisdiction of any court in the State and County of New York in respect of any such proceeding.

8.7    Acknowledgment. By opening this package and/or by accessing the Licensed Content on this Web site, THE END USER ACKNOWLEDGES THAT IT HAS READ THIS AGREEMENT, UNDERSTANDS IT, AND AGREES TO BE BOUND BY ITS TERMS AND CONDITIONS. IF YOU DO NOT ACCEPT THESE TERMS AND CONDITIONS, YOU MUST NOT ACCESS THE LICENSED CONTENT AND RETURN THE LICENSED PRODUCT TO CENGAGE LEARNING (WITHIN 30 CALENDAR DAYS OF THE END USER'S PURCHASE) WITH PROOF OF PAYMENT ACCEPTABLE TO CENGAGE LEARNING, FOR A CREDIT OR A REFUND. Should the End User have any questions/comments regarding this Agreement, please contact Cengage Learning at delmar.help@cengage.com.